SURVEYING

Principles and Applications

BARRY F. KAVANAGH
S. J. GLENN BIRD
Seneca College of Applied Arts and Technology

A RESTON BOOK
PRENTICE-HALL, INC., Englewood Cliffs, New Jersey 07632

Library of Congress Cataloging in Publication Data

Kavanagh, Barry F.
    Surveying : principles and applications.

        1. Surveying.   I. Bird, S. J. Glenn.   II. Title.
TA545.K37   1984   526.9   84-2115
ISBN 0-8359-7405-7

Editorial/production supervision and
interior design by Camelia Townsend

© 1984 by Prentice-Hall, Inc.
A Division of Simon & Schuster
Englewood Cliffs, New Jersey 07632

10   9   8   7   6   5   4   3

Contents

Preface

Surveying: Principles and Applications has been designed for use in Civil Engineering Degree programs as well as in Civil Technology/Technician Diploma programs. It is expected that this text will also be employed for introductory survey courses in allied disciplines such as Survey Engineering, Mining, Resources, Geology, Geography, Earth Sciences, etc.

The material presented in the first eight chapters of Part 1, form the curriculum suitable for most introductory surveying courses. Degree programs will cover the material in one semester using 4 to 6 hours per week. Diploma programs will cover similar material in one semester using 5 to 7 hours per week, the additional time being spent on practical field skills acquisition and theory reinforcement.

The topics in Part 1 are presented in the order normally taught in North American Colleges and Universities. The chapter on survey drafting and computations has been included because of its direct relationship to students taking concurrent drafting programs. The chapter on Electronic Distance Measuring Devices (E.D.M.'s) has been delayed in the order of presentation so that students would first have the opportunity of mastering the traditional levels and theodolites before encountering the more complex and precision-demanding E.D.M.'s. Some instructors may still wish to present E.D.M.'s with the other distance measuring techniques covered in chapter 2.

Discussion of the treatment of random errors has been included in the appendix reflecting the fact that many college students take a concurrent course in statistics or adjustments.

Part 2 has been designed to fill the needs of an advanced Surveying course. The material can be covered in one semester in 4 to 6 hours per week depending on the program. The topics in Part 2 can also be of direct *reference* value to degree program students who do not take a second surveying course,—especially in the areas of *Highway Design, Civil Construction, Marine Engineering and Surveying*, and *Photogrammetry and Airphoto Interpretation*.

In addition to many years of survey education experience, the authors of *Surveying: Principles and Applications*, have both had extensive survey experience in industry. They have been active in the field in legal surveying, route surveying, aerial surveying, hydrographic surveying, pre-engineering surveys and all forms of construction surveying. They have attempted to give the reader the benefit of their actual field experience in all areas of this text. This experience factor, together with the "applications" approach taken by the authors, ensure that this text will also be of immediate use to *practicing surveyors*.

The universal distribution of sophisticated hand held calculators has affected the way survey calculations and survey operations are performed; accordingly, this text is developed ignoring traditional "slide rule" approaches to problem solving. Additionally, since most computations are now performed on calculators or by computer, the tables normally found at the back of survey texts have either been included in the appropriate chapters in the text, or they have been eliminated.

The subject of surveying is studied by a large number of students enrolled in a wide variety of programs; perhaps most of the students who study surveying will never actually work directly in the survey field, but they will use their survey subject experience to advantage in related fields of endeavour. Many students upon graduation, will however work directly in the survey field using their experience to develop a career in surveying, or as an important first step in an engineering or allied career. Surveying as a College or University subject is unique—it is the *only single subject* (together with the experience gained at survey camp) that can prepare a student for *direct entry to the work force*.

The authors can highly recommend surveying either as a career, or as a first step in an allied career. The field work and associated calculations and drawings are challenging and extremely rewarding for any man or woman who enters this field.

Acknowledgements

The authors wish to acknowledge the assistance and support given by the faculty of Seneca College Civil and Resources programs. In particular, thanks are due to Gordon Benson, Peter Crawford and William Habkirk who each contributed advice and many hours of their time reviewing various portions of the manuscript. Thanks are also due to Dr. John van den Berg of the National Swedish Institute for Building Research who shared with us the current work of the international community of surveyors (F.I.G.) in the area of standards for construction surveys; Mr. Don Davis of Davis Canada Engineering Products, Toronto, who provided valuable input for electronic distance position fixing for hydrographic surveys; and Mr. Louis Sebert of the Surveys and Mapping Branch, Ottawa, who provided useful information on original township surveys in Ontario.

The following surveying, engineering, and equipment manufacturing companies have provided us with generous assistance:

AGA Geodimeter of Canada Ltd., Ontario
Alpha Electronics Corp., Colorado
Berntsen Inc. (Survey monuments), Wisconsin
J.D. Barnes Ltd; Surveyors, Ontario
Bird and Hale Limited, Ontario
Cansel Surveying Equipment Co., Ontario
Cooper Tool Group (Lufkin Tapes), Ontario
Davis Canada Engineering Products (Tellurometer), Ontario
Kern and Co. Arau, Switzerland
Keuffel and Esser Co., New Jersey
Laser Alignment Inc., Michigan
Macrometrics Inc., Massachusetts
Marshall, Macklin, Monaghan; Surveyors & Engineers, Ontario
Pentax Corp., Colorado,
Sokkisha Canada Ltd., Ontario
Spectra Physics, California
Topcon Instrument Corporation, New Jersey
Carl Zeiss, Jena, Ontario
Carl Zeiss, Oberkochen, West Germany
Wild Heerbrugg, Switzerland
Wild Leitz, Ontario

The bulk of the typing for the manuscript was performed (flawlessly) by Theresa Hall of Unionville.

Finally, the authors wish to acknowledge the support of their families and wives—Rita Kavanagh and Sue Bird, as well as the staff of Bird and Hale Ltd., during the period of manuscript preparation.

1 SURVEYING PRINCIPLES

1 *Basics of Surveying*

1.1 SURVEYING DEFINED

Surveying is the art of measuring distances and angles on or near the surface of the earth. It is an art in that only a surveyor who possesses a thorough understanding of surveying techniques will be able to determine the most efficient methods required to obtain optimal results over a wide variety of surveying problems.

Surveying is scientific to the degree that rigorous mathematical approaches are used to analyze and adjust the field survey data. The accuracy, and thus reliability, of the survey depend not only on the field expertise of the surveyor, but also on the surveyors' understanding of the scientific principles underlying and affecting all forms of survey measurement.

Figure 1.1 is an aerial photo of undeveloped property. Figure 1.2 is an aerial photo of the same property after development. It can be seen that every line on the development photo is the direct or indirect result of surveying.

3

FIGURE 1.1 Aerial photo of undeveloped property.

FIGURE 1.2 Aerial photo of same property after development.

1.2 TYPES OF SURVEYS

Plane surveying is that type of surveying in which the surface of the earth is considered to be a plane for all X and Y dimensions. All Z dimensions (height) are referenced to the mean spherical surface of the earth (mean sea level). Most engineering and property surveys are plane surveys, although some of these surveys that cover large distances (e.g., highways and railroads) will have corrections applied at regular intervals (e.g., 1 mile) to correct for curvature.

Geodetic surveying is that type of surveying in which the surface of the earth is considered to be spherical (actually an ellipsoid of revolution) for X and Y dimensions. As in plane surveying, the Z dimensions (height) are referenced to the mean surface of the earth (mean sea level). Geodetic surveys are very precise surveys of great magnitude (e.g., national and provincial or state boundaries and control networks).

1.3 CLASSES OF SURVEYS

The *preliminary survey* (data gathering) is the gathering of data (distances and angles) to locate physical features (e.g., trees, rivers, roads, structures, or property markers) so that the data can be plotted to scale on a map or plan. Preliminary surveys also include the determination of differences in elevation (vertical distances) so that elevations and contours may also be plotted.

Layout surveys involve marking on the ground (using wood stakes, iron bars, aluminum and concrete monuments, nails, spikes, etc.) the features shown on a design plan. The layout can be for property lines, as in subdivision surveying, or it can be for a wide variety of engineering works (e.g., roads, pipelines, bridges). The latter group is known as construction surveys. In addition to marking the proposed horizontal (X and Y dimensions) location of the designed feature, reference will also be given to the proposed elevations (Z dimensions).

Control surveys are used to reference both preliminary and layout surveys. Horizontal control can be arbitrarily placed, but it is usually tied directly to property lines, roadway center lines, or coordinated control stations. Vertical control is a series of bench marks—permanent points whose elevation above mean sea level have been carefully determined.

It is accepted practice to take more care in control surveys with respect to precision and accuracy; great care is also taken to ensure that the control used for a preliminary survey can be readily reestablished at a later date, whether it be needed for further preliminary work or for a layout survey.

1.4 DEFINITIONS

1. *Topographic surveys*: preliminary surveys used to tie in the surface features of an area. The features are located relative to one another by tying them all into the same control lines or control grid.
2. *Hydrographic surveys*: preliminary surveys that are used to tie in underwater features to a surface control line. Usually shorelines, marine features, and water depths are shown on the hydrographic map.
3. *Route surveys*: preliminary, layout, and control surveys that range over a narrow, but long strip of land. Typical projects that require route surveys are highways, railroads, transmission lines, and channels.
4. *Property surveys*: preliminary, layout, and control surveys that are involved in determining boundary locations or in laying out new property boundaries (also known as cadastral or land surveys).
5. *Aerial surveys*: preliminary surveys utilizing aerial photography. Photogrammetric techniques are employed to convert the aerial photograph into scale maps and plans.
6. *Construction surveys*: layout surveys for engineering works.
7. *Final ("as built") surveys*: similar to preliminary surveys, final surveys tie in features that have just been constructed to provide a final record of the construction and to check that the construction has proceeded according to the design plans.

1.5 SURVEYING INSTRUMENTATION

The instruments most often used in surveying are (1) the *transit* or *theodolite*—used to establish straight lines, and to measure horizontal and vertical angles; (2) the *level* and *rod*—used to measure differences in elevation; and (3) the *steel tape*—used to measure horizontal and slope distances (Figures 1.3, 1.4 and 1.5).

FIGURE 1.3   Theodolite and transit.

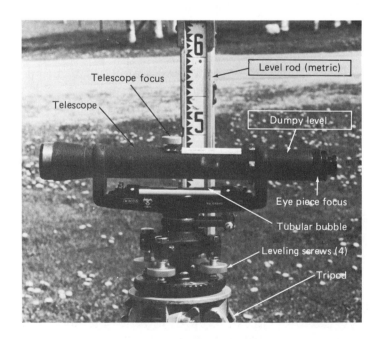

FIGURE 1.4   Level and rod.

8

FIGURE 1.5 Preparing to measure to a stake tack using a plumb bob.

All first courses in surveying enable the student to learn how to operate the equipment just described. The student will surprisingly discover that the steel tape will require more practice to *master* than will the more complex telescopic instruments.

1.6 SURVEY GEOGRAPHIC REFERENCE

It has already been mentioned that surveying involves measuring the location of physical land features relative to one another, and relative to a defined reference on the surface of the earth. In the broadest sense, the earth's reference system is composed of the surface divisions denoted by geographic lines of latitude and longitude. The latitude lines run east–west and are parallel to the equator. The latitude lines are formed by projecting the latitude angle out to the surface of the earth. The latitude angle itself is measured (90° maximum) at the earth's center north or south from the equatorial plane.

The longitude lines all run north–south, converging at the poles. The lines of longitude (meridians) are formed by projecting the longitude angle out to the surface of the earth at the equator. The longitude angle itself is measured at the earth's center east or west (180° maximum) from the plane of 0° longitude, which is arbitrarily placed through Greenwich, England (see Figures 1.6 and 1.7).

Although this system of geographic coordinates is much used in navigation and geodesy, it is not used in plane surveying. Plane surveying utilizes either coordinate grid systems or the original provincial or state township fabric as a basis for referencing.

1.7 SURVEY GRID REFERENCE

All states and provinces have adopted a grid system best suited to their needs. The grid itself is limited in size so that no serious errors will accumulate when the curvature of the earth is ignored.

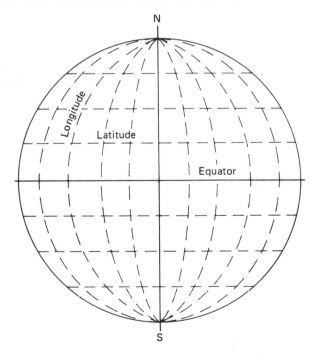

FIGURE 1.6 Sketch of earth showing lines of latitude and longitude.

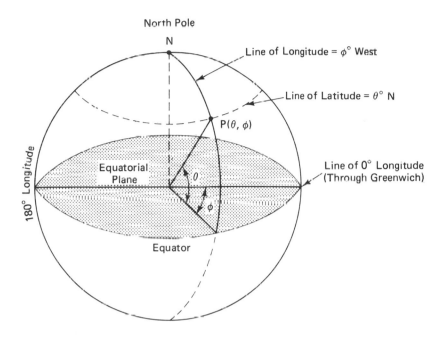

FIGURE 1.7 Sketch showing location of point *P* referenced by geographical coordinates.

Advantages of the grid systems are the ease of calculation (plane geometry and trigonometry) and the availability of one common datum for X and Y dimensions in a large (thousands of square miles) area. The coordinates in most grid systems can be referenced to the central meridian and to the equator so that translation to geographic coordinates is always easily accomplished. This topic is discussed in more detail in Chapter 11.

1.8 SURVEY LEGAL REFERENCE

Public lands in North America were originally laid out for use by the settlers. In the United States and parts of Canada, the townships were laid out in 6-mile squares; however, in the first established areas of Canada, a wide variety of township patterns exists —reflecting both the French and English heritage.

The townships themselves were subdivided into sections (lots) and ranges (concessions), each uniquely numbered. The basic township lots were either 1 mile square or some fraction thereof. Eventually, the township lots were, and still are being, further

subdivided in real estate developments. All developments are referenced to the original township fabric, which has been reasonably
well preserved through ongoing resurveys. This topic is discussed
in detail in Chapter 13.

1.9 SURVEY VERTICAL REFERENCE

The previous sections described how the X and Y dimensions
(horizontal) of any feature could be referenced for plane surveying
purposes. Although vertical dimensions can be referenced to any
datum, the reference datum most often used is that of mean sea
level (MSL). Mean sea level is assigned an elevation of 0.000 feet
(ft) (or meters), and all other points on the earth can be described
by elevations above or below zero. Permanent points whose elevations have been precisely determined (*bench marks*) are available
in most areas for survey use.

1.10 DISTANCE MEASUREMENT

Distances between two points can be *horizontal, slope,* or *vertical,*
and are recorded in feet (Imperial units) or meters (SI units). See
Figure 1.8.
   Horizontal and *slope distances* can be measured with a cloth
or steel tape or with an electronic distance measuring device. In
surveying, the horizontal distance is always required (for plan
plotting purposes); so if a slope distance between two points has
been taken, it must then be converted to its horizontal equivalent.
Slope distances can be trigonometrically converted to horizontal

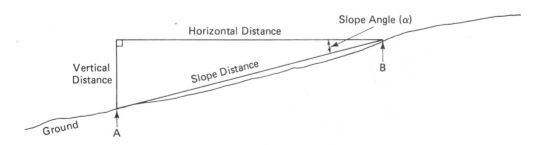

FIGURE 1.8   Horizontal distance measurement.

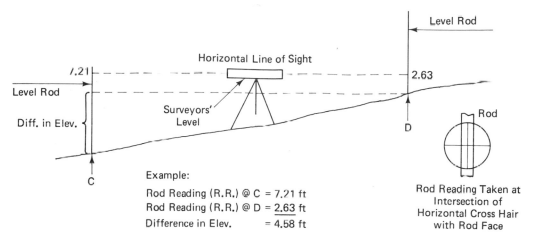

FIGURE 1.9 Leveling technique.

distances by using either the slope angle or the difference in elevation (vertical distance) between the two points.

Vertical distances can be measured with a tape, as in construction work, or, as is more usually the case, with a surveyors' level and leveling rod. See Figure 1.9.

1.11 UNITS OF MEASUREMENT

The English (Imperial) system of measurements has been in use in North America from the early settler days until the present. Canada has recently switched from English units to metric (SI) units, and some agencies (the Geodetic Survey) in the United States have long been working with metric units. SI units are a modernization (1960) of the long-used metric units. This modernization included a redefinition of the meter and the addition of some new units (e.g., newtons; see Table 2.1).

There is a widespread feeling that sooner or later all countries on earth will adopt the meter, the international unit of linear measure. This text will use both units of measure as both units are now being used in North America.

The following tables describe and contrast metric and English units.

TABLE 1.1 *Measurement Definitions and Equivalencies*

Linear Measurements			English Units		
1 mile	=	5280 feet	1 foot	=	12 inches
	=	1760 yards	1 yard	=	3 feet
	=	320 rods	1 rod	=	$16\frac{1}{2}$ feet
	=	80 chains	1 chain	=	66 feet
			1 chain	=	100 links
1 acre	=	43560 ft² = 10 square chains			

Linear Measurement		Metric (SI) Units	
1 kilometer	=	1000	meters
1 meter	=	100	centimeters
1 centimeter	=	10	millimeters
1 decimeter	=	10	centimeters
1 hectare (ha)	=	10,000	m²
1 square kilometer	=	1000000	m²
	=	100	hectares

English to Metric Conversion	
1 ft = 0.3048 m (exactly)	1 inch = 25.4 mm (exactly)*
1 km = 0.62137 miles	
1 hectare (ha) = 2.471 acres	
1 km² = 247.1 acres	

*Prior to 1959, the United States used the relationship 1 m = 39.37 in. This resulted in a U.S. survey foot of .3048006 m.

The most used angle unit is the degree, which is $\frac{1}{360}$ of a circle.

Angular Measurement
1 revolution = 360°
1 degree   =   60'
1 minute   =   60″ (seconds)

Degrees, minutes and seconds are used almost exclusively in both English and metric units; however, in some countries in Europe the circle has been graduated into 400 gon (formerly called grad). Angles are expressed to four decimals (e.g., a right angle = 100.0000 gon).

1.12 LOCATION METHODS

A great deal of surveying time is spent measuring points of interest relative to some reference line so that these points may be later shown on a scaled plan. The illustrations (Figure 1.10) show some common location techniques.

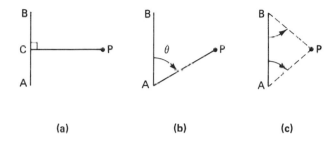

(a)   **(b)**   **(c)**

FIGURE 1.10   Location ties.

Point P in part (a) is located relative to known line AB by determining CB or CA, the right angle at C, and distance CP. This is known as the right-angle offset tie (also known as the *rectangular tie*) and is one of the two most widely used methods of locating a point.

Point P in part (b) is located relative to known line AB by determining the angle (θ) at A and the distance AP. This is known as the angle-distance tie (also known as the *polar tie*) and is the second of the two most widely used methods of locating a point.

Point P in part (c) can also be located relative to known line AB by determining **either** the angles at A and B to P, **or** by determining the distances AP and BP. Both methods are *intersection* techniques.

1.13 ACCURACY AND PRECISION

Accuracy is the relationship between the value of a measurement and the "true" value of the dimension being measured. Precision describes the degree of refinement with which the measurement is made. The concepts of accuracy and precision are illustrated in the following example.

A building wall that is known to be 157.22 ft long is measured by two methods. In the first case the wall is measured very
carefully using a cloth tape graduated to the closest 0.1 ft. The result of this operation is a measurement of 157.2 ft. In the second
case the wall is measured with the same care, but with a more
precise steel tape graduated to the closest 0.01 ft. The result of
this operation is a measurement of 157.23 ft.

	"True" Distance	*Measured Distance*	*Error*
Cloth tape	157.22	157.2	0.02
Steel tape	157.22	157.23	0.01

In this example, the more precise method (steel tape) resulted
in the more accurate measurement.

However, it is conceivable that more precise methods can result in less accurate answers. In the preceding example, if the
steel tape were being used at a very cold temperature (near free-
zing), the results of the measurement could have been 157.26 ft, a
less accurate answer than the value of 157.2 ft given by the cloth
tape. The steel tape would give relatively inaccurate results because at that temperature the steel tape would contract (shorten),
causing most of the significant (0.04 ft) error noted.

1.14 ERRORS

We saw in the previous section how an error due to the effects of
temperature caused a precise measurement to become relatively
inaccurate. It will be shown that temperature is only one of many
factors that can affect the accuracy of distances measured with a
steel tape.

In fact, it can be said that no measurement (except for counting) can be free of error. For every measuring technique used, a
more precise and potentially more accurate method can be found.

For purposes of calculating errors, the "true" value is determined statistically after repeated measurements. In the simplest
case, the "true" value for a distance is the mean value for a series
of repeated measurements. This topic is further discussed in Appendix A.

Systematic errors are defined as being those errors whose
magnitude and algebraic sign can be determined. The fact that

these errors can be determined allows the surveyor to eliminate them from his measurements and thus improve the accuracy. The error due to the effects of temperature on the steel tape noted in the previous section is an example of a systematic error. If the temperature is known, the shortening or lengthening effects on a steel tape can be precisely determined.

Random errors are beyond the control of the surveyor and are associated with the skill and vigilance of the surveyor. Random (also known as accidental) errors are introduced into each measurement mainly because no human can perform perfectly.

Fortunately, random errors, by their very nature, tend to cancel themselves. When surveyors are skilled and careful in measuring, random errors will be of little significance. However, random errors resulting from unskilled or careless work do cause problems. As noted earlier, random errors tend to cancel themselves mathematically, even large random errors—this does not result in accurate work, only in work that appears to be accurate.

1.15 MISTAKES

Mistakes are blunders made by survey personnel. Examples of mistakes include transposing figures (recording a tape value of 68 as being 86), miscounting the number of full tape lengths in a long measurement, measuring to or from the wrong point, and the like.

Students should be aware that mistakes will occur. Mistakes must be discovered and eliminated, preferably by the people who made them. All survey measurements are suspect until they have been verified. Verification may be as simple as repeating the measurement, or verification can result from geometric or trigonometric analysis of related measurements.

As a rule, every measurement is immediately checked or repeated. This immediate repetition enables the surveyor to eliminate most mistakes and, at the same time, improve the precision of the measurement.

1.16 ACCURACY RATIO

The *accuracy ratio* of a measurement or series of measurements is the ratio of error of closure to the distance measured. The *error of closure* is the difference between the measured location and the theoretically correct location. The theoretically correct location can be determined from repeated measurements or mathematical analysis.

Since relevant systematic errors and mistakes can and should be eliminated from all survey measurements, the error of closure will be comprised of random errors.

To illustrate, a distance was measured and found to be 250.56 ft. The distance was previously known to be 250.50 ft. The error is 0.06 ft in a distance of 250.50 ft.

$$\text{accuracy ratio} = \frac{0.06}{250.50} = \frac{1}{4175} = \frac{1}{4200}$$

The accuracy ratio is expressed as a fraction whose numerator is unity and whose denominator is rounded to the closest 100 units.

Survey specifications are discussed in Appendix A. Many land and engineering surveys have in the past been performed at $^{1}/_{5000}$ or $^{1}/_{3000}$ levels of accuracy. With the trend to polar layouts from coordinated control, accuracy ratios in the order of $^{1}/_{10000}$ and $^{1}/_{20000}$ are now often specified. It should be emphasized that *for each of these specified orders of accuracy the techniques and instrumentation used must also be specified.* See Appendix A for survey specifications.

1.17 STATIONING

In surveying, measurements are often taken along a base line and at right angles to the base line. Distances along a survey base line are referred to as stations or chainages, and distances at right angles to the base line (offset distances) are simple dimensions.

The beginning of the survey baseline, the zero end, is denoted by 0+00; a point 100 ft (m) from the zero end is denoted as 1+00; a point 131.26 ft (m) from the zero end is 1+31.26; and so on. If the stationing is extended back of the 0+00 mark, the stations would be 0−50, −1+00, and so on.

In the preceding discussion, the *full stations* are 100 ft (m) and the *half-stations* would be at even 50-ft intervals. Twenty meter intervals are often used as partial stations in the metric system.

With the changeover to metric units, most municipalities have kept the 100-unit station (i.e., 1+00 = 100 meters), whereas highway agencies have adopted the 1000 unit station (i.e., 1+000 = 1000 meters).

Figure 1.11 shows a building tied in to the center-line (℄) of Elm Street, and shows the ℄ (base line) distances as stations and the offset distances as simple dimensions. The sketch also shows that 0+00 is the intersection of the center lines of the two streets.

1.18 FIELD NOTES

One of the most important aspects of surveying is the taking of accurate, neat, legible, and complete field notes. After the survey has been completed, a plan is drawn from the survey notes, and the notes are then filed, often under lock and key.

One aspect of note taking that only comes with practice is the completeness of information being presented. Some details that look too obvious for inclusion in notes by the fieldman will be obscure to a draftsman unfamiliar with the site; these same details could well become obscure even to the surveyor himself within a few weeks.

An experienced surveyor's notes will be complete, without redundancies, well arranged to aid in comprehension, and neat and legible to ensure that the correct information is conveyed. The surveyor will use sketches whenever necessary to aid in comprehension and in the ordering of data.

Students and inexperienced surveyors usually find at first that it is very difficult to make accurate and neat field notes. The first few attempts at note keeping can be quite embarrassing to otherwise gifted students. There is a real temptation to scribble

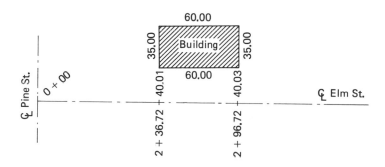

FIGURE 1.11   Base-line stations and offset distances.

the notes on scraps of paper, and then later, in a quiet and peaceful environment, transcribe the scribbled notes neatly onto field note paper. This temptation must be resisted. **Notes cannot be copied without the occurrence of mistakes.**

Copied notes are not field notes and, as such, are outlawed in the surveying profession. It sometimes happens that property and engineering surveyors find themselves in court testifying as to the results of a survey. If the notes being referred to in court are copied notes, the surveyor would no doubt be quickly excused from further participation in the proceedings. It may be a relatively rare occurrence to have to appear in court, but total reliance on the integrity of field notes is a daily requirement for surveyors and their associates. When surveyors are found to be copying or otherwise "cooking" notes, they are soon working elsewhere.

It is sometimes necessary to copy from field notes for other survey purposes; when notes are legitimately copied, they are placed on different-colored note paper or similar note paper with the word "copy" prominently placed on each page.

The field notes themselves are placed in bound field books or in loose-leaf field binders. The pages in bound field books are usually lined and columned on the left leaf and squared on the right leaf. The loose-leaf pages can be lined and columned or squared or in fact in any format required by the surveyor.

A considerable advantage to using loose-leaf note books is that the notes for one project can be filed under that project heading. If a bound book is used for several projects, filing becomes difficult, as several cross references are required just to locate one set of project notes. Bound books are used to advantage on large projects such as highways and other heavy construction operations.

Comments on Field Notes

Bound Books

1. Name, address, phone number, in ink on inside or outside cover.
2. Pages numbered throughout, right leaf only (most bound books have about 80 pages).
3. Room is left at the front of each book for title, index, and diary.

Loose-leaf Books

1. Name, address, phone number, in ink on the binder.

2. Each page must be titled and dated, with identification by project number and surveyors' names.

All Field Notes

1. Entries are to be in pencil in the range of 2H to 4H. The harder pencil (4H) is more difficult to use, but will not smear. The softer pencil (2H) is easy to use for most people but will smear somewhat if care is not exercised. Most surveyors use 2H or 3H lead. Pencils softer than 2H are not used in field notes.

2. All entries are neatly printed. Uppercase lettering can be reserved for emphasis, or it is sometimes used throughout.

3. All arithmetic computations are to be checked and signed.

4. Sketches are used to clarify the field notes. Although the sketches are not scale drawings, they are usually drawn roughly to scale to help order the inclusion of details.

5. Sketches are not freehand. Straight-edges and circle templates are used for all line work.

6. Sketches are properly oriented by the inclusion of a north arrow (preferably pointing up the page or to the left).

7. Avoid crowding the information onto the page. This practice is one of the chief causes of poor looking notes.

8. Mistakes in the entry of *measured data* are to be carefully lined out, *not erased.*

9. Mistakes in entries other than measured data (e.g., descriptions, sums, or products of measured data) may be erased and reentered neatly.

10. Show the word COPY at the top of copied pages.

11. Lettering on sketches is to be read from the bottom of the page or from the right side; any other position is upside down.

12. Measured data are to be entered in the field notes at the time the measurements are taken.

13. The note keeper has all given data verified by repeating it aloud as he is entering it in the notes. The chainman who originally gave the data to the note keeper will listen and respond to the verification call out.

14. If the data on an entire page are to be voided, the word VOID together with a diagonal line is placed on the page. A reference page number is shown for the location of the new relevant data.

1.19 FIELD MANAGEMENT

Survey crews (parties) are comprised of a party chief, an instrument man, and one or two chainmen or rodmen.* The party chief is responsible for the operation of the survey crew and the integrity of the work performed. The instrument man is responsible for the care and operation of the instrument being used. He should be vigilant to ensure that any lack of instrument adjustment is immediately noted and corrected. The chainmen perform the actual taping (chaining) measurements and maintain all equipment. When leveling work is being performed, these personnel are described as rodmen. In three-man crews, the party chief usually acts as head chainman.

It is usual for the least experienced crew member to be involved in the chaining (taping) operation, gradually working up to a position of instrument man.

---

*Women are studying surveying and working at field surveying in ever-increasing numbers. Terms such as rodman, chainman, and instrument man are accordingly becoming less appropriate. Since terms such as "chainperson" have never "caught on" in popular usage, the reader is left to substitute woman for man when warranted.

2 *Tape Measurements*

2.1 METHODS OF LINEAR MEASUREMENT

Throughout recorded history, people have always had some method of measuring distances. Measuring techniques are direct, such as applying a graduated tape against the marks to be measured, or indirect, such as measuring related parameters (e.g., the time required for light waves to be reflected to a source as in EDM work), and then computing the required (horizontal) distance.

2.2 TYPES OF MEASUREMENT

2.2.1 Pacing

Pacing is a very useful (although imprecise) form of measurement. Surveyors can determine the length of a pace that, for them, can be comfortably repeated. Pacing is particularly useful when looking for survey markers in the field; plan distance from a found marker to another marker can be paced off so that the marker can be located. Pacing is also useful as a rough check on construction layouts.

When performed on horizontal or uniformly sloping land, pacing can be performed at an accuracy level of $1/50$ to $1/100$. The accuracy of pacing cannot be relied upon when pacing up or down steep hills. Paces shorten on inclines and lengthen on declines.

2.2.2 Odometer

Odometer readings can be used to measure from one fenceline to another adjacent to a road. Odometer readings are precise enough to enable the surveyor to differentiate fencelines and assist in identification of property lines. This type of measurement is useful when beginning a survey or collecting information in order to begin a survey.

2.2.3 Electronic Distance Measuring Instruments

EDM instruments function by sending a light wave or microwave along the path to be measured and measuring the time involved in traversing the required distance, as with microwaves, or in measuring the time involved in returning the reflected light wave back to source. The instruments and techniques used are described in Chapter 8.

2.2.4 Stadia

Stadia is a form of tacheometry that utilizes a telescopic cross-hair configuration to assist in determining distances. A series of rod readings are taken with a theodolite and the resultant intervals used to determine distances. This technique is described in detail in Chapter 7.

2.2.5 Tacheometry

Tacheometry involves the measurement of a related distance parameter either by means of a fixed angle intercept (see stadia), by means of a timed light-wave signal reflection (see EDM) or by means of a measured angle to a fixed base. The latter type of measurement is obtained by a wide variety of range-finding devices employed by the military. The most popular device used in surveying is the subtense bar.

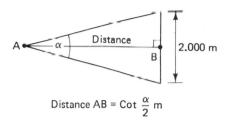

Distance AB = Cot $\frac{\alpha}{2}$ m

FIGURE 2.1 Subtense bar.

Subtense bar targets are held at 2.000 m apart (regardless of fluctuations in temperature) by invar wires under slight tension. The subtense bar, which is mounted on a tripod, is centered and leveled over point B (see Figure 2.1) and aligned roughly toward a theodolite set up over point A. The theodolite (1-second capability) can be used to align the bar exactly perpendicular to the line AB by noting the sighting device on the subtense bar. (The sighting mark can only be clearly seen when the bar is perpendicular to the line of sight.)

The accuracy of this technique over short distances is comparable to careful measuring with a steel tape.

Example 2.1

At 86 m, an error of 1 sec. of arc would result in an error of 0.018 m in distance and an accuracy ratio of $1/4800$:

$$\alpha_1 = 1°20'00'', \quad \frac{\alpha_1}{2} = 0°40'00'', \quad D = 85.940 \text{ m}$$

$$\alpha_2 = 1°20'01'', \quad \frac{\alpha_2}{2} = 0°40'00.5'', \quad D = 85.922 \text{ m}$$

$$E = 0.018$$

$$\text{Accuracy ratio} = \frac{0.018}{85.9} = \frac{1}{4772} = \frac{1}{4800}$$

However, at roughly double the distance, and with the same 1-s error, the accuracy ratio has halved:

$$0°20' \qquad D = 171.885 \text{ m}$$
$$0°20'00.5'' \qquad D = 171.814 \text{ m}$$
$$E = 0.071 \text{ m}$$
$$\text{Accuracy ratio} = \frac{0.071}{171.8} = \frac{1}{2420} = \frac{1}{2400}$$

Since the horizontal angle between the subtense bar targets is independent of any vertical angle (the α angle is being measured between the vertical planes containing the targets), the distance given by cot $(\alpha/2)$ m is always the horizontal distance. This feature makes the subtense bar a valuable aid to surveyors working in hilly and even mountainous country. However, the severe limitations on accuracy for long sights, together with the influx of EDM equipment, has resulted in very restricted use of this technique in modern North American practice.

2.3 GUNTER'S CHAIN

The measuring device in popular use during the time of the settlement of North America was the Gunter's chain, which is 66 ft long, subdivided into 100 links. The length of 66 ft was chosen because of its relationship to other units in the English system of measurements:

$$
\begin{aligned}
80 \text{ chains} &= 1 \text{ mile} \\
10 \text{ square chains} &= 1 \text{ acre } (10 \times 66^2 = 43560 \text{ ft}^2) \\
4 \text{ rods} &= 1 \text{ chain}
\end{aligned}
$$

The original surveys for most of Canada and the United States were performed by surveyors using Gunter's chains. Most of the continent's early legal plans and records contain dimensions in chains and links. The dimensions of the original lots reflect the Gunter's chain (e.g., 20 chains $\times$ 100 chains = 200 acres), and in many parts of Canada the allowance for future roads was routinely set at 1 chain (66 ft) in width. As a matter of fact, in Ontario new road allowances continued to be defined as being 66 ft wide up to the time that the province switched to metric (SI) units (in the late 1970s).

Example 2.2

An old plan shows a dimension of 5 chains 32 links. Convert this value to (a) feet; (b) meters:

$$
\begin{aligned}
\text{(a)} \quad & 5.32 \times 66 = 351.12 \text{ ft} \\
\text{(b)} \quad & 5.32 \times 66 \times 0.3048 = 107.021 \text{ m}
\end{aligned}
$$

The Gunter's chain, which is awkward to use and is relatively imprecise when compared to modern steel tapes, is apparently still being used in England (metric versions); in North America the Gunter's chain is now only of historical interest.

2.4 WOVEN TAPES

Woven tapes made of linen, Dacron, and the like can have fine-gauge copper strands interwoven to provide strength and to limit deformation due to long use and moisture. These so-called metallic tapes can conduct electricity and should not be used near electric installations. Work near electric installations should involve the use of *dry* nonmetallic or fiberglass tapes (see Figure 2.2).

All these tapes come in various lengths, the 100-ft (30 m) tape being the most popular, and are used for many types of measurements where high precision is not required. All woven tapes should be periodically compared with a steel tape to determine their levels of precision.

Many tapes are now manufactured with English units on one side and metric units on the reverse side. English units are in feet, tenths of a foot, and one-fifth or 0.05 ft, or in feet, inches, and quarter-inches. Metric units are in meters, centimeters, and half-centimeters (0.005 m).

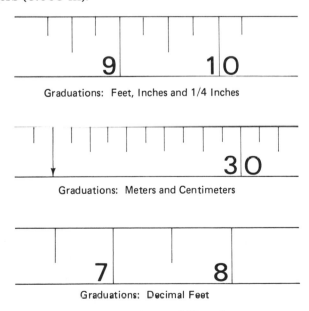

FIGURE 2.2 Metallic and fiberglass tapes.

2.5 STEEL TAPES

Steel tapes are manufactured in both foot and metric units and come in various lengths, markings, and unit weights (see Figure 2.3). In foot units, the steel tapes are manufactured in many lengths, but the 100-, 200-, and 300-foot tapes are the most common, with the 100-foot length being by far the more popular.

In metric units, steel tapes are manufactured in 20-, 30-, 50-, and 100-m lengths. Although metric practice in North America is fairly recent, the 30-m length is already established as the most popular length. It most closely resembles the popular 100-ft length and can be used with a comfortable "normal" tension (see Section 2.13.4).

Steel tapes come in two prevalent cross sections: heavy duty, 8 mm × 0.45 mm ($^5/_{16}"$ × 0.18 in.) or a normal cross section of 6 mm × 0.30 mm ($^1/_4"$ × 0.012 in.). Very lightweight tapes are manufactured in the longer (300 ft or 100 m) lengths, which can be of value for their easier handling characteristics.

Generally, the heavy-duty tapes (drag tapes) are used in route surveying (e.g., highways, railways) and are designed for use off the reel. Leather thongs are tied through the eyelets at both ends of the tape to aid in measuring. The lighter-weight tapes can be used on or off the reel and are found in structural and municipal work.

Invar tapes are composed of 35% nickel and 65% steel. This alloy has a very low coefficient of thermal expansion, making the tapes useful in precise work.

2.5.1 Types of Readouts

Steel tapes are normally marked in one of three ways:

1. Graduated throughout in feet and hundredths (0.01) of a foot or in meters and millimeters (see Figures 2.3 and 2.4).

2. The *cut* tape is marked throughout in feet, with the first and last foot graduated in tenths and hundredths of a foot (see Figure 2.3b). The metric cut tape is marked throughout in meters and decimeters, with the first and last decimeters graduated in millimeters (see Figure 2.3e). Some metric cut tapes have the first and last meters graduated in centimeters and millimeters. A measurement is made with the cut tape by one chainman holding that even foot (decimeter) mark, which will allow the other chainman to read a distance on the first foot graduated in hundredths of a foot (millimeters).

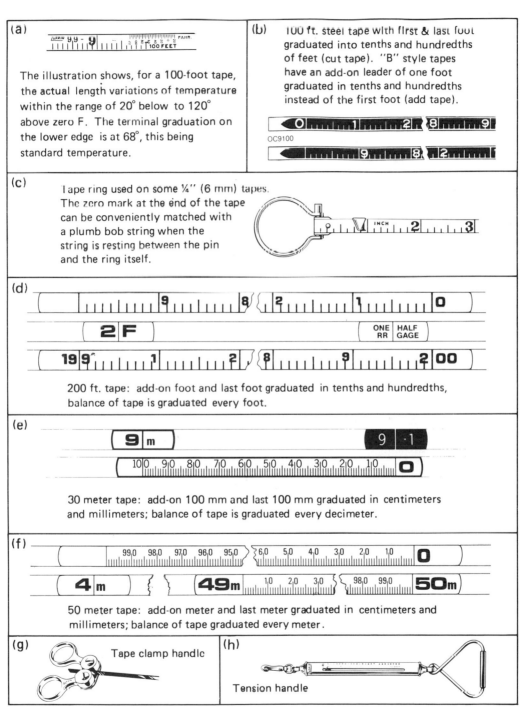

(a)

The illustration shows, for a 100-foot tape, the actual length variations of temperature within the range of 20° below to 120° above zero F. The terminal graduation on the lower edge is at 68°, this being standard temperature.

(b) 100 ft. steel tape with first & last foot graduated into tenths and hundredths of feet (cut tape). "B" style tapes have an add-on leader of one foot graduated in tenths and hundredths instead of the first foot (add tape).

OC9100

(c) Tape ring used on some ¼″ (6 mm) tapes. The zero mark at the end of the tape can be conveniently matched with a plumb bob string when the string is resting between the pin and the ring itself.

(d) 200 ft. tape: add-on foot and last foot graduated in tenths and hundredths, balance of tape is graduated every foot.

(e) 30 meter tape: add-on 100 mm and last 100 mm graduated in centimeters and millimeters; balance of tape is graduated every decimeter.

(f) 50 meter tape: add-on meter and last meter graduated in centimeters and millimeters; balance of tape graduated every meter.

(g) Tape clamp handle

(h) Tension handle

FIGURE 2.3 Lufkin steel tapes and accessories. (Courtesy of the Cooper Tool Group)

For example, the distance from A and B in Figure 2.4a is determined by holding 39 ft at B and reading 0.18 ft at A. Distance AB = 38.82 ft (i.e., 39 ft "cut" 0.18 = 38.82 ft). Each measurement involves this cut subtraction from the even foot (meter) mark being held at the far end. The cut tapes have been with us for generations and are accepted as adequate by most surveyors. However, these tapes, even in the hands of experienced chainmen, can lead to serious blunders. The mental subtraction required for each measurement is sooner or later going to result in an undetected mistake and cost someone considerable expense.

3. The *add* tape is also marked throughout in feet (meters and decimeters) with the last foot (meter or decimeter) being graduated to tenths and hundredths of a foot (cm and mm). An additional graduated foot (decimeter or meter) is included prior to the zero mark (see Figures 2.3b, d, e, and f). The distance from A to B in Figure 2.4b is determined by holding 38 ft at B and reading 0.82 ft at A. Distance AB is 38.82 ft (i.e., 38 "add" 0.82 = 38.82 ft).

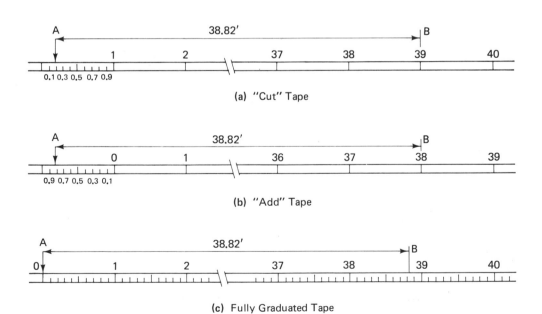

FIGURE 2.4 Various tape markings (hundredth marks not shown): (a) cut tape; (b) add tape; (c) fully graduated tape.

As noted, the cut tapes have the disadvantage of creating opportunities for subtraction mistakes. The add tapes have the disadvantage of forcing the chainman to adopt awkward measuring stances when measuring from the zero mark. The full meter add tape is almost impossible to use correctly consistently. The left (right) hand holding the leather thong at the end of the tape must be fully extended to allow the chainman to properly position the zero tape mark over the ground mark.

It is the authors' belief, based on many years of field experience, that there is no valid reason for employing either cut or add tapes. The small extra cost of fully graduated tapes is a small price to pay to remove the confusion and mistakes associated with add and particularly cut tapes. Tape manufacturers will supply fully graduated tapes in both reel-use tapes and off-the-reel (drag) tapes (Figure 2.4c).

2.6 STANDARD CONDITIONS FOR USE OF STEEL TAPES

Tape manufacturers, noting that steel tapes will behave differently under various temperature, tension, and support situations, specify the accuracy of their tapes under the following standard conditions:

1. Temperature = 68°F (20°C)
2. Tape fully supported throughout
3. Under a tension of 10 lb (50 newtons) (1 lb of force = 4.448 N)

If one or more of these standard conditions cannot be met, suitable corrections or techniques must be used to account for the errors that will result from nonstandard use.

2.7 TAPING ACCESSORIES

2.7.1 Plumb Bob

Plumb bobs are normally made of solid brass and weigh from 8 to 18 oz, with 10 and 12 oz being the most commonly used. Plumb bobs are used in taping to transfer from tape to ground (vice versa) when the tape is being held off the ground to maintain its horizontal alignment (see Figure 2.5).

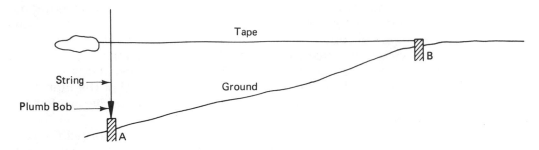

FIGURE 2.5 Plumb bob.

2.7.2 Hand Level

The hand level (see Figure 2.6) can be used to keep the steel tape horizontal, when measuring distances. The hand level is taken by the chainman at the lower elevation and a sight is taken back at the higher elevation chainman (see Figures 2.6a and 2.6b). For example, if the chainman with the hand level is sighting horizon-

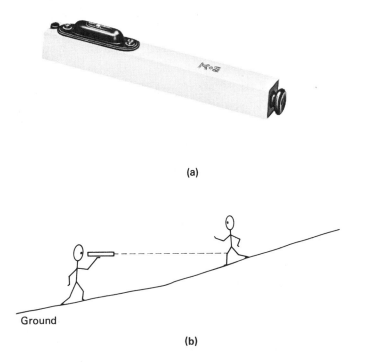

(a)

(b)

FIGURE 2.6 (a) Hand level; (b) hand-level application. (Courtesy of Keuffel & Esser Co.)

tally on his or her partner's waist, and if both are roughly the same height, then the chainman with the hand level is lower by the distance given from his eye to his waist. The low end of the tape is held that distance off the ground (using a plumb bob) with the high end of the tape being held on the mark.

Also shown is an Abney hand level (clinometer) (see Figure 2.7a). This is a hand level that allows for rough vertical angle determination by means of a graduated scale and a movable vernier scale with a level vial reference. This instrument can be used as a hand level, and it can also be used for vertical angle and slope determination.

Example 2.3

For height determination a value of 45° is placed on the scale. The surveyor moves forward or back until he is sighting the point that is being located. The height of the building would be the measured distance that is equal to h_1 (h_1/measured distance = tan 45° = 1), plus the distance from the eye to the ground (h_2) or from a mark on the building (or observer) to the ground (h_2 is also measured).

This technique is particularly useful in determining the height (clearance) of overhead electrical power lines. In that case, one surveyor can line up under the power line by holding up a plumb bob and lining up the power line with the string line. The other surveyor, holding the other end of the tape, backs away until the 45° setting on the clinometer allows him to sight the power line. The height of the power line is $h_1 + h_2$; h_2 can be determined by setting the clinometer to zero (level) and sighting on the surveyor under the power line, and then measuring from that sighted point to the ground.

The Abney hand level (clinometer) is also useful when working on route surveys where extended length (300 ft or 100 m) tapes are being used. The long tape can be used in a slope position (mostly supported) under proper tension. The clinometer can be used to record the slope angle, which will later be used to complete the appropriate horizontal distances.

Course	Slope Angle Line 22–23	Horizontal Distance
300	− 1°14′	299.93
300	− 1°32′	299.89
300	+ 0°52′	299.97
161.72	+ 1°10′	161.69
	Line 22–23 =	1061.48

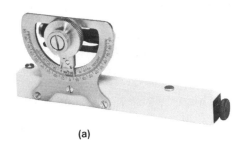

(a)

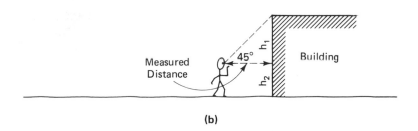

(b)

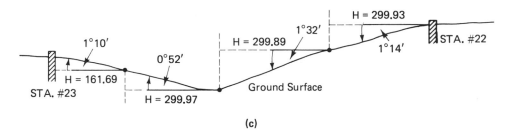

(c)

FIGURE 2.7 (a) Abney hand level; scale graduated in degrees with a vernier reading to 10 minutes (Courtesy Keuffel & Esser Co.); (b) Abney hand-level application in height determination; (c) Abney hand-level typical application in taping.

2.7.3 Additional Taping Accessories

The *clamp handle* (see Figure 2.3g) helps grip the tape at any intermediate point without bending or distorting the tape.

Tension handles (see Figure 2.3h) are used in precise work to ensure that the appropriate tension is being applied. They are graduated to 30 lb in $\frac{1}{2}$-lb graduations (50 N = 11.24 lb).

Chaining pins (marking arrows) come in sets of 11. They are painted alternately red and white and are 14 to 18 in. long. Chaining pins are used to set intermediate marks on the ground. In route surveying work the whole set of 11 pins is used to measure out ℄, the rear chainman being responsible for checking the number of whole tape lengths used by keeping an accurate count of the pins he collects. Eleven pins are used to measure out 1000 ft.

Tape repair kits are available so that broken tapes can be put back into service. The repair kits come in three main varieties: (1) punch pliers and repair eyelets, (2) steel punch block and rivets, and (3) tape repair sleeves. The steel punch block and rivets type is the only method that will give lasting repair. The block and rivets are simple to use, although great care must be exercised to ensure that the repair is precisely accomplished and the integrity of the tape is maintained.

Range poles are 6-ft wood poles with steel points. The poles are usually painted alternately red and white in 1-ft sections. Range poles are used in taping and transit work to provide alignment sights (Figure 2.8).

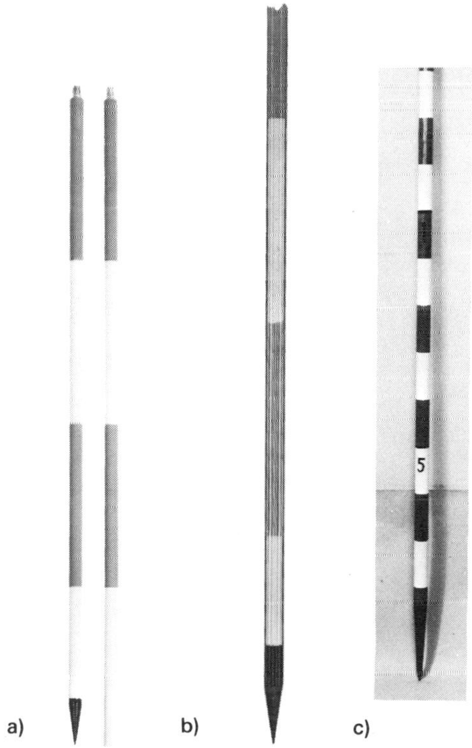

FIGURE 2.8 (a) Eight-foot range pole (two 4-foot sections). Alternate feet are painted red/orange and white. (b) Six- or eight-foot range pole (2½ m). Alternate feet (half-meters) are painted red/orange and white. (c) Wood range pole with steel point decimeters painted alternately red and white. Lengths of 2 or 2.5 m.

Plumb bob targets are designed for use with the plumb bob. The plumb bob string is threaded through the upper and lower notches so that the target ℄ is superimposed on the plumb bob string. The target can be adjusted up or down to aid in sighting. As a sighting device, the target is preferred to the range pole due to its portability (fits in the surveyor's pocket) and to its ability to provide more precise sightings (Figure 2.9).

2.8 TAPING METHODS

Taping (also known as chaining) is a common method either of determining the distance between field points or of establishing points in the field at prescribed distances.

Taping is normally performed with the tape held horizontally. If the distance to be measured is across smooth level land, the tape can be simply laid on the ground, properly aligned and tensioned, and then the end mark on the tape can be marked on the

FIGURE 2.9 Plumb bob cord target used to provide a transit sighting.

ground. If the distance to be measured is across sloping or uneven land, then at least one end of the tape must be raised off the ground to keep the tape horizontal. The raised end of the tape is referenced back to the ground mark with the aid of a plumb bob (see Figure 2.10). Normally, the only occasion when both ends of the tape are plumbed is when the ground rises between the two tapemen (see Figure 2.11).

2.8.1 Taping Procedure

The measurement begins with the head tapeman carrying the zero end of the tape forward toward the final point. He continues until the tape has been unwound, at which point the rear tapeman calls "tape" to alert the head tapeman to stop walking and to prepare for measuring. If a drag tape is being used, the tape is removed from the reel and a leather thong is attached to the reel end to facilitate measuring. If the tape is not designed to come off the reel, the winding handle is folded to the lock position and the reel is used to help hold the tape. The head tapeman is

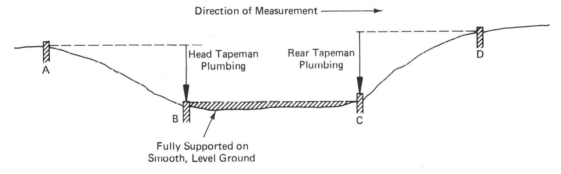

FIGURE 2.10   Horizontal taping; plumb bob used at one end.

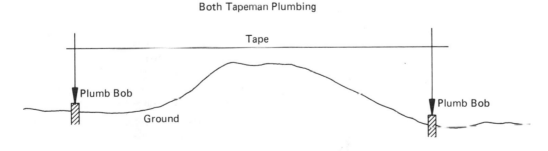

FIGURE 2.11   Horizontal taping; plumb bob used at both ends.

put on line by the rear tapeman, who is sighting forward to a range pole or other target that has been erected at the final mark. In precise work, the intermediate marks can be aligned by transit. The rear tapeman holds the appropriate graduation (e.g., 100.00 ft or 30.000 m) against the mark from which the measurement is being taken. The head tapeman, ensuring that the tape is straight, slowly increases tension to the proper amount and then marks the ground with a chaining pin or other marker. Once the mark has been made, both surveyors repeat the measuring procedure to check the measurement. If necessary, adjustments are made and the check procedure repeated.

If the ground is not level (determined by estimation or by use of a hand level) one or both surveyors must use a plumb bob. When plumbing, the tape is usually held at waist height, although any height between the shoulders and the ground is common. Holding the tape above shoulder height creates more chance for error as the tapeman must move his eyes up and down to include the ground mark and tape graduation in his field of view. The plumb bob string is usually held on the tape with the left thumb (right-handed people), taking care not to completely cover the graduation mark. As the tension is increased, it is common for the tapeman to take up some of the tension with the left thumb, causing it to slide along the tape; if the graduations have been covered with the left thumb, the surveyor is often not aware that the thumb (and string) has moved, resulting in an erroneous measurement. When plumbing, it is advisable to hold the tape close to the body in order to provide good leverage for applying or holding tension, and to accurately transfer from tape to ground, and vice versa.

If the rear tapeman is using a plumb bob, he shouts out "tape," "mark," and so on, at that instant when his plumb bob is steady and over the mark. If the head tapeman is also using a plumb bob, he must wait until both his and the rear tapeman's plumb bobs are simultaneously over their respective marks.

The student will discover that plumbing is a difficult aspect of taping. The student will encounter difficulty in holding the plumb bob steady over the point and at the same time applying appropriate tension. To help steady the plumb bob, the plumb bob is held only a short distance above the mark and is continuously touched down to ground, stake, and so on. This momentary touching down will dampen the plumb bob oscillations and generally steady the plumb bob. The student is cautioned against allowing the plumb bob to actually rest on the ground, stake, or whatever, as this will result in an erroneous measurement.

2.8.2 Taping Summary

Rear Tapeman

1. Aligns the head tapeman by sighting to a range pole or other target placed at the forward station.
2. Holds the tape on the mark, either directly or with the aid of a plumb bob. If a plumb bob is being used the rear tapeman will call out "tape," "mark," and so on, signifying to the head tapeman that, for that instant in time, the plumb bob (tape mark) is precisely over the station.
3. Calls out the station and tape reading for each measurement and listens for verification from the head tapeman.
4. Keeps a count of all full tape lengths included in each overall measurement.
5. Maintains the equipment (e.g., wipes the tape clean at the conclusion of the day's work or as conditions warrant).

Head Tapeman

1. Carries the tape forward, ensuring that the tape is free of loops, which could lead to tape breakage.
2. Prepares the ground surface for the mark (e.g., clears away grass, leaves, etc.).
3. Applies proper tension, after first ensuring that the tape is straight.
4. Places marks (chaining pins, wood stakes, iron bars, nails, rivets, cut crosses, etc.).
5. Takes and records measurements of distances (also temperature and other factors).
6. Supervises the taping work.

2.9 TAPING CORRECTIONS

Section 2.6 outlined the standard conditions that must be met for a steel tape to give precise results. The standard conditions referred to specific temperature and tension and to a condition of full support.

In addition to satisfying the standard conditions, the surveyors must also be concerned with horizontal versus slope distances and with ensuring that their techniques are sufficiently precise.

2.9.1 Taping Errors

The surveyor must make corrections for all significant systematic taping errors, and he must utilize techniques and equipment that will satisfactorily reduce random errors.

Systematic Errors

1. Slope
2. Erroneous tape length
3. Temperature
4. Tension
5. Sag

Random Errors

1. Slope
2. Temperature
3. Tension and sag
4. Alignment
5. Marking and plumbing

The reason for including some factors in both systematic and random categories is that, even when correcting for systematic errors in slope and temperature, there still exists the possibility of error when determining the correction parameters (i.e., the actual temperature or slope angle, etc.). An additional taping correction (reduction to sea level) is covered in Chapter 11.

2.10 SLOPE CORRECTIONS

Survey distances can be measured either horizontally or on a slope. Since survey measurements are normally shown on a plan, if the measurements were taken on a slope, they then must be converted to their horizontal equivalents before they can be plotted. To convert slope distances, either the slope angle or the vertical distance must also be known:

$$\frac{H \text{ (horizontal)}}{S \text{ (slope)}} = \text{Cos } \theta \quad \text{or} \quad H = S \text{ Cos } \theta \qquad (2.1)$$

where θ is the angle of inclination.

$$H^2 = S^2 - V^2, \qquad H = \sqrt{S^2 - V^2} \qquad (2.2)$$

where *V* is the difference in elevation; see part (c).

Example 2.4 *Slope Corrections*
(a) Slope angle: Given the slope distance and slope angle:

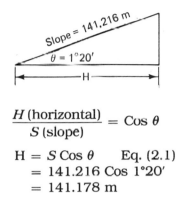

$$\frac{H \text{ (horizontal)}}{S \text{ (slope)}} = \text{Cos } \theta$$

$$H = S \text{ Cos } \theta \qquad \text{Eq. (2.1)}$$
$$= 141.216 \text{ Cos } 1°20'$$
$$= 141.178 \text{ m}$$

(b) Slope gradient: Given the slope distance and gradient (slope):

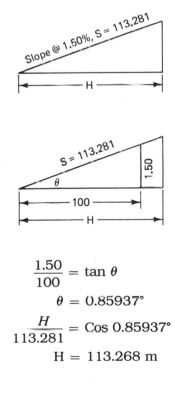

$$\frac{1.50}{100} = \tan \theta$$

$$\theta = 0.85937°$$

$$\frac{H}{113.281} = \text{Cos } 0.85937°$$

$$H = 113.268 \text{ m}$$

(c) Slope and vertical distance: Given the slope distance and difference in elevation (*V*):

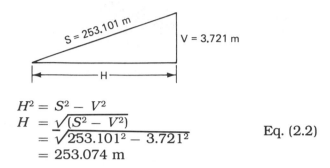

$$H^2 = S^2 - V^2$$
$$H = \sqrt{(S^2 - V^2)}$$
$$= \sqrt{253.101^2 - 3.721^2}$$
$$= 253.074 \text{ m}$$

Eq. (2.2)

(d) Given the slope distance and difference in elevation:

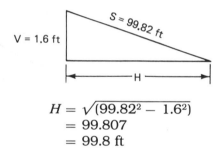

$$H = \sqrt{(99.82^2 - 1.6^2)}$$
$$= 99.807$$
$$= 99.8 \text{ ft}$$

In practice, most measurements are taken with the tape held horizontally. If the slope is too great to allow an entire tape length to be employed, then shorter increments will be measured until the required distance has all been measured. This operation is known as *breaking tape* (Figure 2.12). The sketch shows distance *AB* as being comprised of increments *AL, LM,* and *MB.*

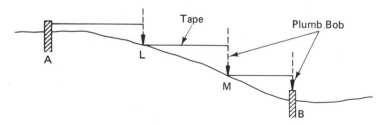

FIGURE 2.12 Breaking tape.

The exception to the foregoing occurs when preliminary route surveys (e.g., hydroelectric transmission lines) are performed using a 300-ft (100-m) steel tape. It is customary to measure slope distances, which allows the chainmen to keep this relatively heavy tape more or less fully supported on the ground. In order to allow for reduction to horizontal, each tape length is accompanied by its slope angle, usually determined by using a clinometer (Abney hand level) (see Section 2.7.2).

2.11 ERRONEOUS TAPE LENGTH CORRECTIONS

For all but precise work, tapes as supplied by the manufacturer are considered to be correct under standard conditions. Through extensive use, tapes do become kinked, stretched, and repaired. The length can become something other than that specified. When this occurs, the tape must be corrected or the measurements taken with the erroneous tape must be corrected.

Example 2.5

A measurement was recorded as 171.278 m with a 30-m tape that was only 29.996 m under standard conditions. What is the corrected measurement?

$$\text{Error per tape length} = -0.004$$
$$\begin{array}{l}\text{Number of times the} \\ \text{tape was used}\end{array} = \frac{171.278}{30}$$
$$\text{Total error} = 0.004 \times \frac{171.278}{30}$$
$$= -0.023 \text{ m}$$
$$\text{Corrected distance} = 171.278 - 0.023$$
$$= 171.255 \text{ m}$$

or

$$\text{Corrected distance} = \frac{29.996}{30} \times 171.278 = 171.255 \text{ m}$$

Example 2.6

It is required to lay out the front corners of a building, a distance of 210.08 ft. The tape to be used is known to be 100.02 ft. under standard conditions.

Error per tape length   = 0.02 ft.
Number of times that
                        = 2.1008
the tape is to be used
Total error   = 0.02 × 2.1008 = +0.04 ft

When the problem involves a *layout* distance, the sign of the correction must be reversed before being applied to the layout measurement. We must find that distance which when corrected by +0.04 will give 210.08 ft.

$$210.08 - 0.04 = 210.04 \text{ ft}$$

This is the distance to be laid out with that tape (100.02 ft) so that the corner points will be exactly 210.08 ft apart.

The student will discover that four variations of this problem are possible: correcting a measured distance while using (1) a long tape or (2) a short tape, or precorrecting a layout distance using (3) a long tape or (4) a short tape. To minimize confusion as to the sign of the correction, the student is urged to consider the problem with the distance reduced to only one tape length (100 ft or 30 m).

In Example 2.5, a recorded distance of 171.278 m was measured with a tape only 29.996 m long. The total correction was found to be 0.023 m. If doubt exists as to the sign of 0.023, ask yourself what would be the procedure for correcting only one tape length. In this example, after one tape length had been measured, it would have been recorded that 30 m had been measured. If the tape were only 29.996 m long, then the field book entry of 30 m must be corrected by −0.004 m.

The magnitude of the tape error is determined by comparing the tape with a tape that has been certified. (National Bureau of Standards, Gaithersburg, Maryland, United States or Ottawa, Canada.) In practice, tapes that require corrections for ordinary work are discarded.

2.12 TEMPERATURE CORRECTIONS

Section 2.6 notes the conditions under which tape manufacturers specify the accuracy of their tapes. One of these standard conditions is that of temperature. In the United States and Canada, tapes are standardized at 68°F or 20°C. Temperatures other than standard result in an erroneous tape length.

The thermal coefficient of expansion of steel is 0.00000645

per unit length per degree Fahrenheit (°F) (0.0000116 per unit length per degree Celsius, °C).

$$C_t = \alpha(T - T_s)L \qquad \text{general formula}$$

English Units

$$C_t = 0.00000645(T - 68)L$$

where C_t = correction due to temperature in feet

T = temperature (°F) of tape during measurement
L = distance measured in feet

Metric Units

$$C_t = 0.0000116(T - 20)L$$

where C_t = correction due to temperature in meters

T = temperature of tape (°C) during measurement
L = distance measured in meters

Example 2.7

A distance was recorded as being 471.37 ft at a temperature of 38°F.

$$C_t = 0.00000645(38 - 68)471.37$$
$$= -0.09$$

Corrected distance = 471.37 − 0.09 = 471.28 ft

Example 2.8

It is required to lay out two points in the field that will be exactly 100.000 m apart. Field conditions indicate that the temperature of the tape will be 27°C. What distance will be laid out?

$$C_t = 0.0000116(27 - 20)100.000$$
$$= +0.008 \text{ m}$$

Since this is a layout (precorrection) problem, the correction sign must be reversed (i.e., we are looking for the distance that when corrected for by +0.008 will give us 100.000 m):

$$\text{Layout distance} = 100.000 - .008$$
$$= 99.992 \text{ m}$$

For most survey work, accuracy requirements do not demand

precision in determining the actual temperature of the tape. Usually, it is sufficient to estimate air temperature. However, for more precise work (say $\frac{1}{10,000}$ and higher), care is required in determining the actual temperature of the tape, which can be significantly different than the temperature of the air.

Invar Steel Tapes

High-precision surveys require the use of steel tapes that have a low coefficient of thermal expansion. Such a tape is composed of a nickel–steel alloy having a thermal expansion ranging from 0.0000002 to 0.00000055 per degree Fahrenheit (3.60×10^{-7} to 5.50×10^{-7} per degree Celsius). Since the temperature of the tape can be significantly different from that of the surrounding air, it is customary to attach thermometers directly to the invar tapes. EDMs have now largely replaced invar tapes for precise distance measurements.

2.13 TENSION AND SAG CORRECTIONS

The three conditions under which tapes are normally standardized are given in Section 2.6. If a tape is fully supported and a tension other than standard is applied, a *tension* (pull) error exists. The tension correction formula is

$$C_p = \frac{(P - P_s)L}{AE}$$

If a tape has been standardized while fully supported and is being used without full support, an error called *sag* will occur.

The force of gravity pulls the center of the unsupported section downward in the shape of a catenary, thus creating an error $B'B$. The sag correction formula is

$$C_s = -\frac{w^2L^3}{24P^2} = \frac{-W^2L}{24P^2}$$

Table 2.1 defines the terms in these two formulas.

TABLE 2.1 *Correction Formula Terms Defined (English, Metric, and SI Units)*

Unit	Description	Foot	Metric	SI
C_p	Correction due to tension per tape length	ft	m	m
C_s	Correction due to sag per tape length	ft	m	m
L	Length of tape under consideration	ft	m	m
P_s	Standard tension Typical Standard Tension	lb (force) 10 lb (f)	kg (force) 4.5–5 kg (f)	N (newtons) 50 N
P	Applied tension	lb (f)	kg (f)	N
A	Cross-sectional area	in.^{2}	cm^2	m^3
E	Average modulus of elasticity of steel tapes	29×10^6 lb (f)/in.^{2}	21×10^5 kg (f)/cm^2	20×10^{10} N/m^2
E	Average modulus of elasticity of invar tapes	21×10^6 lb (f)/in.^{2}	14.8×10^5 kg (f)/cm^2	14.5×10^{10} N/m^2
ω	Weight of tape per unit length	lb (f)/ft	kg (f)/m	N/m
W	Weight of tape	lb (f)	kg (f)	N

Referring to Table 2.1, one newton is the force required to accelerate a mass of 1 kg by 1 meter/s^2.

$$\text{Force} = \text{mass} \times \text{acceleration}$$
$$\text{Weight} = \text{mass} \times \text{acceleration due to gravity (g)}$$
$$g = 32.2 \text{ ft/s}^2 = 9.807 \text{ m/s}^2$$

In SI units, a mass of 1 kg has a weight of 1×9.807 kg $\cdot$ m/s^2 = 9.807 N. That is,

$$1 \text{ kg (f)} = 9.807 \text{ N}$$

Since some tension spring balances are graduated in kilograms and since standard tensions are given in newtons by tape manufacturers, students must be prepared to work in both old metric and SI units.

2.13.1 Examples of Tension Corrections

Example 2.9

Given a standard tension of a 10-lb force for a 100-ft tape that is being used with a 20-lb force pull. If the cross-sectional area of the tape is 0.003 in.^{2}, what is the tension error for each tape length used?

$$C_p = \frac{(20 - 10)100}{29\ 000\ 000 \times 0.003} = +0.011 \text{ ft}$$

If a distance of 421.22 ft had been recorded, the total correction would be $4.2122 \times 0.011 = +0.05$ ft. The corrected distance would be 421.27 ft.

Example 2.10

Given a standard tension of 50 N for a 30-m tape that is being used with a 100-N force. If the cross-sectional area of the tape is 0.02 cm², what is the tension error per tape length?

$$C_p = \frac{(100 - 50)30}{0.02 \times 21 \times 10^5 \times 9.807} = +0.0036 \text{ m}$$

If a distance of 182.716 m had been measured under these conditions, the total correction would be

$$\frac{182.716}{30} \times 0.0036 = +0.022 \text{ m}$$

The corrected distance would be 182.738 m.

2.13.2 Notes on Tension Corrections

The cross-sectional area of the tape can be measured with a micrometer, taken from manufacturer's specifications, or it can be determined by using the following expression:

Tape length × tape area × specific weight of tape steel = weight

or

$$\text{Tape area} = \frac{\text{weight}}{\text{length} \times \text{specific weight}}$$

Example 2.11

A tape is weighed and found to be 1.95 lb. The overall length of the 100-ft tape (end to end) is 102 ft. The specific weight of steel is 490 lb/ft³.

$$\frac{102 \text{ ft} \times 12 \text{ in.} \times \text{area (in.}^2)}{1728 \text{ in.}^3} \times 490 \text{ lb/ft}^3 = 1.95 \text{ lb}$$

$$\text{Area} = \frac{1.95 \times 1728}{102 \times 12 \times 490} = 0.0056 \text{ in.}^2$$

Tension errors are usually quite small and as such have rele-vance only for very precise surveys. Even for precise surveys it is seldom necessary to calculate tension corrections, as availability of a tension spring balance allows the surveyor to apply standard tension and thus eliminate the necessity of calculating a correc-tion.

2.13.3 Sag Corrections

Example 2.12

$$C_s = \frac{-w^2 L^3}{24p^2} = \frac{-W^2 L}{24p^2}$$

$W^2 = w^2 L^2$
w = weight of tape per unit length
W = weight of tape between supports

A 100-ft steel tape weighs 1.6 lb and is supported only at the ends with a force of 10 lb. What is the sag correction?

$$C_s = \frac{-1.6^2 \times 100}{24 \times 10^2} = -0.11 \text{ ft}$$

If the force were increased to 20 lb, the sag is reduced to

$$C_s = \frac{-1.6^2 \times 100}{24 \times 20^2} = -0.03 \text{ ft}$$

Example 2.13

Calculate the length between two supports if the recorded length is 42.071 m, the mass of this tape is 1.63 kg, and the applied ten-sion is 100 N.

$$C_s = \frac{-(1.63 \times 9.807)^2 \times 42.071}{24 \times 100^2}$$

$$= -0.045$$

Therefore, length between supports = 42.071 − 0.045 = 42.026 m

2.13.4 Normal Tension

The error in a measurement due to sag can be eliminated by in-creasing the tension. Although sag cannot be entirely eliminated, the tape can be stretched to compensate for the residual sag.

Tension that will eliminate sag errors is known as *normal tension*. It ranges from 19 lb (light 100-ft tapes) to 31 lb (heavy 100-ft tapes).

$$P_n = \frac{0.204\,W\sqrt{AE}}{\sqrt{P_n - P_s}}$$

This formula will give a value for P_n that will eliminate the error caused by sag. The formula is solved by making successive approximations for P_n until the equation is satisfied. This equation is not much used in practice due to uncertainties associated with individual tape characteristics.

Experiment to Determine Normal Tension

Normal tension can be determined experimentally for individual tapes.

Example 2.14   *For a 100-ft steel tape*

1. Lay out the tape on a flat horizontal surface; an indoor corridor is ideal.
2. Select (or mark) a well-defined point on the surface at which the 100-ft mark is held.
3. Attach a tension handle at the zero end of the tape, apply standard tension, say 10 lb (f), and mark the surface at 0.00 ft.
4. Repeat the process, switching personnel duties, ensuring that the two marks are in fact exactly 100.00 ft apart.
5. Raise the tape off the surface to a comfortable height (waist). While the surveyor at the 100-ft end holds a plumb bob over the point, the surveyor(s) at the zero end slowly increases tension (a third surveyor could perform this function) until the plumb bob is over the zero mark on the surface. The tension is then read off the tension handle. This process is repeated several times, until a set of consistent results is obtained.

The most popular steel tapes (100 ft) now being used require a normal tension of about 24 lb (f). For most 30-m tapes now in use (lightweight), a normal tension of 90 N [20 lb (f) or 9.1 kg (f)] is appropriate.

2.14 RANDOM ERRORS ASSOCIATED WITH SYSTEMATIC TAPING ERRORS

As mentioned in Section 2.9.1, the opportunity exists for random errors to coexist with systematic errors. For example, when dealing with the systematic error caused by variations in temperature, the surveyor can determine the prevailing temperature in several ways:

1. The air temperature can be estimated.
2. The air temperature can be taken from a pocket thermometer.
3. The actual temperature of the tape can be determined by a tape thermometer held in contact with the tape.

For an error of 15°F in temperature, the error in a 100-ft tape would be

$$C_t = 0.00000645(15)100 = 0.01 \text{ ft}$$

Since $0.01/100 = 1/10000$, even an error of 15°F would only be significant for higher-order surveys. If metric equipment were being used, a comparable error would be $8\frac{1}{2}$°C. That is,

$$C_t = 0.0000116(8\frac{1}{2})30 = 0.003 \text{ m}$$

However, for precise work, random errors in determining temperature will be significant. Tape thermometers are recommended for precise work because of difficulties in estimating and because of the large differentials possible between air temperature and the actual temperature of the tape on the ground.

As a second example, consider the treatment of systematic errors dealing with slope versus horizontal dimensions. For a slope angle of 2°40′ read with an Abney handlevel (clinometer) to the closest 10 minutes we can say that an uncertainty of 5 minutes exists.

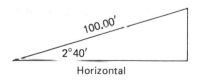

$$\text{Horizontal} = 100 \cos 2°40'$$
$$= 99.89 \text{ ft}$$

In this case an uncertainty of 5 min in the slope angle introduces an uncertainty of only 0.01 ft in the answer.

$$\text{Horizontal} = 100 \text{ Cos } 2°45' = 99.88 \text{ ft}$$
$$= 100 \text{ Cos } 2°35' = 99.90 \text{ ft}$$

Once again this error ($1/10,000$) would be significant for higher-order surveys.

A third example considers the treatment of systematic errors dealing with sag and tension. Consider the sag for a light 100-ft tape weighing 1 lb, with $A = 0.003$ and $E = 30000000$. Normal tension would be

$$\text{Trial 1} \quad 20 \quad = \frac{0.204 \times 1.0 \sqrt{0.003 \times 30000000}}{\sqrt{20 - 10}}$$

$$= \frac{61.2}{3.16} = 19.35$$

$$\text{Trial 2} \quad 19.7 = \frac{61.2}{\sqrt{9.7}} = 19.7 \qquad \text{OK for normal tension}$$

If a tension of 25 lb were exerted instead of 19.7 lb, the following error would occur:

$$C_p = \frac{(P - P_n)L}{AE}$$

$$= \frac{(25 - 19.7)100}{0.003 \times 30000000}$$

$$= 0.006 \text{ ft}$$

Once again, this error ($1/16,700$), by itself, is not significant for ordinary taping.

2.15 RANDOM TAPING ERRORS

In addition to the systematic and random errors already discussed, there are also random errors associated directly with the skill and care of the surveyors. These errors result from the inability of the surveyor to work with perfection in the areas of alignment, plumbing and marking, and estimating horizontal.

Alignment errors exist when the tape is inadvertently aligned off the true path (see Figure 2.13). Under ordinary surveying conditions, the rear tapeman can keep the head tapeman on line by sighting a range pole marking the terminal point. It would take an alignment error of about $1\frac{1}{2}$ ft to produce an error of 0.01 ft in 100 ft. Since it is not difficult to keep the tape aligned by eye to within a few tenths of a foot (0.2 to 0.3 ft), alignment is not usually a major concern. It should be noted that although most random errors are compensating, alignment errors are cumulative (misalignment can randomly occur on the left or on the right but in both cases the result of the misalignment is to make the measured course too long). Alignment errors can be virtually eliminated on precise surveys by using a transit to align all intermediate points.

Marking and plumbing errors are the most significant of all random errors. Even experienced surveyors must exercise great care to accurately place a plumbed mark to within 0.02 ft of true value over a distance of 100 ft. Horizontal measurements taken with the tape fully supported on the ground can be more accurately determined than measurements taken on slope requiring the use of plumb bobs; additionally, rugged terrain conditions that require many breaks in the taping process will cause errors to multiply significantly.

Errors are also introduced when surveyors estimate a horizontal position for a plumbed measurement. The effect of this error is identical to that of the alignment error previously discussed, although the magnitude of these errors is often larger than alignment errors. Skilled surveyors can usually estimate a horizontal position within 1 ft (0.3 m) over a distance of 100 ft (30 m); however, even experienced surveyors can be seriously in error when

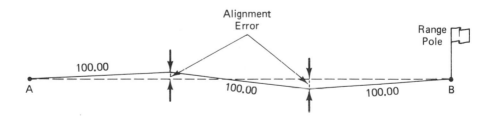

FIGURE 2.13 Alignment errors.

measuring across sidehills where one's perspective with respect to the horizon can be seriously distorted. These errors can be largely eliminated by using a hand level.

2.16 TECHNIQUES FOR ORDINARY TAPING PRECISION

Ordinary taping is referred to as being at the level of $1/5000$ accuracy. The techniques used for ordinary taping, once mastered, can easily be maintained. It is possible to achieve an accuracy level of $1/5000$ with little more effort than is required to attain the $1/3000$ level. Since the bulk of all surveying is either at the $1/3000$ or $1/5000$ level, experienced tapemen will often use $1/5000$ techniques even for the $1/3000$ level work. This practice permits good measuring work habits to be continually reinforced without appreciably increasing survey costs.

Because of the wide variety of field conditions that exist, absolute specifications cannot be prescribed. The specifications in Table 2.2 can be considered as typical for ordinary $1/5000$ taping.

TABLE 2.2   *Specifications for $1/5000$ Accuracy*

	Maximum Effect on One Tape Length	
Source of Error	*100 ft*	*30 m*
Temperature estimated to closest 7°F (4°C)	±0.005 ft	±0.0014 m
Care is taken to apply at least normal tension (lightweight tapes) and tension is known within 5 lb (20 N)	±0.006 ft	±0.0018 m
Slope errors are no larger than 1 ft/100 ft or 0.30 m/30 m	±0.005 ft	±0.0015 m
Alignment errors are no larger than 0.5 ft/100 ft or 0.15 m/30 m	±0.001 ft	±0.0004 m
Plumbing and marking errors are at a maximum of 0.015 ft/100 ft or 0.0046 m/30 m	±0.015 ft	±0.0046 m
Length of tape is known within ±0.005 ft (0.0015 m)	±0.005 ft	±0.0015 m

To determine the total random error (Σe) in one tape length, take the square root of the sum of the squares of the individual maximum errors (see Appendix A):

0.005^2	0.0014^2
0.006^2	0.0018^2
0.005^2	0.0015^2
0.001^2	0.0004^2
0.015^2	0.0046^2
0.005^2	0.0015^2
0.000337	0.000031

$$\Sigma e - \sqrt{0.000337} = 0.018 \text{ ft} \quad \text{or} \quad \sqrt{0.000031} = 0.0056 \text{ m}$$

$$\text{Accuracy} = \frac{0.018}{100} = \frac{1}{5400} \quad \text{or} \quad \frac{0.0056}{30} = \frac{1}{5400}$$

In the foregoing example, it is understood that corrections due to systematic errors had already been applied.

2.17 MISTAKES IN TAPING

If errors can be associated with inexactness, mistakes must be thought of as being blunders. Whereas errors can be analyzed and even, to some degree, predicted, mistakes are totally unpredictable: Since just one undetected mistake can nullify the results of an entire survey, it is essential to perform the work in a manner that will minimize the opportunity for mistakes and allow for verification of the results.

The opportunities for the occurrence of mistakes are minimized by setting up and then rigorously following a standard method of performing the measurement. The more standardized and routine the measurement manipulations, the more likely it is that the surveyor will immediately spot a mistake. The immediate double-checking of all measurement manipulations reduces the opportunities for mistakes to go undetected and at the same time increases the precision of the measurement. In addition to the immediate checking of all measurements, the surveyor is constantly looking for independent methods of verifying the survey results. Gross mistakes can often be detected by comparing the survey results with distances scaled from existing plans. The simple check technique of pacing can be an invaluable tool for rough verification of measured distances, especially construction layout distances. The possibilities for verification are only limited by the surveyor's diligence and imagination.

Common mistakes encountered in taping are the following:

1. Measuring to or from the wrong marker. All members of the survey crew must be vigilant to ensure that measurements begin or end at the appropriate permanent or temporary marker. Markers include legal bars, construction stakes or bars, nails, and the like.

2. Reading the tape incorrectly. It sometimes happens that mistakes are made by reading. Transposing figures is a common mistake (i.e., reading 56 instead of 65).

3. Losing proper count of the full tape lengths involved in a measurement. The counting of full tape lengths is primarily the responsibility of the rear tapeman and can be as simple as counting the chaining pins that have been collected as the work progresses. If the head tapeman is also keeping track of full tape lengths, mistakes, such as failing to pick up all chaining pins, can easily be spotted and corrected.

4. Recording the values in the notes incorrectly. It sometimes happens that the notekeeper will hear the rear tapeman's callout correctly but then transpose the figures as they are being entered in the notes. This mistake can be eliminated if the notekeeper calls out the value as he is recording it. The rear tapeman listens for this callout to ensure that the values called out are the same as the data originally given.

5. Calling out values ambiguously. The rear tapeman can call out 20.27 as twenty (pause) two, seven. This might be interpreted as 22.7. To avoid mistakes this value should be called out as twenty, decimal (point), two, seven.

6. When using cloth or fiberglass tapes, the zero point of the tape is often not identified correctly. This mistake can be avoided if the surveyor checks unfamiliar tapes before use. The tape itself can be used to verify the zero mark.

7. Arithmetic mistakes can exist in sums of dimensions and in error corrections (e.g., for temperature and slope). These mistakes can be identified and corrected if each member of the crew is responsible for checking (and signing) all survey notes.

PROBLEMS

2-1   The following distances were measured with a Gunter's chain – convert these distances to both feet and meters.
      a) 6 chains, 18 links
      b) 12 ch., 81 lk.
      c) 33.27 chains

2-2   Give three examples each of suitable use of the following measuring techniques or instruments;
      a) pacing, b) odometer, c) EDM, d) stadia,
      e) subtense, f) woven tape, g) steel tape.

2-3   A 100-ft "cut" steel tape was used to measure between two property markers. The rear tapeman held 51 ft, while the head tapeman cut 0.94 ft. What was the distance between the markers?

2-4   The slope measurement between two points is 341.236 m and the slope angle is 1°18′. Compute the horizontal distance.

2-5   A distance of 201.97 ft was measured along a 2% slope. Compute horizontal distance.

2-6   The slope distance between two points is 72.777 m and the difference in elevation between the points is 1.33 m. Compute the horizontal distance.

2-7   A 100-ft steel tape known to be only 99.97 ft long (under standard conditions) was used to record a measurement of 861.20 ft. What is the distance corrected for erroneous tape length?

2-8   A 30-m steel tape, known to be 30.009 m (under standard conditions) was used to record a measurement of 117.204 m. What is the distance corrected for erroneous tape length?

2-9   It is required to layout a rectangular commercial building – 80.00 ft wide and 110.00 ft long. If the steel tape being used is 100.05 ft long (under standard conditions), what distances would be laid out?

2-10 A survey distance of 300.00 ft was recorded when the field temperature was 83°F. What is the distance, corrected for temperature?

In Problems 2–11 through 2–15, compute the corrected distance.

	Temperature	Tapelength	Slope Data	Recorded Distance	
				Slope	Horizontal
2-11	18°F	100.00 ft	slope angle = +2°02′	321.68 ft	?
2-12	0°C	30.004 m	slope @ 1.20%	172.193 m	?
2-13	−20°F	99.98 ft	difference in elevation = 6.10 ft	498.98 ft	?
2-14	28°C	29.990 m	slope angle = −3°08′	?	358.621 m.
2-15	92°F	100.03 ft	slope @ −0.80%	?	610.55 ft

In Problems 2–16 through 2–20, compute the required layout distance.

	Temperature	Tapelength	Slope Data	Required Horizontal Distance	Actual Layout Distance
2-16	24°C	30.012 m	slope angle +4°30′	338.666 m	slope layout
2-17	50°F	99.98 ft	horizontal	300.00 ft	horizontal layout
2-18	30°F	100.02 ft	slope @ + 1.5%	500.00 ft	slope layout
2-19	11°C	29.990 m	horizontal	260.000 m	horizontal layout
2-20	92°F	100.04 ft	horizonal	440.00 ft	horizontal layout

2-21   A 50-m tape is used to measure between two points. The average weight of the tape per meter is 0.320 N. If the measured distance is 48.888 m, with the tape supported at ends only and with a tension of 100 N, find the corrected distance.

2-22   A 30-m tape has a mass of 544 g and is supported only at the ends with a force of 80 newtons – what is the sag correction?

2-23   A 100-ft steel tape weighing 1.8 lbs and supported only at the ends with a tension of 24 lbs is used to measure a distance of 471.16 ft, what is the distance corrected for sag?

2-24   A distance of 72.55 ft is recorded using a steel tape supported only at the ends a tension of 15 lbs and weighing 0.016 lbs per foot. Find the distance corrected for sag.

3 *Leveling*

3.1 DEFINITIONS

Leveling is the procedure used when determining differences in elevation between points that are remote from each other. An *elevation* is a vertical distance above or below a reference datum. In surveying, the reference datum that is universally employed is that of *mean sea level* (MSL). In North America, 19 years of observations at tidal stations in 26 locations on the Atlantic, Pacific, and Gulf of Mexico shorelines were reduced and adjusted to provide the national *geodetic vertical datum* (*NGVD*) of 1929. This datum is being further refined to reflect gravimetric and other anomalies in the 1988 general control readjustment. Although, strictly, the *NGVD* datum may not agree with mean sea level at specific points on the earth's surface, the term mean sea level (*MSL*) is generally used to describe the datum. MSL is assigned a vertical value (elevation) of 0.000 ft or 0.000 m. See Figure 3.1.

A **vertical line** is a line from the surface of the earth to the earth's center. It is also referred to as being a plumb line or a line of gravity.

A **level line** is a line in a level surface. A level surface is a curved surface parallel to the mean surface of the earth. A level

59

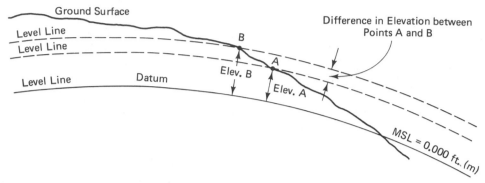

FIGURE 3.1 Leveling concepts.

surface is best visualized as being the surface of a large body of water at rest.

A horizontal line is a straight line perpendicular to a vertical line.

3.2 THEORY OF DIFFERENTIAL LEVELING

Differential leveling is used to determine differences in elevation between points that are remote from each other by using a surveyors' level together with a graduated measuring rod. The surveyors' level consists of a cross-hair equipped telescope and an attached spirit level tube, all of which is mounted on a sturdy tripod.

The surveyor is able to sight through the telescope to a graduated (in feet or meters) rod and determine a measurement reading at the point where the cross hair intersects the rod.

Referring to Figure 3.2, if the rod reading at $A = 6.27$ ft and the rod reading at $B = 4.69$ ft, the difference in elevation between A and B would be $6.27 - 4.69 = 1.58$ ft. Had the elevation of A been 461.27 ft (above MSL), then the elevation of B would be $461.27 + 1.58 = 462.85$ ft. That is,

461.27 (elev. A) + 6.27 (rod reading at A) − 4.69 (rod reading at B) = 462.85 (elev. B)

Referring to Figure 3.3, the student can visualize a potential problem. Whereas elevations are referenced to level lines (surfaces), the line of sight through the telescope of a surveyors' level is in fact a horizontal line. All rod readings taken with a surveyors' level will contain an error c over a distance d. The author has greatly exaggerated the curvature of the level lines shown in

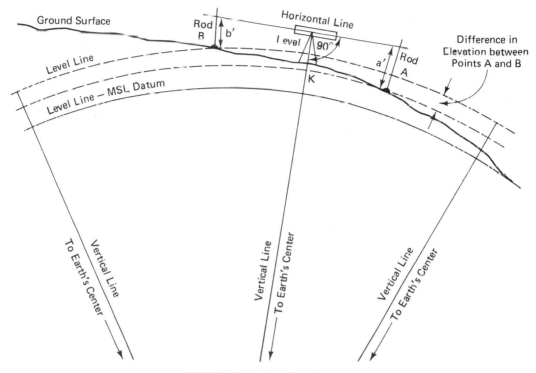

FIGURE 3.2   Leveling terms.

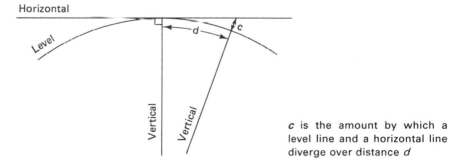

c is the amount by which a level line and a horizontal line diverge over distance *d*

FIGURE 3.3   Relationship between a horizontal line and a level line.

Figures 3.1 through 3.3 for illustrative purposes. In fact, the divergence between a level and a horizontal line is quite small. For example, over a distance of 1000 ft, the divergence will be 0.024 ft, and for a distance of 300 ft the divergence will only be 0.002 ft (0.0008 m in 100 m).

3.3 CURVATURE AND REFRACTION

The previous section introduced the concept of curvature error, that is, the divergence between a level line and a horizontal line over a specified distance. When considering the divergence between level and horizontal lines, one must also account for the fact that all sight lines are refracted downward by the earth's atmosphere. Although the magnitude of the refraction error is dependent on atmospheric conditions, it is generally considered to be about one-seventh of the curvature error. It is seen in Figure 3.4 that the refraction error of AB compensates for part of the curvature error AE, resulting in a net error due to curvature and refraction $(c + r)$ of BE.

From Figure 3.4, the curvature error can be computed:

$$(R + C)^2 = R^2 + KA^2$$
$$R^2 + 2RC + C^2 = R^2 + KA^2$$
$$C(2R + C) = KA^2$$
$$C = \frac{KA^2}{2R + C} \approx \frac{KA^2}{2R} \tag{3.1}$$

Take $R = 6370$ km:

$$C = \frac{KA^2 \times 10^3}{2 \times 6370} = 0.0785 \, KA^2$$

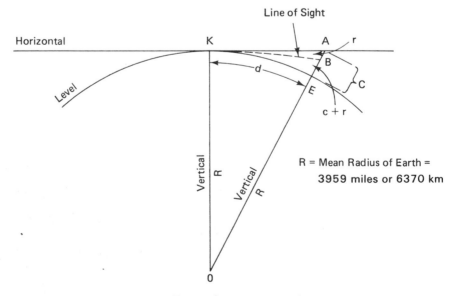

FIGURE 3.4   Effects of curvature and retraction.

Refraction is affected by atmospheric pressure, temperature, and geographic location, but, as noted earlier, it is usually expressed as being roughly equal to one-seventh of C.

$$\text{If } r = 0.14C$$
$$c + r = 0.0675K^2$$

where $K = KA$ (Figure 3.4) and is the length of sight in kilometers. The combined effects of curvature and refraction $(c + r)$ can be determined from the following formulas:

$$\begin{array}{lll}
(c + r)_m = 0.0675K^2, & (c + r)_m \text{ in meters; } K \text{ in kilometers} & (3.2) \\
(c + r)_{ft} = 0.574K^2, & (c + r)_{ft} \text{ in feet;} \quad K \text{ in miles} & (3.3) \\
(c + r)_{ft} = 0.0206M^2, & (c + r)_{ft} \text{ in feet;} \quad M \text{ in thousands} \\
& \qquad\qquad\qquad\qquad\qquad \text{of feet} & (3.4)
\end{array}$$

Example 3.1

Calculate the error due to curvature and refraction for the following distances.

    (a) 2500 ft; $c + r = 0.0206 \times 2.5^2 = 0.13$ ft

    (b) 400 ft; $c + r = 0.0206 \times 0.4^2 = 0.003$ ft

    (c) 2.7 miles; $c + r = 0.574 \quad \times 2.7^2 = 4.18$ ft

    (d) 1.8 km; $c + r = 0.0675 \times 1.8^2 = 0.219$ m

It can be seen from the figures in Table 3.1 that $(c + r)$ errors are relatively insignificant for differential leveling. Even for precise leveling, where distances of rod readings are seldom in excess of 200 ft (60 m), it would seem that this error is only of marginal importance. It will be shown in a later section that the field technique of balancing distances of rod readings effectively cancels out this type of error.

Errors due to curvature and refraction will be discussed further in Sections 3.22, 8.4, and 8.5.

TABLE 3.1 *Selected Values for (c + r) and Distance*

Distance (m)	30	60	100	120	150	300	1 km
$(c + r)_m$	0.0001	0.0002	0.0007	0.001	0.002	0.006	0.152
Distance (ft)	100	200	300	400	500	1000	1 mile
$(c + r)_{ft}$	0.000	0.001	0.002	0.003	0.005	0.021	0.574

3.4 DUMPY LEVEL

The dumpy level (Figure 3.5) was at one time used extensively on all engineering works. Although this simple instrument has, to a large degree, been replaced by more sophisticated instruments, it is shown here in some detail to aid in the introduction of the topic. For purposes of description, the level can be analyzed with respect to its three major components: telescope, level tube, and leveling head (Figure 3.6).

3.5 TELESCOPE

The telescope assembly is illustrated in Figure 3.6b. Rays of light pass through the objective (1) and form an inverted image in the focal plane (4). The image thus formed is magnified by the eyepiece lenses (3) so that the image can be clearly seen. The eyepiece lenses also focus the cross hairs, which are located in the telescope in the principal focus plane. The focusing lens (negative lens) (2) can be adjusted so that images at varying distances can be brought into focus in the plane of the reticle (4). In most telescopes designed for use in North America, additional lenses (prisms) are included in the eyepiece assembly so that the inverted

FIGURE 3.5   Dumpy level. (Courtesy Keuffel & Esser Co.)

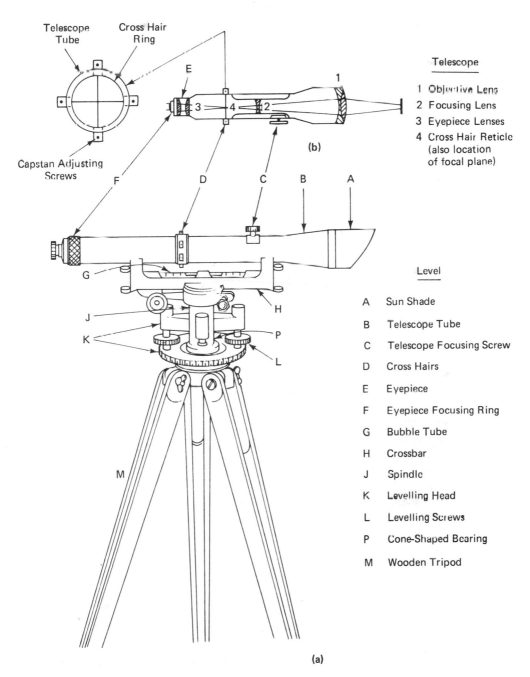

Telescope Tube

Cross Hair Ring

Capstan Adjusting Screws

E

3 4 2

1

(b)

Telescope

1 Objective Lens
2 Focusing Lens
3 Eyepiece Lenses
4 Cross Hair Reticle
(also location
of focal plane)

F D C B A

G

J

K

P

L

H

M

Level

A   Sun Shade
B   Telescope Tube
C   Telescope Focusing Screw
D   Cross Hairs
E   Eyepiece
F   Eyepiece Focusing Ring
G   Bubble Tube
H   Crossbar
J   Spindle
K   Levelling Head
L   Levelling Screws
P   Cone-Shaped Bearing
M   Wooden Tripod

(a)

FIGURE 3.6 (a) Dumpy level; (b) telescope for a dumpy level. (Adapted from *Construction Manual*, Ministry of Transportation and Communications, Ontario)

65

image can be viewed as an erect image. The minimum focusing distance for the object ranges from 1 to 2 m, depending on the instrument.

The line of collimation joins the center of the objective lens to the intersection of the cross hairs. The optical axis is the line taken through the center of the objective lens and perpendicular to the vertical lens axis. The focusing lens (negative lens) moved by focusing screw c (Figure 3.6a) has its optical axis the same as the objective lens.

The cross hairs (Figure 3.6b, no. 4) can be thin wires attached to a cross-hair ring or, as is more usually the case, cross hairs are lines etched on a circular glass plate that is enclosed by a cross-hair ring. The cross-hair ring, which has a slightly smaller diameter than does the telescope tube, is held in place by four adjustable capstan screws. The cross-hair ring (and the cross hairs) can be adjusted left and right or up and down simply by loosening and then tightening the two appropriate opposing capstan screws.

3.6 LEVEL TUBE

The level tube (Figure 3-6a, G) is a sealed glass tube mostly filled with alcohol or a similar substance. The upper (and sometimes lower) surface has been ground to form a circular arc. The degree of precision possessed by a surveyors' level is partly a function of the sensitivity of the level tube; the sensitivity of the level tube is directly related to the radius of curvature of the upper surface of the level tube. The larger the radius of curvature, the more sensitive the level tube.

Sensitivity is usually expressed as the central angle subtending one division (usually 2 mm) marked on the surface of the level tube. The sensitivity of many engineers' levels is 30″; that is, for a 2-mm arc, the central angle is 30″ ($R = 13.75$ m or 45 ft). See Figure 3.7.

The sensitivity of levels used for precise work is usually 10″; that is, $R = 41.25$m or 135 ft.

Precise levels, in addition to possessing more sensitive level tubes, also possess improved optics, including a greater magnification power. The relationship between the quality of the optical system and the sensitivity of the level tube can be simply stated: for any observable movement of the bubble in the level tube, there should be an observable movement of the cross hair on the leveling rod.

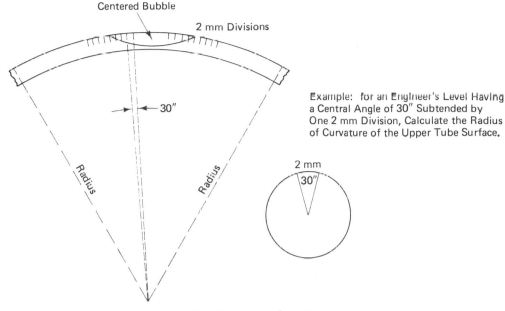

Example: for an Engineer's Level Having a Central Angle of 30″ Subtended by One 2 mm Division, Calculate the Radius of Curvature of the Upper Tube Surface.

Establish an angle/arc ratio

i.e.: $\dfrac{30''}{360°} = \dfrac{0.002}{2\pi R}$, R $= \dfrac{360}{0.00833} \times \dfrac{0.002}{2\pi}$, R = 13.75 m (or 45 ft.)

Note: Curvature shown exaggerated.

FIGURE 3.7 Level tube showing the relationship between the central angle per division and the radius of curvature of the level tube upper surface.

3.7 LEVELING HEAD

In the case of the dumpy level, four leveling foot screws are utilized to set the telescope level. The four foot screws surround the center bearing of the instrument (Figure 3.6) and are used to tilt the level telescope using the center bearing as a pivot.

Figure 3.8 illustrates how the telescope is positioned during the leveling process. The telescope is first positioned directly over two opposite foot screws. The two screws are kept only snugly tightened (overtightening makes rotation difficult and could damage the threads) and rotated in opposite directions until the bubble is centered in the level tube. If the foot screws become loose, it is an indication that the rotations have not proceeded uniformly; at worst, the foot screw pad can rise above the plate, making the telescope wobble. The solution for this condition is to turn one

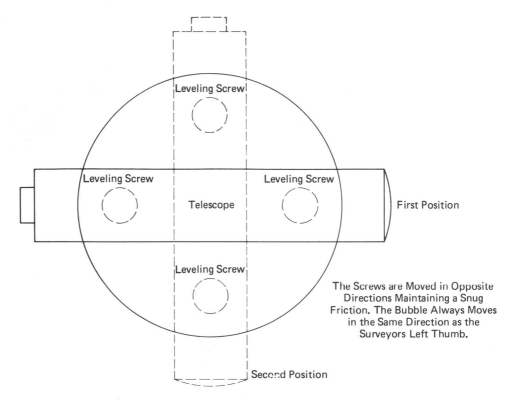

FIGURE 3.8 Telescope positions when leveling a four-screw level tube level.

screw until it again contacts the base plate and provides a snug friction when turned in opposition to its opposite screw.

The telescope is first aligned over two opposing screws; the screws are turned in opposite directions until the bubble is centered in the level tube. The telescope is then turned 90° to the second position, over the other pair of opposite foot screws, and the leveling procedure repeated. This process is repeated until the bubble remains centered. When the bubble remains centered in these two positions, the telescope is then turned 180° to check the adjustment of the level tube.

If the bubble does not remain centered when the telescope is turned 180°, it indicates that the level tube is out of adjustment. The technique used to adjust the level tube is described in Section 5.10 (plate bubbles). However, the instrument can still be used by simply noting the number of divisions that the bubble is off center and by moving the bubble half the number of those divisions. For example, if upon turning the leveled telescope 180° it is noted that the bubble is four divisions off center, the instrument can be leveled by moving the bubble to a position of two divisions off center. It will be noted that the bubble will remain two divisions off center no matter which direction the telescope is pointed. It should be emphasized that the instrument is in fact level if the bubble remains in the same position when the telescope is revolved, regardless of whether or not that position is in the center of the level vial.

3.8 TILTING LEVEL

The tilting level is roughly leveled by observing the bubble in the circular spirit level. Just before each rod reading is to be taken, and while the telescope is pointing at the rod, the telescope is precisely leveled by manipulating a tilting screw, which effectively raises or lowers the eyepiece end of the telescope. The level is equipped with a tube level, which is being precisely leveled by operating the tilting screw. The bubble is viewed through a separate eyepiece or, as is the case shown in Figure 3.10, through the telescope. The image of the bubble is split in two longitudinally and viewed with the aid of prisms. One-half of each end of the bubble is seen (see Figure 3.10), and after adjustment the two half-ends are brought to coincidence and appear as a continuous curve. When coincidence has been achieved, the telescope has been precisely leveled. It has been estimated (Wild Heerbrugg Ltd.) that the accuracy of centering a level bubble with reference to the open tubular scale graduated at intervals at 2 mm is about one-fifth of a division or 0.4 mm. With coincidence-type (split bubble) levels, however, this accuracy increases to about one fortieth of a division or 0.05 mm. As can be seen, these levels are useful where a relatively high degree of precision is required; however, if tilting levels are used on ordinary work (e.g., earthwork), the time (expense) involved in setting the split bubble to coincidence for each rod reading can scarcely be justified.

Most tilting levels (most European surveying instruments) come equipped with a three-screw leveling base. Whereas the support for a four-screw leveling base is the center bearing, the three-screw instruments are supported entirely by the foot screws themselves. This means that adjustment of the foot screws of a three-screw instrument effectively raises or lowers the height of the instrument line of sight. Adjustment of the foot screws of a four-screw instrument does not affect the height of the instrument line of sight, as the instrument is supported by the center bearing. Accordingly, the surveyor should be aware that adjustments made to a three-screw level in the midst of a setup operation will effectively change the elevation of the line of sight and could cause significant errors on very precise surveys (e.g., benchmark leveling or industrial surveying).

The bubble in the circular spirit level is centered by adjusting one or more of the three independent screws. Figure 3.9 shows the positions for a telescope equipped with a tube level when using three leveling foot screws. If this configuration is kept in mind when leveling the circular spirit level, the movement of the bubble is easily predicted. Some manufacturers produce levels and transits equipped with only two leveling screws. Figure 3.9 shows that the telescope positions are identical for both two- and three-screw instruments.

The level shown in Figure 3.10 is a Swiss-made 18″/2-mm precision level. This moderately precise level has the coincidence bubble display in the telescopic field of view, allowing the surveyor to verify coincidence at the instant the rod reading is taken.

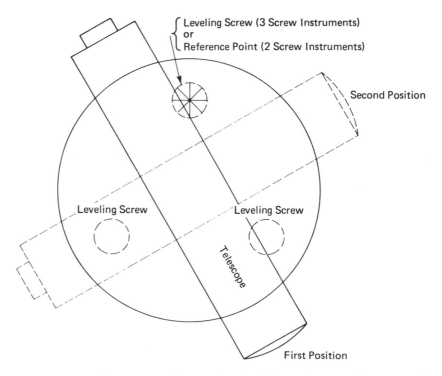

FIGURE 3.9 Telescope positions when leveling a three- or two-screw instrument tube level.

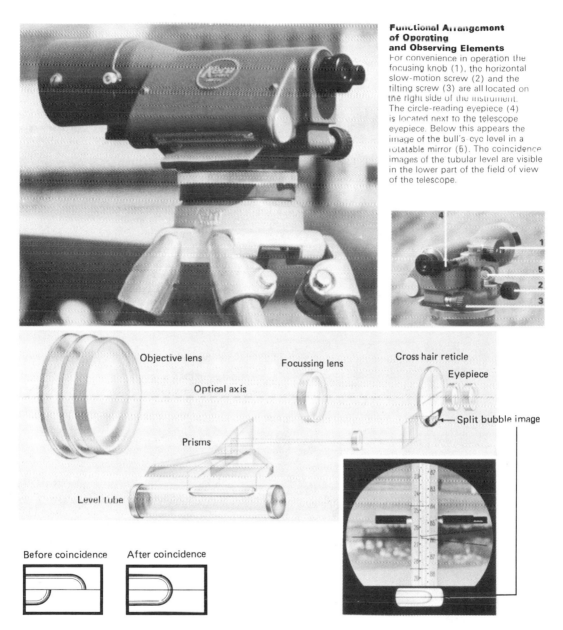

Functional Arrangement of Operating and Observing Elements
For convenience in operation the focusing knob (1), the horizontal slow-motion screw (2) and the tilting screw (3) are all located on the right side of the instrument. The circle-reading eyepiece (4) is located next to the telescope eyepiece. Below this appears the image of the bull's-eye level in a rotatable mirror (5). The coincidence images of the tubular level are visible in the lower part of the field of view of the telescope.

Objective lens

Optical axis

Focussing lens

Cross hair reticle

Eyepiece

Prisms

Split bubble image

Level tube

Before coincidence

After coincidence

FIGURE 3.10 Tilting level with coincidence bubble prisms; Kern Instruments GK 23.

3.9 AUTOMATIC LEVEL

The automatic level employs a gravity-referenced prism or mirror compensator to automatically orient the line of sight (line of collimation).

The instrument is quickly leveled using a circular spirit level; when the bubble has been centered (or nearly so), the compensator takes over and maintains a horizontal line of sight, even if the telescope is slightly tilted (see Figure 3.11).

The instrument shown in Figure 3.11 (a,b,c) is a Sokkisha automatic level, which is used in engineering surveys. The instrument shown in Figure 3.11 (d) is a precise level used in control surveys.

AUTOMATIC LEVELS

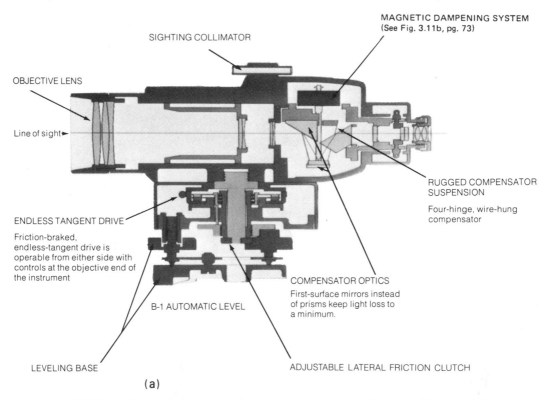

(a)

FIGURE 3.11   (a) Schematic of an engineers automatic level; (b) magnetic damping system for an engineers automatic level; (c) engineers automatic level. This level is mounted on a domed-head tripod, which permits rapid setup (Courtesy Sokkisha Co. Ltd.); (d) precise automatic level. (Courtesy Wild Heerbrugg Co. Ltd.) (*Continued*)

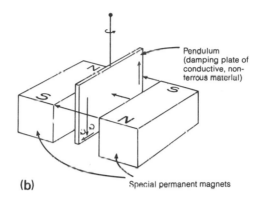

Pendulum
(damping plate of
conductive, non-
ferrous material)

Special permanent magnets

(b)

(c)

S.E. for: 1 mile (double run) = ±0.0011 ft
1 km (double run) = ±0.3 mm
(using a parallel plate micrometer) **(d)**

FIGURE 3.11 (*Continued*)

Automatic levels are extremely popular in present-day surveying operations and are available from most survey instrument manufacturers. They are quick to set up, easy to use, and can be obtained for use at almost any required precision.

A word of caution: all automatic levels employ a compensator referenced by gravity. This operation normally entails freely moving prisms or mirrors, some of which are hung by fine wires. It a wire or fulcrum were to break, the compensator would become inoperative and all subsequent rod readings would be incorrect.

The operation of the compensator can be verified by tapping the end of the telescope or by slightly turning one of the leveling screws (one manufacturer provides a push button), causing the telescopic line of sight to veer from horizontal. If the compensator is operative, the cross hair will appear to deflect momentarily before returning to its original rod reading. The constant checking of the compensator will avoid costly mistakes (see Chapter 12).

3.10 PRECISE LEVEL

Levels used to establish or densify vertical control are manufactured so as to give precise results. These levels can be tilting levels or automatic levels. The magnifying power, setting accuracy of the tubular level or compensator, quality of optics, and so on, are all improved to provide for precise rod readings. The least count on leveling rods is 0.01 ft or 0.001 m. Usually, precise levels are equipped with optical micrometers so that readings can be determined one or two places beyond the least count. Figure 3.12 shows a precise tilting level that utilizes an optical micrometer to measure the deflection of the line of sight resulting from rotation of a plane parallel plate of optical glass inserted just in front of the objective lens.

3.11 LEVELING RODS

Leveling rods are manufactured from wood, metal, or fiberglass and are graduated in feet or meters. The foot rod can be read directly to 0.01 ft, whereas the metric rod can usually only be read to 0.01 m, with millimeters being estimated. Metric rod readings are normally booked to the closest $1/3$ cm (i.e., 0.000, 0.003, 0.005, 0.007, and 0.010); more precise values can be obtained by using an optical micrometer.

One-piece rods are used for more precise work. The most precise work requires the face of the rod to be an invar strip held in place under temperature-compensating spring tension.

Normal leveling utilizes two- or three-piece rods graduated either in feet or meters. The sole of the rod is a metal plate that will withstand the constant wear and tear of leveling. The zero mark is at the bottom of the metal plate. The rods are graduated in a wide variety of patterns, all of which readily respond to logical analysis; the surveyor is well advised to study an unfamiliar rod at close

FIGURE 3.12 (a) Kern high-precision engineering tilting level, GK 23 C; (b) with micrometer and precise rod, reading 253.43 units (126.715 cm); (c) without micrometer (direct reading), 107.7 cm.

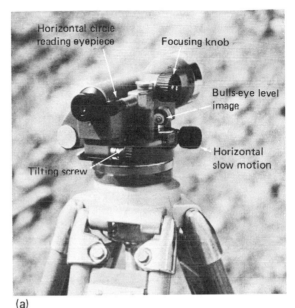

(a)

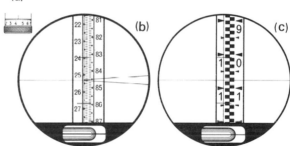

Coincidence image of the tubular level

The coincidence images of the tubular level are projected into the field of view of the telescope so that coincidence is verified at the instant the rod is read.

quarters prior to leveling to ensure that the graduations are thoroughly understood (see Figure 3.13 for a variety of rod cross-sections and graduation markings).

The rectangular sectioned rods are either of the folding (hinged) or sliding variety. Newer fiberglass rods have oval or circular cross sections and fit telescopically together for heights of 3, 5, or 7 m, from a stored height of 1.5 m (equivalent foot rods are also available).

Bench-mark leveling utilizes folding rods or invar rods, both of which have built-in handles and rod levels.

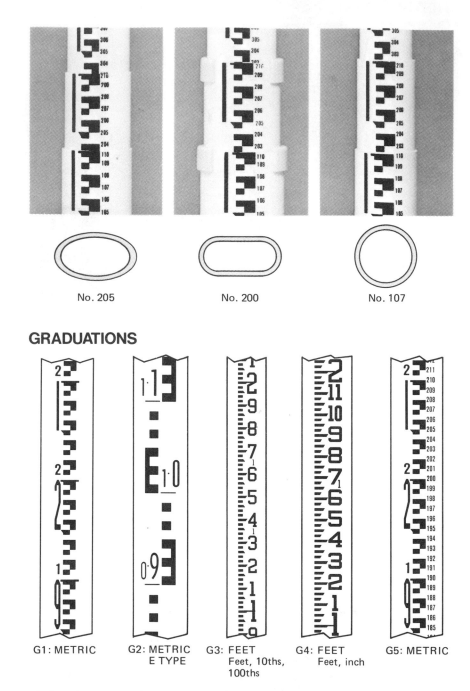

GRADUATIONS

| G1: METRIC | G2: METRIC E TYPE | G3: FEET Feet, 10ths, 100ths | G4: FEET Feet, inch | G5: METRIC |

FIGURE 3.13 (a) Telescopic leveling rods, showing available cross sections and graduation markings. (Courtesy Sokkisha Co. Ltd.) (*Continued*)

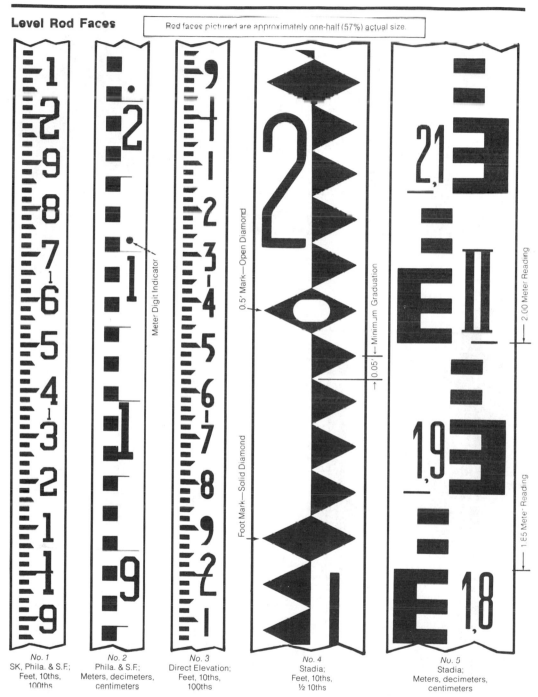

Level Rod Faces

Rod faces pictured are approximately one-half (57%) actual size.

No. 1	No. 2	No. 3	No. 4	No. 5
SK, Phila. & S.F.; Feet, 10ths, 100ths	Phila. & S.F.; Meters, decimeters, centimeters	Direct Elevation; Feet, 10ths, 100ths	Stadia; Feet, 10ths, ½ 10ths	Stadia; Meters, decimeters, centimeters

FIGURE 3.13 (b) Traditional rectangular cross-section leveling rods showing a variety of graduation markings. (Courtesy Lietz Co. Ltd.) (*Continued*)

FIGURE 3.13 (c) Invar rod, also showing the foot plate, which ensures a clearly defined rod position. (Courtesy Kern Instruments Ltd.)

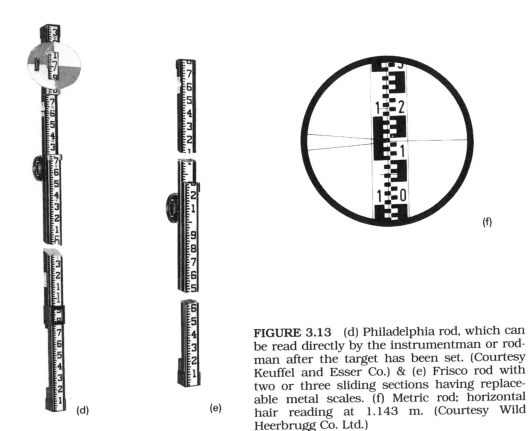

(d)

(e)

(f)

FIGURE 3.13 (d) Philadelphia rod, which can be read directly by the instrumentman or rodman after the target has been set. (Courtesy Keuffel and Esser Co.) & (e) Frisco rod with two or three sliding sections having replaceable metal scales. (f) Metric rod; horizontal hair reading at 1.143 m. (Courtesy Wild Heerbrugg Co. Ltd.)

FIGURE 3.14 Circular rod level. (Courtesy Keuffel & Esser Co.)

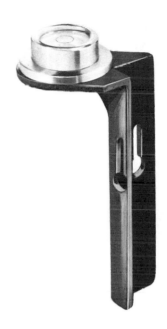

3.12 ROD LEVELS

Precise rods have a built-in circular bull's-eye level. When the bubble is centered, the rod is plumb. All other rods can be plumbed by using a rod level (Figure 3.14).

3.13 LEVELING PROCEDURE

The leveling procedure is used to determine the elevations of selected points with respect to a point of known elevation. Referring to Figure 3.15, the elevation of point A (BM) is known to be 410.26 ft above sea level, and it is desired to know the elevation of point B.

The level is set up midway between A and B, and rod readings are taken at both locations:

To Determine Elevation of B

$$
\begin{aligned}
\text{Elevation pt. } A &= 410.26 \\
\text{Backsight rod reading at } A &= +4.71 \text{ BS} \\
\text{Height (elevation) of instrument line of sight} &= 414.97 \text{ HI} \\
\text{Foresight rod reading at } B &= -2.80 \text{ FS} \\
\text{Elevation pt. } B &= 412.17
\end{aligned}
$$

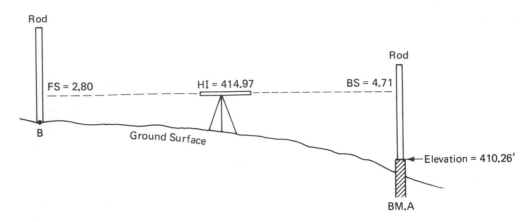

FIGURE 3.15   Leveling procedure: one setup.

After the rod reading of 4.71 is taken at *A*, the elevation of the line of sight of the instrument is known to be 414.97 (410.26 + 4.71). The elevation of point *B* can be determined by holding the rod at *B*, sighting the rod with the instrument, and reading the rod (2.80 ft). The elevation of *B* is therefore 414.97 − 2.80 = 412.17 ft. In addition to determining the elevation of point *B*, the elevations of any other points, lower than the line of sight and visible from the level, can be determined in a similar manner.

The situation depicted in Figure 3.16 shows the technique used when the point whose elevation is to be determined (BM 461) is too far from the point of known elevation (BM 460) for a one-setup solution. The elevation of an intermediate point (TP 1) is determined, allowing the surveyor to move the level to a location where BM 461 can be seen. Real-life situations may require numerous setups and the determination of the elevation of many intermediate points before getting close enough to determine the elevation of the desired point. When the elevation of the desired point has been determined, the surveyor must then either continue the survey to a point (BM) of known elevation, or the surveyor must return (loop) the survey to the point of commencement. The survey must be closed onto a point of known elevation so that the accuracy and acceptability of the survey can be determined. If the closure is not within allowable limits, the survey must be repeated.

In the 1800s and early 1900s, leveling procedures like the one described here were used to survey locations for railroads that traversed North America from the Atlantic Ocean to the Pacific Ocean.

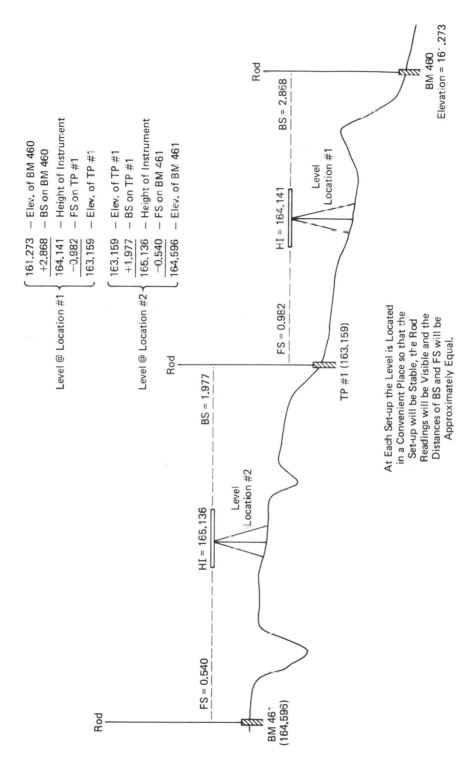

FIGURE 3.16 Leveling procedure: more than one setup.

At Each Set-up the Level is Located in a Convenient Place so that the Set-up will be Stable, the Rod Readings will be Visible and the Distances of BS and FS will be Approximately Equal.

Level @ Location #1
161.273	— Elev. of BM 460
+2.868	— BS on BM 460
164.141	— Height of Instrument
−0.982	— FS on TP #1
163.159	— Elev. of TP #1

Level @ Location #2
163.159	— Elev. of TP #1
+1.977	— BS on TP #1
165.136	— Height of Instrument
−0.540	— FS on BM 461
164.596	— Elev. of BM 461

Rod

HI = 165.136

Level Location #2

BS = 1.977

FS = 0.540

BM 46′ (164.596)

Rod

TP #1 (163.159)

FS = 0.982

HI = 164.141

Level Location #1

BS = 2.868

Rod

BM 460 Elevation = 16′.273

Figure 3.15 shows one complete leveling cycle; actual leveling operations are no more complicated than that shown. Leveling operations typically involve numerous repetitions of this leveling cycle, with some operations requiring that additional (intermediate) rod readings be taken at each instrument setup.

$$\text{Existing elevation} + \text{BS} = \text{HI} \qquad (3.5)$$
$$\text{HI} - \text{FS} = \text{new elevation} \qquad (3.6)$$

These two statements completely describe the differential leveling process.

When leveling between bench marks or turning points, the level is set approximately midway between the BS and FS locations to eliminate (or minimize) errors due to curvature and refraction (Section 3.3) and errors due to a faulty line of sight (Section 3.20).

To ensure that the rod is plumb, either a rod level (Section 3.12) is used, or the rodman gently "waves the rod" toward and away from the instrument. The correct rod reading will be the lowest reading observed. The rodman must ensure that the rod does not sit up on the back edge of the base and effectively raise the zero mark on the rod off the BM (or TP). The instrumentman is sure that the rod has been properly waved if the readings decrease to a minimum value and then increase in value (see Figures 3.17a and b).

3.14 LEVELING OPERATIONS DEFINITIONS

Bench mark (BM) is a permanent point of known elevation. Bench marks are established using precise leveling techniques and instrumentation. Bench marks are bronze disks or plugs set into vertical (usually) wall faces. It is important that the bench mark be placed in a structure that has substantial footings (at least below minimum frost depth penetration). Bench-mark elevations and locations are published by federal, state or provincial, and municipal agencies and are available to surveyors for a nominal fee (see Section 11.12).

Temporary bench mark (TBM) is a semipermanent point of known elevation. TBMs can be flange bolts on fire hydrants, nails in the roots of trees, top corners of concrete culvert headwalls, and so on. The elevations of TBMs are not normally published but are available in the field notes of various surveying agencies.

FIGURE 3.17 (a) Waving the rod; (b) waving the rod slightly to and from the instrument allows the instrumentman to take the most precise (lowest) reading.

(a)

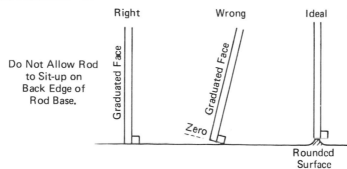

Right   Wrong   Ideal

Do Not Allow Rod
to Sit-up on
Back Edge of
Rod Base.

Graduated Face

Graduated Face

Zero

Rounded
Surface

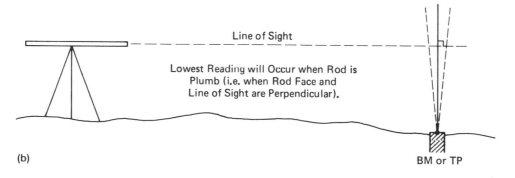

Line of Sight

Lowest Reading will Occur when Rod is
Plumb (i.e. when Rod Face and
Line of Sight are Perpendicular).

(b)

BM or TP

83

Turning point (TP) is a point temporarily used to transfer an elevation (see Figure 3.16).

Backsight (BS) is a rod reading taken on a point of known elevation in order to establish the elevation of the instrument line of sight.

Height of instrument (HI) is the elevation of the line of sight through the level (i.e., elevation of BM + BS = HI).

Foresight (FS) is a rod reading taken on a turning point, bench mark or temporary bench mark in order to determine its elevation (i.e., HI − FS = elevation of TP (BM or TBM)

Intermediate foresight (IS) is a rod reading taken at any other point whose elevation may be required.

$$HI - IS = \text{elevation}$$

Most engineering leveling projects are initiated so as to determine the elevations of intermediate points (as in profiles, cross sections, etc.).

The addition of backsights to elevations to obtain heights of instrument and the subtraction of foresights from heights of instrument to obtain new elevations are known as note reductions.

The arithmetic can be verified by performing the arithmetic check (page check). Since all BS are added, and all FS are subtracted, when the sum of BS are added to the original elevation, and then the sum of FS subtracted from that total, the remainder should be the same as the final elevation calculated (see Figure 3.18). Arithmetic check:

$$\text{Original elevation} + \Sigma\, BS - \Sigma\, FS = \text{new elevation}$$

3.15 LEVELING OPERATIONS

In leveling, as opposed to transit work, the instrument can usually be set up in a relatively convenient location. If the level has to be set up on a hard surface such as asphalt or concrete, the tripod legs will be spread out to avoid problems with wind knocking over the instrument. When the level is to be set up on soft surfaces (e.g., turf), the tripod is first set up so that the tripod top is nearly horizontal, and then the tripod legs are firmly pushed into the earth. The tripod legs are snugly tightened to the tripod top so that a leg, when raised, will just fall back under the force of its own weight. Undertightening can cause an unsteady setup, just as overtightening can cause an unsteady setup due to torque strain.

BENCHMARK VERIFICATION

BM460 to BM461

KANEN-π

Job SOMERVILLE-ROD, LEVEL #C9 19°C, CLOUDY

Date MAY 10 1983 Page 62

STA	B.S. (+)	H.I.		F.S. (−)	ELEV.	DESCRIPTION
BM 460	2.868	164.141			161.273	BRONZE PLATE SET IN --- ETC.
T.P. #1	1.977	165.136		0.982	163.159	NAIL IN ROOF OF MAPLE --- ETC.
BM 461				0.540	164.596	BRONZE PLATE SET IN --- ETC.
					↑	
Σ	4.845			1.522	✓	
ARITHMETIC CHECK:						
	161.273 + 4.845 − 1.522			=	164.596	

FIGURE 3.18 Leveling field notes and arithmetic check (data from Figure 3.16).

On hills it is customary to place one leg uphill and two legs downhill; the instrumentman stands facing uphill while setting up the instrument.

The tripod legs can be adjustable or straight-leg; the straight-leg tripod is recommended for leveling as it contributes to a more stable setup. After the tripod has been set roughly level, with the feet firmly pushed into the ground, the instrument can be leveled.

Four-screw instruments attach to the tripod via a threaded base and are leveled by rotating the telescope until it is directly over a pair of opposite leveling screws, and then by proceeding as described in Section 3.7 and Figure 3.8.

Three-screw instruments are attached to the tripod via a threaded bolt that projects up from the tripod top into the leveling base of the instrument. The three-screw instrument, which usually has a circular bull's-eye bubble level, is leveled as described in Section 3.8 and Figure 3.9. Unlike the four-screw instrument, the three-screw instrument can be manipulated by moving the screws one at a time, although experienced surveyors will usually manipulate at least two at a time; and often all three are manipulated at the same time. After the bubble has been centered, the instrument can be revolved to check that the circular bubble remains centered.

Once the level has been set up, preparation for the rod readings can take place. The eyepiece lenses E (see Figures 3.5 and 3.6) are focused by turning the eyepiece focusing ring (F) until the cross hairs are as black and as sharp as possible (it helps to have the telescope pointing to a light-colored background for this operation). Next, the rod is brought into focus by turning the telescope focusing screw (C in Figure 3.6) until the rod graduations are as clear as possible. If both of these focusing operations have been carried out correctly, it will appear that the cross hairs are superimposed on the leveling rod. If either focusing operation (eyepiece or rod focus) has not been properly carried out, it will appear that the cross hair is moving slightly up and down as the observer's head moves slightly up and down. The apparent movement of the cross hair can result in incorrect readings. The condition where one or both focus adjustments have been improperly made, and the resultant error, is known as *parallax.*

3.16 SIGNALS

Since the instrumentman and the rodman are some distance apart, and often in traffic or construction noise, a series of prearranged arm signals is used to communicate information. If a turning point is being called for (by either the rodman or the instrumentman), the person calling for the turning point holds an arm erect and slowly makes a horizontal circle. If a task has been accomplished, both arms are extended sideways and waved up and down to indicate OK. On some occasions, the rod is being held so close to the instrument that the foot or meter marks do not show up in the field of view. When that happens, the instrumentman asks (or makes an upward motion with one hand) the rodman to "raise for red" (i.e., slowly raise the rod until a red foot or meter mark is visible.) Many rods now have the foot or meter marks repeated on the rod face, thus eliminating the need of "raising for red."

Other signals, often a combination of common sense and tradition, vary from one locality (or agency) to another. Portable radios are used on many noisy construction sites to ensure that the operation goes smoothly.

3.17 BENCH-MARK LEVELING (VERTICAL CONTROL SURVEYS)

Bench-mark leveling is the type of leveling employed when a system of bench marks is to be established or when an existing system of bench marks is to be extended or densified (e.g., perhaps a bench mark is required in a new location, or perhaps an existing bench mark has been destroyed and a suitable replacement is required). Bench-mark leveling is typified by the relatively high level of precision specified, both for the instrumentation and for the technique itself.

Table A-18 (See Appendix A) shows the classification, standards of accuracy and general specifications for vertical control as given by Energy, Mines, and Resources—Canada. Table A.19 (See Appendix A) shows the same data as given by the U.S. Coast and Geodetic Surveys.

The specifications shown in Tables A-18 and A-19 (Appendix A) largely cover the techniques of precise leveling. Precise levels with coincidence tubular bubbles of a sensitivity of 10 seconds per 2-mm division (or equivalent for automatic levels) and equipped with parallel-plate micrometers are used almost exclusively for this type of work. Invar rods (Figure 3.13 d), together with a base plate, rod level, and supports, are used in pairs to minimize the time required for successive readings. Tripods for this type of work are longer than usual, enabling the surveyor to keep the line of sight farther above the ground, thus minimizing interference and errors due to refraction. Ideally, the work is performed on a cloudy windless day, although work can proceed on a sunny day if the instrument is protected from the sun and its possible differential thermal effects on the instrument.

Bench marks, at the national level, are established by federal agencies utilizing first-order methods and first-order instruments. The same high requirements are also specified for state and provincial grids; but as work proceeds from the whole to the part (i.e., down to municipal or regional grids), the rigid specifications are relaxed somewhat. For most engineering works, bench marks are established (from the municipal or regional grid) at third-order specifications. Bench marks established to control isolated construction projects may be at an even lower order of accuracy.

It is customary in bench-mark leveling at all orders of accuracy to first verify that the starting bench mark's elevation is correct. This can be done by two-way leveling to the closest adjacent bench mark.

This check is particularly important when the survey loops back to close on the starting bench mark and no other verification is planned.

3.18 PROFILE AND CROSS-SECTION LEVELING

In engineering surveying, we often consider a route (road, sewer pipeline, channel, etc.) from three distinct perspectives. The *plan view* of route location is the same as if we were in an aircraft looking straight down. The *profile* of the route is a side view or elevation (see Figure 3.19 and 3.20) in which the longitudinal surfaces are highlighted (e.g., of road, top and bottom of pipelines, etc.). The *cross section* shows the end view of a section at a point (0 + 60 in Figure 3.21) and is at right angles to the center line. These three views taken together completely define the route in X, Y, and Z coordinates.

Profile levels are taken along a path that holds interest for the designer. In road work, preliminary surveys often profile the proposed location of the center line (℄) (see Figure 3.19). The proposed ℄ is staked out at an even interval (50 to 100 ft or 20 to 30 m). The level is set up in a convenient location so that the bench mark and as many intermediate points as possible can be sighted. Rod readings are taken at the even station locations and at any other point where the ground surface has a significant change in slope. When the rod is moved to a new location and it cannot be

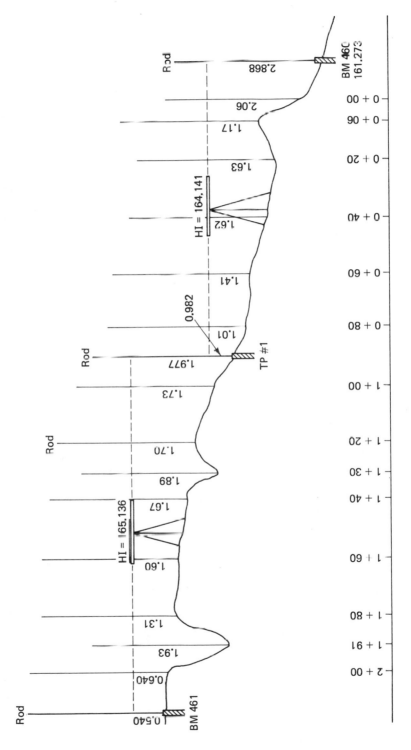

FIGURE 3.19 Example of profile leveling; see Figure 3.20 for survey notes.

89

SMITH—NOTES
BROWN—π
JONES—ROD

PROFILE OF PROPOSED
ROAD 0+00 to 2+00

Job 21°C — SUNNY   LEVEL #L-14

Date ...... AUG. 1 1983 ...... Page 72

STA.	B.S.	H.I.	I.S	F.S.	ELEV.	DESCRIPTION
BM 460	2.868	164.141			161.273	BRONZE PLATE SET IN – – – ETC.
0+00			2.06		162.08	℄
0+06			1.17		162.97	℄ –TOP OF BERM
0+20			1.63		162.51	℄
0+40			1.62		162.52	℄
0+60			1.41		162.73	℄
0+80			1.01		163.13	℄
T.P. #1	1.977	165.136		0.982	163.159	NAIL IN ROOT OF MAPLE – – – ETC.
1+00			1.73		163.41	℄
1+20			1.70		163.44	℄
1+30			1.89		163.25	℄ BOTTOM OF GULLY
1+40			1.67		163.47	℄
1+60			1.60		163.54	℄
1+80			1.31		163.83	℄
1+91			1.93		163.21	℄ BOTTOM OF GULLY
2+00			0.64		164.50	℄
BM 461				-1.522	164.596	BRONZE PLATE SET IN – – – ETC.
	Σ=4.845			Σ=1.522		

ARITHMETIC CHECK: 161.273 + 4.845 – 0.540 = 164.596

Σ=1.522 – 0.540 = 164.596

164.591 – PUBLISHED ELEV.

$$E = \frac{164.596}{\underline{164.591}}$$
$$0.005$$

ALLOWABLE ERROR (3RD ORDER)

= 12 mm √K, = .012 √.2 = .0054 m

ABOVE ERROR (.005) SATISFIES 3RD ORDER.

FIGURE 3.20 Profile field notes.

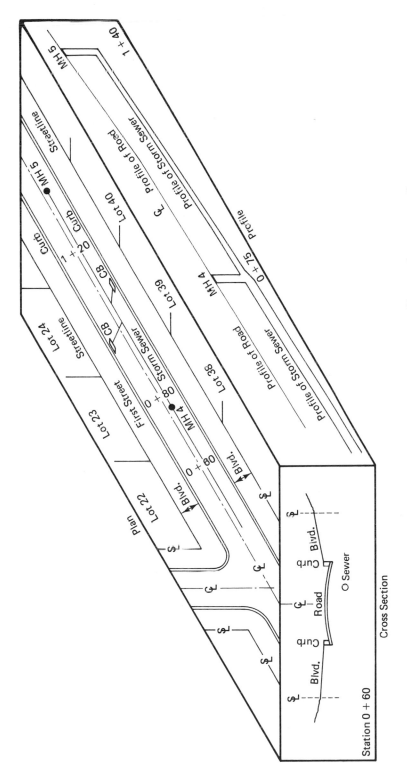

FIGURE 3.21 Relationship between plan, profile, and cross-section views.

91

seen from the instrument, a turning point is called for so that the instrument can be moved ahead and the remaining stations leveled.

The turning point can be taken on a wood stake, corner of concrete step or concrete headwall, a lug on the flange of a hydrant, and so on. The turning point should be a solid, well-defined point that can be precisely described and it is hoped found again at a future date. In the case of leveling across fields, it usually is not possible to find turning point features of any permanance; in that case, stakes are driven in and then abandoned when the survey is finished. In the example shown in Figure 3.20, the survey was closed acceptably to BM 461; had there been no bench mark at the end of the profile, the surveyor would have looped back and closed into the initial bench mark.

The field notes for the profile leveling shown in Figure 3.19 are shown in Figure 3.20. The intermediate sights (IS) are shown in a separate column, and the elevations at the intermediate sights show the same number of decimals as shown in the rod readings. Rod readings on turf, ground, and the like are usually taken to the closest 0.1 ft or 0.01 m. Rod readings taken on concrete, steel, asphalt, and so on are usually taken to the closest 0.01 ft or 0.003 m.

It is a waste of time and money to read the rod more precisely than conditions warrant.

The reader is referred to Chapter 9 for details on plotting the profile.

When the route of the facility has been agreed on, further surveying is required. Once again the ℄ is staked (if necessary) and cross sections are taken at all even stations. In roadwork, rod readings are taken along a line perpendicular to ℄ at each even station. The rod is held at each significant change in surface slope and at the limits (℄) of the job. In uniformly sloping land areas it is often the case that the only rod readings required at each cross-sectioned station are at ℄ and the two ℄'s. Chapter 9 shows how the cross sections are plotted and then utilized to compute volumes of cut and fill (℄ denotes streetline or property line and ℄, the center line).

Figure 3.22 illustrates the rod positions required to suitably define the ground surface at 0 + 60, at right angles to ℄. Figure 3.23 shows typical cross-section note forms employed in many municipalities. Figure 3.24 shows typical cross-section note forms favored by many highways agencies.

The student will note that the HI (178.218) has been rounded to two decimals (178.20) to facilitate reduction of the two-decimal rod readings. The rounded value is placed in brackets to

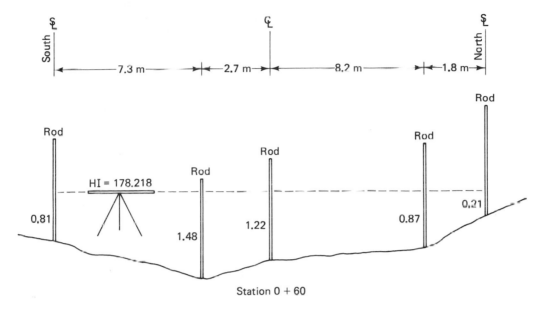

FIGURE 3.22   Cross-section surveying.

distinguish it from the correct HI, from which the next FS will be subtracted.

Road and highway construction often necessitates the location of granular (sand, gravel) deposits for use in the highway road bed. These *borrow pits* (gravel pits) are surveyed to determine the volume of material "borrowed" and transported to the site. Before any excavation takes place, one or more reference baselines are established and a minimum of two bench marks is located in convenient locations. The reference lines are located in secure locations where neither the stripping and stockpiling of topsoil or the actual excavation of the granular material will endanger the stakes (Figure 3.25). Cross sections are taken over (and beyond) the area of proposed excavation. These *original* cross sections will be used as a datum against which interim and final excavations will be measured. The volumes calculated from the cross sections (see Chapter 9) are often converted to tons (tonnes) for payment purposes. In many locations, weigh scales are located at the pit to aid in converting the volumes and as a check on the calculated quantities.

The original cross sections are taken over a grid at 50-ft (20-m) intervals. As the excavation proceeds, additional rod readings (in addition to 50-ft grid readings) for top and bottom of excavation are required. Permanent targets can be established to assist

CROSS-SECTIONS FOR PROPOSED
LOCATION OF DUNCAN ROAD

SMITH—NOTES   JONES—ROD
BROWN—π   TYLER—TAPE
CLOUDY   LEVEL #6
Job 14 °C   Page 23
Date NOV. 24 1983

STA.	B.S.	H.I.	I.S.	F.S.	ELEV.	DESCRIPTION
BM #28	2.011	178.218			176.207	BRONZE PLATE SET IN SOUTH WALL, 0.50m.
		(178.22)				ABOVE GROUND, CIVIC #2242, 23RD AVE.
0 + 60						
10m LT			0.81		177.41	S. ₵
2.7m LT			1.48		176.74	BOTTOM OF SWALE
₵			1.22		177.00	₵
8.2mRT			0.87		177.35	CHANGE IN SLOPE
10 m RT			0.21		178.01	N. ₵
0 + 80						
10m LT			1.02		177.20	S. ₵
3.8mLT			1.64		176.58	BOTTOM OF SWALE
₵			1.51		176.71	₵
7.8mRT			1.10		177.12	CHANGE IN SLOPE
10 m RT			0.43		177.79	N. ₵

FIGURE 3.23 Cross-section notes (municipal format).

CROSS-SECTIONS FOR PROPOSED
LOCATION OF DUNCAN HIGHWAY

SMITH—NOTES
BROWN—R
CLOUDY
NOV. 24 1983

JONES—ROD
TYLER—TAPE
LEVEL #6
Page 23

Job 14 °C

STA.	B.S.	H.I.	I.S.	F.S.	ELEV.					
	2.011	178.218 (178.22)			176.207	BRONZE PLATE SET IN SOUTH WALL
						0.50 ABOVE GROUND, CIVIC #2242, 23RD AVE.
						LEFT		℄	RIGHT	
0 + 60						10.0 / 0.81 / 177.41	2.7 / .48 / 176.74	1.22 / 177.00	8.2 / 0.87 / 177.35	10.0 / 0.21 / 178.01
0 + 80						10.0 / 1.02 / 177.20	3.8 / 1.64 / 176.58	1.51 / 76.71	7.8 / 1.10 / 177.12	10.0 / 0.43 / 177.79

FIGURE 3.24 Cross-section notes (highway format).

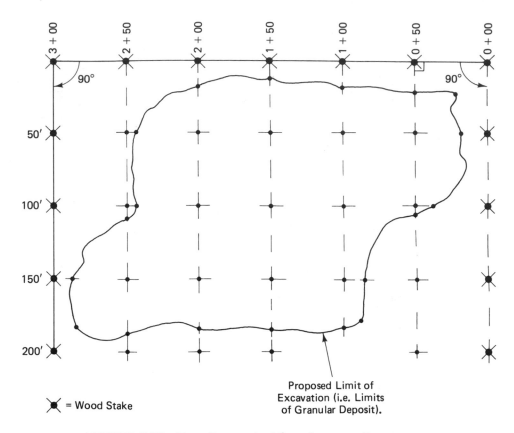

FIGURE 3.25 Base-line control for a borrow pit survey.

in the alignment of the cross-section lines running perpendicular to the base line at each 50-ft station. If permanent targets have not been erected, a surveyor on the base line can keep the rodman on line by using a prism or by estimated right angles.

3.19 RECIPROCAL LEVELING

Section 3.13 advises the surveyor to keep BS and FS distances roughly equal so that instrumental and natural errors will cancel out. In some situations, such as in river or valley crossings, it is not always possible to balance BS and FS distances. The reciprocal leveling technique is illustrated in Figure 3.26. The level is set up and readings are taken on TP 23 and TP 24. (Precision can be improved by taking several readings on the far point, TP 24, and then averaging the results.) The level is then moved to the far side

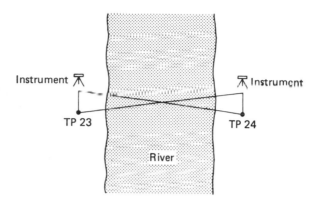

FIGURE 3.26   Reciprocal leveling.

of the river and the process repeated. The differences in elevation thus obtained are averaged to obtain the final result. The averaging process will eliminate instrumental errors and natural errors such as curvature. Errors due to refraction can be minimized by ensuring that the elapsed time for the process is kept to a minimum.

3.20 PEG TEST

The purpose of this test is to check that the line of sight through the level is horizontal (i.e., parallel to the axis of the bubble tube). The line of sight axis is defined by the location of the horizontal cross hair (see Figure 3.27). The reader is referred to Section 5.10 for a description of the horizontal cross-hair orientation adjustment.

To perform the peg test, the surveyor first places two stakes at a distance of 200 to 300 ft (60 to 90 m) apart. The level is set up midway (paced) between the two stakes and rod readings are taken at both locations (see Figure 3.28, first setup).

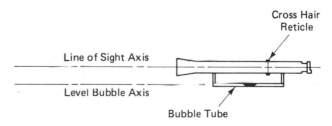

FIGURE 3.27   Optical axis and level bubble axis.

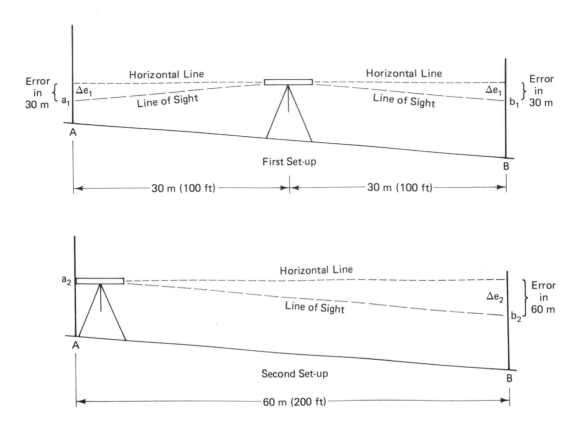

FIGURE 3.28 Peg test.

If the line of sight through the level is not horizontal, the error in rod reading (Δe_1) at both points A and B will be identical as the level is halfway between the points. Since the errors are identical, the calculated difference in elevation between points A and B (difference in rod readings) will be the *true* difference in elevation.

The level is then moved to one of the points (A) and set up so that the eyepiece of the telescope just touches the rod as it is being held plumb at point A. The rod reading (a_2) can be determined by sighting backward through the objective lens at a pencil point that is being moved slowly up and down the rod. The pencil point can be precisely centered, even though the cross hairs are not visible, because the circular field of view is relatively small. Once that reverse rod reading has been determined and booked, the rod is held at B and a normal rod reading obtained. (The reverse rod reading at A will not contain any line-of-sight error because the cross hair was not used to obtain the rod reading.)

Example 3.2

$$\text{First setup: Rod reading at } A, a_1 = 1.075$$
$$\text{Rod reading at } B, b_1 = \underline{1.247}$$
$$\text{True difference in elevations} = 0.172$$

$$\text{Second setup: Rod reading at } A, a_2 = 1.783$$
$$\text{Rod reading at } B, b_2 = \underline{1.946}$$
$$\text{Apparent difference in elevation} = 0.163$$

$$\text{Error } (\Delta\ e_2) \text{ in 60 m} = 0.009$$

This is an error of -0.00015 m/m. Therefore, the collimation correction (C factor) $= +0.00015$ m/m.

In Section 3.13, the reader was advised to try to keep the BS and FS distances equal; the peg test illustrates clearly the benefits to be gained by use of this technique. If the BS and FS distances are kept roughly equal, errors due to a faulty line of sight simply do not have the opportunity to develop.

Example 3.3

If the level used in the peg test of Example 3.2 were used in the field with a BS distance of 80 m and a FS distance of 70 m, the net error in the rod readings would be $10 \times 0.00015 = 0.0015$ (0.002). That is, for a relatively large differential in distances, and a large error in line of sight (0.009 for 60 m), the effect on the survey is negligible for ordinary work.

The peg test can also be accomplished using the techniques of reciprocal leveling (Section 3.19); however, the method described here offers a clearer approach to the problem.

3.21 THREE-WIRE LEVELING

Leveling can be performed by utilizing the stadia cross hairs found on most levels (all theodolites). See Figure 3.29. Each

i.e. 100 X Stadia Hair Interval = Distance

FIGURE 3.29 Reticle cross hairs.

backsight (BS) and foresight (FS) is recorded by reading the stadia hairs in addition to the horizontal cross hair. The three readings thus obtained are averaged to obtain the desired value.

The stadia hairs (wires) are positioned an equal distance above and below the main cross hair and are spaced to give 1.00 ft (m) of interval for each 100 ft (m) of horizontal distance that the rod is away from the level. (Some levels used exclusively for this type of work come equipped with stadia hairs spaced at a 350:1 ratio.)

The recording of three readings at each sighting enables the surveyor to perform a relatively precise survey while utilizing ordinary levels. Readings to the closest thousandth of a foot (mm) are estimated and recorded. The leveling rod used for this type of work should be calibrated to ensure its integrity. Use of an invar rod is recommended.

Figure 3.30 shows typical notes for benchmark leveling. A realistic survey would include a completed loop or a check into another BM of known elevation.

If the collimation correction as calculated in Section 3.20 (+ 0.00015 m/m) were applicable to the survey shown in Figure 3.30, the correction to the elevation would be as follows:

$$C = +0.00015 \times (62.9 - 61.5) = +0.0002$$

Sum of FS corrected to 5.7196 +0.0002 = 5.7198

Elevation of BM 201

$$
\begin{aligned}
\text{Elev. BM 17} &= 186.2830 \\
+\Sigma \text{ BS} &= \underline{+2.4143} \\
& 188.6973 \\
-\Sigma \text{ FS} \quad \text{(corrected)} & \underline{-5.7198} \\
\text{Elev. BM 201} &= 182.9775 \qquad \text{(corrected for collimation)}
\end{aligned}
$$

When levels are being used for precise purposes, it is customary to determine the collimation correction at least once each day.

3.22 TRIGONOMETRIC LEVELING

The difference in elevation between A and B (Figure 3.31) can be determined if the vertical angle (α) and the slope distance (S) are measured.

$$V = S \operatorname{Sin} \alpha \qquad (3.7)$$

Elevation at $\curlywedge$ + HI $\pm$ V − RR = elevation at rod $\qquad$ (3.8)

B.M. LEVELING—3 WIRE
B.M. #17 to B.M. 201
(RETURN RUN ON P.48)

JONES—NOTES
SMITH—K
BROWN—ROD
GREEN—ROD

Job ROD #19; INST. #L.33 8°C CLOUDY
Date MAR. 3 1983 Page 47

STA.	B.S.	DIST.	F.S.	DIST.	ELEV.	DESCRIPTION
BM #17					186.2830	BRONZE PLATE SET IN WALL--- ETC.
	0.825		1.775			
	0.725	10.0	1.673	10.2	+0.7253	
	0.626	9.9	1.572	10.1	187.0083	
	2.176	19.9	5.020	20.3	-1.6733	
	+0.7253		-1.6733			
T.P. #1					185.3350	N. LUG TOP FLANGE FIRE HYD. N/S
	0.698		1.750			MAIN ST. OPP. CIV C #181.
	0.571	12.7	1.620	13.0	+0.5710	
	0.444	12.7	1.490	13.0	185.9060	
	1.713	25.4	4.860	26.0	-1.6200	
	+0.5710		-1.6200			
T.P. #2					184.2860	N. LUG TOP FLANGE FIRE HYD. N/S
	1.199		2.509			MAIN ST. OPP. CIVIC # 63.
	1.118	8.1	2.427	8.2	+1.1180	
	1.037	8.1	2.343	8.4	185.4040	
	3.354	16.2	7.279	16.6	-2.4263	
	+1.1180		-2.4263			
BM. 201					182.9777	BRONZE PLATE SET IN ESTLY FACE
						OF RETAINING WALL --- ETC.
Σ	+2.4143	61.5m	-5.7196	62.9m	182.9777	

ARITHMETIC CHECK: 186.283 + 2.4143 - 5.7196 = 182.9777

FIGURE 3.30 Survey notes for three-wire leveling

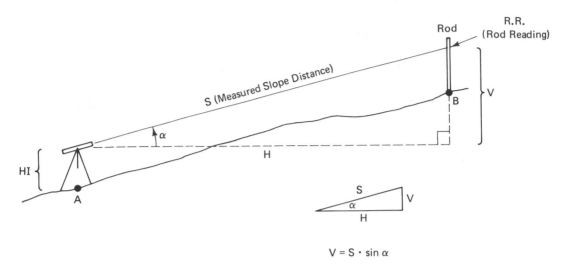

$$V = S \cdot \sin \alpha$$

FIGURE 3.31 Trigonometric leveling.

Note: The HI in this case is not the elevation of the line of sight as it is in differential leveling, but instead HI here refers to the distance from point A up to the optical center of the theodolite measured with a steel tape or rod. See also Chapter 7.

Trigonometric leveling can be used where it is not feasible to use a level. For example, a survey crew running ₵ profile for a route survey comes to a point where the ₵ runs off a cliff; in that case a theodolite can be set up on ₵ with the angle and distance measured to a ₵ station at the lower elevation.

The slope distance can be determined using a steel tape, stadia methods (Chapter 7), or EDM methods (Chapter 8). The angle is normally measured by use of a theodolite, but for lower-order surveys a clinometer (Section 2.7.2) could be used.

For long distances (associated with EDMs), curvature and refraction errors must be eliminated. These matters are discussed in Chapter 8.

Example 3.4 See Figure 3.32

$$\begin{aligned} V &= S \sin \alpha & (3.7)\\ &= 82.18 \sin 30°22'\\ &= 41.54 \text{ ft} \end{aligned}$$

$$\text{Elev. at } \barwedge + \text{HI} \pm \text{v} - \text{RR} = \text{elev. at rod} \qquad (3.8)$$
$$361.29 + 4.72 - 41.54 - 4.00 = \text{elev. at rod}$$
$$320.47 = \text{elev. at rod}$$

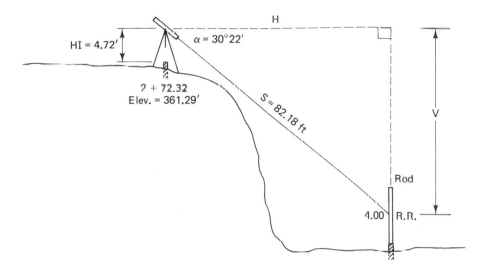

FIGURE 3.32 Example of trigonometric leveling (see Section 3.22).

Note: The RR *could* have been 4.72, the value of the HI. If that reading were visible, the surveyor would have sighted on it in order to eliminate +HI and −RR from the calculation. That is,

$$\text{Elev. at} \; \pitchfork \; \pm \; v = \text{elev. at rod} \qquad (3.8a)$$

In this example the chainage of the rod station could also be determined.

3.23 LEVEL LOOP ADJUSTMENTS

In Section 3.13 it was noted that level surveys had to be closed within acceptable tolerances or the survey would have to be repeated. The tolerances for various orders of surveys are shown in Tables A.18 and A.19 (Appendix A).

If a level survey were performed in order to establish new bench marks, it would be desirable to suitably proportion any acceptable error throughout the length of the survey. Since the error tolerances shown in Tables A.18 and A.19 are based on the distances surveyed, adjustments to the level loop will be based on the relevant distances, or on the number of instrument setups, which is a factor directly related to the distance surveyed (see Appendix A, Section A.8).

Example 3.5

A level circuit is shown in Figure 3.33. The survey, needed for local engineering projects, commenced at BM 20, the elevations of new bench marks 201, 202, and 203 were determined, and then the level survey was looped back to BM 20, the point of commencement (the survey could have terminated at any established BM).

According to Table A-19, the allowable error for a second-order, class II (local engineering projects) survey is $0.008 \sqrt{K}$; thus, $0.008 \sqrt{4.7} = 0.017$m is the permissible error.

The error in the survey was found to be -0.015 m over a total distance of 4.7 km, in this case an acceptable error. It only remains for this acceptable error to be suitably distributed over the length of the survey. The error is proportioned according to the fraction of cumulative distance over total distance, as in the following table.

BM	Loop Distance, Cumulative (km)	Elevation	Correction $\dfrac{\text{Cumulative Distance}}{\text{Total Distance}} \times E$	Adjusted Elevation
20		186.273 (fixed)		186.273
201	0.8	184.242	$+0.8/4.7 \times 0.015 = +0.003 =$	184.245
202	2.4	182.297	$+2.4/4.7 \times 0.015 = +0.008 =$	182.305
203	3.0	184.227	$+3.0/4.7 \times 0.015 = +0.010 =$	184.237
20	4.7	186.258	$+4.7/4.7 \times 0.015 = +0.015 =$	186.273

$E = 186.273 - 186.258 = -0.015$ m

More complex adjustments are normally performed by computer utilizing the adjustment method of least squares.

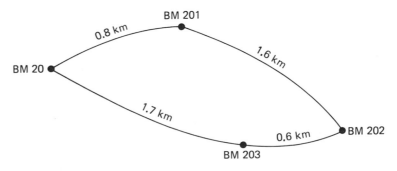

Total Distance Around Loop is 4.7 km.

FIGURE 3.33 Level loop.

3.24 SUGGESTIONS FOR THE RODMAN

1. The rod should be properly extended and clamped; care should be taken to ensure that the bottom of the sole plate does not become encrusted with mud and the like, which could result in mistaken readings. If a rod target is being used, care is exercised to ensure that it is properly positioned and that it cannot slip.

2. The rod should be held plumb for all rod readings. Either a rod level will be used, or the rod will be gently waved to and from the instrument so that the lowest (indicating a plumb rod) reading can be determined. This practice is particularly important for all backsights and foresights.

3. The rodman should ensure that all points used as turning points are suitable (i.e., describable, identifiable, and capable of having the elevation determined to the closest 0.01 ft or 0.001 m). The TP should be nearly equidistant from the two proposed instrument locations.

4. Care should be taken to ensure that the rod is held in precisely the same position for the backsight as it was for the foresight for all turning points.

5. If the rod is being held near to, but not on, a required location, the face of the rod should be turned away from the instrument so that the instrumentman cannot take a mistaken reading. This type of mistaken reading usually occurs when the distance between the rodman and instrumentman is too far to allow for voice communication and sometimes even for good visual contact.

3.25 SUGGESTIONS FOR THE INSTRUMENTMAN

1. Use a straight leg (nonadjustable) tripod, if possible.

2. Tripod legs should be tightened so that when one leg is extended horizontally it falls slowly back to the ground under its own weight.

3. The instrument can be comfortably carried resting on one shoulder; if tree branches or other obstructions (e.g., door frames) threaten the safety of the instrument, it should be carried cradled under one arm with the instrument forward, where it can be seen.

4. When setting up the instrument, gently force the legs into the ground by applying weight on the tripod shoe spurs. On

rigid surfaces (e.g. concrete) the tripod legs should be spread farther apart to increase stability.

5. When the tripod is to be set up on a sidehill, two legs should be placed downhill and the third leg placed uphill. The instrument can be set up roughly leveled by careful manipulation of the third uphill leg.

6. The location of the level setup should be wisely chosen with respect to the ability to "see" the maximum number of rod locations, particularly BS and FS locations.

7. Prior to taking rod readings, the cross hair should be sharply focused; it helps to point the instrument toward a light-colored background.

8. When the instrumentman observes apparent movement of the cross hairs on the rod (parallax), he should carefully check the cross-hair focus adjustment and the objective focus adjustment for consistent results.

9. The instrumentman should consistently read the rod at either the top or the bottom of the cross hair.

10. Never move the level before a foresight is taken; otherwise, all work done from that HI will have to be repeated.

11. Check to ensure that the level bubble remains centered or that the compensating device (in automatic levels) is operating.

12. Rod readings (and the line of sight) should be kept at least 0.5 m above the ground surface to help minimize refraction errors when performing a precise level survey.

3.26 MISTAKES IN LEVELING

Mistakes in level loops can be detected by performing arithmetic checks, and by closing in on the starting BM or any other BM whose elevation is known.

Mistakes in rod readings that do not form part of a level loop, such as intermediate sights taken in profiles, cross sections, or construction grades, are a much more irksome problem. It is bad enough to discover that a level loop contains mistakes and must be repeated, but it is a far more serious problem to have to redesign a highway profile because a key elevation contains a mistake, or to have to break out a concrete bridge abutment (the day after the concrete was poured) because the grade stake elevation contained a mistake.

Since intermediate rod readings cannot be inherently checked, it is essential that the opportunities for mistakes be minimized.

Common mistakes in leveling include the following: misreading the foot (meter) value; transposing figures; not holding the rod in the correct location; resting the hands on the tripod while reading the rod and causing the instrument to go off level; entering the rod readings incorrectly (i.e., switching BS and FS); giving a correct rod reading the wrong station identification; and mistakes in the note reduction arithmetic.

Mistakes in arithmetic can be largely eliminated by having the other crew members check the reductions and initial each page of notes checked. Mistakes in the leveling operation cannot be totally eliminated, but they can be minimized if the crew members are aware that mistakes can (and probably will) occur. All crew members should be constantly alert as to the possible occurrence of mistakes, and all crew members should try to develop rigid routines for doing their work so that mistakes, when they do eventually occur, will be all the more noticeable.

PROBLEMS

3-1   Compute the sighting error, due to curvature and refraction, for the following distances:
a) 50 m;   b) 200 m;   c) 1/2 mile;   d) 5 miles;   e) 800 ft.;
f) 4000 ft.

3-2   An ocean drilling platform is being towed out to sea, on a clear night, to a new drilling position. What is the maximum distance, out to sea, that the lights on top of the platform tower can still be seen by an observer who has an eye height of 5'6" and who is standing at the water's edge? The platform tower lights are 180 ft above the water.

3-3   Reduce the accompanying set of differential leveling notes and perform the arithmetic check.

Station	BS	HI	FS	Elevation
BM 100	2.71			126.13
TP. 1	3.62		4.88	
TP. 2	3.51		3.97	
TP. 3	3.17		2.81	
TP. 4	1.47		1.62	
BM 100			1.21	

3-4   If the distance leveled in Problem 3-3 is 1000 feet, for what order of survey do the results qualify? (See Tables A-18 and A-19).

3-5   Reduce the accompanying set of profile notes and perform the arithmetic check.

Station	BS	HI	IS	FS	Elevation
BM 20	8.27				528.71
TP. #1	9.21			2.60	
0 + 00			11.3		
0 + 50			9.6		
0 + 61.48			8.71		
1 + 00			6.1		
TP. #2	7.33			4.66	
1 + 50			5.8		
2 + 00			4.97		
BM 21				3.88	

3-6   Reduce the accompanying set of profile notes and perform the arithmetic check.

Station	BS	HI	IS	FS	Elevation
BM S.101	0.475				151.283
0 + 000			0.02		
0 + 020			0.41		
0 + 040			0.73		
0 + 060			0.70		
0 + 066.28			0.726		
0 + 080			1.38		
0 + 100			1.75		
0 + 120			2.47		
T.P. #1	0.666			2.993	
0 + 140			0.57		
0 + 143.78			0.634		
0 + 147.02			0.681		
0 + 160			0.71		
0 + 180			0.69		
0 + 200			1.37		
T.P. #2	0.033			1.705	
BM S.102				2.891	

3-7 Reduce the accompanying set of municipal cross-section notes.

Station	BS	HI	IS	FS	Elevation
BM.21	0.711				186.270
0 + 00					
℄			1.211		
10m left, ℄			1.430		
10m right, ℄			1.006		
0 + 20					
10m left, ℄			2.93		
7.3m left			2.53		
4m left			2.301		
℄			2.381		
4m right			2.307		
7.8m right			2.41		
10m right, ℄			2.78		
0 + 40					
10m left, ℄			3.98		
6.2m left			3.50		
4m left			3.103		
℄			3.187		
4m right			3.100		
6.8m right			3.37		
10m right			3.87		
T.P. #1				2.773	

3-8 Reduce the accompanying set of municipal cross-section notes.

Station	BS	HI	IS	FS	Elevation
BM. #41	6.21	443.97			437.76
T.P. 13	4.10	447.18		0.89	443.08
12 + 00					
50 ft left			3.9		
18.3 ft left			4.6		
℄			6.33		
20.1 ft right			7.9		
50 ft right			8.2		
13 + 00					
50 ft left			5.0		
19.6 ft left			5.7		
℄			7.54		
20.7 ft right			7.9		
50 ft right			8.4		
T.P. #14	7.39	453.45		1.12	446.06
BM.S.22				2.41	451.04

17.7   4.42   Check

3-9 Reduce the accompanying set of highways cross-section notes.

Station	BS	HI	FS	Elevation	Left		₵	Right	
BM 37	11.27			480.66					
5 + 50					50	26.7		28.4	50
					4.6	3.8	3.7	3.0	2.7
6 + 00					50	24.1		25.0	50
					4.0	4.2	3.1	2.7	2.9
6 + 50					50	26.4		23.8	50
					3.8	3.7	2.6	1.7	1.1
TP #1			6.71						

3-10 Reduce the accompanying set of differential leveling notes and perform the arithmetic check; (a) determine the order of accuracy (Appendix A) and (b) adjust the elevation of BM #K110. The length of the level run was 780 m with setups equally spaced. The elevation of BM #132 is known to be 199.758.

Station	BS	HI	FS	Elevation
BM #130	0.702			200.786
TP #1	0.970		1.111	
TP #2	0.559		0.679	
TP #3	1.744		2.780	
BM #K110	1.973		1.668	
TP #4	1.927		1.788	
BM #132			0.888	

3-11 A level is set-up midway between two points that are 260 ft apart. The rod reading on point A is 6.27 ft and on point B is 3.78 ft. The level is then moved to point B and set up so that the eyepiece end of the telescope is just touching the rod held on point B. A reading taken by looking backwards through the telescope is 5.21 ft; a rod reading of 7.76 is then taken on point A.

a) What is the correct difference in elevation between points A & B?

b) If the level were in perfect adjustment, what rod reading would have been observed on point A from the second set up?

c) What is the error in the line of sight of the level?

d) How would you eliminate this line-of-sight error from the telescope?

3-12 Assume that the level with the line-of-sight error as indicated in problem 3-11 was used to establish a temporary construction bench mark as follows: the BM elevation was 422.38, the BS was 2.86 at a distance of 30 ft; the FS to the temporary BM was 10.77 at a distance of 180 ft.

a) What is the corrected elevation of the temporary construction bench mark?

b) What effect does the error of curvature and refraction have on the elevation of the temporary construction bench mark?

3-13 It is requires to establish the elevation of point B from point A (elevation 187.298 m); A and B are on opposite sides of a 12-lane urban expressway. Reciprocal leveling is used with the following results: (rod readings are in meters). Set up at A side of expressway: Rod reading A = 0.673, and on B = 2.416 & 2.418. Set up at B side of expressway: Rod reading on B = 2.992, and on A 1.254 & 1.250.

a) What is the elevation of point B?

b) What is the leveling error?

3-14 An EDM at Station A is used to sight stations B, C and D with the heights of instrument, target and reflector all equal for each sighting. The results are as follows:

a) on Station B, EDM distance = 3000.00 ft with a vertical angel of +3° 30′; b) on station C, EDM distance = 2000.00 ft with a vertical angle of −1°30′; and c) on station D, EDM distance = 4000.00 ft with a vertical angle of 0°00′.

Compute the differences in elevation between instrument station A and the other three stations—be sure to correct for curvature and refraction.

3-15 The slope distance between two points, as measured by EDM, is 7280.81 ft and the vertical angle is + 2°45′30″. If the elevation of the instrument station is 630.15, and the height of instrument, height of target and height of EDM reflector are all equal to 5.26 ft, compute the elevation of the target station.

3-16 A line CD was measured at both ends as follows: ⊼ at C, slope distance = 2112.617 m, vertical angle = +1°26′50″. ⊼ at D, slope distance = 2112.601 m, vertical angle = −1°26′38″ The heights strument, reflector and target were all equal for each observation.

a) Compute the horizontal distance CD.

b) If the elevation at C is 150.166 m, what is the elevation at D?

4 *Angles and Directions*

4.1 GENERAL

It was noted in Section 1.11 that the units of angular measurement employed in North American practice are degrees, minutes, and seconds, the sexagesimal system. For the most part, angles in surveying are measured with a theodolite, although angles can be measured using clinometers, sextants (hydrographic surveys), or compasses.

4.2 REFERENCE DIRECTIONS FOR VERTICAL ANGLES

Vertical angles, which are used in slope distance corrections (Section 2.10) or in height determination (Section 3.22), are referenced to (1) the horizon by plus (up) or minus (down) angles, (2) the zenith, or (3) the nadir (see Figure 4.1).

Zenith and nadir are terms describing points on a celestial sphere (i.e., a sphere of infinitely large radius with its center at the center of the earth). The zenith point is directly above the observer and the nadir is directly below the observer; the zenith, nadir, and observer are all on the same vertical line. (See also Figure 11.39)

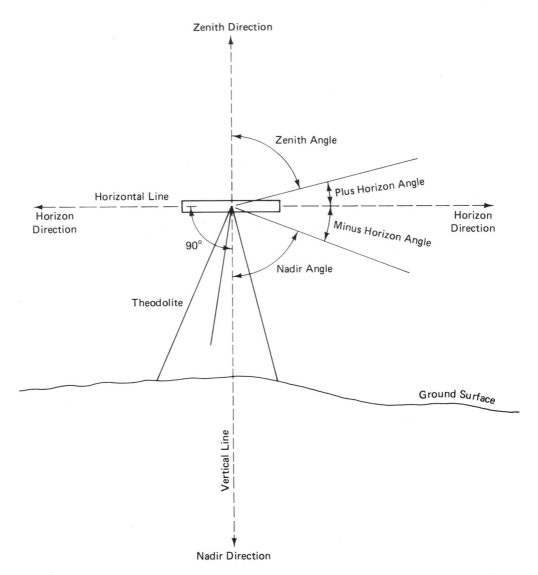

FIGURE 4.1 The three reference directions for vertical angles: horizontal, zenith, and nadir.

4.3 MERIDIANS

A line on the mean surface of the earth joining the north and south poles is called a *meridian*. In Section 1.6, it was noted that surveys could be referenced to lines of latitude and longitude. All lines of longitude are meridians. The term meridian can be more precisely defined by noting that it is the line formed by the intersection with the earth's surface of a plane that includes the earth's axis of rotation. The meridian, as described, is known as the *astronomic meridian*, the *geographic meridian*, or the *true meridian*.

Magnetic meridians are meridians that are parallel to the directions taken by freely moving magnetized needles, as in a compass. Whereas true meridians are fixed, magnetic meridians vary with time and location.

Grid meridians are lines that are parallel to a grid reference meridian (central meridian). The concept of a grid for survey reference was introduced in Section 1.7 and is described in detail in Chapter 11.

Figure 4.2 shows the true meridians, which all converge to meet at the pole, and the grid meridians, which are all parallel to the central (true) meridian. In the case of a small-scale survey of only limited importance, meridians are sometimes assumed, and the survey is referenced to that assumed direction.

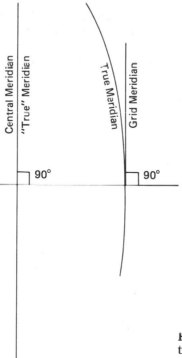

FIGURE 4.2 Relationship between "true" meridians and grid meridians.

We saw in the previous section that vertical directions were referenced to a horizontal line (plus or minus) or to a vertical line (from either the zenith or nadir direction); in contrast, we now see that all horizontal directions are referenced to meridians.

4.4 HORIZONTAL ANGLES

Horizontal angles are measured using a transit (precision of 20 seconds to 1 minute) or a theodolite (precision of 1 second to 20 seconds). These instruments are described in detail in Chapter 5.

Angles can be measured between lines forming a closed traverse, between lines forming an open traverse, or between a line and a point so that the point's location can be determined.

For a closed polygon of n sides, the sum of the interior angles will be $(n - 2)180°$. In Figure 4.3, the interior angles of a five-sided closed polygon have been measured as shown. For a five-sided polygon, the sum of the interior angles must be $(5 - 2)180° = 540°$; the angles shown in Figure 4.3 do, in fact, total 540°. However, in practical field problems the total is usually marginally more or less than $(n - 2)180°$, and it is up to the surveyor to then determine if the error of angular closure is within tolerances as specified for that survey. The adjustment of angular errors is described in Chapter 6.

Note that the exterior angles at each station in Figure 4.3 could have been measured instead of the interior angles as shown.

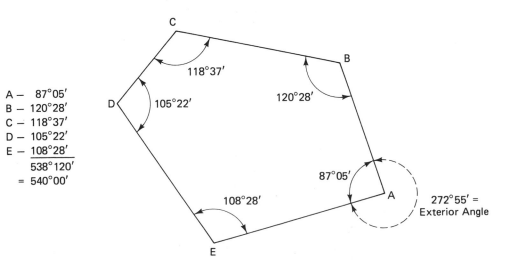

A — 87°05'
B — 120°28'
C — 118°37'
D — 105°22'
E — 108°28'
  538°120'
= 540°00'

FIGURE 4.3 A closed traverse showing the interior angles.

(The exterior angle at A of 272°55′ is shown.) Generally, exterior angles are only measured to occasionally serve as a check on the interior angle. The reader is referred to Chapter 5 for the actual field techniques used in the measurement of angles.

An open traverse is illustrated in Figure 4.4a. The deflection angles shown are measured from the prolongation of the back line to the forward line. The angles are measured either to the left (L) or to the right (R). The direction (L or R) must be shown along with the numerical value.

It is possible to measure the same angle (see Figure 4.4b) by directly sighting the back line and turning the angle left or right to the forward line. This technique is seldom used, as the vast majority of surveying agencies prefer to use deflection angles.

4.5 AZIMUTHS

An azimuth is the direction of a line as given by an angle measured clockwise from the north end of a meridian. In some circumstances, azimuths are measured clockwise from the south end of a meridian (some astronomic and geodetic projects); when that is the case, the directions are clearly labeled as to the angle orientation.

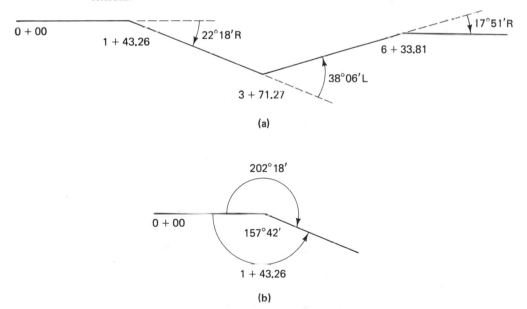

FIGURE 4.4 (a) An open traverse showing deflection angles; (b) same traverse showing angle right (202°18′) and angle left (157°42′).

Azimuths range in magnitude from 0° to 360°. Values in excess of 360°, which are sometimes encountered in computations, are simply reduced by 360° before final listing.

Figure 4.5 illustrates the concept of azimuths by showing four line directions in addition to the four cardinal directions (N, S, E, and W).

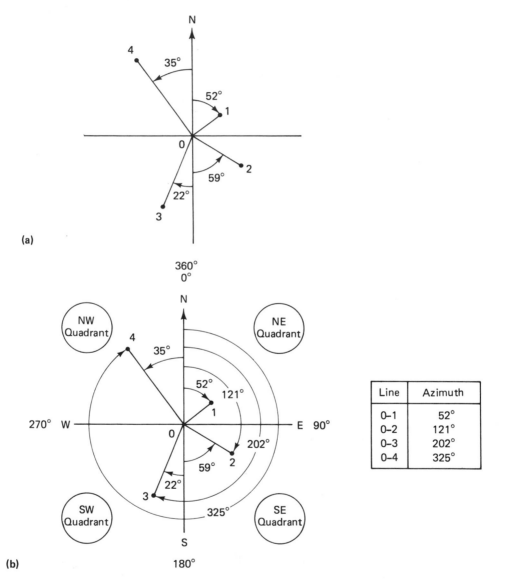

Line	Azimuth
0–1	52°
0–2	121°
0–3	202°
0–4	325°

FIGURE 4.5 (a) Given azimuth data; (b) azimuths calculated from given data.

4.6 BEARINGS

A bearing is the direction of a line as given by the acute angle between the line and a meridian. The bearing angle, which can be measured clockwise or counterclockwise from the north or south end of the meridian, is always accompanied by letters that locate the quadrant in which the line falls (NE, NW, SE, or SW).

Figure 4.6 illustrates the concepts of bearings and shows the proper designation for the four lines shown. In addition, the four cardinal directions are usually designated by the terms Due North, Due South, Due East, and Due West. However, Due West, for example, can also be designated as N 90° W (or S 90° W).

4.7 RELATIONSHIP BETWEEN BEARINGS AND AZIMUTHS

Referring to Figure 4.5, it can be seen that azimuth and bearing directions in the NE quadrant are numerically equal. That is, for line 0–1 the azimuth is 52° and the bearing is N 52° W. Referring to the SE quadrant, it can be seen that 0–2, which has an azimuth of 121°, will have a bearing (acute angle from meridian) of S 59° E. In the SW quadrant, the azimuth of 0–3 is 202° and the bearing will be S 22° W. In the NW quadrant, the azimuth of 0–4 is 325° and the bearing will be N 35° W.

To Convert from Azimuths to Bearings

First the proper quadrant letters are determined (see Figure 4.5):

0 – 90°	NE
90°–180°	SE
180°–270°	SW
270°–360°	NW

Then the numerical value is determined by using the following relationships:

1. NE quadrant : bearing = azimuth
2. SE quadrant : bearing = 180° − azimuth
3. SW quadrant : bearing = azimuth − 180°
4. NW quadrant : bearing = 360° − azimuth

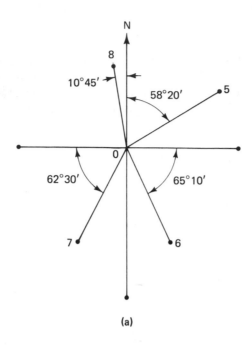

(a)

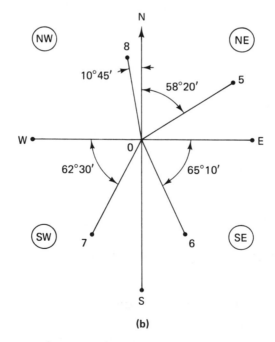

Line	Bearing
0–5	N 58°20′ E
0–6	S 24°50′ E
0–7	S 27°30′ W
0–8	N 10°45′ W

(b)

FIGURE 4.6 (a) Given bearing data; (b) bearings calculated from given data.

To Convert from Bearing to Azimuths

1. NE quadrant : azimuth = bearing
2. SE quadrant : azimuth = 180 − bearing
3. SW quadrant : azimuth = 180 + bearing
4. NW quadrant : azimuth = 360 − bearing

4.8 REVERSE DIRECTIONS

It can be said that every line has two directions. Referring to Figure 4.7, the line shown has direction *AB* or it has direction *BA*. In surveying, a direction is called *forward* if it is oriented in the direction of fieldwork or computation staging. If the direction is the reverse of that, it is called a *back* direction.

The designations of forward and back are often arbitrarily chosen; but if more than one line is being considered, the forward and backward designations must be consistent for all adjoining lines.

In Figure 4.8 the line *AB* has a bearing of N 62°30′ E, whereas the line *BA* has a bearing of S 62°30′ W. That is, *to reverse a bearing, simply reverse the direction letters;* in this case N and E become S and W, and the numerical value (62°30′) remains unchanged.

In Figure 4.9, the line *CD* has an azimuth of 128°20′; analysis of the sketch leads quickly to the conclusion that the azimuth of *DC* is 308°20′; that is, *to reverse an azimuth, simply add 180° to the original direction.* If the original azimuth is greater than 180°, 180° can be subtracted from it in order to reverse its direction. The *key factor to remember is that a forward and back azimuth must differ by 180°.*

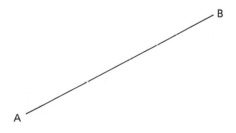

FIGURE 4.7 Reverse directions.

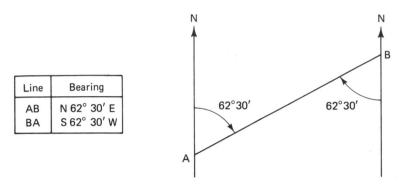

Line	Bearing
AB	N 62° 30′ E
BA	S 62° 30′ W

FIGURE 4.8   Reverse bearings.

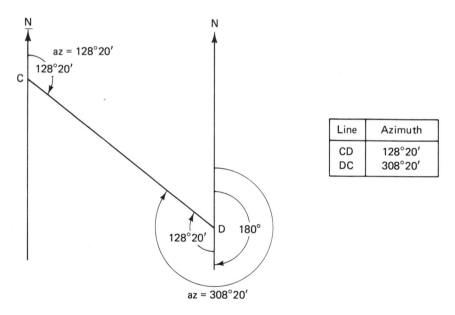

Line	Azimuth
CD	128°20′
DC	308°20′

FIGURE 4.9   Reverse azimuths.

4.9 AZIMUTH COMPUTATIONS

The data in Figure 4.10 will be utilized to illustrate the computation of azimuths. **Before azimuths or bearings are computed, it is important to check that the figure is geometrically closed: $(n - 2)180$.**

In Figure 4.10 the five angles do add to 540°00′ and *AB* has a given azimuth of 330°00′. At this point a decision must be made

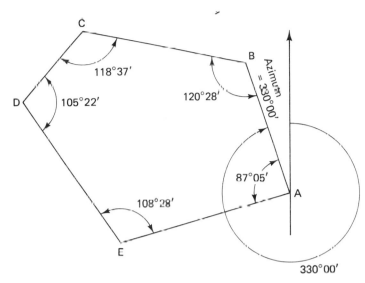

FIGURE 4.10 Sketch for azimuth calculations.

as to the direction that the computation will proceed. Using the given azimuth and the angle at B, the azimuth of BC can be computed (counterclockwise direction); or using the given azimuth and the angle at A, the azimuth of AE can be computed (clockwise direction). Once a direction for solving the problem has been established, the computed directions must all be consistent with that general direction. It is strongly advised that a neat, well-labeled sketch accompany each step of the computation.

Counterclockwise Solution

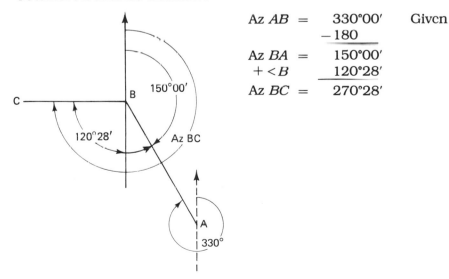

Az AB =	330°00′	Given
	− 180	
Az BA =	150°00′	
+ <B	120°28′	
Az BC =	270°28′	

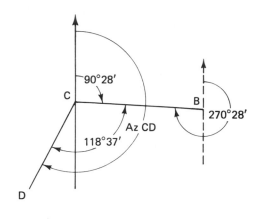

$$
\begin{aligned}
\text{Az } BC &= & 270°28' \\
& & -180 \\
\hline
\text{Az } CB &= & 90°28' \\
+ <C & & 118°37 \\
\hline
\text{Az } CD &= & 208°65' \\
\text{Az } CD &= & 209°05'
\end{aligned}
$$

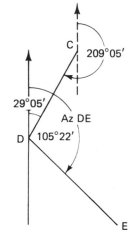

$$
\begin{aligned}
\text{Az } CD &= & 209°05' \\
& & -180 \\
\hline
\text{Az } DC &= & 29°05' \\
+ <D & & 105°22' \\
\hline
\text{Az } DE &= & 134°27'
\end{aligned}
$$

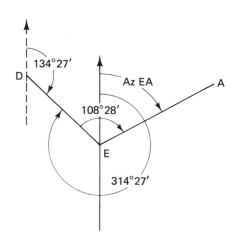

$$
\begin{aligned}
\text{Az } DE &= & 134°27' \\
& & +180° \\
\hline
\text{Az } ED &= & 314°27' \\
+ <E & & 108°28' \\
\hline
\text{Az } EA &= & 422°55' \\
& & -360 \\
\hline
\text{AZ } EA &= & 62°55'
\end{aligned}
$$

124

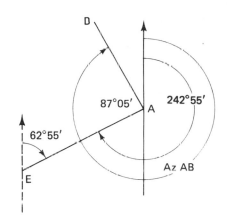

$$
\begin{aligned}
\text{Az } EA \ - \quad & 62°55' \\
& \underline{+ \ 180} \\
\text{Az } AE \ = \quad & 242°55' \\
+ <A \quad & \underline{87\ 05} \\
\text{Az } AB \ = \quad & 329°60' \\
\text{Az } AB \ = \quad & 330°00' \qquad \text{Check}
\end{aligned}
$$

Clockwise Solution

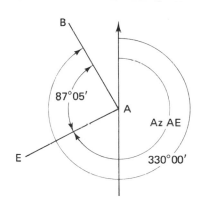

$$
\begin{aligned}
\text{Az } AB \ = \quad & 330°00' \qquad \text{Given} \\
- <A \quad & \underline{87°05'} \\
\text{Az } AE \ = \quad & 242°55'
\end{aligned}
$$

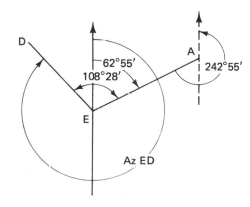

$$
\begin{aligned}
\text{Az } AE \ = \quad & 242°55' \\
& \underline{- 180°} \\
\text{Az } EA \ = \quad & 62°55' \\
& \underline{+ 360°} \\
\text{Az } EA \ = \quad & 422°55' \\
< E \quad & \underline{108°28'} \\
\text{Az } ED \ = \quad & 314°27'
\end{aligned}
$$

125

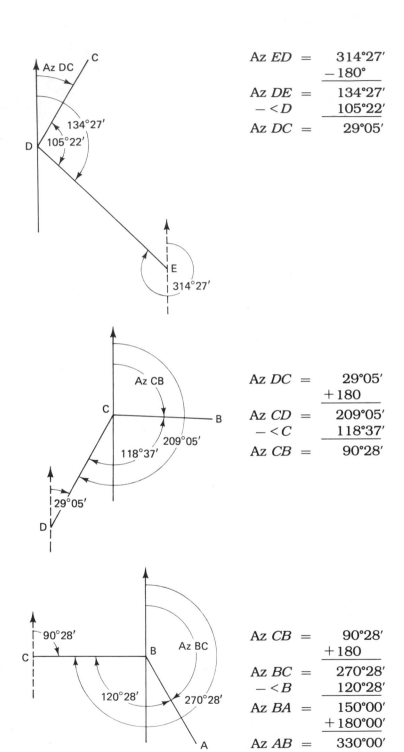

Az ED =   314°27′

          −180°

Az DE =   134°27′

  − $<D$ 105°22′

Az DC =   29°05′

Az DC =   29°05′

          +180

Az CD =   209°05′

  − $<C$ 118°37′

Az CB =   90°28′

Az CB =   90°28′

          +180

Az BC =   270°28′

  − $<B$ 120°28′

Az BA =   150°00′

        +180°00′

Az AB =   330°00′   Check

Analysis of the preceding azimuth computations gives the following observations: To compute the azimuths of the sides of a closed figure:

1. If the computation is proceeding in a *counterclockwise* manner, *add the interior angle to the back azimuth of the previous course.*
2. If the computation is proceeding in a *clockwise* manner, *subtract the interior angle from the back azimuth of the previous course.*

If the bearings of the sides are also required, they can now be derived from the computed azimuths.

Counterclockwise Solution

Course	Azimuth	Bearing
BC	270°28'	N 89°32' W
CD	209°05'	S 29°05' W
DE	134°27'	S 45°33' E
EA	62°55'	N 62°55' E
AB	330°00'	N 30°00' W

Clockwise Solution

Course	Azimuth	Bearing
AE	242°55'	S 62°55' W
ED	314°27'	N 45°33' W
DC	29°05'	N 29°05' E
CB	90°28'	S 89°32' E
BA	150°00'	S 30°00' E

Notes

1. It will be noted that for reversal of direction (i.e., clockwise versus counterclockwise) azimuths for the same side differ by 180°, whereas bearings for the same side remain numerically the same and have the letters (N/S, E/W) reversed.
2. It should be emphasized that the bearings calculated from azimuths have no built-in check. The only way that the correctness of the calculated bearings can be verified is to double-check the computation. Constant reference to a good problem diagram will help reduce the incidence of mistakes.

4.10 BEARING COMPUTATIONS

As with azimuth computations, the solution can proceed in a clockwise or counterclockwise manner. Referring to Figure 4.11, side *AB* has a given bearing of N 30°00′ W, and the bearing of either *BC* or *AE* may be computed first.

Since there is no systematic method of directly computing bearings, each bearing computation will be regarded as a separate problem; it is essential that a neat, well-labeled diagram accompany each computation.

The sketch of each individual bearing computation will show the appropriate interior angle together with one bearing angle. The required bearing angle should also be clearly shown.

In Figure 4.12.a, the interior angle (*B*) is shown; the bearing angle (30°) for side *BA* is shown; and the required bearing angle for side *BC* is shown as a question mark. Analysis of the sketch clearly shows that the required bearing angle (?) = 180 − (120°28′ − 30°00′) = 89°32′, and that the quadrant is NW. The bearing of *BC* = N 89°32′ W.

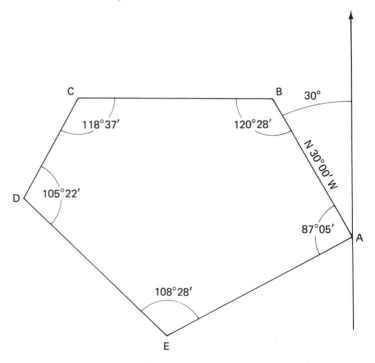

FIGURE 4.11 Sketch for bearing computations.

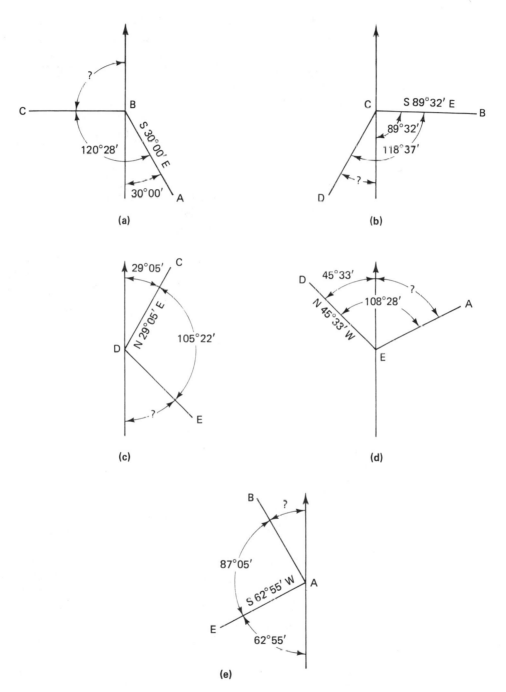

FIGURE 4.12 Sketches for each bearing calculation for the example of Section 4.10.

In Figure 4.12b, the bearing for *CB* is shown as S 89°32′ E. This is the reverse of the bearing that was just calculated for side *BC*. When the meridian line is moved from *B* to *C* for this computation, it necessitates the reversal of direction. Analysis of the completely labeled sketch clearly shows that the required bearing angle for *CD* (?) = (118°37′ − 89°32′) = 29°05′ and that the direction is SW. The bearing of *CD* is S 29°05′ W.

Analysis of Figure 4.12c clearly shows that the bearing angle of line *DE* (?) = 180° − (105°22′ + 29°05′) = 45°33′, and that the direction of *DE* is SE. The bearing of *DE* is S 45°33′ E.

Analysis of Figure 4.12d clearly shows that the bearing angle of line *EA* (?) is (108°28′ − 45°33′) = 62°55′, and that the direction is NE. The bearing of *EA* is N 62°55′ E.

This problem's original data included the bearing of *AB* as being N 30°00′ W. The bearing of *AB* will now be computed using the interior angle at *A* and the bearing just computed for the previous course (*EA*). The bearing angle of *AB* = 180° − (62°55′ + 87°05′) = 30°00′ and the direction is NW. The bearing of *AB* is N 30°00′ W (see Figure 4.12e). This last computation serves as a check on all our computations.

4.11 COMMENTS ON BEARINGS AND AZIMUTHS

It has been shown that both bearings and azimuths may be used to give the direction of a line. North American tradition in this regard favors the use of bearings over azimuths; most legal plans (plats) show directions in bearings.

In the previous sections, it was shown that bearings can be derived from computed azimuths (Section 4.9), or bearings can be computed directly from the given data (Section 4.10). The advantage of computing bearings directly from the given data is that the final computation (of the given bearing) provides a check on all the problem computations, ensuring (normally) the correctness of all the computed bearings. In contrast, if bearings are to be derived from computed azimuths, there is no intrinsic check on the correctness of the derived bearings.

The disadvantage associated with computing bearings directly from the given data is that there is no systematic approach to the overall solution. Each bearing computation is unique, requiring individual analysis. It is sometimes difficult persuading people to prepare neat, well-labeled sketches for computations involving only intermediate steps in a problem; and without neat, well-labeled sketches for each bearing computation, the potential for mistakes is quite large. Additionally, when mistakes do occur in the computation, the lack of a systematic approach to the solution often means that much valuable time is lost before the mistake is found and corrected.

In contrast, the computation of azimuths involves a highly systematic routine: *add (subtract) the interior angle from the back azimuth of the previous course.* If the computations are arranged as shown in Section 4.9, mistakes that may be made in the computation will be found in very short order.

With the advent of the universal use of computers and sophisticated hand-held calculators, it is expected that azimuths will be used more and more to give the direction of a line. It is easier to deal with straight numeric values rather than the alphanumeric values associated with bearings, and it is more efficient to have the algebraic sign generated by the calculator or computer rather

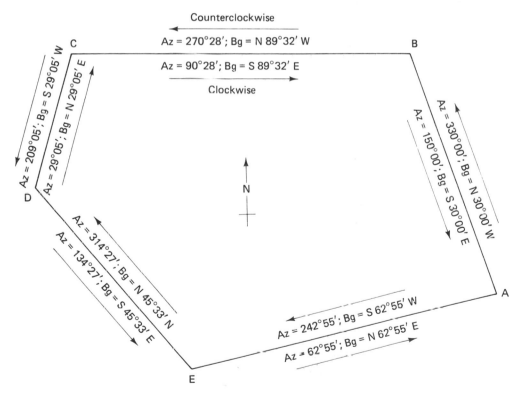

FIGURE 4.13 Summary of results from clockwise and counterclockwise approaches.

than trying to remember if the direction was north, south, east, or west.

We will see, in Chapter 6, that bearings or azimuths are used in calculating the geometric closure of a closed survey. The absolute necessity of having mistakes eliminated from the computation of line directions will become more apparent in that chapter; it will be seen that the computation of direction (bearings or azimuths) is only the first step in what can be a very involved computation. (See Figure 4.13 for a summary of results.)

4.12 MAGNETIC DIRECTION

A freely moving magnetized needle (as in a compass) will always point in the direction of magnetic north. The magnetic north pole is located about 1000 miles south of the geographic pole, near Bathurst Island in the Canadian Arctic.

Because the magnetic north pole is some distance from the geographic north pole, the magnetized needle does not, for the most part, point to geographic north. The horizontal angle between the direction taken by the compass needle and geographic north is the *magnetic declination.* In North America, the magnetic declination ranges from about 25° east on the west coast to about 25° west on the east coast. The magnetic declination, which, as seen, varies with location, also varies with time.

Careful records are kept over the years so that, although magnetic variations are not well understood, it is possible to predict magnetic declination over a short span of time. Accordingly, many countries issue *isogonic charts,* usually every five years, on which lines are drawn (isogonic lines) that join points of the earth's surface having equal magnetic declination (see Figure 4.14). Additional lines are shown on the chart that join points on the earth's surface that are experiencing equal annual changes in magnetic declination. Due to the uncertainties of determining magnetic declination and the effects of local attraction (e.g., ore bodies) on the compass needle, magnetic directions are not employed for any but the lowest order of surveys.

Isogonic charts are a valuable aid when original surveys (magnetic) are to be retraced. Most original township surveys in North America were magnetically referenced. When magnetically referenced surveys are to be retraced, it is necessary to determine the magnetic declination for that area at the time of the survey and at the time of the retracement survey.

FIGURE 4.14 (a) Isogonic map showing the magnetic declination in Canada. (Source: Department of Energy, Mines and Resources, Canada.

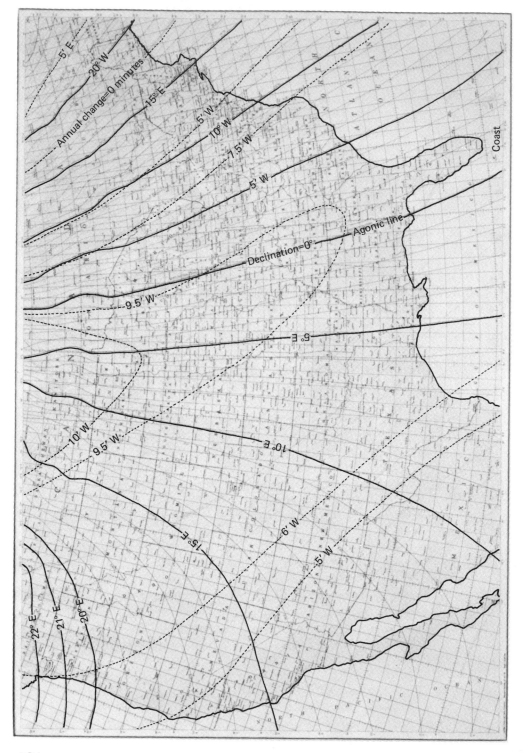

FIGURE 4.14 (b) Isogonic chart showing magnetic declination in the United States. (Source: U.S. Geological Surveys)

134

Example 4.1

A magnetic bearing was recorded for a specific lot line on an original township plan as being N 10°30′ E. The date of the original survey was June 25, 1889. It is desired to retrace the survey from the original notes during the first week of September 1983. What compass bearing will be used for the same specific lot line during the retracement survey?

Using an isogonic chart, the following data are scaled:

$$
\begin{aligned}
\text{Declination (1980.0)} &= 10°\ \text{W} \\
\text{Annual change} &= 06′\ \text{W} \\
\text{Change in } 90\tfrac{1}{2} \text{ years} = 90.5 \times 06′ &= 543′\ \text{or}\ 9°03′ \\
\text{Declination June 25, 1889} = 10° - 9°03′ &= 0°57′\ \text{W} \\
\text{Geographic bearing} &= 10°30′ - 0°57′ \\
&= \text{N } 9°33′ \text{ E}
\end{aligned}
$$

$$
\begin{aligned}
\text{Declination (1980.0)} &= 10°\ \text{W} \\
\text{Annual change} &= 06′\ \text{W} \\
\text{Change in } 3\tfrac{3}{4} \text{ years} = 3.75 \times 6 &= 22\tfrac{1}{2}′ \\
\text{Declination September 1983} &= 10°22\tfrac{1}{2}′\ \text{W} \\
\text{Magnetic bearing September 1983} &= \text{true bearing } + \\
& \quad \text{magnetic declination} \\
&= 9°33′ + 10°22.5′ \\
&= 19°55.5′ \\
&= \text{N } 19°55.5′ \text{ E}
\end{aligned}
$$

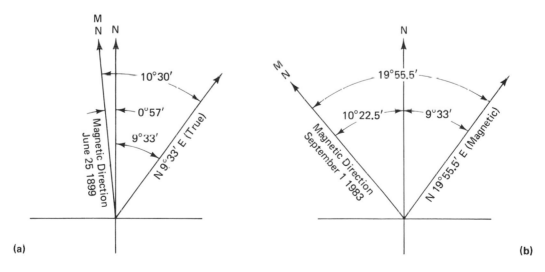

(a)   **(b)**

FIGURE 4.15   Magnetic declination determination.

FIGURE 4.16 Pocket transit, combining the features of a sighting compass, prismatic compass, hand level, and clinometer. Can be staff mounted for more precise work. (Courtesy Keuffel & Esser Co.)

Another area in which modern-day surveyors have need of a compass is the area of meridian determination by observation on the North Star, Polaris (see Chapter 11).

There is a wide variety of compasses available, from the simple to the more complex. Figure 4.16 shows a *Brunton* compass, which is popular with many surveyors and geologists. It can be used hand-held or mounted on a tripod. In addition to providing magnetic direction, this instrument can also be used as a clinometer, with vertical angles being read to the closest five minutes.

PROBLEMS

4-1 A closed five-sided field traverse has the following interior angles:

A 110°27′00″

B 130°52′30″

C   88°17′00″

D   97°33′30″

E ?

Find the angle at E.

4-2   Convert the following azimuths to bearings:
a) 241°16'; b) 145°02'; c) 167°50'; d) 280°19';
e) 21°46'; f) 333°33'; g) 191°14'.

4-3   Convert the following bearings to azimuths:
a) N71°50'W; b) N1°03'E; c) S14°53'E; d) S89°29'W;
e) N89°08'E; f) S10°10'W; g) S70°40'E.

4-4   Convert the azimuths given in Problem 4-2 to reverse (back) azimuths.

4-5   Convert the bearings given in Problem 4-3 to reverse (back) bearings.

4-6   An open traverse that runs from A through H has the following deflection angles:
B) 8°13'R.; C) 2°21'R.; D) 14°41'R.; E) 21°08'L.;
F) 6°32'L.; G) 1°15'R.
If the bearing of AB is N41°21'E, compute the bearings of the remaining sides.

4-7   Closed traverse ABCD has the following bearings:
AB = N.60°38'E.; BC = S.49°49'E.; CD = S.17°13'W.;
DA = N.58°49'W.
Compute the interior angles and provide a geometric check on your work.

Use the following sketch and interior angles for Problems 4-8 to 4-11.

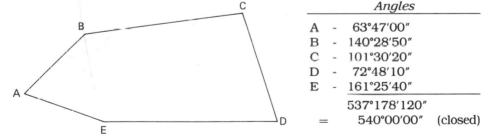

		Angles
A	-	63°47'00"
B	-	140°28'50"
C	-	101°30'20"
D	-	72°48'10"
E	-	161°25'40"
		537°178'120"
	=	540°00'00" (closed)

4-8   If Bearing AB is N.30°38'10"E., compute the bearings of the remaining sides. Provide two solutions: one solution proceeding clockwise and the other solution proceeding counterclockwise.

4-9   If the azimuth of AB is 30°38'10", compute the azimuths of the remaining sides. Provide two solutions: one solution proceeding clockwise and the other proceeding counterclockwise.

4-10   If the bearing of AB is N.51°44′E., compute the bearings of the remaining sides proceeding in a clockwise direction.

4-11   If the azimuth of AB is 51°44′, compute the azimuths of the remaining sides proceeding in a counterclockwise direction.

4-12   On January 2, 1984, the compass reading on a survey line in the Seattle, Washington area was N.40°20′E. Use Figure 4-14a or 4-14b to determine:

a) What the compass reading would have been for the same survey line on July 2, 1968.

b) What is the astronomic bearing of the survey line?

5 *Transits and Theodolites*

5.1 INTRODUCTION

As noted in Section 1.5, the term transit or theodolite (transiting theodolite) can be used to describe those survey instruments designed to precisely measure horizontal and vertical angles. In addition to measuring horizontal and vertical angles, transits and theodolites are used to establish straight lines, to establish horizontal and vertical distances through the use of stadia (Chapter 7), and to establish elevations when used as a level.

To avoid confusion, this text will, from this point on, refer to as *transits* (Figure 5.1) those American-style instruments using a four-screw leveling base, metallic (silvered) horizontal and vertical circles read with the aid of vernier scales, and equipped with plumb bobs for setting over specific points, whereas *theodolites* (Figure 5.2) are those European-style instruments using a three-screw leveling base, glass horizontal and vertical circles read either directly or with the aid of an optical micrometer, and equipped with right-angle prisms serving as optical plummets for setting over specific points.

The verniers used on the transit circles permit angles to be read to 1 minute, 30 seconds or 20 seconds of arc. The direct

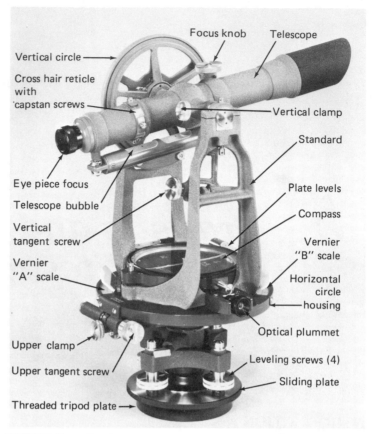

FIGURE 5.1 Engineers transit. (Courtesy Keuffel & Esser Co.)

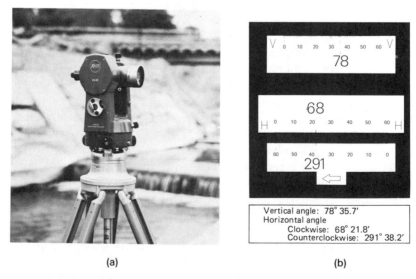

(a)

(b)

FIGURE 5.2 (a) Kern K1-S half-minute engineers theodolite with optical plummet and direct reading scales; (b) scale readings for the K1-S theodolite. (Courtesy Kern Instruments, Inc.)

reading theodolites can be read to 1 minute of arc, whereas those theodolites equipped with optical micrometers can be read to 20 seconds, 10 seconds, and 1 second of arc. Theodolites used in very precise survey control can be read to fractions of 1 second.

5.2 THE TRANSIT

Figure 5.3 shows the three main assemblies of the transit. The upper assembly, called the *alidade*, includes the standards, telescope, vertical circle and vernier, two opposite verniers for reading the horizontal circle, plate bubbles, compass, and upper tangent (slow motion) screw.

The spindle of the alidade fits down into the hollow spindle of the circle assembly. The circle assembly includes the horizontal circle that is covered by the alidade plate except at the vernier windows, the upper clamp screw, and the hollow spindle previously mentioned.

THE ALIDADE ASSEMBLY, WHICH INCLUDES

> TELESCOPE
> VERTICAL CIRCLE
> STANDARDS
> VERNIERS
> VERNIER COVER
> PLATE LEVELS
> INNER CENTRE
> UPPER TANGENT

THE CIRCLE ASSEMBLY, WHICH INCLUDES

> HORIZONTAL LIMB
> OUTER CENTRE
> UPPER CLAMP

THE LEVELING HEAD ASSEMBLY, WHICH INCLUDES

> LEVELING HEAD
> LEVELING SCREWS
> SHIFTING PLATE
> FRICTION PLATE
> HALF BALL
> TRIPOD PLATE
> LOWER CLAMP
> LOWER TANGENT

THE INNER CENTRE OF THE ALIDADE ASSEMBLY FITS INTO THE OUTER CENTRE OF THE CIRCLE ASSEMBLY AND CAN BE ROTATED IN THE OUTER CENTRE. THE OUTER CENTRE FITS INTO THE LEVELING HEAD AND CAN BE ROTATED IN THE LEVELING HEAD.

FIGURE 5.3 Three major assemblies of the transit. (Courtesy Sokkisha Co. Ltd.)

The hollow spindle of the circle assembly fits down into the leveling head. The leveling head includes the four leveling screws, the half-ball joint about which opposing screws are manipulated to level the instrument, a threaded collar that permits attachment to a tripod, the lower clamp and slow-motion screw, and a chain with an attached hook for attaching the plumb bob.

The upper clamp tightens the alidade to the circle, whereas the lower clamp tightens the circle to the leveling head. These two independent motions permit angles to be accumulated on the circle for repeated measurements. Transits that have these two independent motions are called *repeating* instruments.

Instruments with only one motion (upper) are called *direction* instruments. Since the circle cannot be precisely zeroed (older instruments), angles are usually determined by subtracting the initial setting from the final value. It is not possible to accumulate or repeat angles with a direction theodolite.

5.3 CIRCLES AND VERNIERS

The horizontal circle is usually graduated into degrees and half-degrees or 30 minutes (Figure 5.4), although it is not uncommon to find the horizontal circle graduated into degrees and one-third degrees (20 minutes). To determine the angle value more precisely than the least count of the circle (i.e., 30 or 20 minutes), vernier scales are employed.

Figure 5.5 shows a double vernier scale alongside a transit circle. The left vernier scale is used for clockwise circle readings (angles turned to the right), and the right vernier scale is used for counterclockwise circle readings (angles turned to the left). To avoid confusion as to which vernier (left or right) scale is to be

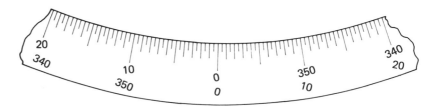

FIGURE 5.4   Part of a transit circle showing a least count of 30 minutes. The circle is graduated in both clockwise and counterclockwise directions permitting the reading of angles turned to both the right and left.

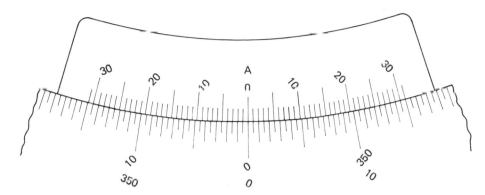

FIGURE 5.5 Double vernier scale set to zero on the horizontal circle.

used, one can recall that the vernier to be used is the one whose graduations are increasing in the same direction as are the circle graduations.

The vernier scale is constructed so that 30 vernier divisions cover the same length of arc as do 29 divisions (half-degrees) on the circle. The length of one vernier division is $(29/30) \times 30' = 29'$ on the circle. Therefore, the space difference between one division on the circle and one division on the vernier represents $01'$; referring to Figure 5.5, the first division on the vernier (left or right of the index mark) fails to exactly line up with the first division on the circle (left or right) by $01'$. The second division on the vernier fails to line up with the corresponding circle division by $02'$, and so on. If the vernier were moved so that its first division exactly lined up with the first circle division (30' mark), the reading would be $01'$; if the vernier again were moved the same distance at arc ($1'$), the second vernier mark would now line up with the appropriate circle division line, indicating a vernier reading of $02'$.

Generally, the vernier is read by finding which vernier division line exactly coincides with any circle line, and by then adding the value of that vernier line to the value of the angle obtained from reading the circle to the closest 30' (in this example).

In Figure 5.6a the circle is divided into degrees and half-degrees (30'). Before even looking at the vernier, we know that its range will be 30' (left or right) to cover the least count of the circle. Inspection of the vernier shows that 30 marks cover the range of 30', indicating that the value of each mark is $01'$. (Had each of the minute marks been further subdivided into two or three intervals, the angle could then have been read to the closest 30" or 20".)

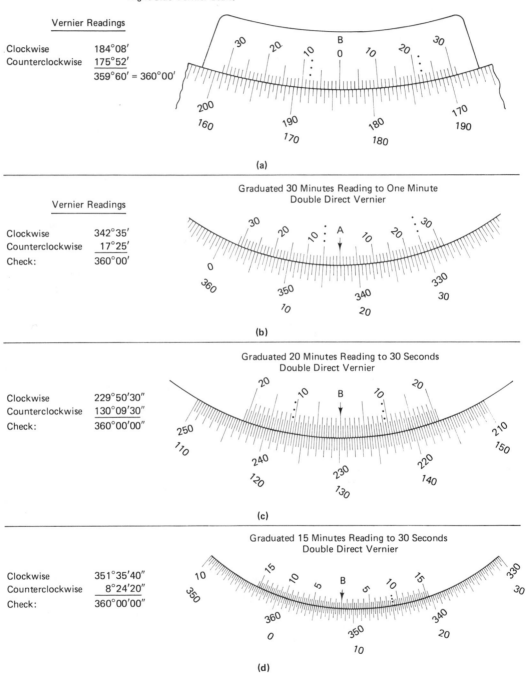

FIGURE 5.6 Sample vernier readings. (Triple dots identify aligned vernier graduations.)

If we consider the clockwise circle readings (field angle turned left to right), we see that the zero mark is between 184° and 184°30'; the circle reading is therefore 184°. Now to find the value to the closest minute we use the left side vernier and moving from the zero mark, we look for the vernier line that exactly lines up with a circle line. In this case, the 08' mark lines up; this is confirmed by noting that both the 07' and 09' marks do not line up with their corresponding circle mark, both by the same amount. The angle for this illustration is 184° + 08' = 184°08'.

If we consider the counterclockwise circle reading in Figure 5.6a, we see that the zero mark is between 175°30' and 176°; the circle reading is therefore 175°30, and to that value we will add the right side vernier reading of 22' to give an angle of 175°52'. As a check, the sum of the clockwise and counterclockwise readings should be 360°00'.

All transits are equipped with two double verniers (A and B) located 180° apart. Although increased precision can theoretically be obtained by reading both verniers for each angle, usually only one vernier is employed. Furthermore, to avoid costly mistakes, most surveying agencies favor use of the same vernier, the A vernier, at all times.

As noted earlier, the double vernier permits angles to be turned to the right (left vernier) or to the left (right vernier). However, by convention, field angles are normally turned only to the right. The exceptions to this occur when deflection angles are being employed, as in route surveys, or when construction layouts necessitate angles to the left, as in some curve deflections. There are a few more specialized cases (e.g., star observations) where it is advantageous to turn angles to the left, but as stated earlier, the bulk of surveying experience favors angles turned to the right. This type of consistency provides the routine required to foster a climate where fewer mistakes occur, and where mistakes that do occur can be readily recognized and eliminated.

The graduations of the circles and verniers as illustrated are in wide use in the survey field. However, there are several variations to both circle graduations and vernier graduations. Typically, the circle is graduated to the closest 30' (as illustrated), 20', or 10' (rarely). The vernier will have a range in minutes covering the smallest division on the circle (30', 20', or 10'), and could be further graduated to half-minute (30") or one-third minute (20") divisions. A few minutes spent observing the circle and vernier graduations of an unfamiliar transit will easily disclose the proper technique required for reading. (See also Figs. 5.6b, c, and d.)

The use of a magnifying glass (5×) is recommended for reading the scales, particularly for the 30" and 20" verniers.

5.4 TELESCOPE

The telescope (see Figures 5.7 and 5.8) in the transit is somewhat shorter than in a level with a reduced magnifying power (26×). The telescope axis is supported by the standards, which are of sufficient height so as to permit the telescope to be revolved (transited) 360° about the axis. A level vial tube is attached to the telescope so that, if desired, it may be used as a level.

The telescope level has a sensitivity of 30″ to 40″ per 2-mm graduation, compared to a level sensitivity of about 20″ for a dumpy level. When the telescope is positioned so that the level tube is under the telescope, it is said to be in the *direct* (normal) position; when the level tube is on top of the telescope, the telescope is said to be in a *reversed* (inverted) position. The eyepiece focus is always located at the eyepiece end of the telescope, whereas the object focus can be located on the telescope barrel just ahead of the eyepiece focus, midway along the telescope, or on the horizontal telescope axis at the standard.

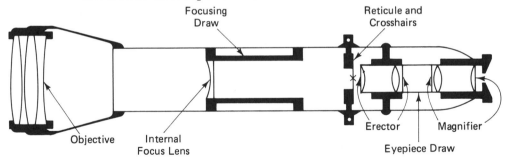

FIGURE 5.7   Transit telescope. (Courtesy Sokkisha Co. Ltd.)

Diagram of Optical System

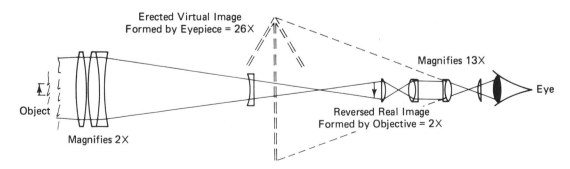

FIGURE 5.8   Diagram of optical system. (Courtesy Sokkisha Co. Ltd.)

5.5 LEVELING HEAD

The leveling head supports the instrument, and proper manipulation of the leveling screws (Section 3.7) allows the horizontal circle and telescope axis to be placed in a horizontal plane, which forces the alidade and circle assembly spindles to be placed in a vertical direction.

When the leveling screws are loosened, the pressure on the tripod plate is removed, permitting the instrument to be shifted laterally a short distance (⅜ in.). This shifting capability permits the surveyor to precisely position the transit center over the desired point.

5.6 PLATE LEVELS

Transits come equipped with two plate levels set at 90° to each other. Plate levels have a sensitivity range of 60″ to 80″ per 2-mm division on the level tube, depending on the overall precision requirements of the instrument.

5.7 TRANSIT SETUP

The transit is removed from its case, held by the standards or leveling base (never by the telescope), and placed on a tripod by screwing the transit snugly to the threaded tripod top. When carrying the transit indoors or near obstructions (e.g., tree branches), it is carried cradled under the arm, with the instrument forward, where it can be seen. Otherwise, the transit and tripod can be carried on the shoulder.

The instrument is placed roughly over the desired point and the tripod legs adjusted so that (1) the instrument is at a convenient height and (2) the tripod plate is nearly level. Usually, two legs are placed on the ground, and the instrument is roughly leveled by manipulation of the third leg. If the instrument is to be set up on a hill, the instrumentman faces uphill and places two of the legs on the lower position; the third leg is placed in the upper position and then manipulated to roughly level the instrument. The wing nuts on the tripod legs are tightened and a plumb bob is attached to the plumb bob chain, which hangs down from the leveling head. The plumb bob is attached by means of a slip knot, which allows placement of the plumb bob point immediately over the mark. If it appears that the instrument placement is reasonably close to its final position, the tripod legs are pushed into the ground, taking care not to jar the instrument.

If necessary, the length of plumb bob string is adjusted as the setting up procedure advances. If after pushing in the tripod legs the instrument is not centered, one leg is either pushed in farther or pulled out and repositioned until the plumb bob is very nearly over the point or until it becomes obvious that manipulation of another leg would be more productive. When the plumb bob is within $1/4$ in. of the desired location, the instrument is then leveled.

Now, two adjacent leveling screws are loosened so that pressure is removed from the tripod plate and the transit can be shifted laterally until it is precisely over the point. If now the same two adjacent leveling screws are retightened, the instrument will return to its level (or nearly so) position. Final adjustments to the leveling screws at this stage will not be large enough to displace the plumb bob from its position directly over the desired point.

The actual leveling procedure is a faster operation than that for a level. The transit has two plate levels, which means that the transit can be leveled in two directions, 90° opposed, without rotating the instrument. When both bubbles have been carefully centered, it only remains for the instrument to be turned through 180° to check the adjustment of the plate bubbles. If one (or both) bubble(s) do not center, after turning 180°, the discrepancy is noted and the bubble brought to half the discrepancy by means of the leveling screws. If this has been done correctly, the bubbles will remain in the same position as the instrument is revolved.

5.8 MEASURING ANGLES BY REPETITION

Assuming that the instrument is over the point and level, the following procedure is used to turn and "double" an angle. *Turning the angle at least twice permits the elimination of mistakes, and increases precision owing to the elimination of most instrumental errors.*

It is recommended that only the A vernier scale be used.

1. *Set the scales to zero.* Loosen both the upper and lower motion clamps. While holding the alidade stationary, revolve the circle by pushing on the circle underside with the fingertips.

When the zero on the scale is close to the vernier zero, tighten (snug) the upper clamp. Then, with the aid of a magnifying glass, turn the upper tangent screw (slow motion screw) until the zeros are precisely set. It is good practice to make the last turn of the tangent screw against the tangent screw spring so that spring tension is assured.

2. *Sight the initial point* (see Figure 5.9 and assume the instrument at *A*, and an angle from *B* to *E*). With the upper clamp tightened and the lower clamp loose, turn and point at station *B*, and then tighten the lower clamp. At this point check the eyepiece focus and object focus to eliminate parallax (Section 3.15). If the sight is being given by a range pole, or even a pencil, always sight as close to the ground level as possible to eliminate plumbing errors. If the sight is being given with a plumb bob, sight high on the plumb bob string to minimize the effect of plumb bob oscillations. Using the lower tangent screw, position the vertical cross hair on the target, once again making the last adjustment motion against the tangent screw spring.

3. *Turn the angle.* Loosen the upper clamp and turn clockwise to the final point (*E*). When the sight is close to *E*, tighten the upper clamp. Using the upper tangent screw set the vertical cross hair precisely on the target using techniques already described. Read the angle by utilizing (in this case) the left side vernier, and book the value in the appropriate column in the field notes (see Figure 5.9).

4. *Repeat the angle.* After the initial angle has been booked, *transit (plunge) the telescope*, loosen the lower motion and sight at the initial target, station *B*. The simple act of transiting the telescope removes nearly all the potential instrumental errors associated with the transit.

The procedure described in the first three steps is repeated, the only difference being that the telescope is now inverted and that the initial horizontal angle setting is 101°24′ instead of 0°00′. The angle that is read as a result of this repeating procedure should be approximately double the initial angle. This "double" angle is booked, and then divided by two to find the mean value, which is also booked.

If the procedure has been properly executed, the mean value should be the same as the direct reading or half the least count (30″) different. In practice, a discrepancy equal to the least count (01′) is normally permitted. Although doubling the angle is sufficient for most engineering projects, precision can be increased by

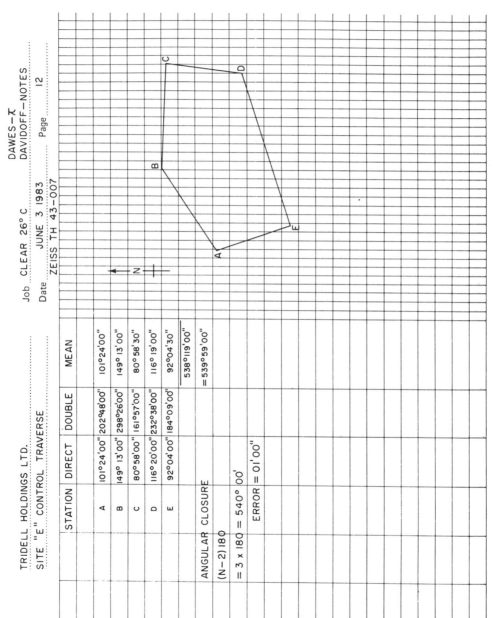

FIGURE 5.9 Field notes from angles by repetition.

150

repeating the angle a number of times. Due to personal errors of sighting and scale reading, this procedure has practical constraints as to improvement in precision. It is generally agreed that repetitions beyond six or perhaps eight times will not further improve the precision

When multiple repetitions are being used, only the first angle and the final value are recorded. The final value is divided by the number of repetitions to arrive at the mean value. It may be necessary to augment the final reading by 360° or multiples of 360° prior to determining the mean. The proper value can be roughly determined by multiplying the first angle recorded by the number of repetitions.

5.9 GEOMETRY OF THE TRANSIT

The vertical axis of the transit goes up through the center of the spindles and is oriented over a specific point on the earth's surface. The circle assembly and alidade revolve about this axis. The horizontal axis of the telescope is perpendicular to the vertical axis, and the telescope and vertical circle revolve about it. The line of sight (line of collimation) is a line joining the intersection of the reticle cross hairs and the center of the objective lens. The line of sight is perpendicular to the horizontal axis and should be truly horizontal when the telescope level bubble is centered and when the vertical circle is set at zero.

5.10 ADJUSTMENT OF THE TRANSIT

Figure 5.10 shows the geometric features of the transit. The most important relationships are as follows:

1. The vertical cross hair should be perpendicular to the horizontal axis (tilting axis).
2. The axis of the *plate bubble* should be in a plane perpendicular to the vertical axis.
3. The *line of sight* should be perpendicular to the horizontal axis.
4. The horizontal axis should be perpendicular to the vertical axis (*standards adjustment*).

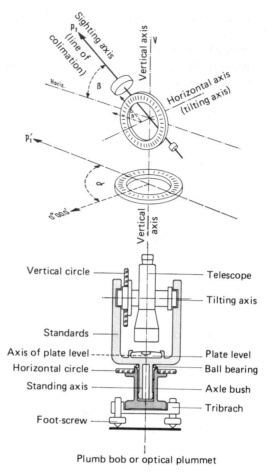

FIGURE 5.10 Geometry of the transit. (Courtesy Wild Heerbrugg Co. Ltd.)

In addition to the above, the following secondary features must be considered:

5. The axis of the telescope and the axis of the *telescope bubble* should be parallel.

6. The *vertical circle vernier* zero mark should be aligned with the vertical circle zero mark when the plate bubbles and the telescope bubble are centered.

These features are discussed in the following paragraphs.

5.10.1 Vertical Cross Hair

If the vertical cross hair is perpendicular to the horizontal axis, all parts of the vertical cross hair can be used for line and angle sightings. This adjustment can be checked by sighting a well-defined distant point and then clamping the horizontal movements. The telescope is now moved up and down so that the point sighted appears to move on the vertical cross hair. If the point appears to move off the vertical hair, an error exists and the cross-hair reticle must be rotated slightly until the sighted point appears to stay on the vertical cross hair as it is being revolved. The reticle can be adjusted slightly by loosening two adjacent capstan screws, rotating the reticle, and then retightening the same two capstan screws.

This same cross-hair orientation adjustment is performed on the dumpy level, but in the case of the level, the horizontal cross hair is of prime importance. The horizontal cross hair is checked by sighting a distant point on the horizontal cross hair with the vertical clamp set, and then moving the telescope left and right, checking to see that the sighted point remains on the horizontal cross hair; the adjustment for any maladjustment of the reticle is performed as described previously.

5.10.2 Plate Bubbles

It has been previously noted (Section 5.7) that after a bubble has been centered, its position is checked by rotating the instrument through 180°; if the bubble does not remain centered, it can be properly set by bringing the bubble halfway back using the foot screws. For example, if, when checking a bubble position, it is out by four division marks, the bubble can now be properly set by turning the foot screws until the bubble is only two division marks off center.

The bubble should remain in this off-center position as the telescope is rotated, indicating that the instrument is, in fact, level. Although the instrument can now be safely used, it is customary to remove the error by adjusting the bubble tube.

The bubble tube can now be adjusted by turning the capstan screws at one end of the bubble tube until the bubble becomes precisely centered. The entire leveling and adjusting procedure is repeated until the bubble remains centered as the instrument is rotated and checked in all positions. All capstan screw adjustments are best done in small increments. That is, if the end of the bubble tube is to be lowered, first loosen the lower capstan screw a slight turn (say one-eighth); then tighten (snug) the top capstan screw to close the gap. This incremental adjustment is continued until the bubble is precisely centered.

5.10.3 Line of Sight

Vertical Cross Hair. The vertical line of sight should be perpendicular to the horizontal axis so that a vertical plane is formed by the complete revolution of the telescope on its axis. The technique for testing for this adjustment is known as double centering (Section 5.17).

The transit is set over a point A (Figure 5.11a) and leveled. A backsight is taken on any distant well-defined point B (a well-defined point on the horizon is best); the telescope is transited (plunged) and a point C is set on the opposite side of the transit 300 to 400 ft away, and at roughly the same elevation as the transit station. Now loosen either the upper or lower plate clamp and revolve the transit back toward point B. With the telescope still inverted, sight B and then transit the telescope and sight toward point C previously set. It is highly probable that the new line of sight will not fall precisely on point C, and that a new point, D, will be set adjacent to C. The correct location of line AB produced is at B^1, which is located midway between points C and D. Since the distance CB^1 or B^1D is double the sighting error, the line of sight correction is accomplished by setting point E midway between B and D (or one-quarter of the way from D to C) and by moving the vertical cross hair onto point E.

The vertical cross hair is adjusted laterally by first loosening the left or right reticle capstan screw and then by tightening the opposite screw. This procedure is repeated until the spread CD is eliminated or, as is more usually the case, CD is reduced to a manageable size, say 0.5 ft at 1000 ft for normal engineering work. (Some surveyors, knowing that this error cannot be totally eliminated, prefer to have an error large enough to warrant continual correction.)

Horizontal Cross Hair. The horizontal cross hair must be adjusted so that it lies on the optical axis of the telescope. To test this relationship, set up the transit at point A (Figure 5.11b), and place two stakes B and C in a straight line, B about 25 ft away and C about 300 ft away. With the vertical motion clamped, take a reading first on C (K) and then on B (J). Transit (plunge) the telescope and set the horizontal cross hair on the previous rod reading (J) at B and then take a reading at point C. If the transit is in perfect

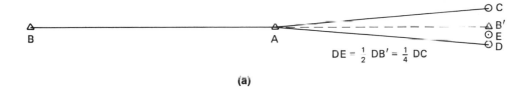

$$DE = \tfrac{1}{2} DB' = \tfrac{1}{4} DC$$

(a)

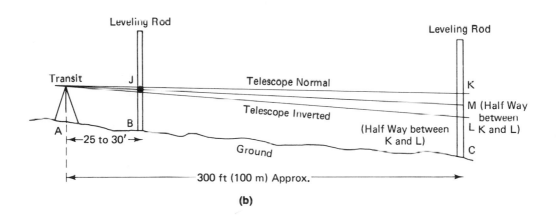

(b)

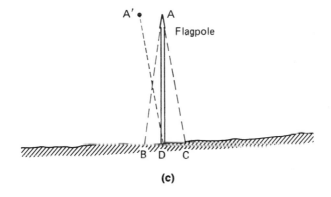

(c)

FIGURE 5.11 Instrument adjustments: (a) line of sight perpendicular to horizontal axis; (b) horizontal cross-hair adjustment; (c) standards adjustment.

adjustment, the two rod readings at C will be the same; if the cross hair is out of adjustment, the rod reading at C will be at some reading L instead of K. To adjust the cross hair, adjust the cross-hair reticle up or down by first loosening and then tightening the appropriate opposing capstan screws until the cross hair lines up with the average of the two readings (M).

5.10.4 Standards Adjustment

The horizontal axis should be perpendicular to the vertical axis. The standards are checked for proper adjustment by first setting up the transit and then sighting a high (at least 30° altitude) point (point A in Figure 5.11c). After clamping the instrument in that position, the telescope is depressed and point B is marked on the ground. The telescope is then transited (plunged), a lower clamp is loosened, and the transit is turned and once again set precisely on point A. The telescope is again depressed, and if the standards are properly adjusted, the vertical cross hair will fall on point B; if the standards are not in adjustment, a new point C is established. The discrepancy between B and C is double the error resulting from the standards maladjustment. Point D, which is now established midway between B and C, will be in the same vertical plane as point A. The error is removed by sighting point D and then elevating the telescope to A^1, adjacent to A. The adjustable end of the horizontal axis is then raised or lowered until the line of sight falls on point A. When the adjustment is complete, care is taken in retightening the upper friction screws so that the telescope revolves with proper tension.

5.10.5 Telescope Bubble

If the transit is to be used for leveling work, the axis of the telescope bubble and the axis of the telescope must be parallel. To check this relationship, the bubble is centered with the telescope clamped, and the peg test (Section 3.20) is performed. When the proper rod reading has been determined at the final setup, the horizontal cross hair is set on that rod reading by moving the telescope with the vertical tangent (slow motion) screw. The telescope bubble is then centered by means of the capstan screws located at one (or both) end(s) of the bubble tube.

5.10.6 Vertical Circle Vernier

When the transit has been carefully leveled (plate bubbles), and the telescope bubble has been centered, the vertical circle should read zero. If a slight error (index error) exists, the screws holding the vernier are loosened, the vernier is tapped into its proper position, and then the screws are retightened so that the vernier is once again just touching, without binding, the vertical circle.

5.11 REPEATING THEODOLITES

As noted in Section 5.1, theodolites (now also manufactured in Japan and the United States) are European-style instruments characterized by three-screw leveling heads, optical plummets, light weight, with glass circles being read either directly or with the aid of a micrometer (see Figures 5.2 and 5.12). In contrast with the American engineers transit, most theodolites do not come equipped with compasses or telescope levels. Instead of a telescope level, theodolite telescopes can be "leveled" by means of a coincidence-type collimation level used in conjunction with a horizontal setting of the vertical circle. Most theodolites are now equipped with a compensating device that automatically indexes the horizontal direction when the vertical circle has been set to the horizontal setting of of 90°.

The horizontal setting for theodolites is 90° (270°), whereas for the transit it is 0°. A word of caution: although all theodolites have a horizontal setting of 90° direct or 270° inverted, some theodolites have their zero set at the nadir, while others have the zero set at the zenith. The method of graduation can be quickly ascertained in the field by simply setting the telescope in an upward (positive) direction and noting the scale reading. If the reading is less than 90°, the zero has been referenced to the zenith direction; if the reading is more than 90°, the zero has been referenced to the nadir direction.

The graduations of both the vertical and horizontal circles are, by means of prisms and lenses, projected to one location just adjacent to the telescope eyepiece where they are read by means of a microscope. Light, which is necessary in the circle reading procedure, is controlled by an adjustable mirror located on one of the standards. Light required for underground or night work is directed through the mirror window by attached battery packs.

The theodolite tripod, unlike the transit tripod with its threaded top, has a flat base through which a bolt is threaded up into the three-screw leveling base, thus securing the instrument to the tripod.

Most theodolites have a leveling head release feature that permits the alidade and circle assemblies to be lifted from the leveling head and interchanged with a target or prism (see Figure 5.12d). When the theodolite (minus its leveling head) is placed on the leveling head vacated by the target or prism, it will be instantly over the point and nearly level. This system, called *forced centering,* speeds up the work and reduces centering errors associated with multiple setups.

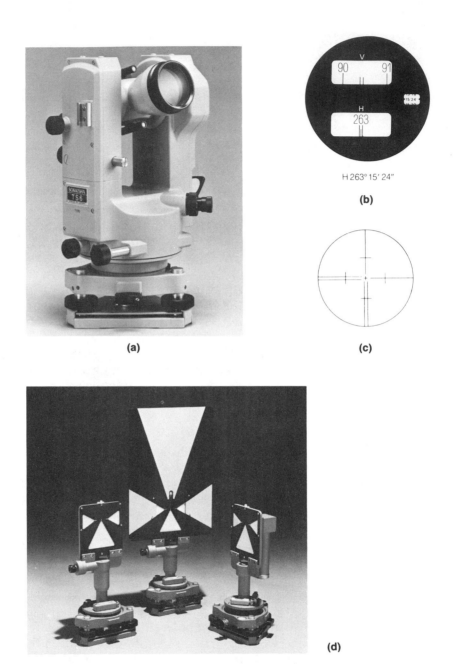

H 263° 15′ 24″

(b)

(a)

(c)

(d)

FIGURE 5.12 (a) Six-second repeating theodolite; (b) horizontal circle and micrometer reading; (c) cross-hair reticle pattern; (d) a variety of tribrach-mounted traverse targets. Targets and theodolites can be easily interchanged to save setup time (forced centering system). (Courtesy Sokkisha Co. Ltd.)

Typical Specifications for Repeating Micrometer Theodolites

Magnification: 30×
Clear objective aperture: 1.6 in. (42 mm)
Field of view at 100 ft (100 m): 2.7 ft (2.7 m)
Shortest focusing distance: 5.6 ft (1.7 m)
Stadia multiplication constant: 100
Bubble sensitivity
    Circular bubble: 8' per 2 mm
    Plate level: 30" per 2 mm
Direct circle reading: 01" to 06" (20" in older versions)

Similar to the engineers transit, the repeating theodolite has two independent motions (upper and lower), which necessitates upper and lower clamps with their attendant tangent screws. However, some theodolites come equipped with only one clamp and one tangent screw; these instruments have a lever or switch that transfers clamp operation from upper to lower motion and thus probably reduces the opportunity for mistakes due to wrong-screw manipulation.

5.12 THEODOLITE SETUP

The theodolite can be set up in much the same manner as described in Section 5.7 for the transit. The difference in setups is related to the use of optical plummets instead of the more traditional plumb bobs to position the instrument over the point. Although the optical plummet results in a more precise positioning, it is, for the beginner, more difficult to use. Since the optical plummet can realistically only give position when the instrument is level (or nearly so), the beginning surveyor is often unaware as to the relative location of the instrument until the setup procedure is almost complete, and if the instrument has not been properly positioned, the entire procedure must then be repeated. To reduce setup times, a systematic approach is recommended.

Setup Procedure

1. Place the instrument over the point with the tripod plate as level as possible and with two tripod legs on the downhill side, if applicable.

2. Stand back a pace or two and see if the instrument appears to be over the station; if it does not, adjust the location, and check again from a pace or two away.

3. Move to a position 90° opposed to the original inspection location and repeat step 2.

   Note: This simple act of "eyeing-in" the instrument from two directions, 90° opposed, takes only seconds but could save a great deal of time in the long run.

4. Check to see that the station point can be seen through the optical plumb and then firmly push in the tripod legs by pressing down on the tripod shoe spurs.

5. While looking through the optical plumb, manipulate the leveling screws (one, two, or all three at a time) until the cross hair (bull's-eye) of the optical plumb is directly on the station mark.

6. Now, level the theodolite circular bubble by adjusting the tripod legs up or down. This is accomplished by noting which leg, when slid up or down, will move the circular bubble into the bull's-eye. Upon adjusting the leg, the bubble will either move into the circle (the instrument is level), or it will slide around until it is exactly opposite another tripod leg. That leg would then be adjusted up or down until the bubble moved into the circle. If the bubble does not move into the circle, adjust the leg until the bubble is directly opposit another leg and repeat the process. If this manipulation has been done correctly, the bubble will be centered after the second leg has been adjusted; it is seldom necessary to adjust the legs more than three times. Comfort can be taken from the fact that these manipulations take less time to perform than they do to read about.

7. A check through the optical plumb will now confirm that its cross hairs (bull's-eye) are still quite close to being over the station mark.

8. The circular bubble is exactly centered (if necessary) by turning one (or more) leveling screws.

9. The tripod clamp bolt is loosened a bit and the instrument is slid on the flat tripod top until the optical plummet cross hairs (bull's-eye) are exactly centered on the station mark. The tripod clamp bolt is retightened and the circular bubble reset, if necessary. When sliding the instrument on the tripod top, it is advised not to twist the instrument, but to move it in a rectangular fashion; this will ensure that the instrument

will not go seriously off level if the tripod top itself is not close to being level.

10. The instrument can now be precisely leveled by centering the tubular bubble. The tubular bubble is set so that it is aligned in the same direction as two of the foot screws. These two screws can be turned (together or independently) until the bubble is centered. The instrument is then turned 90°, at which point the tubular bubble will be aligned with the third leveling screw. That screw is then turned to center the bubble. The instrument now should be level, although it is always checked by turning the instrument through 180°.

5.13 ANGLE MEASUREMENT WITH A THEODOLITE

The technique for turning and doubling (repeating) an angle is the same as that described for a transit (Section 5.10). The only difference in procedure is that of zeroing and reading the scales. In the case of the direct reading optical scale (Figure 5.2b), zeroing the circle is simply a matter of turning the circle until the zero degree mark lines up approximately with the zero minute mark on the scale. Once the upper clamp has been tightened, the setting can be precisely accomplished by manipulation of the upper tangent screw. The scale is read directly as illustrated in Figure 5.2b.

In the case of the optical micrometer instruments (see Figures 5.12 and 5.13), it is important to first set the micrometer to zero, and then to set the horizontal circle to zero. When the angle has been turned, it will be noted that the horizontal (or vertical) circle index mark is not directly over a degree mark. The micrometer knob is turned until the circle index mark is set to a degree mark. Movement of the micrometer knob also moves the micrometer scale; the reading on the micrometer scale is then added to the even degree reading taken from the circle. Figure 5.12 shows a micrometer graduated to the closest 06″, which is a refinement of an instrument that formerly was graduated to the closest 20″ (Figure 5.13).

The vertical circle is read in the same way using the same micrometer scale. If the vertical index is not automatically compensated (as it is for most repeating theodolites), the vertical index

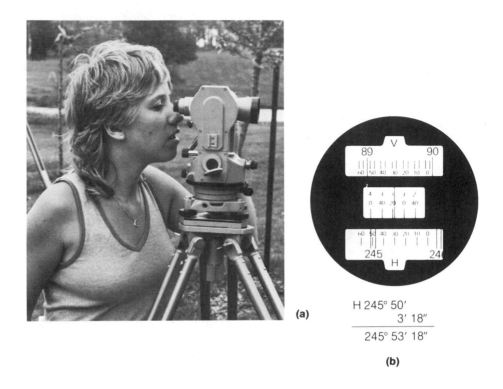

H 245° 50'
—————— 3' 18"
245° 53' 18"

(b)

FIGURE 5.13   (a) Twenty-second theodolite, Sokkisha TM 20-C; (b) horizontal and vertical scales with micrometer scale.

coincidence bubble must be centered by rotating the appropriate screw when vertical angles are being read.

5.14 DIRECTION THEODOLITES

The essential difference between a direction theodolite and a repeating theodolite is that the direction theodolite has only one motion (upper), whereas the repeating theodolite has two motions (upper and lower). Since it is difficult to precisely set angle values on this type of instrument, angles are usually determined by reading the initial direction and the final direction, and by then determining the difference between the two.

Direction theodolites are generally more precise; for example, the wild T2 shown in Figure 5.14 reads directly to 01″ and by estimation to 0.5″, whereas the wild T3 shown in Figure 5.15 reads directly to 0.2″ and by estimation to 0.1″ (see also Figure 5.16).

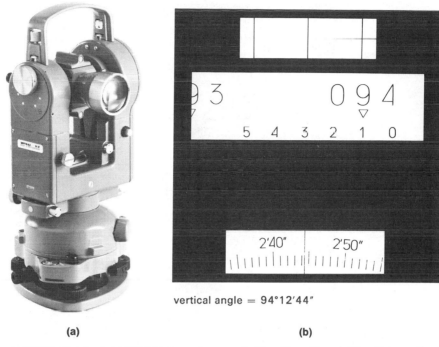

vertical angle = 94°12′44″

(a)   (b)

FIGURE 5.14 (a) Wild T-2, a 1-second direction theodolite; (b) vertical circle reading. (Courtesy Wild Heerbrugg Co. Ltd.)

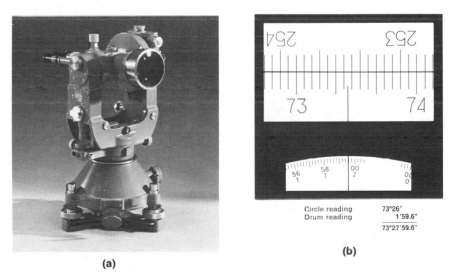

Circle reading   73°26′
Drum reading   1′59.6″
                       73°27′59.6″

(b)

(a)

FIGURE 5.15 (a) Wild T-3, precise theodolite for first-order surveying; (b) circle reading (least graduation is 0.2 second). (Courtesy Wild Heerbrugg Co. Ltd.)

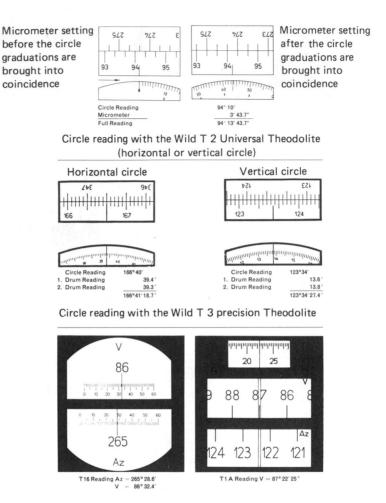

FIGURE 5.16 A variety of scale and micrometer readings. (Courtesy Wild Heerbrugg Co. Ltd.)

Figures 5.17, 5.18 and 5.19 show additional 1-second theodolites. It will be noted that only one circle can be seen in some figures. These instruments have a switch knob that permits either the vertical circle or the horizontal circle to be seen separately with the micrometer scale.

In the case of the Wild T2, (and the other 1-second theodolites), the micrometer is turned to force the index to read an even 10″ [the grid lines shown above (beside) the scale are brought to coincidence], and then the micrometer scale reading (02′44″) is added to the circle reading (94° 10′) to give a result of 94° 12′44″.

In the case of the T3 (Figure 5.15), both sides of the circle are viewed simultaneously; one reading is shown erect, the other in-

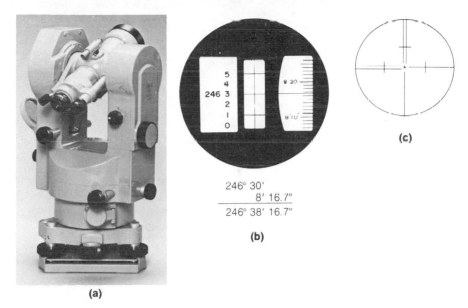

246° 30'
8' 16.7"
246° 38' 16.7"

(b)

FIGURE 5.17 (a) TM-1A one-second thodolite, a direction instrument with detachable tribrach, automatic vertical circle indexing, and optical plummet; (b) horizontal circle reading; (c) reticle pattern. (Courtesy Sokkisha Co. Ltd.)

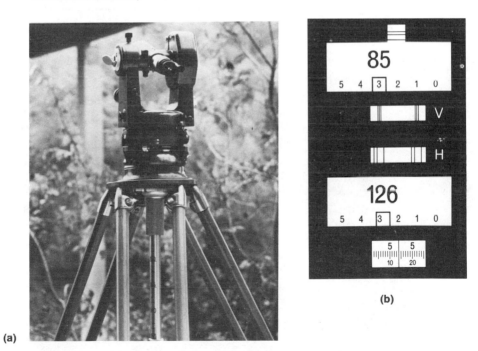

FIGURE 5.18 (a) Kern DKM2-A one-second direction theodolite; (b) scale reading for the DKM2-A; in this illustration the micrometer has been set for the vertical circle and reading is 85°35'14". (Courtesy Kern Instruments, Inc.)

165

(a)

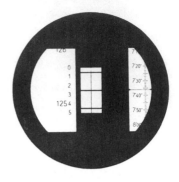

Microscope reading:
125° 47′ 36″
Approx. ⅓ actual size

(b)

FIGURE 5.19 (a) Zeiss Th 2 one-second direction theodolite featuring a ball-joint base for rapid coarse leveling; (b) horizontal circle reading. (Courtesy Carl Zeiss—Oberkochen)

verted. The micrometer knob is used to precisely align the erect and inverted circle markings. Each division on the circle is 04′; but if the lower scale is moved half a division, the upper also moves half a division, once again causing the markings to align, that is, *a movement of only 02′*. The circle index line is between the 73° and 74° mark, indicating that the value being read is 73°. Minutes can be read on the circle by counting the number of divisions from the erect 73° to the inverted value that is 180° different than 73° (i.e., 253°). In this case, the number of divisions between these two numbers is 13, each having a value of 02′ (i.e, 26′).

On the micrometer, a value of 01′59.6″ can be read. The reading is therefore 73°27′59.6″.

5.15 ANGLES MEASURED WITH A DIRECTION THEODOLITE

As noted earlier, since it is not always possible to precisely set angles on a direction theodolite scale, directions are observed and then subtracted, one from the other, in order to determine angles. Furthermore, if several sightings are required for precision purposes, it is customary to distribute the initial settings around the circle to minimize the effect of circle graduation distortions. For example, using a directional theodolite where both sides of the circle are viewed simultaneously, the initial settings (positions) would be distributed per 180/n, where n is the number of settings required by the precision specifications (specifications for precise surveys are published by the National Geodetic Survey—United States, and the Geodetic Surveys of Canada). To be consistent, not only should the initial settings be uniformly distributed around the circle, but the range of the micrometer scale should be noted as well and appropriately apportioned.

For example, using the scales shown in Figure 5.14 the initial settings for four positions would be near 0°, 45°, 90°, and 135° on the circle and near 00'00", 02'30", 05'00", and 07'30" on the micrometer.

For the instruments shown in Figures 5.14 and 5.15, the settings would be as given in Table 5.1.

TABLE 5.1   *Approximate Initial Scale Settings for Four Positions*

10-Minute Micrometer, Wild T-2	2-Minute Micrometer, Wild T-3
0°00'00"	0°00'00"
45°02'30"	45°00'30"
90°05'00"	90°01'00"
135°07'30"	135°01'30"

These initial settings are accomplished by setting the micrometer to zero (02'30", 05'00", 07'30"), and then setting the circle as closely to zero as is possible using the tangent screw. Precise coincidence of the zero (45, 90, 135) degree mark is achieved using the micrometer knob, which moves the micrometer scale slightly off zero (2'30", 5'00", 7'30", etc.).

The direct readings are taken first in a clockwise direction; the telescope is then transited (plunged), and the reverse readings are taken counterclockwise. In Figure 5.20, the last entry at position 1 is 180°00'12" (R). If the angles (shown in the abstract) do not meet the required accuracy, the procedure is repeated while the instrument still occupies that station.

FOXLEA SUBDIVISION
DIRECTIONS FOR CONTROL EXTENSION

Job CLEAR 17°C

CRAWFORD–K
HABKIRK–NOTES

Date MARCH 14 1983   Page 28

WILD T–2 #4128–B

K @ STATION #481

STATION SIGHTED	D/R	READING	MEAN D/R	REDUCED DIRECTION
POSITION #1				
1001	D	0° 00' 08"	10"	0° 00' 00"
	R	180° 00' 12"		
778	D	40° 37' 44"	46"	40° 37' 36"
	R	220° 37' 47"		
779	D	78° 52' 19"	16"	78° 52' 06"
	R	258° 52' 13"		
POSITION #2				
1001	D	45° 02' 22"	24"	0° 00' 00"
	R	225° 02' 26"		
778	D	85° 40' 02"	04"	40° 37' 40"
	R	265° 40' 05"		
779	D	123° 54' 30"	33"	78° 52' 09"
	R	303° 54' 36"		
POSITION #3				
1001	D	90° 05' 03"	05"	0° 00' 00"
	R	270° 05' 07"		
778	D	130° 42' 44"	44"	40° 37' 39"
	R	310° 42' 45"		
779	D	168° 57' 10"	12"	78° 52' 07"
	R	348° 57' 14"		
POSITION #4				
ETC.				

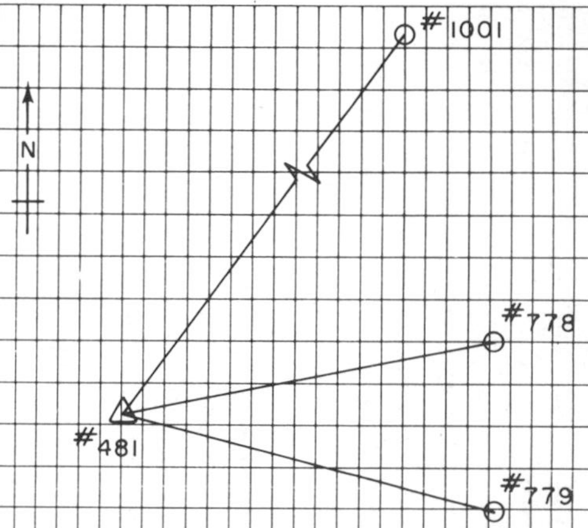

ABSTRACT OF ANGLES

POSITION	ANGLE 1001–778	ANGLE 778–779
1	40° 37' 36"	38° 14' 30"
2	40° 37' 40"	38° 14' 33"
3	40° 37' 39"	38° 14' 28"
4		

FIGURE 5.20 Field notes for directions.

168

5.16 LAYING OFF ANGLES

Case 1.   The angle is to be laid out no more precisely than the least count of the transit or theodolite.

Assume a route survey where a deflection angle (31°12′R) has been determined photogrammetrically to position the ℄ clear of natural obstructions (see Figure 5.21a). The surveyor sets the instrument at the PI (point of intersection of the tangents), and sights the back line with the telescope reversed and the horizontal circle set to zero. Next the surveyor transits (plunges) the telescope and turns off the required deflection angle and sets a point on line. The deflection angle is 31°12′R and a point is set at *a′*.

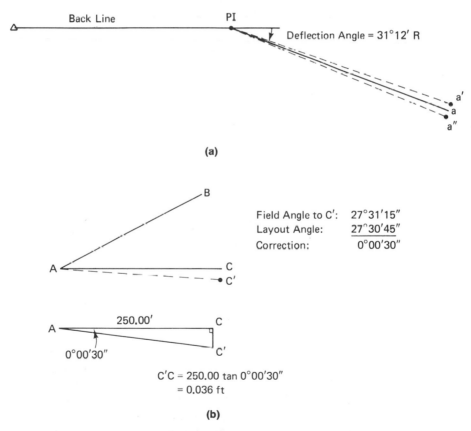

(a)

Field Angle to C′:   27°31′15″
Layout Angle: 27°30′45″
Correction: 0°00′30″

C′C = 250.00 tan 0°00′30″
    = 0.036 ft

(b)

FIGURE 5.21   Laying off an angle: (a) case I, angle laid out to least count of instrument; (b) case II, angle to be laid out more precisely than the least count of the instrument will permit.

The surveyor then loosens the lower motion (repeating instruments) sights again at the back line, transits the telescope, and turns off the required value (31°12′ × 2 = 62°24′). It is very likely that this line of sight will not precisely match the first sighting at a'; if the line does not cross a', a new mark a'' is made and the instrumentman given a sighting on the correct point a, which is midway between a' and a''.

Case 2. The angle is to be laid out more precisely than the least count of the instrument will directly permit.

Assume that an angle of 27°30′45″ is required in a heavy construction layout, and that a 01-minute transit is being used. In Figure 5.21b, the transit is set up at A zeroed on B with an angle of 27°31′ turned to set point C'. The angle is then repeated to point C' a suitable number of times so that an accurate value of that angle can be determined. Let's assume that the scale reading after four repetitions is 110°05′, giving a mean angle value of 27°31′15″ for angle BAC'.

If the layout distance of AC is 250.00 ft, point C can be precisely located by measuring from C' a distance $C'C$:

$$C'C = 250.00 \tan 0°00′30″$$
$$= .036 \text{ ft}$$

After point C has been located, its position can be verified by repeating angles to it from point B.

5.17 PROLONGING A STRAIGHT LINE

Prolonging a straight line (also known as *double centering*) is a common survey procedure used every time a straight line must be prolonged. The best example of this requirement would be in route surveying, where straight lines are routinely prolonged over long distances and often over very difficult terrain. The technique of reversion (the same technique as used in repeating angles) is used to ensure that the straight line is properly prolonged.

Referring to Figure 5.22, the straight line AB is to be prolonged to C (see also Figure 5.11a). With the instrument at B, a sight is carefully made on station A. The telescope is transited, and a temporary point is set at C'. The transit is revolved back to station A, and a new sighting is made (the telescope is in a position reversed to the original sighting). The telescope is transited, and a temporary mark is made at C'', adjacent to C'.

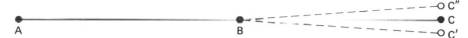

FIGURE 5.22 Double centering to prolong a straight line.

Note: Over short distances, well-adjusted transits will show no appreciable displacement between points C' and C''; however, over the longer distances normally encountered in this type of work, all transits will display a displacement between direct and reversed sightings; the longer the forward sighting, the greater the displacement.

The correct location of station C is established midway between C' and C'' by measuring with a steel tape.

5.18 INTERLINING (BALANCING IN)

It is sometimes necessary to establish a straight line between two points that themselves are not intervisible (i.e., a transit set up at one point cannot, because of an intervening hill, be sighted at the other required point). It is usually possible to find an intermediate position from which both points can be seen.

In Figure 5.23, points A and B are not intervisible, but point C is in an area from which both A and B can be seen. The interlining procedure is as follows: the transit is set up in the area of C (at C_1) and as close to line AB as is possible to estimate. The transit is roughly leveled and a sight is taken on point A; then the telescope is transited and a sight taken toward B. The line of sight will, of course, not be on B but on point B', some distance away.

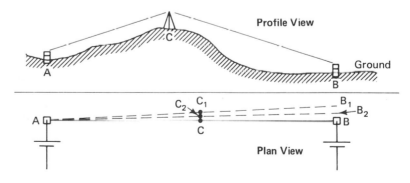

FIGURE 5.23 Interlining (balancing in).

Noting roughly the distance $B'B$ and the position of the transit between A and B (e.g., halfway, one-third, or one-quarter of the distance AB), an estimate is made as to proportionately how far the instrument is to be moved in order to be on the line AB. The transit is once again roughly leveled (position C_2), and the sighting procedure is repeated.

This trial-and-error technique is repeated until, after sighting A, the transited line of sight falls on point B or close enough to point B so that it can be precisely set by shifting the transit on the leveling head shifting plate. When the line has been established, a point is set at or near point C so that the position can be saved for future use.

The entire procedure of interlining can be accomplished in a surprisingly short period of time. All but the final instrument setups are only roughly leveled, and at no time does the instrument have to be set up over a point.

5.19 INTERSECTION OF TWO STRAIGHT LINES

The intersection of two straight lines is also a very common survey technique. In municipal surveying, street surveys usually begin (0 + 00) at the intersection of the ℄'s of two streets, and the chainage and angle of the intersections of all subsequent street ℄'s are routinely determined.

Figure 5.24a illustrates the need for intersecting points on a municipal survey, and Figure 5.24b illustrates just how the intersection point is located. Referring to Figure 5.24b, with the instrument set on a Main St. station and a sight taken also on Main St. ℄ (the longer the sight, the more precise the sighting), two points (2 to 4 ft apart) are established on Main St. ℄, one point on either side of where the surveyor estimates that 2nd Ave. ℄ will intersect. The instrument is then moved to a 2nd Ave. station, and a sight is taken some distance away, on the far side of Main St. ℄. The surveyor can stretch a plumb bob string over the two points (A and B) established on Main St. ℄, and the instrumentman can note where on the string the 2nd Ave. ℄ intersects. If the two points (A and B) are reasonably close together (2 to 3 ft), the surveyor can use the plumb bob itself to take line from the instrumentman on the plumb bob string; otherwise the instrumentman can take line with a pencil or any other suitable sighting target.

The intersection point is then suitably marked (e.g., nail and

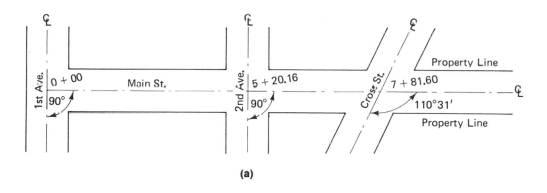

(a)

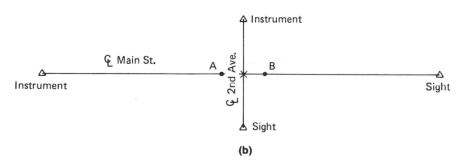

(b)

FIGURE 5.24 Intersection of two straight lines: (a) example of the intersection of ℄'s of streets; (b) intersecting technique.

flagging on asphalt, wood stake with tack on ground), and then the angle of intersection and the chainage of the point can be determined. After marking the intersection point, temporary markers *A* and *B* are removed.

5.20 PROLONGING A MEASURED LINE BY TRIANGULATION OVER AN OBSTACLE

In route surveying, obstacles such as rivers or chasms must be traversed. Whereas the alignment can be conveniently prolonged by double centering, the chainage may be deduced from the construction of a geometric figure. In Figure 5.25, the distance from 1 + 121.271 to the station established on the far side of the river can be determined by solving the constructed triangle (triangulation).

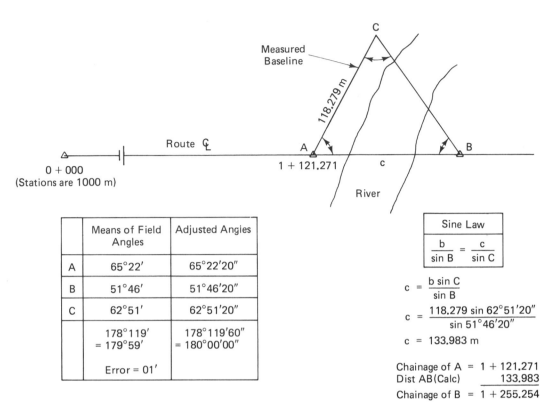

FIGURE 5.25 Prolonging a measured line over an obstacle by triangulation.

The ideal (strongest) triangle is one having angles close to 60° (equilateral), although angles as small as 20° may be acceptable. The presence of rugged terrain and heavy tree cover adjacent to the river often result in a less than optimal geometric figure.

The base line and a minimum of two angles are measured so that the missing distance can be calculated. The third angle (on the far side of the river) should also be measured to check for mistakes and to reduce errors.

Using the sine law,

$$\frac{b}{\text{Sin } B} = \frac{c}{\text{Sin } C}$$

$$\frac{118.279}{\text{Sin } 51°46'20''} = \frac{c}{\text{Sin } 62°51'20''}$$

$$C = 133.983 \text{ m}$$

5.21 PROLONGING A LINE PAST AN OBSTACLE

It often occurs in property surveying that obstacles, such as trees, block the path of the survey. Whereas in route surveying, it is customary for the surveyor to cut down the offending trees (later construction will require them to be removed in any case), in property surveying the owner would be quite upset to find valuable trees destroyed just so the surveyor could establish a boundary line. Accordingly, the surveyor must find an alternate method of providing distances and/or locations for blocked survey lines.

In Figure 5.26a the technique of right-angle offset is illustrated. Boundary line AF cannot be run because of the wooded area. The survey continues normally to point B just clear of the wooded area. At B a right angle is turned (and doubled), and point C is located a sufficient distance away from B to provide a clear parallel line to the boundary line. The transit is set at C and sighted at B (great care must be exercised because of the short sighting distance); an angle of 90° is turned to locate point D. Point E is located on the boundary line using a right angle and the offset distance used for BC. The survey then can continue to F. If distance CD is measured, then the required boundary distance (AF) is $AB + CD + EF$.

If intermediate points are required on the boundary line between B and E (e.g., fencing layout), a right angle can be turned from a convenient location on CD and the offset distance (BC) is used to measure back to the boundary line. Use of a technique like this will minimize the destruction of trees and other obstructions.

In Figure 5.26b, trees are scattered over the area, preventing the establishment of a right-angle offset line. In this case, a random line (open traverse) is run (by deflection angles) through the scattered trees. The distance AF is the sum of AB, EF, and the resultant of BE (see Chapter 6 for appropriate computation techniques for problems such as this).

In Figure 5.26c, the line must be prolonged past an obstacle, a large tree. In this case, a triangle is constructed with the three angles and two distances measured as shown. As noted earlier, the closer the constructed triangle is to equilateral, the stronger will be the calculated distance (BD). Also, as noted earlier, the optimal equilateral figure cannot always be constructed due to topographic constraints, and angles as small as 20° are acceptable for many surveys.

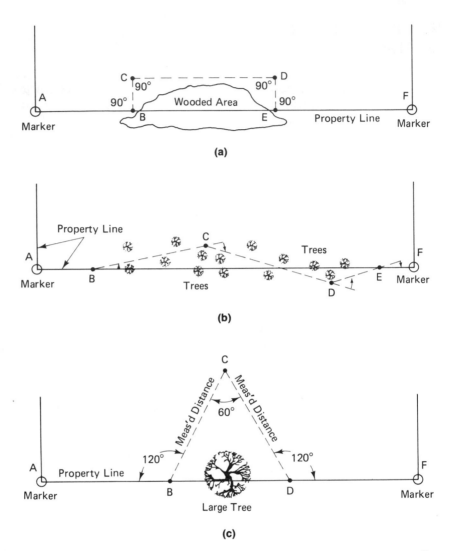

FIGURE 5.26 Prolonging a line past an obstacle: (a) right-angle offset method; (b) random-line method; (c) triangulation method.

It was noted that the technique of right-angle offsets has larger potential for error due to the weaknesses associated with several short sightings, but, at the same time, this technique gives a simple and direct method for establishing intermediate points on the boundary line. In contrast, the random line and triangulation methods provide for stronger geometric solutions to the missing property line distances, but they also require less direct and much more cumbersome calculations for the placement of intermediate line points.

6 *Traverse Surveys*

6.1 GENERAL

A traverse is usually a control survey and is employed in all forms of legal and engineering work. Essentially, traverses are a series of established stations that are tied together by angle and distance. The angles are measured using transits or theodolites, whereas the distances can be measured using steel tapes or electronic distance measurement instruments (EDMI), or by tacheometric devices such as the subtense bar or stadia (see Chapter 7). Traverses can be open, such as in route surveys, or closed, as in a closed geometric figure (see Figures 6.1 and 6.2).

Boundary surveys, which constitute a consecutive series of established (or laid out) stations, are usually described as being traverse surveys (e.g., retracement of the distances and angles of a piece of property).

In engineering work, traverses are used as control surveys to (1) locate topographic detail for the preparation of plans, (2) to lay out (locate) engineering works, and (3) for the processing and ordering of earthwork and other engineering quantities. Traverses also provide horizontal control for aerial surveys in the preparation of photogrammetric mapping (see Chapter 15).

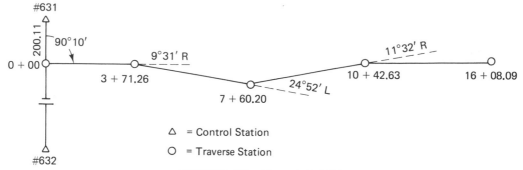

FIGURE 6.1 Open traverse.

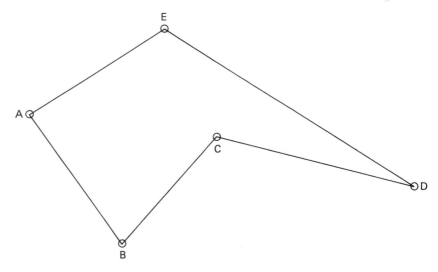

FIGURE 6.2 Closed traverse (loop).

6.2 OPEN TRAVERSE

Simply put, an open traverse is a series of measured straight lines (and angles) that do not geometrically close. This lack of geometric closure means that there is no geometric verification possible with respect to the actual positioning of the traverse stations. Accordingly, the measuring technique must be refined to provide for field verification. As a minimum, distances are measured twice (once in each direction) and angles are doubled.

In route surveys, positions can be verified by computation from available property plans or by scale from existing topographic plans. Directions can be verified by scale from existing plans or by observation on the sun or Polaris. Many states and provinces are now providing densely placed third- and fourth-order horizontal control monuments as an extension to their coordinate grid systems. It is now often possible to tie in the initial and terminal stations of a route survey to a coordinate grid monument whose position has been precisely determined; when this is the case, the route survey becomes a closed traverse and is subject to geometric verification and analysis.

As was noted in Section 4.4, open traverses are tied together angularly by deflection angles, and distances are shown in the form of stations that are cumulative measurements referenced to the initial point of the survey (0 + 00). See Figure 6.3.

6.3 CLOSED TRAVERSE

A closed traverse is one that either begins and ends at the same point or one that begins and ends at points whose positions have been previously determined (Section 6.2). In both cases the angles can be closed geometrically, and the position closure can be determined mathematically.

A closed traverse that begins and ends at the same point is known as a *loop* traverse (see Figure 6.2). In this case the distances are measured from one station to the next (and verified) by using a steel tape or EDMI; the interior angle is measured at each station (and doubled). The loop distances and angles can be obtained by proceeding consecutively around the loop in a clockwise or counterclockwise manner; in fact, the data can be collected in any order convenient to the surveyor. However, as noted in the previous chapter, the angles themselves (by convention) are always measured from left to right.

6.4 BALANCING ANGLES

In Section 4.4, it was noted that the geometric sum of angles in an n-sided closed figure is $(n - 2)180$. For example, a five-sided figure would have $(5 - 2)180 = 540°$; a seven-sided figure would have $(7 - 2)180 = 900°$.

When all the interior angles of a closed field traverse are summed, they may or may not total the number of degrees required for geometric closure. As a general rule, the more precise the instrument being used to measure the angles, the less likely will it be that the sum of the angles will **precisely** agree with the correct geometric total.

PRELIMINARY SURVEY FOR CLEAR LAKE
ACCESS ROAD–TOMLIN TOWNSHIP

Job CLEAR 68° F
Date MAY 17 1982 Page 14

STATION	DIRECT	DOUBLE	MEAN	L/R	
0+00	90° 11'	180° 20'	90° 10'	SEE SKETCH.	
	DEFLECTION ANGLES				
3+71.26	9° 31'	19° 02'	9° 31'	R	
7+60.20	24° 51'	49° 44'	24° 52'	L	
10+42.63	11° 32'	23° 04'	11° 32'	R	
		CHAINAGES			
	DIRECT	REVERSE	MEAN		CHAINAGES
					0+00.00
	371.24	371.28	371.26		3+71.26
	388.93	~~388.80~~			
		388.95	388.94		7+60.20
	282.43	282.43	282.43		10+42.63
	565.44	565.49	565.46		16+08.09
			1608.09		

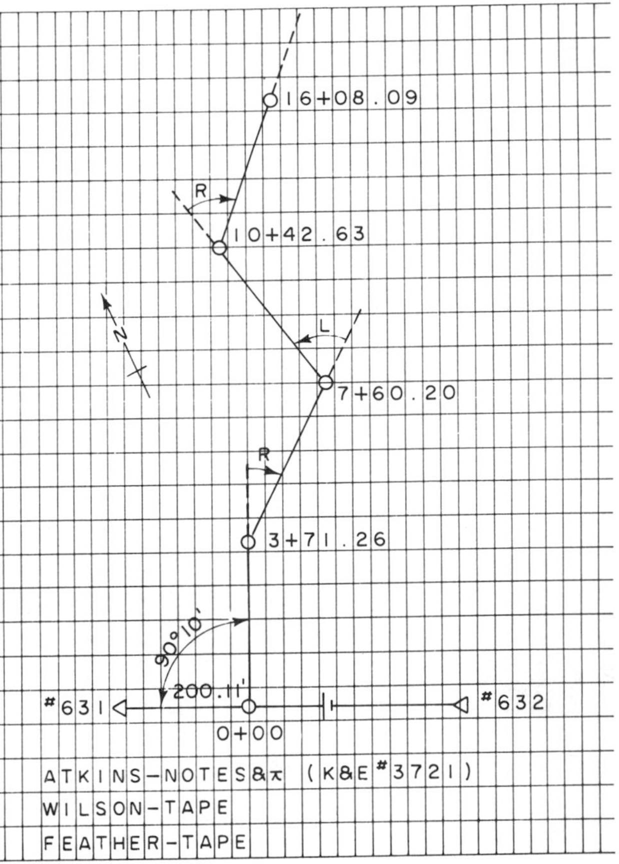

FIGURE 6.3 Field notes for open traverse.

Before mathematical analysis of the traverse can begin, before even the bearings can be calculated, the field angles must be adjusted so that their sum equals the correct geometric total. The angles can be balanced by distributing the angular error evenly to each angle, or one or more angles can be adjusted to force the closure.

The acceptable total error of angular closure is usually quite small (i.e., <03′); otherwise, the fieldwork will have to be repeated. The actual size of the allowable angular error is governed by the specifications being used for that specific traverse.

The angles for traverse Example 6.1 (Figure 6.4) are shown in Table 6.1. If one of the traverse stations had been in a particu-

TABLE 6.1 *Two Methods of Adjusting Field Angles*

Station	Field Angle	Arbitrarily Balanced	Equally Balanced
A	101° 24′ 00″	101° 24′ 00″	101° 24′ 12″
B	149° 13′ 00″	149° 13′ 00″	149° 13′ 12″
C	80° 58′ 30″	80° 59′ 00″	80° 58′ 42″
D	116° 19′ 00″	116° 19′ 00″	116° 19′ 12″
E	92° 04′ 30″	92° 05′ 00″	92° 04′ 42″
	538°119′ 00″	538°120′ 00″	538°118′120″
	= 539° 59′ 00″	= 540° 00′ 00″	= 540° 00′ 00″
	Error = 01′	Balanced	Balanced

$$\text{Correction/angle} - \frac{60}{5} - 12''$$

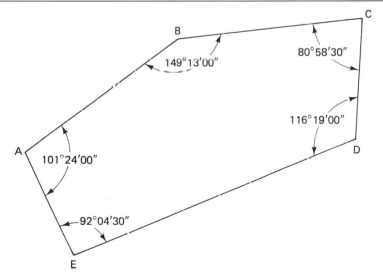

FIGURE 6.4 Example 6.1 (closed traverse problem) field angles.

larly suspect location (e.g., a swamp), a larger proportion of the angle correction could be assigned to that one station. In all balancing operations, however, a certain amount of guesswork is involved, as we don't know if any of the balancing procedures give us values closer to the "true" value than did the original field angle. The important thing is that the overall angular closure is no larger than that specified.

6.5 LATITUDES AND DEPARTURES

In Section 1.12 it was noted that a point could be located by polar ties (direction and distance) or by rectangular ties (two distances at 90°). In Figure 6.5a, point *A* is located, with respect to point *D*, by direction (bearing) and distance. In Figure 6.5b, point *A* is located, with respect to point *D*, by a distance north (Δy) and a distance east (Δx).

By definition, *latitude is the north–south rectangular component of a line,* and to differentiate direction, north is considered plus, whereas south is considered minus. Similarly, *departure is the east–west rectangular component of a line,* and to differentiate direction, east is considered plus, whereas west is considered minus. When working with azimuths, the plus–minus designation is directly given by the appropriate trigonometric function

$$\text{Latitude } \Delta Y = \text{distance(S) Cos } \alpha \qquad (6.1)$$
$$\text{Departure } \Delta X = \text{distance(S) Sin } \alpha \qquad (6.2)$$

where α is the bearing or azimuth of the traverse course, and distance(S) is the distance of the traverse course.

Latitudes (lats) and departures (deps) can be used to calculate the accuracy of a traverse by noting the plus–minus closure of

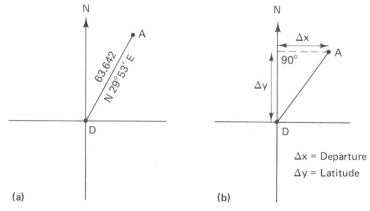

FIGURE 6.5   Location of a point: (a) polar ties; (b) rectangular ties.

both latitudes and departures. If the survey has been perfectly performed (angles and distances), the plus latitudes will equal the minus latitudes, and the plus departures will equal the minus departures.

In Figure 6.6, the figure has been approached in a counterclockwise mode (all signs would simply be reversed for a clockwise approach). Latitudes *CD*, *DA*, and *AB* are all positive and should precisely equal (if the survey were perfect) the latitude of *BC*, which is negative. Similarly, the departures of *CD* and *DA* are positive, and they should equal the departures of *AB* and *BC*, which are both negative.

As noted earlier when using bearings for directions, the north and east directions are positive, whereas the south and west directions are negative. Figure 6.7 shows these relationships and, in addition, that *for azimuth directions the algebraic sign is governed by the algebraic sign of the appropriate trigonometric function (cos or sin)*.

In Section 4.11, it was noted that the algebraic sign of trigonometric functions was automatically generated by computers and hand-held calculators. The remaining parts of this section involve traverse computations that use the trigonometric functions of directions (bearings or azimuths) to calculate latitudes and departures; the student is asked to consider the relative advantages of using azimuths in computations instead of the more traditional use of bearings. However, bearings are still universally used to show direction on legal plans (plats).

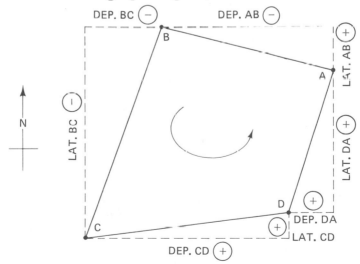

FIGURE 6.6 Closure of latitudes and departures (counterclockwise approach).

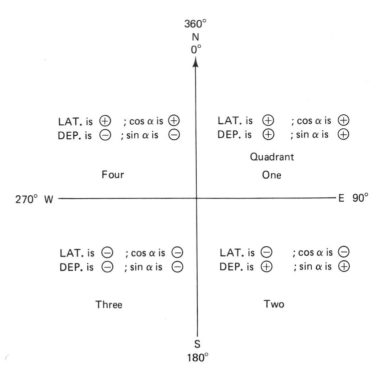

FIGURE 6.7   Algebraic signs of latitudes and departures by trigonometric functions (where α is the azimuth).

6.6 COMPUTATION OF LATITUDES AND DEPARTURES TO DETERMINE THE ERROR OF CLOSURE AND THE ACCURACY OF A TRAVERSE

Step 1

Balance the Angles. Using the examples introduced in Section 6.4, for this computation, use the angles resulting from equal distribution of the error among the field angles. These balanced angles are shown in Figure 6.8.

Step 2

Compute the Bearings (Azimuths). Starting with the given direction of *AB* (N 51°22'00" E), the directions of the remaining sides are computed. The computation can be solved going clockwise (Figure 6.10) or counterclockwise (Figure 6.9) around the figure. Use the techniques developed in Section 4.9 (counterclockwise approach) for this example.

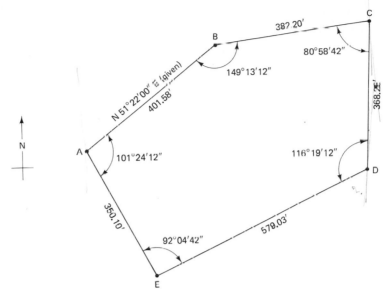

FIGURE 6.8 Distances and balanced angles for Example 6.1.
Refer to Figure 6.9:

Bg AB =	N 51°22′00″	E
Az AB =	51°22′00″	
+ < A	101°24′12″	
Az AE =	152°46′12″	
	+ 180°	
Az EA =	332°46′12″	
+ < E	92°04′42″	
Az ED =	424°50′54″	
	− 360°	
Az ED =	64°50′54″	
	+ 180°	
Az DE =	244°50′54″	
+ < D	116°19′12″	
Az DC =	361°10′06″	
	− 360°	
Az DC =	1°10′06″	
	+ 180°	
Az CD =	181°10′06″	
+ < C	80°58′42″	
Az CB =	262°08′48″	
	− 180°	

$$\text{Az } BC = \quad\quad 82°08'48''$$
$$+ < B \quad\quad\quad 149°13'12''$$
$$\text{Az } BA = \quad\quad 231°22'00''$$
$$-180°$$
$$\text{Az } AB = \quad\quad\; 51°22'00''$$
$$\text{Bg } AB = \; \text{N} \;\; 51°22'00'' \; \text{E} \quad\quad \text{Check}$$

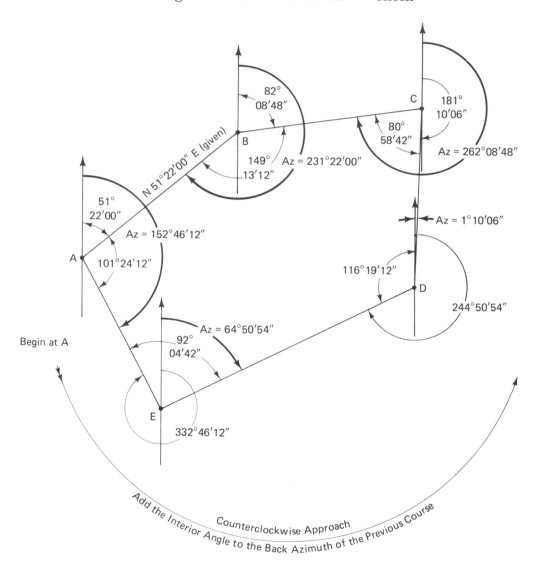

FIGURE 6.9 Example 6.1: azimuth computation, counterclockwise approach.

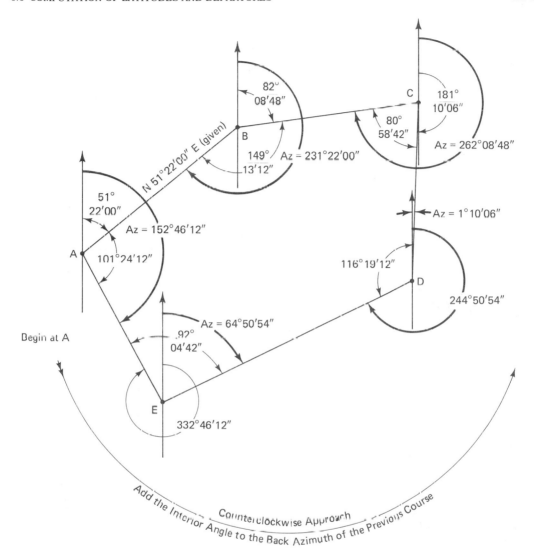

Kavanagh & Bird 6.10

FIGURE 6.10 Example 6.1: azimuth computation, clockwise approach.

For comparative purposes, the clockwise approach is also shown: (Refer to Figure 6.10)

$$
\begin{array}{lll}
\text{Bg } AB & = & \text{N } 51°22'00'' \text{ E} \\
\text{Az } AB & = & 51°22'00'' \\
& & +\,180° \\ \hline
\text{Az } BA & = & 231°22'00'' \\
-\angle B & & 149°13'12''
\end{array}
$$

$$
\begin{array}{rl}
\text{Az } BC = & 82°08'48'' \\
 & +180° \\
\hline
\text{Az } CB = & 262°08'48'' \\
 -<C & 80°58'42'' \\
\hline
\text{Az } CD = & 181°10'06'' \\
 & +180° \\
\hline
\text{Az } DC = & 361°10'06'' \\
 -<D & 116°19'12'' \\
\hline
\text{Az } DE = & 244°50'54'' \\
 & -180° \\
\hline
\text{Az } ED = & 64°50'54'' \\
 *+ & 360°00'00'' \\
\hline
\text{Az } ED = & 424°50'54'' \\
 -<E & 92°04'42'' \\
\hline
\text{Az } EA = & 332°46'12'' \\
 & -180° \\
\hline
\text{Az } AE = & 152°46'12'' \\
 -<A & 101°24'12'' \\
\hline
\text{Az } AB = & 51°22'00'' \\
\end{array}
$$

*To provide an angle large enough to permit subtraction of the interior angle.

Check

Summary of Azimuths and Related Bearings

	Counterclockwise			Clockwise	
Course	Azimuth	Bearing	Course	Azimuth	Bearing
AE	152°46'12"	S 27°13'48" E	EA	332°46'12"	N 27°13'48" W
ED	64°50'54"	N 64°50'54" E	DE	244°50'54"	S 64°50'54" W
DC	1°10'06"	N 1°10'06" E	CD	181°10'06"	S 1°10'06" W
CB	262°08'48"	S 82°08'48" W	BC	82°08'48"	N 82°08'48" E
BA	231°22'00"	S 51°22'00" W	AB		N 51°22'00" E
					(Given)

Step 3

Compute the Latitudes and Departures. Reference to Table 6.2 will show the format for a typical traverse computation. The table shows both the azimuth and the bearing for each course (usually only the azimuth or the bearing is included); it is noted that the algebraic sign for both latitude and departure is given directly by the calculator (computer) for each azimuth angle. In the case of bearings, as noted earlier, latitudes are plus if the bearing is north and minus if the bearing is south; similarly, departures are positive if the bearing is east and negative if the bearing is west.

TABLE 6.2 *Latitudes and Departure Computations, Counterclockwise Approach*

Course	Distance	Azimuth (α_a)	Bearing (α_B)	Latitude	Departure
AE	350.10'	152°46'12"	S 27°13'48" E	−311.30	+160.19
ED	579.03'	64°50'54"	N 64°50'54" E	+246.10	+524.13
DC	368.28'	1°10'06"	N 1°10'06" E	+368.20	+ 7.51
CB	382.20'	262°08'48"	S 82°08'48" W	− 52.22	−378.62
BA	401.58'	231°22'00"	S 51°22'00" W	−250.72	−313.70
	P = 2081.19 ft			Σ lat = +0.06	Σ dep = −0.49

$$E = \sqrt{\Sigma_{lat}{}^2 + \Sigma_{dep}{}^2}$$
$$= \sqrt{0.06^2 + 0.49^2}$$
$$= 0.494$$

Accuracy ratio $= \dfrac{E}{P} = \dfrac{0.49}{2081.19} = \dfrac{1}{4247} = \dfrac{1}{4200}$

For example, course *AE* has an azimuth of 152°46'12"(cos is negative; sin is positive; Figure 6.7), meaning that the latitude will be negative and the departure positive. Similarly, *AE* has a bearing of S 27°13'48" E, which expectedly results in a negative latitude (south) and a positive departure (east).

Table 6.2 shows that the latitudes fail to close by +0.06 (Σ lat) and the departures fail to close by −0.49 (Σ dep). Figure 6.11 shows graphically the relationship between the *C* lat and *C* dep. *C* lat and *C* dep are opposite in sign to Σ lat and Σ dep and reflect the direction consistent (in this example) with the counterclockwise approach to the problem.*

The traverse computation began at *A* and concluded at *A'*. The error of closure is given by the line *A'A*, and the *C* lat and *C* dep are in fact the latitude and departure of the *linear error of closure*.

The length of *A'A* is the square root of the sum of the squares at the *C* lat and *C* dep:

$$A'A = \sqrt{C\,lat^2 + C\,dep^2} = \sqrt{0.06^2 + 0.49^2} = 0.494$$

It is sometimes advantageous to know the bearing of the linear error of closure. Reference to Figure 6.11 shows that *C* dep/*C* lat = tan Bg: Bg *A'A* = S 83°01' E.

---

*Σ lat = error in latitudes
Σ dep = error in departures
C lat = required correction in latitudes
C dep = required correction in departures

Traverse Computation Began at A, and Terminated at A'

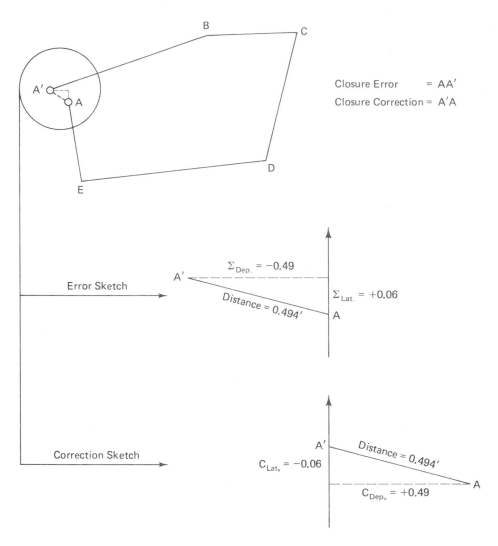

Closure Error   = AA'
Closure Correction = A'A

$\Sigma_{Dep.} = -0.49$

$\Sigma_{Lat.} = +0.06$

Error Sketch

Distance = 0.494'

Correction Sketch

$C_{Lat.} = -0.06$

Distance = 0.494'

$C_{Dep.} = +0.49$

$A'A = \sqrt{C_{Lat.}^2 + C_{Dep.}^2} = 0.494'$

Bearing of A'A can be Computed from the Relationship:

$$\tan \text{ Bearing} = \frac{C_{Dep.}}{C_{Lat.}} = \frac{0.49}{-0.06}$$

Bearing Angle $= 83.0189° = 83°01'$
Bearing A'A $= S\ 83°01'\ E$

FIGURE 6.11 Closure error and closure correction.

190

The *error of closure* (linear error of closure) is the net accumulation of the random errors associated with the measurement of the traverse angles and traverse distances. In this example the total error is showing up at *A* simply because the computation began at *A*. If the computation had commenced at any other station, the identical linear error of closure would have shown up at that station.

The error of closure is compared to the perimeter (*P*) of the traverse to determine the *accuracy ratio.*

In the example of Table 6.2, the closure accuracy is *E/P* = 0.49/2081 (rounded).

$$\text{Accuracy ratio} = \frac{1}{4247} = \frac{1}{4200}$$

The fraction of *E/P* is always expressed so that the numerator is 1, and the denominator is rounded to the closest 100 units. In the example shown, both the numerator and denominator are divided by the numerator (0.49).

The concept of an accuracy ratio was introduced in Section 1.13. Most states and provinces have these ratios legislated for minimally acceptable surveys for boundaries. The values vary from one area to another, but usually the minimal values are $^1/_{5000}$ to $^1/_{7500}$. It is logical to assign more precise values (e.g., $^1/_{10000}$) to high-cost urban areas. Engineering surveys are performed at levels of $^1/_{3000}$ to $^1/_{10000}$ depending on the importance of the work and the materials being used. For example, a gravel highway could well be surveyed at a $^1/_{3000}$ level of accuracy, whereas overhead rails for a monorail facility could well be surveyed in at levels of $^1/_{7500}$ to $^1/_{10000}$. As noted in the earlier section, control surveyors for both legal and engineering projects must locate their control points at a much higher level of precision than is necessary for the actual location of the legal or engineering project markers that are surveyed in from those control points.

If the accuracy computed using latitudes and departures is not acceptable as determined by the survey specifications, additional fieldwork must be performed to improve the level of accuracy. (Fieldwork is never undertaken until all calculations have been double-checked.) When fieldwork is to be checked, usually the distances are checked first, as the angles have already been verified [i.e., $(n - 2)180$].

If a large error (or mistake) has been made on one side, it will significantly affect the bearing of the linear error of closure. Accordingly, if a check on fieldwork is necessary, the surveyor first computes the bearing of the linear error of closure and checks that bearing against the course bearings. If a similarity exists

($\pm 5°$), that course is the first course remeasured in the field. If a difference is found in that course measurement (or any other course), the new course distance is immediately substituted into the computation to check whether or not the required accuracy level has then been achieved.

Summary of Initial Traverse Computations

1. Balance the angles.
2. Compute the bearings and/or the azimuths.
3. Compute the latitudes and departures, the linear error of closure, and the accuracy ratio of the traverse.

If the accuracy ratio is satisfactory, further treatment of the data is possible (e.g., coordinates and area computations).

If the accuracy ratio is unsatisfactory (e.g., an accuracy ratio of only $1/4000$ when a ratio of $1/5000$ was specified), complete the following steps:

1. Double-check all computations.
2. Double-check all field-book entries.
3. Compute the bearing of the linear error of closure and check to see if it is similar to a course bearing ($\pm 5°$).
4. Remeasure the sides of the traverse beginning with a course having a similar bearing to the linear error of closure bearing (if there is one).
5. When a correction is found for a measured side, try that value in the latitude–departure computation to determine the new level of accuracy.

6.7 TRAVERSE PRECISION AND ACCURACY

The actual accuracy of a survey as given by the accuracy ratio of a closed traverse can be misleading. The opportunity exists for significant errors to cancel out; this can result in "high-accuracy" closures from relatively imprecise field techniques. Some closed traverses do not begin and end on the same point; instead, they begin and end on points whose position has been previously determined (e.g., coordinate grid monuments). When this is the case, in addition to the random errors already noted, the linear error of closure will also contain systematic errors. In the case of a closed-loop traverse, the systematic errors will have been largely balanced during the computation of latitudes and departures. For example, in a square-shaped traverse, systematic taping errors (long or short tape) will be completely balanced and beyond mathematical detection. Therefore, in order for high precision to also reflect favorably on accuracy, control of field practice and techniques is essential.

If a traverse has been closed to better than $1/5000$, the taping should have conformed to the specifications shown in Table 2.2. Reference to Figure 6.12 will show that, for consistency, the survey should be designed so that the maximum allowable error in angle (Ea) should be roughly equal to the maximum allowable error in distance (Ed). If the linear error is $1/5000$, the angular error should be consistent:

$$1/5000 = \tan \theta$$
$$\theta = 0°00'41''$$

See Table 6.3 for additional linear and angular error relationships. The overall allowable angular error in an n-angled closed

Point Y is to Be Set Out from Fixed Points X and Z

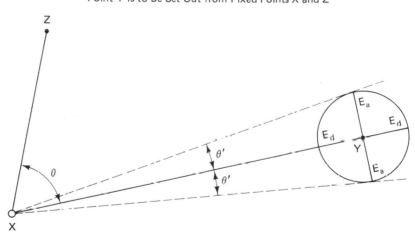

For the Line XY

E_d is the Possible Error in Distance Measurement and E_a is the Position Error Resulting from a Possible Angle Error of θ' in an Angle of θ.

FIGURE 6.12 Relationship between errors in linear and angular measurements.

TABLE 6.3   *Linear and Angular Error Relationships*

Linear Accuracy Ratio	*Maximum Angular Error*	*Least Count of Transit or Theodolite Scale*
1/1000	0°03′26″	01′
1/3000	0°01′09″	01′
1/5000	0°00′41″	30″
1/7500	0°00′28″	20″
1/10000	0°00′21″	20″
1/20000	0°00′10″	10″

NOTE: Most theodolites equipped with optical micrometers are now (1980's) capable of reading angles directly to 6 seconds.

traverse would be $Ea\sqrt{n}$. (Random errors accumulate as the square root of the number of observations.)

Thus for a five-sided traverse with a specification for accuracy of $^1/_{3000}$, the maximum angular misclosure would be $01′\sqrt{5} = 02′$ (to the closest minute), and for a five-sided traverse with a specification for precision of $^1/_{5000}$, the maximum misclosure of the field angles would be $30″ \sqrt{5} = 01′$ (to the closest 30 seconds).

6.8 TRAVERSE ADJUSTMENTS

As was noted in Section 6.6, latitudes and departures can be used for the computation of coordinates or the computation of the area enclosed by the traverse. In addition, the station coordinates can then by used to establish control for further survey layout. However, before any further use can be made of these latitudes and departures, they must be adjusted so that their errors are suitably distributed, and the algebraic sums of the latitudes and departures are each zero. These adjustments, if properly done, will ensure that the final position of each traverse station, as given by the station coordinates, is optimal with respect to the "true" station location.

Current surveying practice favors either the compass rule (Bowditch) adjustment or the least squares adjustment. The compass rule (an approximate method) is applied in most cases, with the least squares method being reserved for large, precise traverses (e.g., extension of a state or province coordinate grid). Although the least squares method requires extensive computations, advances in computer technology are now enabling the surveyor to use the least squares technique with the help of a desk-top computer or even a hand-held calculator.

6.9 COMPASS RULE ADJUSTMENT

The compass rule is used in most survey computations. The compass rule distributes the errors in latitude and departure for each traverse course in the same proportion as the course distance is to the traverse perimeter. That is, generally,

$$\frac{C \operatorname{lat} AB}{\Sigma \operatorname{lat}} = \frac{AB}{P} \quad \text{or} \quad C \operatorname{lat} AB = \Sigma \operatorname{lat} \times \frac{AB}{P} \tag{6.3}$$

where $C \operatorname{lat} AB$ = correction in latitude AB
$\Sigma \operatorname{lat}$ = error of closure in latitude
AB = distance AB
P = perimeter of traverse

and

$$\frac{C \operatorname{dep} AB}{\Sigma \operatorname{dep}} = \frac{AB}{P} \quad \text{or} \quad C \operatorname{dep} AB = \Sigma \operatorname{dep} \times \frac{AB}{P} \tag{6.4}$$

where $C \operatorname{dep} AB$ = correction in departure AB
$\Sigma \operatorname{dep}$ = error of closure in departure
AB = distance AB
P = perimeter of traverse

Referring to traverse example 6.1, Table 6.2 has been expanded in Table 6.4 to provide space for traverse adjustments. The magnitudes of the individual corrections are shown next:

$$C \operatorname{lat} AE = \frac{0.06 \times 350.10}{2081.81} = 0.01$$

$$C \operatorname{lat} ED = \frac{0.06 \times 579.03}{2081.81} = 0.02$$

$$C \operatorname{lat} DC - \frac{0.06 \times 368.28}{2081.81} = 0.01$$

$$C \operatorname{lat} CB = \frac{0.06 \times 382.20}{2081.81} = 0.01$$

$$C \operatorname{lat} BA = \frac{0.06 \times 401.58}{2081.81} = \underline{0.01}$$

$$\text{Check } C \operatorname{lat} = 0.06$$

$$C \operatorname{dep} AE = \frac{0.49 \times 350.10}{2081.81} = 0.08$$

$$C \operatorname{dep} ED = \frac{0.49 \times 579.03}{2081.81} = 0.14$$

$$C \operatorname{dep} DC = \frac{0.49 \times 368.28}{2081.81} = 0.09$$

$$C \operatorname{dep} CB = \frac{0.49 \times 382.20}{2081.81} = 0.09$$

TABLE 6.4 *Traverse Adjustments: Compass Rule, Example 6.1*

Course	Distance	Bearing	Latitude	Departure	C Lat	C Dep	Balanced Latitudes	Balanced Departures
AE	350.10	S 27°13′48″ E	−311.30	+160.19	−0.01	+0.08	−311.31	+160.27
ED	579.03	N 64°50′54″ E	+246.10	+524.13	−0.02	+0.14	+246.08	+524.27
DC	368.28	N 1°10′06″ E	+368.20	+7.51	−0.01	+0.09	+368.19	+7.60
CB	382.20	S 82°08′48″ W	−52.22	−378.62	−0.01	+0.09	−52.23	−378.53
BA	401.58	S 51°22′00″ W	−250.72	−313.70	−0.01	+0.09	−250.73	−313.61

$P = 2081.19$ ft   Σ lat = +0.06   Σ dep = −0.49 C_{lat} = −0.06 C_{dep} = +0.49   0.00   0.00

$$C \text{ dep } BA = \frac{0.49 \times 401.58}{2081.81} = \underline{0.09}$$

$$\text{Check } C \text{ dep } = 0.49$$

These computations are normally performed on a hand-held calculator with the constants 0.06/2081.81 and 0.49/2081.81 entered into storage for easy retrieval and thus quick computations.

It now only remains for the algebraic sign to be determined. Quite simply, the corrections are opposite in sign to the errors. Therefore, for this example, the latitude corrections are negative and the departure corrections are positive. The corrections are now added algebraically to arrive at the balanced values. For example, in Table 6.4, the correction for latitude AE is -0.01, which is to be "added" to latitude AE, 311.30. Since the correction is the same sign as the latitude, the two values are added to get the answer. In the case of course ED, the latitude correction (-0.02) and the latitude ($+246.10$) have opposite signs, indicating that the difference between the two values is the desired value (i.e., subtract to get the answer).

To check the work, the balanced latitudes and balanced departures are totaled to see if their respective sums are zero. It sometimes happens that the balanced latitude or balanced departure totals fail to equal zero by one last-place unit (0.01 in this example). This discrepancy is probably caused by rounding off and is normally of no consequence; the discrepancy is removed by arbitrarily changing one of the values to force the total to zero.

It should be noted that when the error (in latitude or departure) to be distributed is quite small, the corrections can be arbitrarily assigned to appropriate courses. For example, if the error in latitude (or departure) were only 0.03 ft in a five-sided traverse, it would be appropriate to apply corrections of 0.01 ft to the latitude of each of the three longest courses. Similarly, in Example 6.1, where the error in latitude for that five-sided traverse is $+0.06$, it would have been appropriate to apply a correction of -0.02 to the longest course latitude and a correction of -0.01 to each of the remaining four latitudes, the same solution provided by the compass rule.

See Appendix A for further treatment of error adjustments.

6.10 EFFECTS OF TRAVERSE
ADJUSTMENTS ON THE ORIGINAL DATA

Once the latitudes and departures have been adjusted, the original polar coordinates (distance and direction) will no longer be valid. In most cases, the adjustment required for the polar coordinates is too small to warrant consideration; but if the data are to be used for layout purposes, the *corrected* distances and directions should be used.

The following relationships are inferred from Figures 6.5 and 6.11:

$$\text{Distance } AD = \sqrt{\text{lat } AD^2 + \text{dep } AD^2} \tag{6.5}$$

$$\text{Tan Bg } AD = \frac{\text{dep } AD}{\text{lat } AD} \tag{6.6}$$

Using the values from Table 6.4,

Example   *Course AE*

$$\text{Distance } AE = \sqrt{311.31^2 + 160.27^2} = 350.14 \text{ ft}$$

$$\text{Tan Bg } AE = \frac{+160.27}{-311.31}, \qquad \text{Bg} = \text{S } 27°14'26'' \text{ E}$$

The remaining corrected bearings and distances are shown in Table 6.5.

6.11 OMITTED MEASUREMENTS

The techniques developed in the computation of latitudes and departures can be used to solve for missing course information on a closed traverse, and, as well, these techniques can be utilized to solve any surveying problem that can be arranged in the form of a closed traverse.

Example 6.2

A missing course in a closed traverse is illustrated in Figure 6.13 and tabulated in Table 6.6. The data can be treated in the same manner as in a closed traverse. When the latitudes and departures are totaled, they will not balance. Both the latitudes and departures will fail to close by the amount of the latitude and departure of the missing course, *DA.*

The length and direction of *DA* can be simply computed:

$$\text{Distance } DA = \sqrt{\text{lat } DA^2 + \text{dep } DA^2} \tag{6.5}$$
$$= \sqrt{202.82^2 + 276.13^2}$$
$$= 342.61 \text{ ft}$$

$$\text{Tan Bg } DA = \frac{\text{dep } AD}{\text{lat } AD} \tag{6.6}$$
$$= \frac{-276.13}{+202.82}$$
$$\text{Bg } DA = \text{N } 53°42' \text{ W} \quad \text{(rounded to closest minute)}$$

TABLE 6.5   Adjustment of Bearings and Distances Using Balanced Latitudes and Departures, Example 6.1

Course	Balanced Latitude	Balanced Departure	Adjusted Distance	Adjusted Bearing	Original Distance	Original Bearing
AE	−311.31	+160.27	350.14	S 27°14′26″ E	350.10	S 27°13′48″ E
ED	+246.08	+524.27	579.15	N 64°51′21″ E	579.03	N 64°50′54″ E
DC	+368.19	+7.60	368.27	N 1°10′57″ E	368.28	N 1°10′06″ E
CB	−52.23	−378.53	382.12	S 82°08′38″ W	382.20	S 82°08′48″ W
BA	−250.73	−313.61	401.52	S 51°21′28″ W	401.58	S 51°22′00″ W
	0.00	0.00	P = 2081.20′		P = 2081.19′	

199

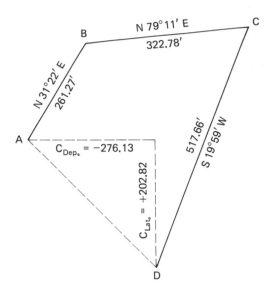

FIGURE 6.13 Example 6.2, missing course computation.

TABLE 6.6 *Example 6.2, Missing Course*

Course	Distance	Bearing	Latitude	Departure
AB	261.27	N 31°22′ E	+223.09	+135.99
BC	322.78	N 79°11′ E	+60.58	+317.05
CD	517.66	S 19°59′ W	−486.49	−176.91
			Σ lat = −202.82	Σ dep = +276.13
DA			C lat = +202.82	C dep = −276.13

It will be noted that this technique does not permit a check on the accuracy ratio of the fieldwork. Since this is the closure course, the computed value will also contain *all* accumulated errors.

Example 6.3

Figure 6.14 illustrates an intersection jog-elimination problem that occurs in many municipalities. For a variety of reasons, some streets do not directly intersect other streets. They jog a few feet to a few hundred feet before continuing. In Figure 6.14, the sketch indicates that, in this case, the entire jog will be taken out on the north side of the intersection. The designer has determined that if the Springfield Rd. ℄ is produced northerly 300 ft north of the ℄ of Finch Ave. E., it can then be joined to the existing Springfield Rd. ℄ at a distance of 1100 ft northerly from the ℄ of Finch Ave. E. Presumably, these distances will allow for the insertion of curves that will satisfy the geometric requirements for the design speed

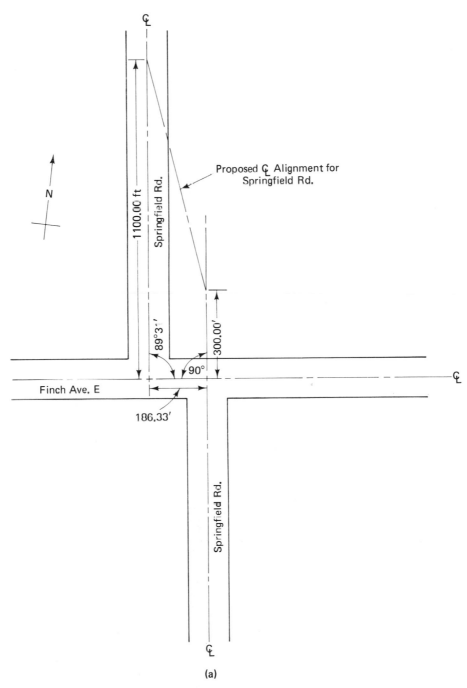

FIGURE 6.14 (a) JOG elimination, missing course problem; (b) distances and bearings; (c) determination of deflection angles. (*Continued*)

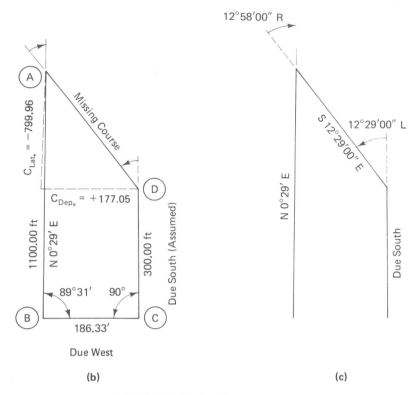

FIGURE 6.14 *(Continued)*

and traffic volumes. The problem here is to compute the length of *AD* and the deflection angles at *D* and *A* (see Figure 6.14b and Table 6.7).

Assume a bearing for line *DC* of due south. The bearing for *CB* is therefore due west, and obviously the bearing of *BA* is N 0° 29′ E. The problem is then set up as a missing course problem, and the corrections in latitude and departure are the latitude and departure of *AD*.

$$\text{Distance } AD = \sqrt{799.96^2 + 177.05^2} = 819.32 \text{ ft}$$
$$\text{Tan Bg } AD = \frac{177.05}{-799.96}$$
$$\text{Bg } AD = \text{S } 12°29'00'' \text{ E} \quad \text{(to the closest 30}'')$$

The bearings are shown in Figure 6.14c, which leads to the calculation of the deflection angles as shown.

There are obviously many other situations where the missing course techniques can be used. If one bearing and one distance

TABLE 6.7 *Example 6.3, Missing Course*

Course	Distance	Bearing	Latitude	Departure
DC	300.00	S 0°00′ E(W)	−300.00	0.00
CB	186.33	S 90°00′ (W)	0.00	−186.33
BA	1100.00	N 0°29′ E	+1099.96	+9.28
			Σ lat = 799.96	Σ dep = −177.05
AD			C lat = −799.96	C dep = +177.05

(not necessarily on the same course) are omitted from a traverse, the missing data can be solved. In some cases it may be necessary to compute intermediate cutoff lines and utilize cosine and sine laws in conjunction with the latitude–departure solution.

6.12 RECTANGULAR COORDINATES OF TRAVERSE STATIONS

Rectangular coordinates define the position of a point with respect to two perpendicular axes. Analytic geometry uses the concepts of a y axis (north–south) and an x axis (east–west), concepts that are obviously quite useful in surveying applications.

In state and provincial coordinate grid systems, the x axis is often the equator, and the y axis is a central meridian through the middle of the zone in which the grid is located (see Chapter 11). For surveys of a limited nature, where a coordinate grid has not been established, the coordinate axes can be assumed.

If the axes are to be assumed, values are chosen such that the coordinates of all stations will be positive (i.e., all stations will be in the northeast quadrant).

The traverse tabulated in Table 6.4 will be used for illustrative purposes. Values for the coordinates of station A are assumed to be 1000.00 north and 1000.00 east.

To calculate the coordinates of the other traverse stations, it is simply a matter of applying the **balanced** latitudes and departures to the previously calculated coordinates. In Figure 6.15 the balanced latitude and departure of course AE are applied to the assumed coordinates of station A to determine the coordinates of station E, and so on. These simple computations are shown in Table 6.8.

A check on the computation is possible by using the last latitude and departure (BA) to recalculate the coordinates of station A.

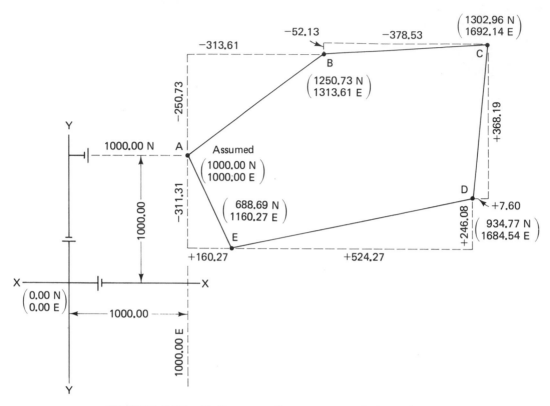

FIGURE 6.15 Station coordinates, using balanced latitudes and departures.

TABLE 6.8 *Computation of Coordinates Using Balanced Latitudes and Departures*

Course	Balanced Latitude	Balanced Departure	Station	Northing		Easting	
			A	1000.00	(assumed)	1000.00	(assumed)
AE	−311.31	+160.27		−311.31		+160.27	
			E	688.69		1160.27	
ED	+246.08	+524.27		+246.08		+524.27	
			D	934.77		1684.54	
DC	+368.19	+7.60		+368.19		+7.60	
			C	1302.96		1692.14	
CB	−52.23	−378.53		−52.23		−378.53	
			B	1250.73		1313.61	
BA	−250.73	−313.61		−250.73		−313.61	
			A	1000.00	Check	1000.00	Check

If a scale drawing of the traverse is required, it can be accomplished by methods of rectangular coordinates, where each station is located independently of the other stations by scaling the appropriate north and east distances from the axes; or the traverse can be drawn by the direction and distance method of polar coordinates (i.e., by scaling the interior angle between courses and by scaling the course distances, the stations are located in counterclockwise or clockwise sequence).

Whereas in the rectangular coordinate system, plotting errors are isolated at each station, in the polar coordinate layout method, all angle and distance scale errors are accumulated and only show up when the last distance and angle are scaled to theoretically relocate the starting point. The resultant plotting error is similar in type to the linear error of traverse closure ($A'A$) illustrated in Figure 6.11.

The use of coordinates to define the positions of boundary markers has been steadily increasing over the years (see Chapter 11). The storage of property-corner coordinates in large-memory civic computers will, in the not so distant future, permit lawyers, municipal authorities, and others to have instant retrieval of current land registration information, assessment, and other municipal information, such as concerning census and the level of municipal services. Although such use is important, the truly impressive impact of coordinate use will result from the coordination of topographic detail (digitization) so that plans can be prepared by computer-assisted plotters (see Chapter 16), and the coordination of all legal and engineering details so that not only will the plans be produced by computer-assisted plotters, but the survey layout will be accomplished by sets of computer-generated coordinates (rectangular and polar) fed either manually or automatically through electronic tacheometer instruments (see Chapter 8). Complex layouts can then be accomplished quickly by a few surveyors from one or two centrally located control points with a higher level of precision and a lower incidence of mistakes.

6.13 GEOMETRY OF RECTANGULAR COORDINATES

Figure 6.16 shows two points P_1 (x_1, y_1) and P_2 (x_2, y_2) and their rectangular relationships to the x and y axes.

$$\text{length } P_1 P_2 = \sqrt{(x_2 - x_1)^2 + (y_2 - y_1)^2} \tag{6.7}$$

$$\text{Tan } \alpha \qquad = \frac{x_2 - x_1}{y_2 - y_1} \tag{6.8}$$

Where α is the bearing or azimuth of $P_1.P_2$.

Also:

$$\text{length } P_1P_2 = \frac{x_2 - x_1}{\sin \alpha} \tag{6.9}$$

$$\text{length } P_1P_2 = \frac{y_2 - y_1}{\cos \alpha} \tag{6.10}$$

Use the equation having the larger numerical value of $(x_2 - x_1)$ or $(y_2 - y_1)$.

It will be clear from Figure 6.16 that $(x_2 - x_1)$ is the departure of P_1P_2 and that $(y_2 - y_1)$ is the latitude of P_1P_2. In survey work, the y value (latitude) is known as the northing, and the x value (departure) is known as the easting.

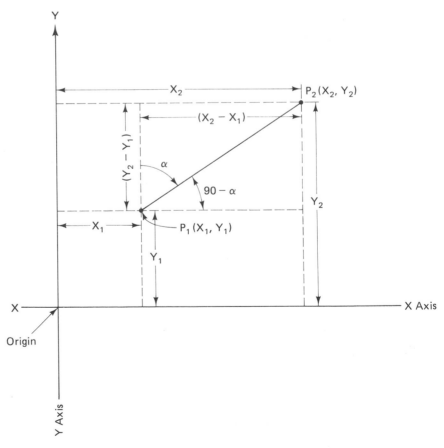

FIGURE 6.16 Geometry of rectangular coordinates.

From analytic geometry, the slope of a straight line is $M = \tan(90 - \alpha)$, where $(90 - \alpha)$ is the angle of the straight line with x axis (Figure 6.16); that is,

$$M = \cot \alpha \qquad \text{(Figure 6.16.)}$$

From coordinate geometry, the equation of straight line $P_1 P_2$, when the coordinates of P_1 and P_2 are known, is

$$\frac{y - y_1}{y_2 - y_1} = \frac{x - x_1}{x_2 - x_1} \qquad (6.11)$$

This can be written

$$y - y_1 = \frac{y_2 - y_1}{x_2 - x_1}(x - x_1)$$

where $(y_2 - y_1)(x_2 - x_1) = \cot a = m$ (from analytic geometry), the slope.

When the coordinates of one point (P_1) and the bearing or azimuth of a line are known, the equation becomes

$$y - y_1 = \cot \alpha \, (x - x_1) \qquad (6.12)$$

Where α is the azimuth or bearing of the line through P_1 (x_1, y_1). Also from analytical geometry,

$$y - y_1 = \frac{-1}{\cot \alpha}(x - x_1) \qquad (6.13)$$

which represents a line perpendicular to the line represented by Eq. (6.12); that is, the slopes of perpendicular lines are negative reciprocals.

Equations for circular curves are quadratics in the form of

$$(x - H)^2 + (y - K)^2 = r^2 \qquad (6.14)$$

where r is the curve radius, (H, K) are the coordinates of the center, and (x, y) are the coordinates of point P, which locates the circle (see Figure 6.17). When the circle center is at the origin, the equation becomes

$$x^2 + y^2 = r^2 \qquad (6.15)$$

6.14 ILLUSTRATIVE PROBLEMS IN RECTANGULAR COORDINATES

Example 6.4

From the information shown in Figure 6.18, calculate the coordinates of the point of intersection of lines EC and DB (K_1). From Eq. (6.11), the equation of EC is

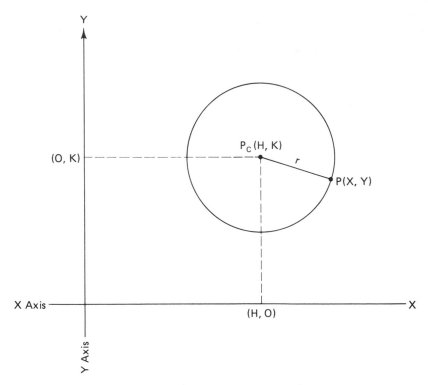

FIGURE 6.17   Circular curve coordinates.

$$\frac{y - 688.69}{1302.96 - 688.69} = \frac{x - 1160.27}{1692.14 - 1160.27} \tag{1}$$

The equation of *DB* is

$$\frac{y - 934.77}{1250.73 - 934.77} = \frac{x - 1684.54}{1313.61 - 1684.54} \tag{2}$$

Simplifying, these equations become

$$EC: \quad 614.27x - 531.87y = 346\ 425.50 \tag{1A}$$
$$DB: \quad 315.96x + 370.93y = 878\ 981.50 \tag{2A}$$

Multiply equation (1A) by $\dfrac{370.93}{531.87}$ to get equation (1B).

$$EC: \quad 428.40x - 370.93y = 241\ 599.66 \tag{1B}$$
$$ 1120581.16 \qquad (1B + 2A)$$
$$ 1505.429$$

Substitute the value of x in Eq. (2A) and check the results in Eq. (1B):

$$y = 1087.34$$

Therefore, the coordinates of point of intersection $K1$ are (1087.34 N, 1505.43 E).

Example 6.5

From the information shown in Figure 6.18, calculate (a) the coordinates K_2, the point of intersection of line ED and a line *perpendicular to ED* running through station B; and (b) distances K_2D and K_2E.

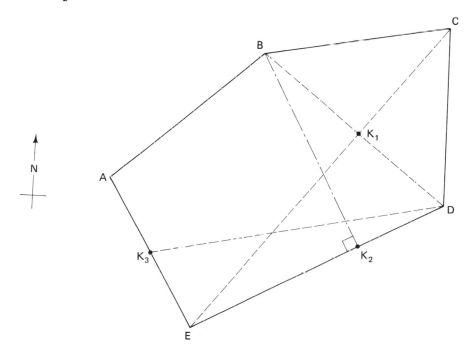

	Coordinates	
Station	North	East
A	1000.00	1000.00
B	1250.73	1313.61
C	1302.96	1692.14
D	934.77	1684.54
E	688.69	1160.27

FIGURE 6.18 Coordinates for traverse problem, Example 6.1 (see Table 6.8).

(a) From Eq. (6.11), the equation of ED is

$$\frac{y - 688.69}{934.77 - 688.69} = \frac{x - 1160.27}{1684.54 - 1160.27} \qquad (1)$$

$$y - 688.69 = \frac{246.08}{524.27}(x - 1160.27) \qquad (1A)$$

From Eq. (6.12), the equation of BK_2 is

$$y - 1250.73 = -\frac{524.27}{246.08}(x - 1313.61) \qquad (2)$$

Simplifying, these equations become (use five decimals to avoid rounding errors)

$$\begin{array}{llr}
ED: & 0.46938x - y = -144.09 & (1B) \\
BK_2: & 2.13049x + y = +4049.36 & (2B) \\
& 2.59987x = 3905.27 & (1B + 2B) \\
& x = 1502.102 &
\end{array}$$

Substitute the value of x in Eq. (1A) and check the results in Eq. (2):

$$y = 849.15$$

Therefore, the coordinates of K_2 are (849.15 N, 1502.10 E).

(b) Figure 6.19 shows the coordinates for stations E and D and intermediate point K_2 from Eq. (6.7):

$$\text{length } K_2D = \sqrt{85.62^2 + 182.44^2} = 201.53$$

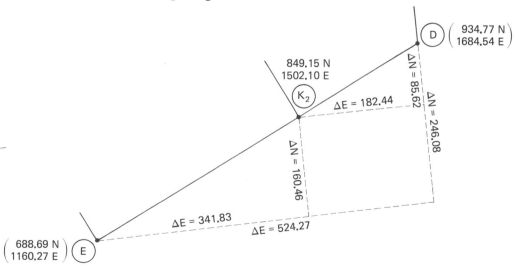

FIGURE 6.19 Sketch for Example 6.5.

and

$$\text{length } K_2E = \sqrt{160.46^2 + 341.83^2} = 377.62$$
$$K_2D + K_2K = \quad ED = 579.15$$

Check

$$\text{length } ED = \sqrt{246.08^2 + 524.27^2} = 579.15$$

Example 6.6

From the information shown in Figure 6.18, calculate the coordinates of the point of intersection (K_3) of a *line parallel to CB* running from station D to line EA. From Eq. (6.11),

$$CB: \quad \frac{y - 1302.96}{1250.73 - 1302.96} = \frac{x - 1692.14}{1313.61 - 1692.14}$$

$$y - 1302.96 = \frac{-52.23}{-378.53}(x - 1692.14)$$

$$\text{Slope (cot } \alpha) \text{ of } CB = \frac{-52.23}{-378.53}$$

Since DK_3 is parallel to BC,

$$\text{slope (cot } \alpha) \text{ of } DK_3 = \frac{-52.23}{-378.53}$$

$$DK_3: \quad y - 934.77 = \frac{52.23}{378.53}(x - 1684.54) \qquad (1)$$

$$EA: \quad \frac{y - 688.69}{1000.00 - 688.69} = \frac{x - 1160.27}{1000.00 - 1160.27} \qquad (2)$$

$$DK_3: \quad 0.13798x - y = -702.34 \qquad (1A)$$
$$EA: \quad 1.94241x + y = +2942.41 \qquad (2A)$$
$$2.08039x = +2240.07 \qquad (1A + 2A)$$
$$x = 1076.7547$$

Substitute the value of x in Eq. (1A) and check the results in Eq. (2A):

$$y = 850.91$$

Therefore, the coordinates of K_3 are (850.91 N, 1076.75 E).

Example 6.7

From the information shown in Figure 6.20, calculate the coordinates of the point of intersection (L) of the ℄ of Fisher Road with the ℄ of Elm Parkway.

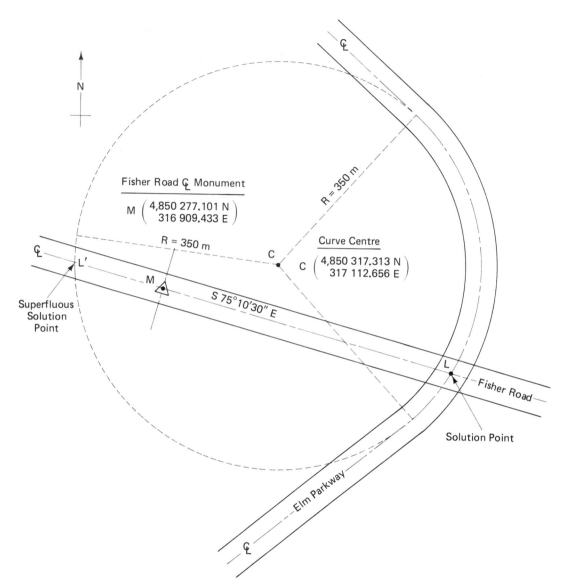

FIGURE 6.20 Intersection of a straight line with a circular curve, Example 6.7.

The coordinates of station *M* on Fisher Road are (4,850277.101 N, 316909.433 E) and the bearing of the Fisher Road ℄ (*ML*) is S 75°10'30" E. The coordinates of the center of the 350 *M* radius highway curve are (4,850 317.313 N, 317 112.656 E).

The coordinates here are referred to a coordinate grid system having 0.000 *M* north at the equator, and 304,800.000 east at longitude 79°30' W.

The coordinate values are, of necessity, very large and would cause significant rounding errors if they were used in the computations. Accordingly, an auxiliary set of coordinate axes will be used, allowing the values of the given coordinates to be greatly reduced for the computations; the amount reduced will later be added to give the final coordinates. The summary of coordinates is shown next:

Station	Grid Coordinates		Reduced Coordinates	
	y	*x*	*y'(y − 4850000)*	*x'(x − 316500)*
M	4850277.101	316909.433	277.101	409.433
C	4850317.313	317112.656	317.313	612.656

From Eq. (6.12), the equation of Fisher Road $\text{\L}$ (*ML*) is

$$y' - 277.101 = \text{Cot } 75°10'30'' \ (x' - 409.433) \tag{1}$$

From Eq. (6.14), the equation of Elm Parkway $\text{\L}$ is

$$(x' - 612.656)^2 + (y' - 317.313)^2 = 350.000^2 \tag{2}$$

Simplify Eq. (1) to

$$y' - 277.101 = -0.2646782x' + 108.368$$
$$y' = -0.2646782x' + 385.469 \tag{1A}$$

Substitute the value of y' into Eq. (2):

$$(x' - 612.656)^2 + (-0.2646782x' + 385.469 - 317.313)^2$$
$$- 350.000^2 = 0$$
$$(x' - 612.656)^2 + (-2646782x' + 68.156)^2 - 350.000^2 = 0$$
$$1.0700545x'^2 - 1261.391x' + 257492.61 = 0$$

This quadratic of the form $ax^2 + bx + c = 0$ has roots

$$x = \frac{-b \pm \sqrt{b^2 - 4ac}}{2a}$$

$$x' = \frac{1261.3908 \pm \sqrt{1\,591\,107.30 - 1\,102\,124.50}}{2.140\,109}$$

$$= \frac{1261.3908 \pm 699.27305}{2.140\,109}$$

$$= 916.1514 \quad \text{or} \quad x' = 262.658$$

Solve for y' by substituting in Eq. (1A):

$$y' = 142.984 \quad \text{or} \quad y' = 315.949$$

When these coordinates are now enlarged by the amount of the original axes reduction, the following values are obtained:

| Station | Reduced Coordinates | | Grid Coordinates | |
	y'	x'	$y(y' + 4850000)$	$x(x' + 316500)$
L	142.984	916.151	4850142.984	317416.151
L'	315.949	262.658	4850315.949	316762.658

Analysis of Figure 6.20 is required in order to determine which of the two solutions is the correct one. The sketch shows that the desired intersection point L is south and east of station M. That is, L(4 850 142.984 N, 317 416.151 E) is the set of coordinates for the intersection of the ℄'s of Elm Parkway and Fisher Road.

The other intersection point (L') solution is superfluous.

Example 6.8 Review Problem

Figure 6.21 shows the field data for a five-sided closed traverse:

(a) Balance the angles.

(b) Compute the azimuths and bearings.

(c) Compute the linear error of closure and the accuracy ratio of the traverse.

(d) If the accuracy ratio is equal to or better than $1/4000$, balance the latitudes and departures using the compass rule.

(e) Assuming that the coordinates of station B are (1000.000 N, 1000.000 E), compute the coordinates of the remaining stations.

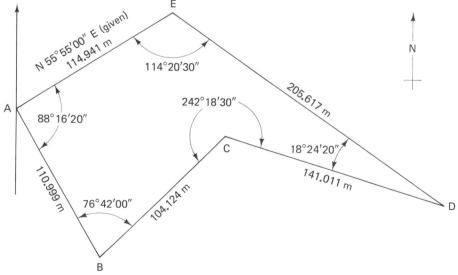

FIGURE 6.21 Traverse sketch for Example 6.8, showing mean field distances and interior angles.

The solutions are as follows:
  (a) Balance the angles:

Field Angles		Balanced Angles	
A	88°16′20″	A	88°16′00″
B	76°42′00″	B	76°41′40″
C	242°18′30″	C	242°18′10″
D	18°24′20″	D	18°24′00″
E	114°20″30″	E	114°20″10″

$$538°120′100″ = 540°00′100″ \qquad 538°119′60″ = 540°00′00″$$

$$(n - 2)180 = 540°$$
$$\text{Error in angular closure} = 100″$$
$$\text{Correction per angle} = -20″$$

(b) Compute the azimuths and bearings (see Figure 6.22):

Azimuths

Az AE =	55°55′00″	See Figure 6.21
+ $<A$	88°16′00″	
Az AB =	144°11′00″	
+ 180°		
Az BA =	324°11′00″	
+ $<B$	76°41′40″	
Az BC =	400°52′40″	
− 360°		
Az BC =	40°52′40″	
+ 180°		
Az CB =	220°52′40″	
+ $<C$	242°18′10″	
Az CD =	463°10′50″	
− 360°		
Az CD =	103°10′50″	
+ 180°		
Az DC =	283°10′50″	
+ $<D$	18°24′00″	
Az DE =	301°34′50″	
− 180°		
Az ED −	121°34′50″	
+ $<E$	114°20′10″	
Az EA −	235°55′00″	
− 180°		
Az AE =	55°55′00″	Check

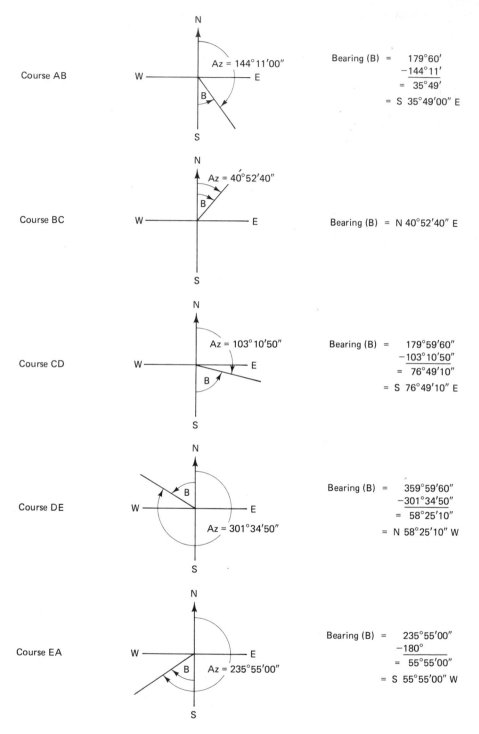

FIGURE 6.22 Computation of bearings from computed azimuths, Example 6.8.

Course	Azimuths (Clockwise)	Course	Bearings (Clockwise) Computed from Clockwise Azimuths
AB	144°11′00″	AB	S 35°49′00″ E
BC	40°52′40″	BC	N 40°52′40″ E
CD	103°10′50″	CD	S 76°49′10″ E
DE	301°34′50″	DE	N 58°25′10″ W
EA	235°55′00″	EA	S 55°55′00″ W

(c) Compute the linear error of closure and the accuracy ratio of the traverse: Referring to Table 6.9, the latitudes and departures are computed by using either the azimuths or the bearings and the distances. (Both azimuths and bearings are shown here only for comparative purposes.)

Since the accuracy ratio (1/4000) meets the specification noted, the computations can continue.

(d) Using the compass rule, balance the latitudes and departures.

$$C \text{ lat } AB = \Sigma \text{ lat} \times \frac{\text{distance } AB}{\text{perimeter}} \qquad (6.3)$$

and

$$C \text{ dep } AB = \Sigma \text{ dep} \times \frac{\text{distance } AB}{\text{perimeter}} \qquad (6.4)$$

For the example problem

$$C \text{ lat } AB = 0.164 \times \frac{111}{677} \text{ (values rounded)} = 0.027 \text{ m}$$

$$C \text{ dep } AB - 0.033 \times \frac{111}{677} = 0.005 \text{ m}$$

As a check, the algebraic sum of the balanced latitudes (and balanced departures) will be zero.

TABLE 6.9 *Computation of Linear Error of Closure (E) and the Accuracy Ratio E/P of the Traverse in Example 6.8*

Course	Azimuth	Bearing	Distance	Latitude	Departure
AB	144°11′00″	S 35°49′00″ E	110.999	−90.008	+64.956
BC	40°52′40″	N 40°52′40″ E	104.124	+78.729	+68.144
CD	103°10′50″	S 76°49′10″ E	141.011	−32.153	+137.296
DE	301°34′50″	N 58°25′10″ W	205.617	+107.681	−175.166
EA	235°55′00″	S 55°55′00″ W	114.941	−64.413	−95.197
			$P = 676.692$	$E_L = -0.164$	$E_D = +0.033$

$$E = \sqrt{E \text{ lat}^2 + E \text{ dep}^2} = \sqrt{0.164^2 + 0.033^2} = 0.167 \text{ m}$$

$$\text{Accuracy ratio of traverse} = \frac{0.167}{676.692} = \frac{1}{4052} = \frac{1}{4100}$$

TABLE 6.10   *Computation of Balanced Latitudes and Departures Using the Compass Rule*

Course	Latitude	Departure	CL	CD	Balanced Latitude	Balanced Departure
AB	−90.008	+64.956	+0.027	−0.005	−89.981	+64.951
BC	+78.729	+68.144	+0.025	−0.005	+78.754	+68.139
CD	−32.153	+137.296	+0.034	−0.007	−32.119	+137.289
DE	+107.681	−175.166	+0.050	−0.010	+107.731	−175.176
EA	−64.413	−95.197	+0.028	−0.006	−64.385	−95.203
	$E_L = -0.164$	$E_D = +0.033$	+0.164	+0.033	0.00	0.00

(e) Assuming that the coordinates of station B are (1000.000 N, 1000.000 E), compute the coordinates of the remaining stations: The *balanced* latitudes and departures (Table 6.10) are used to compute the coordinates (see Figure 6.23). The axes are selected so that B is 1000.000 N, 1000.000 E.

If the computation returns to B with values of 1000.000 N and 1000.000 E, the computation is verified.

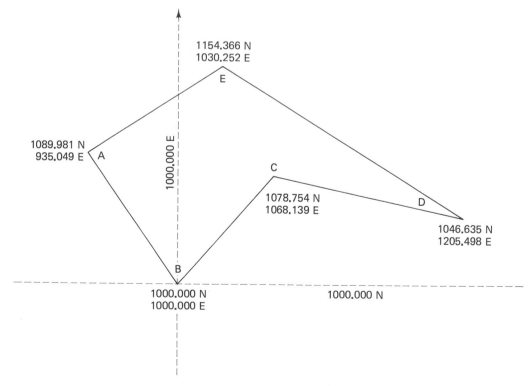

FIGURE 6.23   Station coordinates, Example 6.8.

Station	Coordinates			
	North		East	
B	1000.000		1000.000	
	+78.754	lat BC	+68.139	dep BC
C	1078.754		1068.139	
	−32.119	lat CD	+137.289	dep CD
D	1046.635		1205.498	
	+107.731	lat DE	−175.176	dep DE
E	1154.366		1030.252	
	−64.385	lat EA	−95.203	dep EA
A	1089.981		935.049	
	−89.981	lat AB	+64.951	dep AB
B	1000.000	Check	1000.000	Check

6.15 AREA OF A CLOSED TRAVERSE BY THE COORDINATE METHOD

When the coordinates of the stations of a closed traverse are known, it is a simple matter to then compute the area within the traverse, either by computer or hand-held calculator. Figure 6.24a shows a closed traverse 1, 2, 3, 4 with the appropriate x and y coordinate distances. Figure 6.24b illustrates the technique used to compute the traverse area.

With reference to Figure 6.24b, it can be seen that the desired *area of the traverse is, in effect, area 2 minus area 1.* Area 2 is the sum of the areas of trapezoids 4′433′ and 3′322′. Area 1 is the sum of trapezoids 4′411′ and 1′122′.

$$\text{Area 2} = \tfrac{1}{2}(x_4 + x_3)(y_4 - y_3) + \tfrac{1}{2}(x_3 + x_2)(y_3 - y_2)$$
$$\text{Area 1} = \tfrac{1}{2}(x_4 + x_1)(y_4 - y_1) + \tfrac{1}{2}(x_1 + x_2)(y_1 - y_2)$$
$$2A = [(x_4 + x_3)(y_4 - y_3) + (x_3 + x_2)(y_3 - y_2)]$$
$$- [(x_4 + x_1)(y_4 - y_1) + (x_1 + x_2)(y_1 - y_2)]$$

Expand this expression and collect the remaining terms:

$$2A = x_1(y_4 - y_2) + x_2(y_1 - y_3) + x_3(y_2 - y_4) + x_4(y_3 - y_1) \quad (6.16)$$

Stated simply, the double area of a closed traverse is the algebraic sum of each *x coordinate multiplied by the difference between the y values of the adjacent stations.*

The double area is divided by 2 to determine the final area. The final area can be positive or negative, reflecting only the direction of computation approach (clockwise or counterclockwise). The area is, of course, positive; there is no such thing as a negative area.

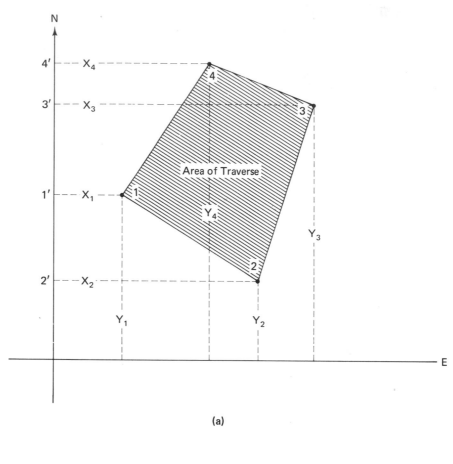

(a)

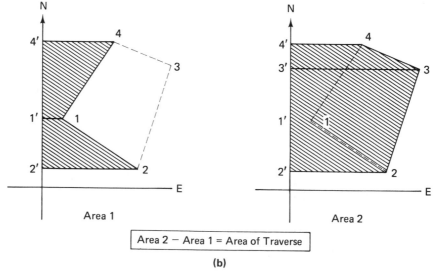

Area 2 − Area 1 = Area of Traverse

(b)

FIGURE 6.24 Area by rectangular coordinates.

Example 6.9   *Area Computation by Coordinates*

Refer to the traverse example, 6.8, and the computed coordinates shown in Figure 6.23, which are summarized next:

Station	North	East
A	1089.981	935.049
B	1000.000	1000.000
C	1078.754	1068.139
D	1046.635	1205.498
E	1154.366	1030.252

$$2A = x_1(y_1 - y_2) + x_2(y_1 - y_3) + x_3(y_2 - y_4) + x_4(y_3 - y_1) \quad (6.16)$$

The *double area computation* (to the closest m²) is as follows:

$$XA(YB - YE) = 935.049(1000.000 - 1154.366) = -144\ 340$$
$$XB(YC - YA) = 1000.000(1078.754 - 1089.981) = -11\ 227$$
$$XC(YD - YB) = 1068.139(1046.635 - 1000.000) = +49\ 813$$
$$XD(YE - YC) = 1205.498(1154.366 - 1078.754) = +91\ 150$$
$$XE(YA - YD) = 1030.252(1089.981 - 1046.635) = +44\ 657$$

$$2A = +30\ 053\ \text{m}^2$$

$$\text{Area} = 15027\ \text{m}^2$$
$$= 1.503\ \text{hectares}$$

Example 6.10   *Area Computation by Coordinates*

Referring to the traverse example, 6.1, as shown in Figures 6.15 and 6.18, the coordinates are summarized next:

Station	North	East
A	1000.00 ft	1000.00 ft
B	1250.73	1313.61
C	1302.96	1692.14
D	934.77	1684.54
E	688.69	1160.27

The double area computation (to the closest ft²) using the relationships shown in Eq. (6.16) is as follows:

$$2A = x_1(y_4 - y_2) + x_2(y_1 - y_3) + x_3(y_2 - y_4) + x_4(y_3 - y_1) \quad (6.16)$$

$$XA(YB - YE) = 1000.00(1250.73 - 688.69) = +562\ 040$$
$$XB(YC - YA) = 1313.61(1302.96 - 1000.00) = +397\ 971$$
$$XC(YD - YB) = 1692.14(934.77 - 1250.73) = -534\ 649$$
$$XD(YE - YC) = 1684.54(688.69 - 1302.96) = -1\ 034\ 762$$
$$XE(YA - YD) = 1160.27(1000.00 - 934.77) = +75\ 684$$

$$-533\ 716\ \text{ft}^2$$
$$2A = \quad 533\ 716\ \text{ft}^2$$

$$\text{Area} = 266\ 858\ \text{ft}^2$$
$$1\ \text{Acre} = 43\ 560\ \text{ft}^2$$
$$\text{Area} = \frac{266\ 858}{43\ 560} = 6.126\ \text{acres}$$

6.16 AREA OF A CLOSED TRAVERSE BY THE DOUBLE MERIDIAN DISTANCE METHOD

This method utilizes balanced latitudes and departures to directly calculate the area within a closed traverse. By definition, the meridian distance of a line is the distance from the midpoint of the line to some meridian. The meridian being used may be some distance from the line (Figure 6.25a) or the meridian could be placed so that it goes through one end of the line (Figure 6.25b).

In Figure 6.25a, the meridian distance of line AB is the mean of the distances from the ends of the line to the meridian.

$$\text{Meridian distance (MD) of } AB = \frac{B'B + A'A}{2}$$

In Figure 6.25b, the meridian distance of line CD is half the distance from the end of the line not on the meridian.

i.e.:
$$\text{Meridian distance (MD) of } CD = \frac{D'D}{2}$$

Area of figure (trapezoid) $B'BAA'$ = meridian distance $AB \times$ lat AB

$$= \frac{B'B + A'A}{2} \times \text{lat. } AB$$

Area of figure (triangle) $D'DC$ = meridian distance $CD \times$ lat CD

$$= \frac{D'D}{2} \times \text{lat } CD$$

Generally,

$$\text{Area} = \text{meridian distance (MD)} \times \text{latitude}$$

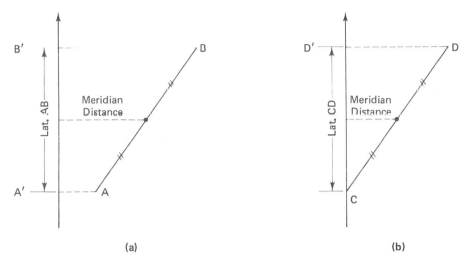

FIGURE 6.25 Meridian distances and areas: (a) meridian is located some distance from straight line (*AB*); (b) meridian is through one end of straight line (*CD*).

or

$$\text{Double area} = \text{double meridian distance (DMD)} \times \text{latitude}$$
$$2A = \text{DMD} \times \text{lat} \qquad (6.17)$$

It is customary to calculate the double areas to avoid the multiple divisions by 2. The total double area is first computed and then finally divided by 2 to get the required area.

Analysis of Figure 6.26 will show the following:

1. The desired area within the traverse can be determined by subtracting the areas of trapezoid *C'CBB'* and triangles *B'BA* and *ADD'* from the large area of trapezoid *C'CDD'*.

2. The areas of the three geometric figures to be subtracted will all have the same algebraic sign as their latitudes all have the same algebraic sign. Their latitudes will be positive (northerly) if the computation approach is clockwise, or the latitudes will be negative (southerly) if the computation approach is counterclockwise.

3. The area of the large geometric figure will be opposite in sign to the three areas mentioned above, resulting in a net double area for the actual traverse boundary.

Figure 6.27 shows a four-sided traverse with a meridian drawn through the most westerly point (*A*). The area computation

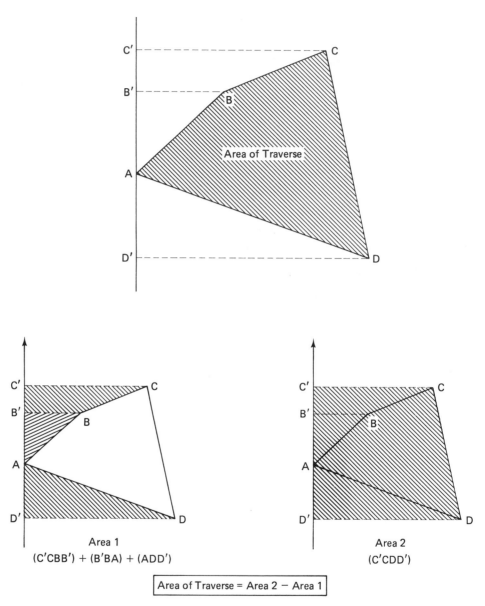

FIGURE 6.26 Area by double meridian distances.

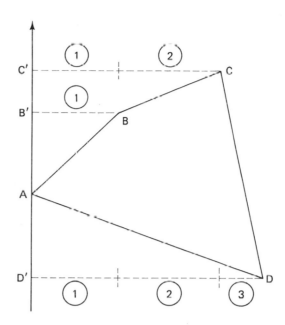

FIGURE 6.27 Double meridian distances.

can proceed clockwise or counterclockwise, although to avoid confusion the area computation always proceeds in the same direction as did the computation for latitudes and departures.

Generally, the computation for double area would proceed as follows [see Eq. (6.17)]:

	DMD	*Double Area*
DMD *AB* =	Dep *AB* (1)	DMD *AB* × lat *AB*
DMD *BC* =	(1) + (1 + 2)	
	— DMD *AB* (1)	DMD *BC* × lat *BC*
	+ dep. *AB* (1)	
	+ dep. *BC* (2)	
DMD *CD* =	(1 + 2) + (1 + 2 + 3)	
	= DMD *BC* (1 + 1 + 2)	DMD *CD* × lat *CD*
	+ dep *BC* (2)	
	+ dep *CD* (3)	
DMD *DA* =	(1 + 2 + 3)	
	= Dep *DA*	DMD *DA* × lat *DA*

Note: The accumulation of distances in the preceding steps would have been algebraically reduced to the simple sum shown (1 + 2 + 3) in a numerical problem.

The preceding analysis leads to the following observations:

1. The DMD of the first course is equal to the departure of the first course.
2. The DMD of each succeeding course is equal to the DMD of the previous course + the departure of the previous course + the departure of the course itself.
3. The DMD of the last course will turn out to be equal to the departure of the last course, but opposite in sign.

It was suggested that the meridian be placed through the most westerly station; this practice will ensure that all DMDs are positive. With all DMDs positive, the algebraic sign of the individual double areas will be given by the sign of the latitude of the individual traverse courses.

The algebraic sign of the resultant double area can be plus or minus depending only on the direction (clockwise or counterclockwise) taken for the computation. The actual area is obviously positive. In the general example shown in Figures 6.26 and 6.27, the resultant area (area 2 − area 1) will be negative, as area 2 will have the minus sign of the latitude of *CD* (course *CD* is SE).

Some things take much longer to describe than to actually perform; areas by double meridian distances is one such thing. The following two examples will illustrate just how simple and quick the actual computations are.

Example 6.11 *Area Computation by DMDs*

Refer to Table 6.10, which shows the balanced latitudes and departures for the traverse example, 6.8 (Figure 6.21):

Course	Balanced Latitude	Balanced Departure
AB	−89.981	+64.951
BC	+78.754	+68.139
CD	−32.119	+137.289
DE	+107.731	−175.176
EA	−64.385	−95.203
	0.000	0.000

Course	DMD	Latitude	Double Area +	Double Area −
AB	64.951	−89.981		5 844
	+64.951			
	+68.139			
BC	198.041	+78.754	15 597	
	+68.139			
	+137.289			
CD	403.469	−32.119		12 959
	+137.289			
	540.758			
	−175.176			
DE	365.582	+107.731	39 385	
	−175.176			
	190.406			
	−95.203			
EA	95.203 Check	−64.385		6 130
			54 982	24 933

$$2A = 54\ 982 - 24\ 933 = 30\ 049\ \text{m}^2$$
$$A = 15\ 025\ \text{m}^2$$
$$= 1.503\ \text{ha (hectares)}$$

See Example 6.9 and Figure 6.21.

Example 6.12 *Area Computation by DMDs*

Refer to the traverse example, 6.1 (Figures 6.8 and 6.15) and Tables 6.4 and 6.8; the balanced latitudes and departures are as follows:

Course	Balanced Latitude	Balanced Departure
AE	−311.31	+160.27
ED	+246.08	+524.27
DC	+368.19	+7.60
CB	−52.23	−378.53
BA	−250.73	−313.61
	0.00	0.00

Course	DMD	Latitude	Double Area +	−
AE	160.27	−311.31		49 894
	160.27			
	524.27			
ED	844.81	+246.08	207 891	
	524.27			
	7.60			
DC	1376.68	+368.19	506 880	
	+7.60			
	1384.28			
	−378.53			
CB	1005.75	−52.23		52 530
	−378.53			
	627.22			
	−313.61			
BA	313.61	−250.73		78 631
			714 771	181 055

$$2A = 714\ 771 - 181\ 055 = 533\ 716\ \text{ft}^2$$
$$A = 266\ 858\ \text{ft}^2$$
$$= \frac{266\ 858}{43\ 560} = 6.126\ \text{acres}$$

6.17 SUMMARY OF TRAVERSE COMPUTATIONS

1. Balance the field angles.
2. Correct (if necessary) the field distances (e.g., for temperature).
3. Compute the bearings and/or azimuths.
4. Compute the linear error of closure and the accuracy ratio of the traverse.
5. Compute the balanced latitudes and balanced departures.
6. Compute the coordinates.
7. Compute the area by coordinates *or* compute the area by DMDs.

Allied topics, such as plotting by rectangular coordinates, and the computation of land areas by other methods are covered in Chapter 9.

PROBLEMS

6-1   A five-sided closed field traverse has the following angles:

A - 110°27'00"
B - 130°52'30"   Determine the angular error
C - 88°17'00"   of closure and balance
D - 97°33'30"   the angles by applying equal
E - 112°52'30"   corrections to each angle

6-2   A four-sided closed field traverse has the following angles:
      A = 81°22'30"; B = 72°32'30"; C = 89°39'30"; D = 116°23'30"
      The lengths of the sides are as follows:
      AB = 636.45ft; BC = 654.49ft; CD = 382.85ft; DA = 512.77ft.
      The bearing of AB is S 23°06'W. BC is in the S.E. quadrant
      a) Balance the field angles; b) compute the bearings; c) compute the latitudes and departures; and d) determine the linear error of closure and the accuracy ratio.

6-3   Using the data from Problem 6-2, a) balance the latitudes and departures by use of the compass rule and b) compute the coordinates of stations A, C and D if the coordinates of station B are 1000.00N, 1000.00E.

6-4   Using the data from Problems 6-2 and 6-3 compute the area enclosed by the traverse:

      a) compute area using coordinate method
      b) compute area using the DMD method

6-5   A five-sided closed field traverse has the following distances in meters: AB = 50.276; BC = 26.947; CD = 37.090; DE = 35.292; EA = 20.845. The adjusted angles are as follows: A = 101°28'26"; B = 102°10'42"; C = 104°42'06"; D = 113°04'42"; and E = 118°34'04". The bearing of AB is N 71°10'34"E. BC is in the S. E. quadrant

      a) Compute the bearings; b) compute the latitudes and departures; c) determine the linear error of closure and the accuracy ratio.

6-6   Using the data from Problem 6-5, a) balance the latitudes and departures by using the Compass rule and b) compute the coordinates of the traverse stations using coordinates of station A as 1000.000N, 1000.000E.

6-7   Using the data from Problems 6-5 and 6-6 compute the area enclosed by the traverse:

a) compute area using coordinate method
b) compute area using DMD method

6-8   The two frontage corners of a large tract of land were joined by the following open traverse:

Course	Distance	Bearing
AB	80.32ft	N 70°10′07″E
BC	953.83ft	N 74°29′00″E
CD	818.49ft	N 70°22′45″E

Compute the distance and bearing of the property frontage *AD*.

6-9   Given the following data for a closed property traverse, a) compute the missing data (i.e., distance *CD* and bearing *DE*); b) compute the area of the property bounded by the traverse.

Course	Bearing	Distance (ft)
AB	N 37°10′49″	537.144
BC	N 79°29′49″E	1109.301
CD	S 18°56′31″W	?
DE	?	953.829
EA	N 26°58′31″W	483.669

6-10   A six-sided traverse has the following station coordinates: A (559.319 N, 207.453 E); B (738.562 N, 666.737 E); C (541.742 N, 688.350 E); D (379.861 N, 839.008 E); E (296.099 N, 604.048 E); and F (218.330 N, 323.936 E). Compute the distance and bearing of each side.

6-11   Using the data from Problem 6-10, compute the area enclosed by the traverse (hectares).

6-12   Use the data from Problem 6-10 to solve the following: If the intersection point of lines *AD* and *BF* is *K*, and if the intersection point of lines *AC* and *BE* is *L*, compute the distance and bearing of line *KL*.

6-13   A five-sided field traverse has the following balanced angles: A = 101°28′26″; B = 102°10′42″; C = 104°42′06″; D = 113°04′42″ and E = 118°34′04″. The lengths of the sides are as follows: AB = 50.276 m; BC = 26.947 m; CD = 37.090 m; DE = 35.292 m; and EA = 20.854 m.

The bearing of EA is N 7°21'00" W, and AB is oriented northeasterly.

a) Compute the bearings; b) compute the latitudes and departures; c) compute the linear error of closure and the accuracy ratio.

6-14 From the data in Problem 6-13, balance the latitudes and departures employing the compass rule.

6-15 From the data in Problems 6-13 and 6-14; a) if the coordinates of station A are (1000.000 N, 1000.000 E), compute the coordinates of the other stations.

6-16 From the data in Problems 6-14, and 6-15, use either the DMD method, or the coordinate method to compute the area of the enclosed figure.

6-17 A theodolite with EDM was set up at control station K which is within the limits of a five-sided property. Coordinates of control station K are 1990.000 N, 2033.000 E. Azimuth angles and distances to the five property corners were determined as follows:

Direction	Azimuth	Horizontal Distance (meters)
KA	286°51'30"	34.482
KB	37°35'28"	31.892
KC	90°27'56"	38.286
KD	166°26'49"	30.916
KE	247°28'43"	32.585

Compute the coordinates the property corners A, B, C, D, and E.

6-18 Using the data from Problem 6-17, compute the area of the property.

6-19 Using the data from Problem 6-17, compute the bearings and distances of the five sides of the property.

7 *Topographic Surveys*

7.1 GENERAL

Topographic surveys are performed in order to determine the position of natural and man-made features (e.g., trees, shorelines, roads, sewers, and buildings). These natural and man-made features can then be drawn to scale on a plan or map. In addition, topographic surveys also include the determination of ground elevations, which can later be plotted on plans for the construction of contours or plotted in the form of profiles or cross sections. Chapter 9 covers these topics in detail. In engineering work, topographic surveys are often called preliminary or pre-engineering surveys.

Section 1.12 lists the techniques that can be used to locate (tie in) a point (feature or elevation) for the purpose of plan preparation. The vast majority of topographic surveys utilize rectangular or polar ties. Chapter 15 describes how topographic features and elevations can be shown on plans and maps utilizing aerial surveys and photogrammetric plotting techniques, whereas Chapter 9 describes standard plotting and drafting techniques.

7.2 SCALES AND PRECISION

Maps and plans are drawn so that a distance on the map or plan conforms to a set distance on the ground. The ratio (called *scale*) between plan distance and ground distance is consistent throughout the plan. The scales can be stated as equivalencies, for example, 1″ = 50′ or 1″ = 1000′, or the same scales can be stated as representative fractions: 1:600 (1″:50 × 12″) or 1:12000 (1″:1000 × 12″). When representative fractions are used, all units are valid; that is, 1:500 is the same scale for inches, feet, meters, and so on. Only representative fractions are used in the metric (SI) system.

Maps and plans used in Canada are now utilizing metric scales, whereas in the United States the conversion to metric appears to be underway.

Table 7.1 shows recommended map and plan scales and the replaced scales in the foot–inch system. Almost all surveying required for the production of intermediate and small-scale maps is done by aerial surveying, with the maps being produced photogrammetrically (see Chapter 15).

Even in municipal areas where services (roads, sewers, water) plans for subdivisions are drawn at 1:1000 and plans and profiles for municipal streets are drawn at 1:500, it is not uncommon to have the surveys flown and the maps produced photogrammetrically. For street surveys, the surveying manager will have a good idea as to the cost per kilometer or mile for various orders of urban density and will arrange for field or aerial surveys depending on which method is deemed cost effective.

Before a field survey is undertaken, a clear understanding of the reason for the survey is necessary so that appropriately precise techniques can be employed. If the survey is required to locate points that will later be shown on a small-scale map, the precision of the survey will be of a very low order. Generally, points are located in the field with a precision that will at least be compatible with the plotting precision possible at the designated plan (map) scale.

For example, if we can assume that points can be plotted to the closest 0.5 mm ($1/50$ in.) at a scale of 1:500, this represents a plotting capability to the closest ground distance of 0.25 m (i.e., 0.0005×500), whereas at a scale of 1:20000 the plotting capability is $(20000 \times 0.0005) = 10$ m of ground distance.

The latter example is rarely encountered since most surveys for maps at 1:20000 are aerial surveys. In the former example, although plotting capabilities indicate that a point should be tied in

TABLE 7.1   *Metric and Foot–Inch Scales*

Category	Recommended Scales, Metric (SI) System	Replaced Scales, Foot–Inch System
Large scale	1:100	1" = 10' and 1" = 8'
	1:200	1" = 20'
	1:500	1" = 40' and 1" = 50'
	1:1000	1" = 80' and 1" = 100'
Intermediate scale	1:2000	1" = 200'
	1:5000	1" = 400'
	1:10000	1" = 800'
Small scale	1:20000	1:25000 ($2^{1}/_{2}$" = 1 mile)
	1:50000	1:63360 (1" = 1 mile)
	1:100000	1:126720 ($^{1}/_{2}$" = 1 mile)
	1:200000	1:250000 ($^{1}/_{4}$" = 1 mile, approx.)
	1:500000	1:625000 ($^{1}/_{10}$" = 1 mile, approx.)
	1:1000000	1:1000000 ($^{1}/_{16}$" = 1 mile, approx.)

to the closest 0.25 m, in reality the point probably would be tied in to a higher level of precision (e.g., 0.1 m).

In this regard, the following points should be considered:

1. Some detail can be precisely defined and located (e.g., building corners, subway tracks, bridge beam seats).
2. Some detail cannot be precisely defined or located (e.g., stream banks, edges of a gravel road, $\mathcal{L}$ of ditches, limits of a wooded area, rock outcrops).
3. Some detail can be located with only moderate precision, using normal techniques (e.g., large single trees, manholes and catchbasins, curbs, walks, culverts, docks).

Usually, the detail that is fairly well defined is located with more precision than is required just for plotting. The reasons for this are as follows:

1. As in the preceding example, it takes little (if any) extra effort to locate detail to 0.1 m than it would to locate it to 0.25 m.
2. By using the same techniques to locate all detail, as if all detail were precisely defined, the survey crew is able to develop uniform practices, which will reduce mistakes and increase efficiency.
3. Some location measurements taken in the field, and design parameters that may be based on those field measurements,

are also shown on the plan as layout dimensions (i.e., levels of precision are required that greatly supersede the precision required simply for plotting).

Most natural features are themselves not precisely defined, so if a topographic survey is required in an area having only natural features (e.g., stream or watercourse surveys, site development surveys, large-scale mapping surveys), a relatively imprecise survey method (e.g., stadia) can be employed.

All topographic surveys are tied into both horizontal and vertical (bench marks) control. The horizontal control for topographic surveys can be closed-loop traverses, traverses from a coordinate grid monument closed to another coordinate grid monument, route center line ($\mathcal{L}$), or some assumed base line. The survey measurements used to establish the horizontal and vertical control are always taken more precisely than are the location ties.

Surveyors are conscious of the need for accurate and well-referenced survey control. If the control is inaccurate, the survey and resultant design will also be inaccurate; if the control is not well referenced, it will be costly (perhaps impossible) to precisely relocate the control in the field once it is lost. In addition to providing control for the original survey, the survey control must be used if additional survey work is required to supplement the original survey, and of course the survey control must be used for any construction layout resulting from designs based on the original survey.

7.3 LOCATION BY RIGHT-ANGLE OFFSETS

Many topographic surveys, excluding mapping surveys, but including most pre-engineering surveys, utilize the right-angle offset technique to locate detail. This techique not only provides the location of plan detail but also provides location for area elevations taken by cross sections (see Section 7.4).

Plan detail is located by measuring the distance perpendicularly from the baseline to the object and, in addition, measuring along the base line to the point of perpendicularity (Figure 1.11). The base line is laid out in the field with stakes (nails in pavement) placed at appropriate intervals, usually 100 ft or $^{20}/_{30}$ m. A sketch is entered in the field book before the measuring commences. If the terrain is smooth, a tape can be laid on the ground between the station marks. This will permit the surveyor to move along the tape (toward the forward chainage), noting and booking the chainages of the sketched detail on both sides of the base line. The right angle for each location tie can be established using a pentaprism (Figure 7.1), or a right angle can be approximately established in the following manner: the surveyor stands on the base line facing the detail to be tied in and then points one arm down the base line in one direction and then the other arm down the base line in the opposite direction; after checking both arms (pointed index fingers) for proper alignment, the surveyor closes his eyes while he swings his arms together in front of him, pointing (presumably) at the detail. If he is not pointing at the detail, the surveyor moves slightly along the base line and repeats the procedure until the detail has been correctly sighted in. The chainage is then read off the tape and booked in the field notes. This approximate method is used a great deal in route surveys and municipal surveys. This technique provides good results over short offset distances (50 ft or 15 m). For longer offset distances or for very important detail, a pentaprism or even a transit can be used to determine the chainage.

Once all the chainages have been booked for the interval (100 ft or 20 to 30 m), it only remains to measure the offsets left and right of the base line. If the steel tape has been left lying on the ground during the determination of the chainages, it is usually left in place to mark the base line while the offsets are measured with another tape (e.g., a cloth tape).

Figure 7.2a illustrates topographic field notes that have been booked using a single base line and in Figure 7.2b using a split

FIGURE 7.1 Double right-angle prism. (Courtesy Keuffel & Esser Co.)

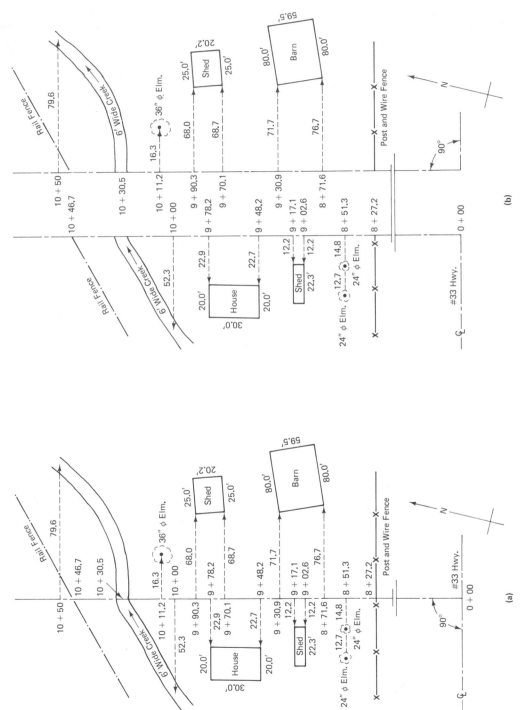

FIGURE 7.2 Topographic field notes: (a) single base line; (b) split base line.

238

base line. In Figure 7.2a, the offsets are shown on the dimension lines and the chainages are shown opposite the dimension line or as close as possible to the actual tic point on the base line.

In Figure 7.2b, the base line has been "split"; that is, two lines are drawn representing the base line, leaving a space of zero dimension between them for the inclusion of chainages. The split-base line technique is particularly valuable in densely detailed areas where single base-line notes would tend to become crowded and difficult to decipher. The earliest topographic surveyors in North America used the split-base-line method of note keeping (see Figure 7.3).

7.4 CROSS SECTIONS AND PROFILES

Cross sections are a series of elevations taken at *right angles to a base line* at specific stations, whereas profiles are a series of elevations taken *along a base line* at some specified repetitive station interval. The elevations thus determined can be plotted on maps and plans either as spot elevations or as contours, or they can be plotted as end areas for construction quantity estimating.

As in offset ties, the base line interval is usually 100 ft (20 to 30 m), although in rapidly changing terrain the interval is usually smaller (e.g., 50 ft or 10 to 15 m). In addition to the regular intervals, cross sections are also taken at each abrupt change in the terrain (top, bottom of slopes, etc.).

Figure 7.4 illustrates how the rod readings are used to define the ground surface. In Figure 7.4(a) the uniform slope permits a minimum (℄ and both limits of the survey) number of rod readings. In Figure 7.4(b) the varied slope requires several more (than the minimum) rod readings to adequately define the ground surface.

Figure 7.4c illustrates how cross sections are taken before and after construction. Chapter 9 covers how to calculate the end area (lined section) at each station and then the volumes of cut and fill.

The profile consists of a series of elevations along the base line. If cross sections have been taken, the necessary data for plotting a profile will also have been taken. If cross sections are not planned for an area for which a profile is required, the profile elevations can be determined by simply taking rod readings along the line at regular intervals and at all points where the ground slope changes (see Figure 3.19).

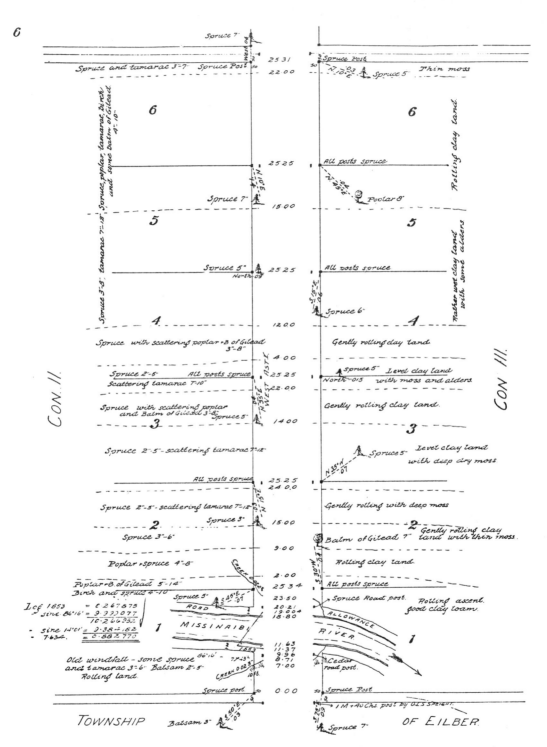

FIGURE 7.3 Original topographic field notes (1907), Province of Ontario (distances shown are in chains).

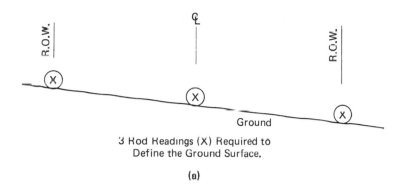

3 Rod Readings (X) Required to
Define the Ground Surface.

(a)

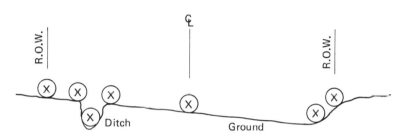

7 Rod Readings (X) Required to
Define this Ground Surface.

(b)

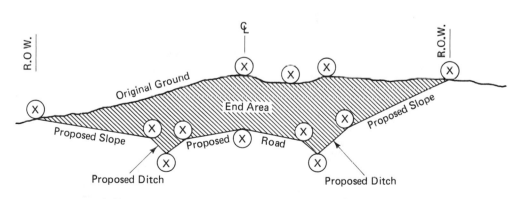

5 Rod Readings (X) Required to Define Original Ground Surface.
9 Rod Readings (X) Required to Define the Constructed Ground Surface.

(c)

FIGURE 7.4 Cross sections used to define ground surface: (a) uniform slope; (b) varied slope; (c) ground surface before and after construction.

241

Typical field notes for profile leveling are shown in Figure 3.20. Cross sections are booked in two different formats. Figure 3.23 shows cross sections booked in standard level note format. All the rod readings for one station (that can be "seen" from the HI) are booked together. In Figure 3.24, the same data are entered in a format popular with highway agencies. The latter format is more compact and thus takes less space in the field book; the former format takes more space in the field book but allows for a description for each rod reading, an important consideration for municipal surveyors.

The assumption in this and the previous section is that the data are being collected by conventional offset ties and cross-section methods (i.e., steel and cloth tapes for the tie-ins and a level and rod for the cross sections). In municipal work, a crew of four surveyors can be efficiently utilized. While the party chief is making sketches for the detail tie-ins, the instrumentman (rod readings and bookings) and two rodmen (one on the tape, the other on the rod) can perform the cross sections.

In cases where the terrain is very rugged, making the level and rod work very time-consuming (many instrument setups), the survey can be performed by stadia (see Section 7.5), keeping all the elevation rod shots on cross-section station lines. If stadia precision is sufficient, the plan detail location can also be accomplished using stadia.

Chapter 8 describes electronic tacheometer instruments (ETI), also known as *total stations*. These instruments can very quickly measure distances and differences in elevation; the rodman holds a reflecting prism mounted on a range pole instead of holding a rod. Many of these instruments have the distance and elevation data recorded automatically for future computer processing, while others require that the data be manually entered into the data recorder. Either method could be used to advantage on a multitude of surveying projects, including right-angle offset tie-ins and cross sections.

7.5 STADIA PRINCIPLES

Stadia is a tacheometric form of distance measurement that relies on a fixed-angle intercept. Stadia is used on topographic surveys where a limiting accuracy of $1/400$ will be acceptable. As noted earlier, stadia is ideally suited for the location of natural features that themselves cannot be precisely defined or located.

Cross Hair Reticle

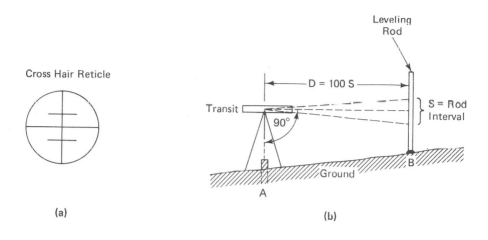

(a)

(b)

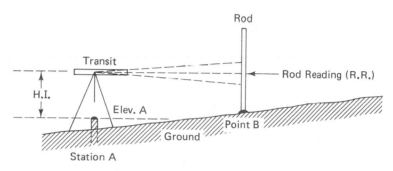

Elevation A. + H.I. − R.R. = Elevation B.

(c)

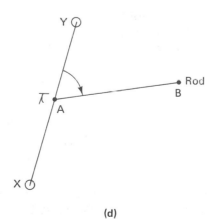

(d)

FIGURE 7.5 Stadia principles: (a) stadia hairs; (b) distance determination; (c) elevation determination; (d) angle determination.

The transit (some levels) cross-hair reticle has, in addition to the normal cross hairs, two additional horizontal hairs (see Figure 7.5a), one above and the other below the main horizontal hair. The stadia hairs are positioned in the reticle so that, if a rod were held 100 ft (m) away from the transit (telescope level), the difference between the upper and lower stadia hair readings (rod interval) would be exactly 1.00 ft (m). It can be seen (Figure 7.5b) that distances can be determined simply by sighting a rod with the telescope level and determining the rod interval; the rod interval is then multiplied by 100 to get the horizontal distance.

$$D = 100S \qquad (7.1)$$

Elevations can be determined by stadia in the manner illustrated in Figure 7.5c. The elevation of the instrument station (A) is usually determined using a level and rod. When the transit is set up on the station in preparation for a stadia survey, the height of the optical center of the instrument above the top of stake (HI) is measured with a tape and noted in the field book. A rod reading can then be taken on the rod with the telescope level. The elevation of the point (B) where the rod is being held is

Elevation of station A ($\curlywedge$) + HI − RR =
$$\text{elevation of point } B \text{ (rod)} \qquad (7.2)$$

Note: HI in stadia work is the vertical distance from the station to the optical center of the transit, whereas in leveling work HI is the elevation of the line of sight through the level.

The optical center of the transit is at the center of the telescope at the horizontal axis. The exact point is marked with a cross, colored dot, or screw. Measuring the HI (height of instrument) with a tape is not exact because the tape must be bent over the circle assembly when measuring; however, the error encountered using this method will not significantly affect the stadia results.

Figure 7.5d illustrates that the location of point B (rod) can be tied in by angle to a reference base line X–A–Y. The plan location of point B can now be plotted using the angle from the reference line and the distance as determined by Eq. (7.1). The elevation of point B can also be plotted, either as a spot elevation or as a component of a series of contours. It can be seen that the stadia method permits the surveyor to determine the three-dimensional location of any point with just one set of observations.

7.6 INCLINED STADIA MEASUREMENTS

The discussion thus far has assumed that the stadia observations were taken with the telescope level; however, the stadia method is particularly well suited for the inclined measurements required by rolling topography.

When the telescope is inclined up or down, the computations must be modified to account for the effects of the sloped sighting. Inclined sights require consideration in two areas: (1) the distance from the instrument to the rod must be reduced from slope to horizontal, and (2) the rod interval of a sloped sighting must be reduced to what the interval would have been if the line of sight had been perpendicular to the rod.

Figure 7.6 illustrates these two considerations. The value of HI and the rod reading (RR) have been made equal to clarify the sketch. The geometric relationships are as follows: (1) S is the rod interval when the line of sight is horizontal, and (2) S' is the rod interval when the line of sight is inclined by angle θ.

$D = 100\ S$	(7.1)	Figure 7.5b
$S = S' \cos \theta$	(7.3)	Figure 7.6
$D = 100\ S' \cos \theta$	(7.4)	From Eqs. (7.3) and (7.1)
$H = D \cos \theta$	(7.5)	Figure 7.6
$H = 100\ S' \cos^2 \theta$	(7.6)	From Eqs. (7.4) and (7.5)
$V = D \sin \theta$	(7.7)	Figure 7.6
$D = 100\ S' \cos \theta$	(7.4)	
$V = 100\ S' \cos \theta \sin \theta$	(7.8)	From Eqs. (7.7) and (7.4)

Equations (7.6) and (7.8) can be used in computing the horizontal distance and difference in elevation for any inclined stadia measurement.

In the past, special slide rules and/or tables were used to compute H and V. However, with the universal use of hand-held calculators, slide rules and stadia tables have become less popular. The computations can be accomplished just as quickly working with Eqs. (7.6) and (7.8). Stadia reduction tables are given in Table 7.2. The table is entered at the value of the VCR (vertical circle reading) with the horizontal distance factor and the difference in elevation factor both being multiplied by the rod interval to give H and V. See Examples 7.1 and 7.2.

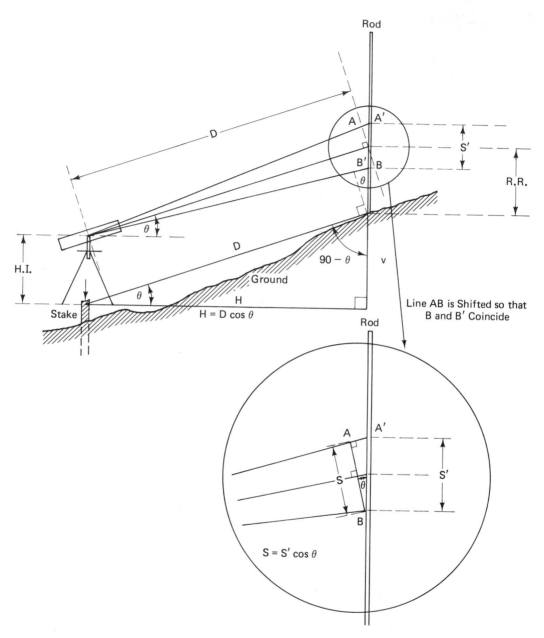

FIGURE 7.6 Geometry of an inclined stadia measurement.

246

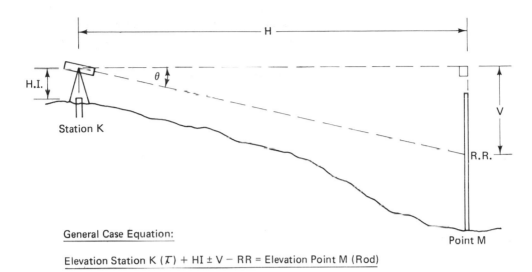

General Case Equation:

Elevation Station K ($\mathcal{T}$) + HI ± V − RR = Elevation Point M (Rod)

FIGURE 7.7 General case of an inclined stadia measurement.

Figure 7.7 shows the general case of an inclined stadia measurement, which can be stated as follows:

$$\text{Elevation } (\pi) \text{ station } K + \text{HI} \pm V - \text{RR} =$$
$$\text{elevation (rod) point } M \qquad (7.9)$$

The relationship is valid for every stadia measurement. If the HI and RR are equal, Eq. (7.9) becomes

$$\text{Elevation } (\pi) \text{ station } K \pm V = \text{elevation (rod) point } M \quad (7.10)$$

as the HI and RR cancel out each other. In practice, the surveyor will read the rod at the value of the HI unless that value is obscured (e.g., tree branch, vehicle, rise of land). If the HI value cannot be sighted, usually a value an even foot (decimeter) above or below is sighted, allowing for a mental correction to the calculation. Of course, if an even foot (decimeter) above or below the desired value cannot be read, *any value* can be read and Eq. (7.9) is used.

7.7 EXAMPLES OF STADIA MEASUREMENTS

There are three basic variations to a standard stadia measurement:

1. The rod reading is taken to be the same as the HI.
2. The rod reading is not the same as the HI.
3. The telescope is horizontal.

Example 7.1

This example, where the rod reading has been made to coincide with the value of the HI, is typical of 90% of all stadia measurements (see Figures 7.8 and 7.11). Here the VCR (vertical circle reading) is + 1°36′ and the rod interval is 0.401. Both the HI and the rod reading are 1.72 m.

From Eq. (7.6),

$$H = 100 \ S' \cos^2 \theta$$
$$= 100 \times 0.401 \times \cos^2 1°36′$$
$$= 40.1 \text{ m}$$

From Eq. (7.8),

$$V = 100 \ S' \cos \theta \sin \theta$$
$$= 100 \times 0.401 \times \cos 1°36′ \times \sin 1°36′$$
$$= +1.12 \text{ m}$$

From Eq. (7.10)

$$\text{Elev. } (\pi) \pm V = \text{elev. (rod)}$$
$$185.16 + 1.12 = 186.28$$
$$\text{See STA. } 0+00 \text{ on Figure } 7.11$$

From Table 7.2, VCR = + 1°36′, rod interval = 0.401
H = 99.92 × 0.401 = 40.1 m.
V = 2.79 × 0.401 = +1.12 m

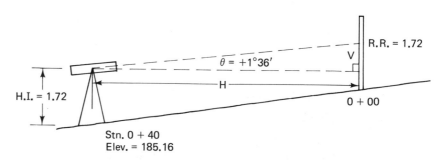

FIGURE 7.8   Example 7.1: RR = HI.

Example 7.2

This example illustrates the case where the value of the HI cannot be seen on the rod due to some obstruction (see Figures 7.9 and 7.11). In this case a rod reading of 2.72 with a vertical angle of $-6°37'$ was booked, along with the HI of 1.72 and a rod interval of 0.241.

From Eq. (7.6),

$$H = 100\ S'\cos^2\theta$$
$$= 100 \times 0.241 \times \cos^2 6°37'$$
$$= 23.8 \text{ m}$$

From Eq. (7.8),

$$V = 100\ S'\cos\theta\sin\theta$$
$$= 100 \times 0.241 \times \cos 6°37' \times \sin 6°37'$$
$$= -2.76 \text{ m}$$

From Eq. (7.9),

$$\text{Elev. (\Lambda) + HI} \pm V - \text{RR} = \text{elev. (rod)}$$
$$185.16 + 1.72 - 2.76 - 2.72 = 181.40$$
$$\text{See STA. three on Figure 7.11}$$

From Table 7.2, VCR $= -6°37'$, rod interval $= 0.241$
H $= 98.675$ (interpolated) $\times 0.241 = 23.8$ m.
V $= 11.445$ (interpolated) $\times 0.241 = -2.76$ m

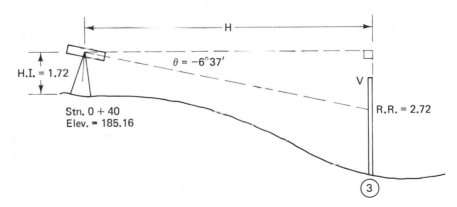

FIGURE 7.9   Example 7.2: RR $\neq$ HI.

TABLE 7.2 *Stadia Tables*

EXAMPLE: VCR = −3°21'
Rod interval = 0.123 m
From tables: V = 5.83 × 0.123 = 0.72 m
H = 99.66 × 0.123 = 12.3 m

Minutes	0° Hor. Dist.	0° Diff. Elev.	1° Hor. Dist.	1° Diff. Elev.	2° Hor. Dist.	2° Diff. Elev.	3° Hor. Dist.	3° Diff. Elev.
0	100.00	.00	99.97	1.74	99.88	3.49	99.73	5.23
2	100.00	.06	99.97	1.80	99.87	3.55	99.72	5.28
4	100.00	.12	99.97	1.86	99.87	3.60	99.71	5.34
6	100.00	.17	99.96	1.92	99.87	3.66	99.71	5.40
8	100.00	.23	99.96	1.98	99.86	3.72	99.70	5.46
10	100.00	.29	99.96	2.04	99.86	3.78	99.69	5.52
12	100.00	.35	99.96	2.09	99.85	3.84	99.69	5.57
14	100.00	.41	99.95	2.15	99.85	3.89	99.68	5.63
16	100.00	.47	99.95	2.21	99.84	3.95	99.68	5.69
18	100.00	.52	99.95	2.27	99.84	4.01	99.67	5.75
20	100.00	.58	99.95	2.33	99.83	4.07	99.66	5.80
22	100.00	.64	99.94	2.38	99.83	4.13	99.66	5.86
24	100.00	.70	99.94	2.44	99.82	4.18	99.65	5.92
26	99.99	.76	99.94	2.50	99.82	4.24	99.64	5.98
28	99.99	.81	99.93	2.56	99.81	4.30	99.63	6.04
30	99.99	.87	99.93	2.62	99.81	4.36	99.63	6.09
32	99.99	.93	99.93	2.67	99.80	4.42	99.62	6.15
34	99.99	.99	99.93	2.73	99.80	4.47	99.61	6.21
36	99.99	1.05	99.92	2.79	99.79	4.53	99.61	6.27
38	99.99	1.11	99.92	2.85	99.79	4.59	99.60	6.32
40	99.99	1.16	99.92	2.91	99.78	4.65	99.59	6.38
42	99.99	1.22	99.91	2.97	99.78	4.71	99.58	6.44
44	99.98	1.28	99.91	3.02	99.77	4.76	99.58	6.50
46	99.98	1.34	99.90	3.08	99.77	4.82	99.57	6.56
48	99.98	1.40	99.90	3.14	99.76	4.88	99.56	6.61
50	99.98	1.45	99.90	3.20	99.76	4.94	99.55	6.67
52	99.98	1.51	99.89	3.26	99.75	4.99	99.55	6.73
54	99.98	1.57	99.89	3.31	99.74	5.05	99.54	6.79
56	99.97	1.63	99.89	3.37	99.74	5.11	99.53	6.84
58	99.97	1.69	99.88	3.43	99.74	5.17	99.52	6.90
60	99.97	1.74	99.88	3.49	99.73	5.23	99.51	6.96

Minutes	4° Hor. Dist.	4° Diff. Elev.	5° Hor. Dist.	5° Diff. Elev.	6° Hor. Dist.	6° Diff. Elev.	7° Hor. Dist.	7° Diff. Elev.
0	99.51	6.96	99.24	8.68	98.91	10.40	98.51	12.10
2	99.51	7.02	99.23	8.74	98.90	10.45	98.50	12.15
4	99.50	7.07	99.22	8.80	98.88	10.51	98.49	12.21
6	99.49	7.13	99.21	8.85	98.87	10.57	98.47	12.27
8	99.48	7.19	99.20	8.91	98.86	10.62	98.46	12.32
10	99.47	7.25	99.19	8.97	98.85	10.68	98.44	12.38
12	99.46	7.30	99.18	9.03	98.83	10.74	98.43	12.43
14	99.46	7.36	99.17	9.08	98.82	10.79	98.41	12.49
16	99.45	7.42	99.16	9.14	98.81	10.85	98.40	12.55
18	99.44	7.48	99.15	9.20	98.80	10.91	98.39	12.60
20	99.43	7.53	99.14	9.25	98.78	10.96	98.37	12.66
22	99.42	7.59	99.13	9.31	98.77	11.02	98.36	12.72
24	99.41	7.65	99.11	9.37	98.76	11.08	98.34	12.77
26	99.40	7.71	99.10	9.43	98.74	11.13	98.33	12.83
28	99.39	7.76	99.09	9.48	98.73	11.19	98.31	12.88
30	99.38	7.82	99.08	9.54	98.72	11.25	98.30	12.94
32	99.38	7.88	99.07	9.60	98.71	11.30	98.28	13.00
34	99.37	7.94	99.06	9.65	98.69	11.36	98.27	13.05
36	99.36	7.99	99.05	9.71	98.68	11.42	98.25	13.11
38	99.35	8.05	99.04	9.77	98.67	11.47	98.24	13.17
40	99.34	8.11	99.03	9.83	98.65	11.53	98.22	13.22
42	99.33	8.17	99.01	9.88	98.64	11.59	98.20	13.28
44	99.32	8.22	98.99	9.94	98.63	11.64	98.19	13.33
46	99.31	8.28	98.99	10.00	98.61	11.70	98.17	13.39
48	99.30	8.34	98.98	10.05	98.60	11.76	98.16	13.45
50	99.29	8.40	98.97	10.11	98.58	11.81	98.14	13.50
52	99.28	8.45	98.96	10.17	98.57	11.87	98.13	13.56
54	99.27	8.51	98.94	10.22	98.56	11.93	98.11	13.61
56	99.26	8.57	98.93	10.28	98.54	11.98	98.10	13.67
58	99.25	8.63	98.92	10.34	98.53	12.04	98.08	13.73
60	99.24	8.68	98.91	10.40	98.51	12.10	98.06	13.78

TABLE 7.2 Continued

Minutes	8° Hor. Dist.	8° Diff. Elev.	9° Hor. Dist.	9° Diff. Elev.	10° Hor. Dist.	10° Diff. Elev.	11° Hor. Dist.	11° Diff. Elev.
0	98.06	13.78	97.55	15.45	96.98	17.10	96.36	18.73
2	98.05	13.84	97.53	15.51	96.96	17.16	96.34	18.78
4	98.03	13.89	97.52	15.56	96.94	17.21	96.32	18.84
6	98.01	13.95	97.50	15.62	96.92	17.26	96.29	18.89
8	98.00	14.01	97.48	15.67	96.90	17.32	96.27	18.95
10	97.98	14.06	97.46	15.73	96.88	17.37	96.25	19.00
12	97.97	14.12	97.44	15.78	96.86	17.43	96.23	19.05
14	97.95	14.17	97.43	15.84	96.84	17.48	96.21	19.11
16	97.93	14.23	97.41	15.89	96.82	17.54	96.18	19.16
18	97.92	14.28	97.39	15.95	96.80	17.59	96.16	19.21
20	97.90	14.34	97.37	16.00	96.78	17.65	96.14	19.27
22	97.88	14.40	97.35	16.06	96.76	17.70	96.12	19.32
24	97.87	14.45	97.33	16.11	96.74	17.76	96.09	19.38
26	97.85	14.51	97.31	16.17	96.72	17.81	96.07	19.43
28	97.83	14.56	97.29	16.22	96.70	17.86	96.05	19.48
30	97.82	14.62	97.28	16.28	96.68	17.92	96.03	19.54
32	97.80	14.67	97.26	16.33	96.66	17.97	96.00	19.59
34	97.78	14.73	97.24	16.39	96.64	18.03	95.98	19.64
36	97.76	14.79	97.22	16.44	96.62	18.08	95.96	19.70
38	97.75	14.84	97.20	16.50	96.60	18.14	95.93	19.75
40	97.73	14.90	97.18	16.55	96.57	18.19	95.91	19.80
42	97.71	14.95	97.16	16.61	96.55	18.24	95.89	19.86
44	97.69	15.01	97.14	16.66	96.53	18.30	95.86	19.91
46	97.68	15.06	97.12	16.72	96.51	18.35	95.84	19.96
48	97.66	15.12	97.10	16.77	96.49	18.41	95.82	20.02
50	97.64	15.17	97.08	16.83	96.47	18.46	95.79	20.07
52	97.62	15.23	97.06	16.88	96.45	18.51	95.77	20.12
54	97.61	15.28	97.04	16.94	96.42	18.57	95.75	20.18
56	97.59	15.34	97.02	16.99	96.40	18.62	95.72	20.23
58	97.57	15.40	97.00	17.05	96.38	18.68	95.70	20.28
60	97.55	15.45	96.98	17.10	96.36	18.73	95.68	20.34

Minutes	12° Hor. Dist.	12° Diff. Elev.	13° Hor. Dist.	13° Diff. Elev.	14° Hor. Dist.	14° Diff. Elev.	15° Hor. Dist.	15° Diff. Elev.
0	95.68	20.34	94.94	21.92	94.15	23.47	93.30	25.00
2	95.65	20.39	94.91	21.97	94.12	23.52	93.27	25.05
4	95.63	20.44	94.89	22.02	94.09	23.58	93.24	25.10
6	95.61	20.50	94.86	22.08	94.07	23.63	93.21	25.15
8	95.58	20.55	94.84	22.13	94.04	23.68	93.18	25.20
10	95.56	20.60	94.81	22.18	94.01	23.73	93.16	25.25
12	95.53	20.66	94.79	22.23	93.98	23.78	93.13	25.30
14	95.51	20.71	94.76	22.28	93.95	23.83	93.10	25.35
16	95.49	20.76	94.73	22.34	93.93	23.88	93.07	25.40
18	95.46	20.81	94.71	22.39	93.90	23.93	93.04	25.45
20	95.44	20.87	94.68	22.44	93.87	23.99	93.01	25.50
22	95.41	20.92	94.66	22.49	93.84	24.04	92.98	25.55
24	95.39	20.97	94.63	22.54	93.82	24.09	92.95	25.60
26	95.36	21.03	94.60	22.60	93.79	24.14	92.92	25.65
28	95.34	21.08	94.58	22.65	93.76	24.19	92.89	25.70
30	95.32	21.13	94.55	22.70	93.73	24.24	92.86	25.75
32	95.29	21.18	94.52	22.75	93.70	24.29	92.83	25.80
34	95.27	21.24	94.50	22.80	93.67	24.34	92.80	25.85
36	95.24	21.29	94.47	22.85	93.65	24.39	92.77	25.90
38	95.22	21.34	94.44	22.91	93.62	24.44	92.74	25.95
40	95.19	21.39	94.42	22.96	93.59	24.49	92.71	26.00
42	95.17	21.45	94.39	23.01	93.56	24.55	92.68	26.05
44	95.14	21.50	94.36	23.06	93.53	24.60	92.65	26.10
46	95.12	21.55	94.34	23.11	93.50	24.65	92.62	26.15
48	95.09	21.60	94.31	23.16	93.47	24.70	92.59	26.20
50	95.07	21.66	94.28	23.22	93.45	24.75	92.56	26.25
52	95.04	21.71	94.26	23.27	93.42	24.80	92.53	26.30
54	95.02	21.76	94.23	23.32	93.39	24.85	92.49	26.35
56	94.99	21.81	94.20	23.37	93.36	24.90	92.46	26.40
58	94.97	21.87	94.17	23.42	93.33	24.95	92.43	26.45
60	94.94	21.92	94.15	23.47	93.30	25.00	92.40	26.50

Example 7.3

This example illustrates the situation where the ground is level enough to permit horizontal rod sightings (see Figures 7.10 and 7.11). The computations for this observation are quite simple; the horizontal distance is simply 100 times the rod interval [$D = 100S$, Eq. (7.1)]. Since there is no vertical angle, there is no triangle to solve (i.e., $V = 0$). The difference in elevation is simply $+\text{HI} - \text{RR}$.

If the survey is in a level area where many observations can be taken with the telescope level, this technique will speed up the survey computations and the field time. (Vertical angles are not read.) However, if the survey is in typical rolling topography, the surveyor will not normally spend the time necessary to see if a single horizontal observation can be made; the surveyor will instead continue sighting the rod at the HI value to maintain the momentum of the survey (a good instrumentman can keep two rodmen busy).

In this example, a rod interval of 0.208 was booked, together with the HI of 1.72 and a rod reading of 2.06. From Eq. (7.1),

$$D = 100S$$
$$= 100 \times 0.208$$
$$= 20.8 \text{ m} \quad \text{(horizontal distance)}$$

From Eq. (7.9),

$$\text{Elev. } (\bar{\wedge}) + \text{HI} + V - \text{RR} = \text{elev. (rod)}$$
$$185.16 + 1.72 + 0 - 2.06 = 184.82$$
$$\text{See STA. four on Figure 7.11}$$

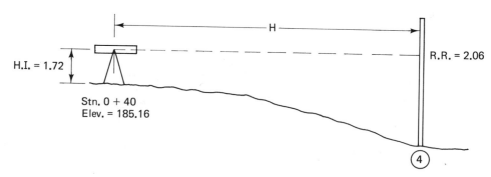

FIGURE 7.10   Example 7.3, telescope horizontal.

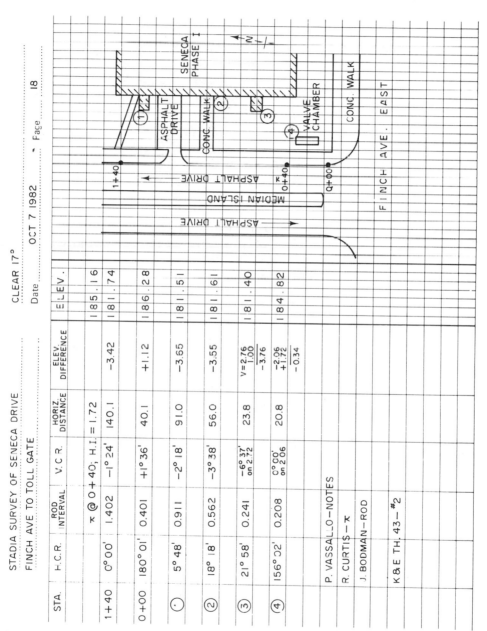

FIGURE 7.11 Stadia field notes.

253

7.8 PRECISION OF STADIA MEASUREMENTS

Normal field practice permits accurate rod readings of 0.01 ft or 0.003 m for distances of 300 ft or 100 m. If rod intervals are read accordingly, horizontal distances (100S) can be computed to the closest 1 ft or 0.3 m. This results in a maximum relative accuracy of $1/300$ to $1/400$ (400 ft being the longest sight for which realistically accurate rod intervals can be determined).

Consistent reasoning (including recognition of the effect of significant figures) indicates that differences in elevation (V) can be realistically computed to the closest 0.1 ft or 0.03 m. (The established field practice of showing V to the closest 0.01 m causes little harm.)

If the length of horizontal distances can only be given to the closest 1 ft (i.e., ±0.5 ft), it follows that angles need only be read to a sufficient level of precision to give the desired results. In 300 ft, a lateral uncertainty of 0.5 ft permits an angular precision to the closest $5\frac{1}{2}$ minutes (i.e., tan θ = 0.5′/300′, θ = 0°05.7′).

In 300 ft, a vertical angle uncertainty of 01 minutes in a 2°30′ slope permits differences in elevation to be computed to the closest 0.1 ft (0.03 m).

Established field practice dictates that both the HCR (horizontal circle reading) and the VCR (vertical circle reading) be read and booked to the closest 1 minute of arc. Although the HCR need only be booked to the closest 5 minutes, it takes no longer to read to the closest minute and does permit a consistency of field operation. Reading a stadia angle to seconds is a waste of time.

7.9 STADIA FIELD PRACTICE

In stadia work, the transit is set on a point for which the horizontal location and elevation have been determined. If necessary, the elevation of the transit station can be determined after setup by sighting on a point of known elevation and working backward through Eq. (7.9).

The horizontal circle is zeroed and a sight is taken on another control point (1 + 40 in Figure 7.11). All stadia pointings are accomplished by working on the circle utilizing the upper clamp and upper tangent screw. It is a good idea to periodically check the zero setting by sighting back on the original backsight; this will ensure that the setting has not been invalidated by inadvertent use of the lower clamp or lower tangent screw. At a bare minimum, a zero check is made just before the instrument is moved from the transit station; if the check proves that the setting has been inadvertently moved from zero, all the work must be repeated.

Before any observations are made, the HI is measured with a steel tape (sometimes with the leveling rod) and the value booked, as shown in Figure 7.11.

The actual observation proceeds as follows: with the upper clamp loosened, the rod is sighted; precise setting can be accomplished by the upper tangent screw after the upper clamp has been locked. The main cross hair is sighted approximately to the value of the HI, and then the telescope is revolved up or down until the lower stadia hair is on the closest even foot (decimeter) mark (see Figure 7.12). The upper stadia hair is then read and the rod interval determined mentally by simply subtracting the lower hair reading from the upper hair reading. After the rod interval is booked (see Figure 7.11), the main cross hair is then moved to read the value of the HI on the rod. When this has been accomplished, the rodman is waved off, and while the rodman is walking to the next point, the instrumentman reads and books the VCR and the HCR (see Figure 7.11). Usually, the calculations for horizontal distance and elevation are performed after field hours.

The technique of temporarily moving the lower stadia hair to an even value to facilitate the determination of the rod interval introduces errors in the readings, but these errors are not large enough to significantly affect the results. The alternative to this technique would be to initially lock the main hair on the value of the HI and then read and book the upper and lower hair readings. When the lower hair is subtracted from the upper hair, the result can then be booked as the rod interval. This alternative technique is more precise, but it is far too cumbersome and time-consuming for general use.

If the value of the HI cannot be seen on the rod, another value (e.g., even foot or decimeter above or below the HI) can be sighted and that value is booked along with the vertical angle in the VCR column (see Figure 7.11, STA. three).

If the telescope is level, the rod reading is booked in the VCR column alone or together with 0°00′ (see Figure 7.11, STA. four).

It sometimes happens that the entire rod interval cannot be seen on the rod (e.g., tree branches, intervening ground, extra-long shots); in this case, half the rod interval can be read and that val-

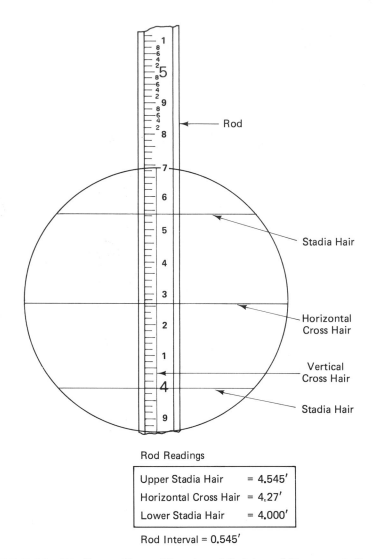

FIGURE 7.12 Stadia readings. (Courtesy Ministry of Transportation and Communications—Ontario)

ue doubled and then entered in the rod interval column. Reading only half the rod interval reduces the precision considerably, so extra care should be taken when determining the half-interval.

Generally, if the relative accuracy ratio of $1/300$ to $1/400$ is to be maintained on long sights and on steeply inclined sights, extra care is required (particularly in reading the rod, plumbing the rod, etc.).

Two areas that seem to contribute the most to the occurrence of mistakes in stadia work are as follows:

1. Recording the vertical angle incorrectly. With American transits, the mistake is simply switching plus for minus, and vice versa. With European-style theodolites that directly give the zenith angle, a mistake can be made in converting from the zenith angle to its complement, the vertical angle. That is, a zenith angle of 88°29′ is a vertical angle of +1°31′ or a zenith angle of 92°30′ is a vertical angle of −2°30′. To help reduce mistakes, some surveyors prefer to book the zenith angle instead of the vertical angles, performing the angle conversions later when the notes are being reduced.

2. Recording the observation location incorrectly. This type of mistake occurs when the rodman is some distance from the instrument and confusion exists due to poor communications, or the mistake could occur because of an erroneous or sloppy sketch. Detail locations are shown (Figure 7.11) as numbered entries directly on an unambiguous sketch; in the case of cross sections or other well-ordered surveys, the detail locations can be clearly described by chainage and offset.

7.10 ESTABLISHING CONTROL BY STADIA METHODS

Stadia methods can be used to establish secondary control points or even to establish closed traverses that will be used for topographic stadia control. The technique essentially consists of taking stadia observations from both ends of each line and then averaging the results to obtain horizontal distances and differences in elevation. The double readings provide an increase in precision, which permits stations so established to be used as control for further stadia work.

Figure 7.13 illustrates an extension of primary control to the secondary control point K.

With the transit at $0 + 40$, a horizontal angle is turned (and doubled) on to point K. The HI, VCR and rod interval are determined in the usual manner.

Given the following field data:

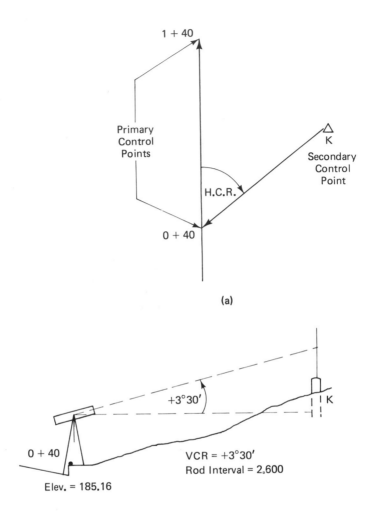

(a)

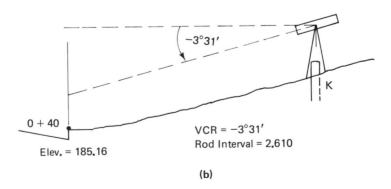

(b)

FIGURE 7.13 Establishing control by stadia: (a) plan view; (b) profile views.

⌅ at 0 + 40   VCR = + 3°30′, rod int. = 2.600, RR = HI
                Horiz. dist. by Eq. (7.6) = 259.0
                Elevation difference by Eq. (7.8) = 15.84
⌅ at K   VCR = − 3°31′, rod int. = 2.610, RR = HI
           Horiz. dist. by Eq. (7.6) = 260.0
           Elevation difference by Eq. (7.8) = 15.98

Horizontal distance 0 + 40 to K = 259.5
Difference in elevation 0 + 40 to K = +15.91
Elevation of station K = 185.16
                                      $\underline{+\,15.91}$
                                        201.07 m

A further example is provided in Figure 7.14, where control is extended to secondary station A from primary station 854 so that additional stadia observations (1 to 4) can be taken.

7.11 SELF-REDUCING STADIA THEODOLITE

The self-reducing instrument (also known as a reduction tacheometer) is a theodolite designed specifically for stadia observations. Instead of the reticle with two stadia hairs and a main horizontal hair, the self-reducing stadia theodolite comes equipped with three curved lines of varying radius. As the telescope is moved up or down, the interval between the hairs changes. For example, the interval between the main (lower) hair and the upper hair moves in such a way that 100 times the rod interval automatically gives the horizontal distance, regardless of the vertical angle.

The middle hair is also moving so that the interval between the middle hair (elevation curve) and the main hair, when multiplied by some factor, gives the difference in elevation. In some instruments the elevation factor is 100. In other instruments the elevation factor is shown in the telescopic field of view; in this case, the algebraic sign and magnitude of the elevation factor (as viewed in the telescope) change as the telescope is revolved (see Figure 7.15).

Some systems (e.g., Wild-Leitz) employ an extension to the stadia rod that permits the decimal portion of the HI value to be extended down from the zero mark of the rod. This facilitates the sighting of the HI value on the rod. For example, using this rod, if the HI were 1.201 m, the graduated extension to the rod would be

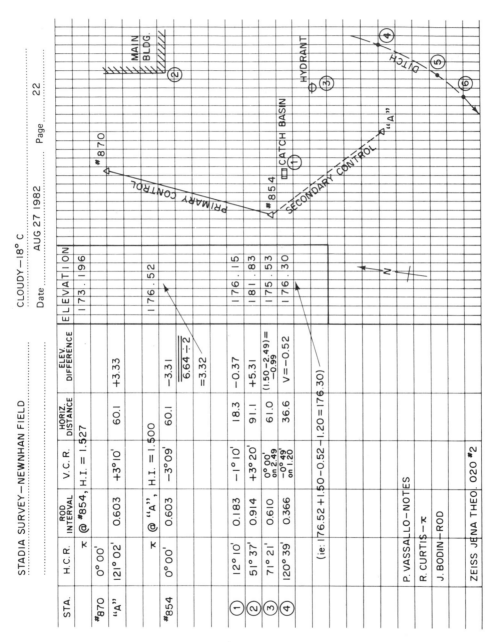

FIGURE 7.14 Stadia notes, including extension of control.

260

lowered 0.201 m, permitting the stadia surveyor to quickly sight 1.000 on the rod each time the HI value is to be sighted.

Figure 7.16 illustrates field notes for a self-reducing stadia theodolite. For comparison purposes, the detail is the same as that shown in Figure 7.14, in which the detail is being obtained by conventional stadia methods. Essentially, the advantage to the self-reducing technique is that the time-consuming after-hours computations have been eliminated.

The additional cost for a self-reducing stadia theodolite can be recouped quickly if many stadia surveys are required. When not being used for stadia work, the self-reducing stadia theodolite can be employed in the same manner as any 1-minute theodolite.

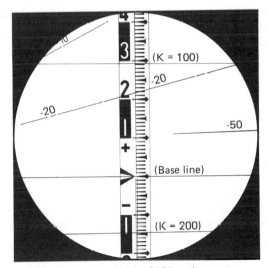

Reticule of the Dahlta 010A tacheometer.

Example:
1. Horizontal distance
   @ K = 100, interval = 0.292, distance = 29.2 m.
   @ K = 200, interval = 0.146, distance = 29.2 m.
2. Difference in elevation
   Elevation interval = 0.218
   Difference in elevation = 0.218 x (-20) = -4.36 m.

FIGURE 7.15 Self-reducing stadia theodolite rod readings. (Courtesy Zeiss JENA Instruments, Ltd.)

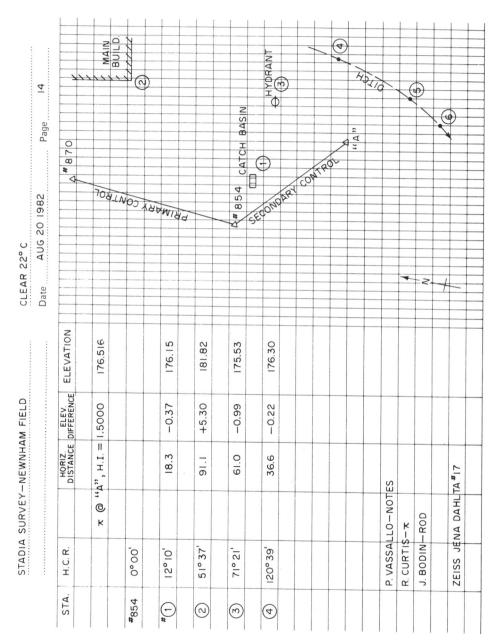

FIGURE 7.16 Stadia notes using a self-reducing stadia theodolite (see Figure 7.14).

262

7.12 SUMMARY OF STADIA FIELD PROCEDURE: CONVENTIONAL INSTRUMENT

1. Set theodolite over station.
2. Measure HI with a steel tape.
3. Set horizontal circle to zero.
4. Sight the reference station at 0°00'.
5. Sight stadia point by loosening upper clamp (lower clamp is tight).
6. Sight main horizontal hair roughly on value of HI. Then move lower hair to closest even foot (decimeter) mark.
7. Read upper hair; determine rod interval and enter that value in the notes.
8. Sight main horizontal hair precisely on the HI value.
9. Wave off the rodman.
10. Read and book the horizontal angle (HCR) and the vertical (VCR) angles.
11. Check zero setting for horizontal angle before moving the instrument.
12. Reduce the notes (compute horizontal distances and elevations) after field hours; check the reductions.

Notes

1. If the HI cannot be sighted, any point on the rod can be sighted, and that rod reading is then booked along with the vertical angle to that rod reading.
2. For European theodolites, the zenith angle may be booked instead of the vertical angle, with the angle conversion taking place with the field notes reductions.
3. For self-reducing stadia theodolites, use of an extendable-foot rod permits all main-hair sightings to be at an even foot (decimeter) and greatly speeds up this method.

PROBLEMS

7-1   The data shown below are stadia rod intervals and vertical
       angles taken to locate points on a topographic survey. The
       elevation of the transit station is 348.78 ft and all vertical
       angles were read with the cross hair on the rod at the value
       of the height of instrument (HI).

Point	Rod Interval, Ft	Vertical Angle
1	3.22	+ 0°58′
2	0.29	− 3°38′
3	1.17	− 1°30′
4	2.86	+ 0°20′
5	0.08	+ 2°41′

       Compute the horizontal distances and the elevations using
       Table 7-2.

7-2   The data shown below are stadia rod intervals and vertical
       angles taken to locate points on a watercourse survey.
       Where the HI value could not be sighted on the rod, the
       sighted rod reading is booked along with the vertical angle.
       The HI was 1.83 and the elevation of the transit station is
       174.74 m.

Point	Rod Interval, M	Vertical Angle
1	0.018	+ 2°19′
2	0.089	+ 1°57′on 1.43
3	0.626	0°00′on 2.71
4	1.238	− 2°13′
5	1.002	− 4°55′on 1.93
6	0.119	+ 0°30′

       Compute the horizontal distances and the elevations using
       Equations 7-6 and 7-8.

7-3   A transit is set up on station K with a HI of 5.36 ft. Read-
       ings are taken at station L as follows: rod interval = 1.31
       ft, cross hair at 4.27 with a vertical angle of + 3°12′. The
       elevation of station L = 486.70ft. a) What is the elevation
       of station K? b) What is distance KL?

7-4   Reduce the following set of stadia notes using either Table
       7-2 or Equations 7-6 and 7-8.

Station Point	Rod Interval, Ft	Horizontal Angle	Vertical Angle	Horizontal Distance	Elevation Difference	Elevation
⋏ @ MON. #36; HI = 5.06'						471.36
Mon. #37	3.56	0°00'	+ 3°46'			
1	2.33	2°37'	+ 2°52'			
2	1.71	8°02'	+ 1°37'			
3	1.26	27°53'	+ 2°18' on 4.06			
4	0.81	46°20'	+ 0°18'			
5	1.96	81°32'	0°00' on 8.11			
6	2.30	101°17'	− 1°38'			
Mon. #38	3.48	120°20'	− 3°41'			

7-5   Reduce the following cross-section stadia notes.

Station Point	Rod Interval, m	Horizontal Angle	Vertical Angle	Horizontal Distance	Elevation Difference	Elevation
⋏ @ CTRL. POINT "K"; H.I. = 1.82 m						143.78
CTRL.PT."L"		0°00'				
0+00 ₡	0.899	34°15'	− 18°10'			
south ditch	0.851	33°31'	− 21°08'			
north ditch	0.950	37°08'	− 19°52'			
0+50 ₡	0.622	68°17'	− 15°07'			
south ditch	0.503	64°10'	− 19°32'			
north ditch	0.687	70°48'	− 18°44'			
1+00 ₡	0.607	113°07'	− 12°51'			
south ditch	0.511	109°52'	− 15°58'			
north ditch	0.710	116°14'	− 13°47'			
1+50 ₡	0.852	139°55'	− 9°00'			
south ditch	0.800	135°11'	− 10°15'			
north ditch	0.932	144°16'	− 9°20'			
2+00 ₡	1.228	152°18'	− 5°49'			
south ditch	1.148	155°43'	− 7°05'			
north ditch	1.263	147°00'	− 6°32'			

7-6   A traverse required as control for a mapping survey was measured using stadia techniques. The elevation of Sta. A is 200.50; all vertical angle readings were taken with the cross hair set to the HI value.

Transit Station	Rod Station	Rod Interval	Vertical Angle
A	B	0.551	+ 2°12'
	E	1.137	+ 1°58'
B	A	0.551	− 2°11'
	C	0.826	+ 4°27'
C	B	0.827	− 4°25'
	D	0.733	− 2°33'
D	C	0.732	+ 2°33'
	E	0.606	− 1°44'
E	D	0.606	+ 1°46'
	A	1.137	− 1°57'

a) Determine the horizontal distances of the traverse sides.
b) determine the elevations of the traverse stations.

8 *Electronic Surveying Instruments*

8.1 GENERAL

Advances in electronics technology have resulted in rapid advances in surveying methodology. Distances are now measured electronically with electronic distance measurement instruments (EDM), angles are now being read electronically with digital theodolites, and positions are being determined by satellite receivers.

The EDM, first introduced in the late 1950s (Geodimeter) has undergone continual refinements. Although the early EDMs were known for their high precision and long-range capabilities, they were also known for their sheer bulk and relatively cumbersome operating procedures. Surveyors who have back-packed heavy EDMs long distances into the bush appreciate the compact size of the recently designed EDMs. Short-range (2 mi or 3 km) EDMs have found wide acceptance with engineering and legal surveyors who seldom require longer-range capabilities. These infrared, laser, and microwave instruments, when used in conjunction with theodolites, can provide horizontal and vertical position of one point relative to another (the slope distance provided by the EDM can be reduced to its horizontal equivalent when the slope angle is determined with the theodolite). Newer EDMs have built-in or

267

add-on calculator-processors, which can be utilized to provide slope reduction to horizontal, atmospheric, sea level, and scale corrections (see Chapter 11), and reflector constants. The electronic digitized theodolite, first introduced in the late 1960s (Zeiss-Oberkochen), set the stage for modern field data collection and processing. When the electronic theodolite is used with a built-in EDM (e.g., Zeiss Elta 3, Figure 8.1) or an interfaced EDM (e.g., K & E Vectron with Autoranger EDM, Figure 8.2), the surveyor has a very powerful instrument. Add to that instrument an onboard microprocessor that automatically records and processes field data and you have what is known as an electronic tacheometer instrument (ETI). The name "Total Station" was introduced by the Hewlett-Packard Co. to describe its H-P 3820A instrument, which had the features described previously. Hewlett-Packard discontinued its line of survey instruments in 1982, and the name *total station* is now widely used to describe any ETI having automatic data-collection capability. Some instruments automatically record the distance (EDM) value; others automatically record distance and slope angle values, with the other (HCR) data being manually entered to the microprocessor; these instruments are called *semitotal stations.*

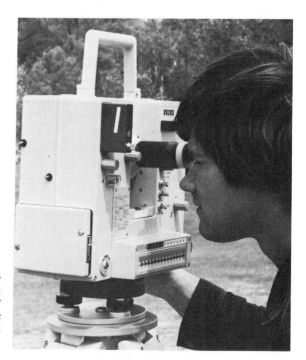

FIGURE 8.1 Total station: Zeiss ELTA-3 featuring automatic data entry; coordinate computation; angle accuracy of 02 seconds; triple prism distance range of 1.6 km; SE = ±5 mm + 2 ppm.

FIGURE 8.2 Keuffel & Esser vectron the-odolite, shown with the autoranger EDM and the vectron field computer and data collector.

8.2 ELECTROMAGNETIC MEASUREMENT

EDM instruments utilize either microwaves (radio waves) or light waves (including laser and infrared). The microwave systems require a transmitter-receiver at both ends of the measured line, whereas light-wave systems use a transmitter at one end and a reflector at the other end. Microwave systems are often used in hydrographic surveying and have a usual upper limit of 60 km in measuring range. Although microwave systems can be used in poorer weather conditions (fog, rain, etc.) than light-wave systems, the uncertainties caused by varying humidity conditions over the line result in lower accuracy expectations.

Long-range land measurements can be taken with laser instruments that have a distance capability of up to 60 km at nighttime and a somewhat reduced capability during daylight hours. The power source can be low powered, as with the helium neon laser; although low-powered lasers present little or no danger to the operator or to the public, care should be taken to ensure that the laser beam does not directly strike the eye.

Most short-range (0 to 3000 m) instruments now in use are infrared instruments (gallium arsenide diode). These instruments are lightweight and relatively inexpensive ($4000 to $8000). There are, however, several short-range laser EDMs that are price competitive. Laser instruments have one advantage over infrared in that the beam is visible; the visibility of the beam can be helpful in some difficult sighting situations.

8.3 PRINCIPLES OF EDM MEASUREMENT

Figure 8.3 shows a wave of wavelength λ. The wave is traveling along the x axis with a velocity of 299 792.5 5±0.4 km/s (in vacuum). The frequency of the wave is the time taken for one complete wavelength.

$$\lambda = \frac{v}{f} \qquad (8.1)$$

where λ = wavelength in meters
v = velocity of km/s
f = frequency in hertz (one cycle per second)

Figure 8.4 shows the electromagnetic spectrum with wave frequencies shown in kilohertz (kilo = 1000) along with relevant wavelengths.

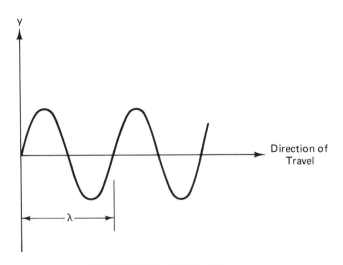

FIGURE 8.3 Lightwave.

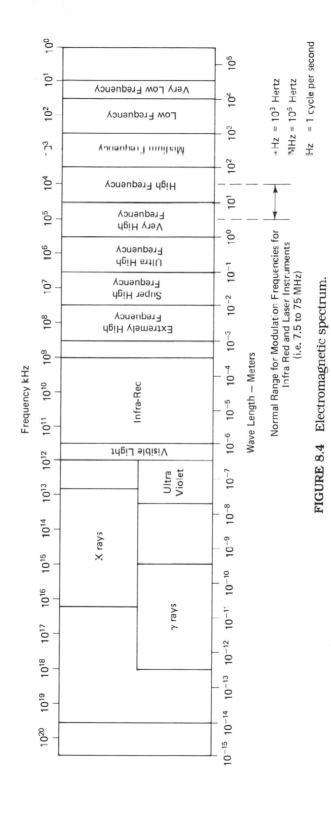

FIGURE 8.4 Electromagnetic spectrum.

Figure 8.5 shows the modulated electromagnetic wave leaving the EDM and being reflected (light waves) or retransmitted (microwaves) back to the EDM. It can be seen that the double distance (2L) is equal to a whole number of wavelengths ($n\,\lambda$), plus the partial wavelength ($\bar\phi$) occurring at the EDM

$$L = \frac{n\,\lambda + \phi}{2} \text{ meters} \tag{8.2}$$

The partial wavelength (ϕ) is determined in the instrument by noting the phase delay required to precisely match up the transmitted and reflected or retransmitted waves. The instrument (e.g., Wild Distomat) can count the number of full wavelengths ($n\lambda$), or, instead, the instrument can send out a series (three or four) of modulated waves at different frequencies. (The frequency is typically reduced each time by a factor of 10, and of course the wavelength is increased each time also by a factor of 10.) By substituting the resulting values of λ and ϕ into Eq. (8.2), the value of n can be found. The instruments are designed to carry out this procedure in a matter of seconds and then to display the value of L in digital form.

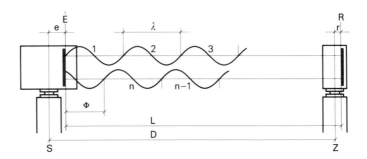

S   Station
Z   Target
E   Reference plane within the distance meter for phase comparison between transmitted and received wave
R   Reference plane for the reflection of the wave transmitted by the distance meter
a   Addition constant
e   Distance meter component of addition constant
r   Reflector component of addition constant
λ Modulation wave length
φ Fraction to be measured of a whole wave length of modulation $(\Delta \lambda)$

The addition constant a applies to a measuring equipment consisting of distance meter and reflector. The components e and r are only auxiliary quantities.

FIGURE 8.5 Principles of EDM measurement. (Courtesy Kern Instruments, Inc.)

8.4 ATMOSPHERIC CORRECTIONS

The velocity (299 792.5 km/s) of light shown in the previous section is reduced somewhat due to atmospheric conditions.

$$V = \frac{V_0}{n} \tag{8.3}$$

where V = velocity of light in the atmosphere
V_0 = velocity of light in vacuum
n = refractive index (> 1)

Most instruments utilize crystals oscillating at fixed frequencies in order to produce the measurement waves. Since the atmospheric velocity V is a variable, the wavelength λ must also vary:

$$F = \frac{V}{\lambda}, \quad \text{a fixed constant} \qquad \text{(from 8.1)}$$

The value of the refractive index n varies with the following:

1. Temperature
2. Pressure
3. Water vapor content
4. Wavelength (light-wave instruments)

The refractive index for standard air conditions (0°C, 760 mm Hg, and 0.03 CO_2) for the group velocity of light waves is given by the Barrel and Sears formula:

$$n_s = 1 + \left(287.604 + \frac{4.8864}{\lambda^2} + \frac{0.068}{\lambda^4} \right) 10^{-6} \tag{8.4}$$

where λ is the wavelength of the carrier light wave being used in micrometers (μ). Some typical values for λ are given:

Carrier	$\lambda (\mu m)$
Mercury vapor	0.5500
Incandescent	0.5650
Red laser	0.6328
Infrared	0.8600 to 0.9300

For nonstandard air conditions (the norm), the correction formula for light waves becomes

$$n = 1 + \frac{0.359474(n_s - 1)p}{273.2 + t} - \frac{1.5026e \times 10^{-5}}{273.2 + t} \tag{8.5}$$

where p = atmospheric pressure in mm of mercury (torr)
t = temperature in degrees Celsius
e = vapor pressure in torr

Since water vapor has little effect on light waves, the second term, which contains e, is usually omitted. The working formula then becomes

$$n = 1 + \frac{0.35947(n_s - 1)p}{273.2 + t} \qquad (8.6)$$

The working formula for microwaves, where the vapor pressure exerts a significant effect, is as follows:

$$(n - 1)10^6 = \frac{103.49}{273.2 + t}(p - e) + \frac{86.26}{273 + t}\left(1 + \frac{5748}{273 + t}\right)^e \quad (8.7)$$

This is a modified form of the Essen and Froome formula for the refractive index of the atmosphere, which was provisionally adopted by the International Association of Geodesy in 1960.

In practice, the foregoing formulas are seldom used; most EDMs now have the capability of reducing the atmospheric corrections after the parameters (t, P, and e) have been entered directly into the instrument. Older versions of EDMs come equipped with nomographs and the like, which the manufacturer supplies, to facilitate the application of atmospheric corrections (see Figure 8.6). The wise surveyor will, however, check this operation of the instrument at regular intervals to ensure that the instrument is functioning correctly.

For short distances using light-wave EDMs, atmospheric corrections have a relatively small significance. For long distances using light-wave instruments and particularly for microwave instruments, atmospheric corrections can become quite important. The following chart shows the comparative effects of the atmosphere on both light waves and microwaves.

		Error (parts per million)	
Parameter	Error	Light Wave	Microwave
t, temperature	+1°C	−1.0	−1.25
p, pressure	+1 mm Hg	+0.4	+0.4
e, partial water-vapor pressure	1 mm Hg	−0.05	+ 7 at 20°C
			+17 at 45°C

At this point it is also worth noting that several studies of general EDM use show that more than 90% of all distance determinations involve distances of 1000 m or less, and more than

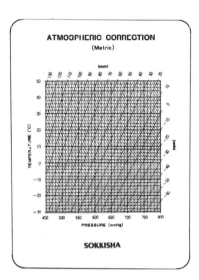

FIGURE 8.6 Atmospheric correction graph. (Courtesy Sokkisha Co. Ltd.)

95% of all layout measurements involve distances of 400 m or less. The values in the preceding chart would seem to indicate that, for the type of measurements normally encountered in the civil field, instrumental errors and centering errors hold more significance than do the atmosphere-related errors.

8.5 EDM TYPES

EDMs are designed for tribrach mount, instrument mount, or are actually built into the theodolite (see the illustrations). Since the technology is improving so rapidly, no attempt will be made here to describe any of the instruments in detail. Instrument manufacturers will gladly provide this information.

Essentially, there are now two types of EDM. The first type measures only the slope distance; if this instrument is mounted on a theodolite, a vertical angle can also be measured so that the horizontal distance can be quickly computed. Some instruments make provision for manual entry of the vertical angle into the instrument so that an onboard calculator can compute both the horizontal and vertical distances (see Figures 8.7 and 8.8).

The other type of instrument, the electronic tacheometer instrument (Figure 8.9), has provision for automatic recording of the horizontal angle, the vertical angle, and the slope distance. A touch of a button can also provide the horizontal or vertical dis-

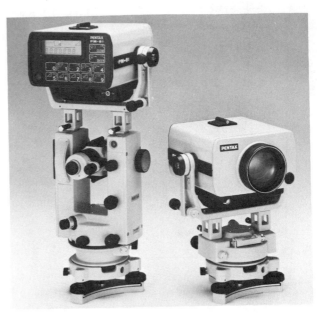

FIGURE 8.7 Pentax PM 81 EDM mounted on a 6-second Pentax theodolite and also shown tribrach mounted. EDM has a triple prism range of 2 km (6600 ft) with a SE = ±(5 mm + 5 ppm). (Courtesy Pentax Corporation—Colorado)

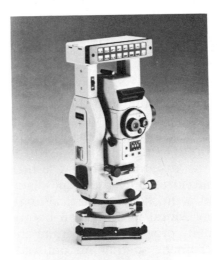

FIGURE 8.8 SDM 3E "total station": built-in calculator permits entry of distance and angle data for horizontal distance computations. Angles read directly to 10 seconds, distance range of up to 1800 m with a triple prism. Both distance and angle measurement accomplished with one target sighting. (Courtesy Sokkisha Co. Ltd.)

FIGURE 8.9   Omni I total station featuring automatic acquisition of data, generation of *X*, *Y*, and *Z* coordinates, and direct off-load to computer. (Courtesy Alpha Electronics, Littleton, Colorado)

tance, station coordinates, and so on, and all the data can be temporarily held in storage and then electronically transferred to an interfaced computer back at the office; or the data can be transmitted over telephone lines at any hour of the day.

8.6 RANGE OF EDM MEASUREMENTS

EDMs are designed for short-range (up to 3 km), long-range (up to 60 km), and intermediate distances. The measurement capability of any specific electro-optical instrument can be affected by the quality and number of prisms; and both electro-optical and microwave instruments are affected by atmospheric factors.

Generally, the range capability of an electro-optical EDM can be doubled by squaring the number of prisms used. That is, if a cluster of nine prisms were used instead of a cluster of three, the range of the instrument would be roughly doubled. The upper limit of the number of prisms that can realistically be employed is nine to twelve, depending on the manufacturer. If a surveyor discovers that he needs more than a dozen prisms to take measurements, he should be using a longer-range EDM.

Both electro-optical and microwave EDMs are affected by atmospheric factors. The visibility factor is affected by the absorption and scattering effects of particles of rain, snow, fog, smoke, and dust. Poor visibility can reduce the range achievable under good visibility conditions by a factor of 2 or 3. Range capability is also affected by refraction caused by the shimmering phenomenon found close to the ground surface; this can be reduced by keeping the EDM beam as high as possible off the ground.

Background radiation in electro-optical instruments can reduce range and cause inconsistent results; this can be eliminated by keeping the instrument pointed away from the sun (occupy the other end of the line or take the measurement at a different time).

8.7 PRISMS

Prisms are used with electro-optical EDMs (light, laser, and infrared) to reflect the transmitted signals (see Figure 8.10). A single reflector is a cube corner prism that has the characteristic of reflecting light rays back precisely in the same direction as they are received. This retro-direct capability means that the prism can be somewhat misaligned with respect to the EDM and still be effective. Cube corner prisms are formed by cutting the corners off a solid glass cube; the quality of the prism is determined by the flatness of the surfaces and the perpendicularity of the 90° surfaces.

FIGURE 8.10 Various reflector and target systems in tribrach mounts. (Courtesy Topcon Instrument Corporation of America)

Prisms can be tribrach mounted on a tripod, centered by optical plummet, or they can be attached to a prism pole held vertical on a point with the aid of a bull's-eye level.

In control surveys, tribrach-mounted prisms can be detached from their tribrachs and then interchanged with a theodolite (and EDM) similarly mounted at the other end of the line being measured. This interchangeability of prism and theodolite (also targets) speeds up the work, as the tribrach mounted on the tripod is only centered and leveled one time. Equipment that can be interchanged and mounted on already set up tribrachs is known as forced-centering equipment.

Prisms mounted on adjustable-length prism poles are very portable and, as such, are particularly suited for stakeout surveys. Figure 8.11 shows the prism pole being steadied with the aid of an additional target pole.

FIGURE 8.11 Steadying the EDM reflector with the aid of a target pole.

8.8 EDM ACCURACIES

EDM accuracies are stated in terms of a constant instrumental error and a measuring error proportional to the distance being measured.

Typically, accuracy is claimed as $\pm$ [5 mm + 5 parts per million (ppm)] or $\pm$(0.02 ft + 5 ppm). The $\pm$5 mm (0.02 ft) is the instrument error that is independent of the length of the measurement, whereas the 5 ppm (5 mm/km) denotes the distance-related error.

Most instruments now on the market have claimed accuracies in the range from $\pm$(3 mm + 1 ppm) to $\pm$(10 mm + 10 ppm). The proportional part error (ppm) is insignificant for most work, and the constant part of the error assumes less significance as the distances being measured lengthens: at 100 m an error of $\pm$5 mm represents $1/20000$ accuracy, whereas at 1000 m the same instrumental error represents $1/200000$ accuracy.

When dealing with accuracy, it should be noted that both the EDM and the prism reflectors must be corrected for off-center characteristics. The measurement being recorded goes from the electrical center of the EDM to the back of the prism (allowing for refraction through glass) and then back to the electrical center of the EDM. The difference between the electrical center of the EDM and the plumb line through the tribrach center is compensated for by the EDM manufacturer at the factory. The prism constant (30 to 40 mm) is eliminated either by the EDM manufacturer at the factory or in the field. That is, if the EDM is designed to be used with a prism having a 30-mm constant and instead is being used with a 40-mm constant prism, the surveyor could simply enter 40 mm in the appropriate microprocessor register.

The EDM/prism constant value can be field checked in the following manner: A line is laid out with end stations and an intermediate station (see Figure 8.12). The overall distance AC is measured along with partial lengths AB and BC. The constant value will be present in all measurements; therefore,

$$AC - AB - BC = \text{instrument/prism constant}$$

Alternately, the constant can be determined by measuring a known base line, if one can be conveniently accessed.

FIGURE 8.12   Method of determining the instrument-reflector constant.

8.9 EDM OPERATION

All EDMs follow the same basic operating steps: (1) setup, (2) aim, (3) measure, and (4) record.

8.9.1 Setup

Tribrach-mounted EDMs are simply inserted into the tribrach (forced centering) after the tribrach has been set over the point (by means of optical plummet) and leveled. Telescope or theodolite yoke-mounted EDMs are simply attached to the theodolite either before or after the theodolite has been set over the point and leveled. Prisms are set over the remote station either by insertion into an already set up tribrach (forced centering), or prisms can be held vertically over the remote station on a prism pole. The EDM is turned on, and a quick check is made to ensure that the EDM is in working condition (battery, display, etc.).

8.9.2 Aim

The EDM is aimed at the prism by utilizing either built-in sighting devices on the EDM or the theodolite telescope. Telescope or yoke-mounted EDMs will have the optical line of sight a bit lower than the electronic signal; however, if the theodolite telescope is clamped when positioned on the prism, the electronic fine adjusting can be easily accomplished using either the theodolite tangent screws or the EDM tangent screws (vertical movement on yoke-mounted EDMs must be accomplished using the EDM vertical fine adjustment screw).

The operator can set the electronic signal precisely on the prism by adjusting vertical and horizontal fine adjustment screws until a maximum intensity return signal is displayed on a signal scale. Some EDMs also have an audible prism locator whose variable-tone indicator helps to properly align the electronic signal to the prism. Some EDMs have a signal attenuator that must be used to adjust the strength of the signal to the distance being measured and the atmospheric conditions encountered. (Newer EDMs have automatic attenuation.)

8.9.3 Measure

The measurement of the distance is accomplished simply by pressing the "measure" button and waiting the few seconds for the result to appear in the display. The displays are either liquid crystal (LCD) or light emitting diodes (LED). The measurement is shown to two decimals of a foot, or three decimals of a meter; a foot/meter switch is used to switch from one system to the other.

Some EDMs have a tracking mode that continuously updates and displays the distances as the prism is moved closer to its final layout position. Usually, the tracking mode display is shown to one decimal less than the normal measurement display. The more precise measurement mode can be used when the tracked prism is very close to its final layout position. All microwave EDMs provide two-way communication on the measuring wave itself; some electro-optical EDMs (e.g., Geodimeter) provide one-way communication from the EDM; obviously, voice communications are a great help when long distances are being measured or when points are being laid out. Two-way field radios are also used for these purposes. Figure 8.13 shows a remote EDM display device that is particularly useful in layout surveys and in high-noise data-gathering surveys.

Some EDM instruments have the internal capability of providing corrections to measured distances, whereas other EDMs (older versions usually) require manual corrections to the displayed measurement. Modern EDMs automatically correct for curvature and refraction (c & r) and instrument/prism constants,

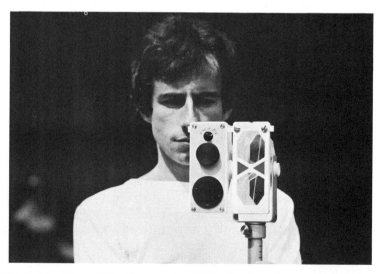

FIGURE 8.13 Kern RD 10 remote EDM display shown attached to the EDM reflecting prism. Slope, horizontal, and vertical distances (from the EDM to the reflecting prism) are displayed on the RD 10. Maximum range of 1300 ft (400 m).

and can internally correct for atmospheric factors when tempera-ture and pressure are entered. Instrument/prism constants other than that for which the EDM has been calibrated can be entered; and vertical angles can be entered to reduce the slope distance to its horizontal equivalent. Sea level and scale factors (see Chapter 11) can be similarly treated.

8.9.4 Recording

The displayed data can be recorded conventionally in field notes or they can be manually entered in an electronic data collector. Total station instruments have the capability of automatically recording all the data collected by the electronic tacheometer. For older EDMs, in addition to the displayed distance, all other correction-related data (e.g., temperature, prism constants) must be booked for each measurement. Figures 8.14 to 8.19 show a variety of cur-rent EDM equipment.

FIGURE 8.14 Sokkisha Red Mini; lightweight (0.9 kg or 2 lb) and short-range (800 m-triple prism) EDM, mounted on a 10-second theodolite; SE = ±(5 mm + 5 ppm).

FIGURE 8.15 Red1A shown mounted on a theodolite and on a tribrach. Triple prism range = 2500 m; SE = ±(5 mm + 5 ppm). (Courtesy Sokkisha Co. Ltd.)

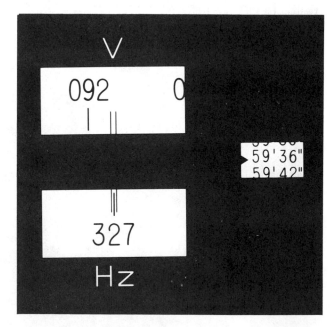

FIGURE 8.16 (a) Wild DI4 Distomat mounted on a Wild T-1 repeating theodolite. Keyboard calculator is used for trigonometric reductions; Distomat measures 1.6 km (triple prism) with one-key operation; (b) circle reading (least count 6 seconds); horizontal angle = 327°59′36″. (Courtesy Wild Heerbrugg Ltd.)

284

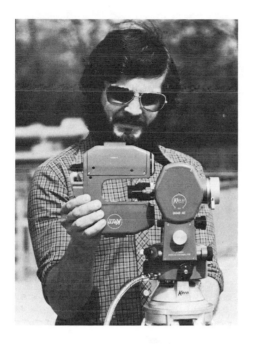

FIGURE 8.17 Kern DM 502 (infrared EDM) attaches readily to a theodolite, a DMK2 1-second theodolite in this case. U-shaped design permits one sighting for both distance and angle measurement. Distance range is 2000 m with a single prism. (Courtesy Kern Instruments, Inc.)

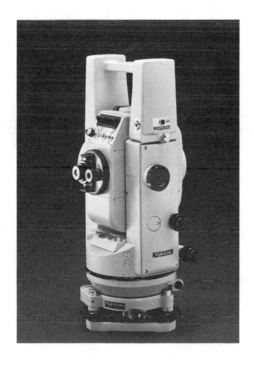

FIGURE 8.18 Topcon GTS-2 Geodetic "Total Station" has a triple prism range of 1700 m with a SE = ±(5mm + 5 ppm). Angles are read to 10 seconds (6 seconds on some models). Optional slope reduction calculator permits keyboard entry of slope distance and vertical angle to compute horizontal distance.

FIGURE 8.19 Wild long-range infrared EDM, DI 20 mounted on a Wild T-2 directional theodolite, featuring a triple prism range of 7 km, with a SE of ±(5 mm + 1 ppm). Built-in keyboard calculator permits on-site data reduction.

8.10 GEOMETRY OF AN EDM MEASUREMENT

Figure 8.20 shows the general case of an EDM measurement. The slope distance (s) is measured by EDM; the slope angle (α) is measured by an accompanying theodolite. The height of the EDM and theodolite (HI) is measured with a steel tape, and the height (HR) of the prism (reflector) is also measured with a steel tape. Adjustable-length prism poles permit the surveyor to adjust the height of the prism (HR) to equal the height of the instrument (HI) thus simplifying the computations.

From Figure 8.20, if the elevation of STA A is known, and the elevation of STA B is required,

Elevation STA B (reflector) =
$$\text{elevation STA A (inst.)} + \text{HI} \pm V - \text{HR} \qquad (8.8)$$

When the EDM is mounted on a theodolite and the target is located beneath the prism on the prism pole, the geometric relationship can be as shown in Figure 8.21.

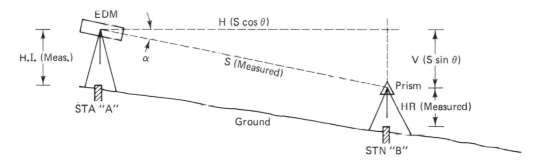

FIGURE 8.20 Geometry of an EDM measurement, general case.

The additional problem encountered here is the computation of the correction to the vertical angle ($\Delta\alpha$), which occurs when ΔHI and ΔHR are different. The precise size of the vertical angle is important as it is used in conjunction with the measured slope distance to compute the horizontal and vertical distances.

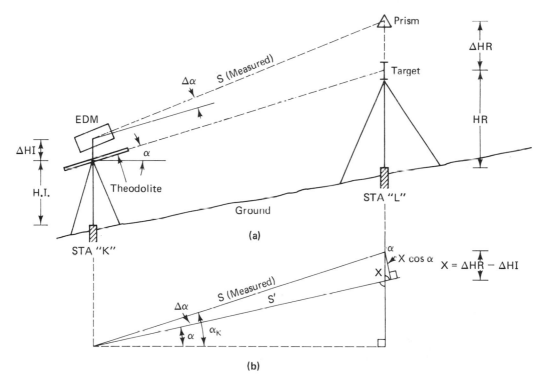

FIGURE 8.21 Geometry of an EDM measurement, usual case.

In Figure 8-21, the difference between ΔHR and ΔHI is x (i.e., ΔHR − ΔHI = x). The small triangle formed by extending S' (Figure 8.21b) has hypotenuse equal to x and an angle of α. This permits determination of the side $x \cos \alpha$, which can be used together with S to determine $\Delta\alpha$:

$$\frac{x \cos \alpha}{S} = \text{Sin} \ \Delta\alpha$$

Example 8.1

An EDM slope distance AB is determined to be 561.276 m. The EDM is 1.820 m above its station (A), and the prism is 1.986 m above its station (B). The EDM is mounted on a theodolite whose optical center is 1.720 m above the station. The theodolite was used to measure a vertical angle (+6°21'38") to a target on the prism rod; the target is 1.810 above station B. Compute (a) the horizontal distance AB, and (b) the elevation of station B, given that the elevation of station A is 186.275.

Solution: The given data are shown in Figure 8.22a and the resultant figure is shown in Figure 8.22b. The X value introduced in Figure 8.21b is in this case determined as follows;

$$X = (1.986 - 1.810) - (1.820 - 1.720)$$
$$= 0.176 - 0.100$$
$$= 0.076 \quad \text{(i.e., } \Delta HR - \Delta HI)$$
$$\text{Sin } \Delta\alpha = \frac{0.076 \cos 6°21'38''}{561.276}$$

$$\Delta\alpha = 28''$$
$$\alpha_k = 6°22'06''$$
(a) $\quad H = 561.276 \cos 6°22'06''$
$$= 557.813 \text{ m}$$

If H had been computed using the field vertical angle of 6°21'38", the result would have been 557.821 m, not a significant difference in the example.

(b) Elev. B = elev. A + 1.820 + 561.276 sin 6°22'06" − 1.986
$$= 186.275 + 1.820 + 62.257 - 1.986$$
$$= 248.366$$

If V had been computed using 6°21'38", the result would have been 62.181 instead of 62.257 m, a more significant discrepancy.

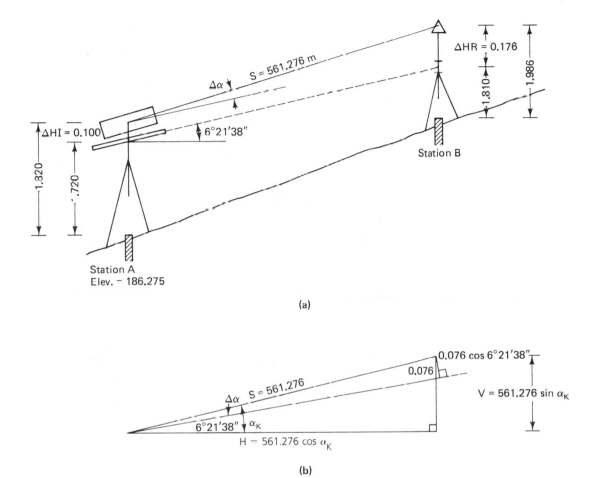

FIGURE 8.22 Illustration for Example 8.1.

8.11 ELECTRONIC TACHEOMETER INSTRUMENTS (ETI)

Electronic tacheometer instruments automatically record distances, horizontal angles, and vertical angles. Early models of electronic tacheometer instruments (e.g., Hewlett-Packard 3820A and the Zeiss Elta 2) used an *absolute* method of angle measurement. For example, the Zeiss Elta 2 is essentially an optical coincidence instrument with photoelectric sensors being used to electronically scan and read the angles.

Later ETI models (e.g., Wild TCI, K & E Vectron, Geodimeter 140, Kern E1, Zeiss Elta 4, and Alpha Omni) employ an *incremental* method of angle measurement. These instruments have glass circles that are graduated into unnumbered gratings. The number of gratings involved in a measurement is determined from whole-circle electronic scanning. Circle imperfections are thus compensated for, permitting higher precision with only one circle setting. The distance measurement is obtained using electro-optical range finders (infrared Ga As diode). Many ETIs have coaxial electronic and optical systems, thus permitting simultaneous electronic and optical pointing.

The onboard microprocessor in the ETI monitors the instrument status (e.g., level or plumb orientation) and controls the angle and distance data acquisition and processing. In addition to computing horizontal distances and differences in elevation, most ETI microprocessors will also compute coordinates. The K & E Vectron field computer is also programmed to compute traverse and curve problems.

ETIs that have automatic data collection capability for angle and distance measurement are called total stations (see Figures 8.23, and 8.24). In addition to the fully automatic instruments described here, there is a wide variety of instruments that have some of the total station characteristics. That is, some instruments have automatic distance recording and others have automatic distance and vertical angle recording. In each case, the nonrecorded data must be entered manually into the electronic keyboard. These instruments are sometimes called semitotal stations.

Most ETIs are designed so that data stored in the data collector can be automatically unloaded to the computer via a RS 232 C or IEEE interface, with appropriate transfer software. The data can be adjusted by the computer (mainframe, mini or micro) and can then be printed out or graphically portrayed by an interfaced digital plotter. The chief attributes of an ETI system are the speed and ease of data collection and processing, and the elimination of many of the usual opportunities for mistakes and errors.

The only disadvantage the authors see in the use of ETIs is the lack of hard-copy field notes that can be scanned and checked in the field. Although individual lines of data can be recalled in the field from the data collector, the overall sense of the survey must wait for the computer printout or digital plot. Recognizing this danger, ETI surveyors carefully design their survey implementation in order to minimize errors and mistakes (e.g., use of rigidly specific techniques and redundant measurements); and

FIGURE 8.23 Kern E-1 electronic theodolite (total station), shown with the Kern DM 502 EDM and the Hewlett-Packard 41CV as a data collector and calculator. The HP 41CV can accept measured values from the E-1 and transmit computed values via the DM 502 to the remote receiver (RD 10 Figure 8.13).

FIGURE 8.24 Wild T-2E electronic theodolite with GRE data recorder can be used with a Distomat DI4 for total station capabilities. Angles read with SE = 0.5 second.

they design the computer system output so as to highlight possible discrepancies (e.g., "extra" cross sections, profiles, and plan views). Any apparent output inconsistencies are quickly investigated in the field; the computer graphics and printouts can in some cases be available within 24 hours of the completion of the fieldwork.

8.12 ELECTRONIC TACHEOMETER OPERATION

ETIs are designed for use with standard forced-centering tribrachs and standard tripods. The optical plummet in the tribrach is used to center the instrument over the instrument station. In addition to the tribrach circular bubble, most ETIs have a tubular bubble that is utilized for fine leveling.

The ETI operation begins with the entry of the initial data:

Typical ETI Operation Procedure

A. Initial data entry (can be in imperial or metric units):
  1. x, y, and z coordinates of instrument station (if required).
  2. HI (height of instrument) ft or m.
  3. HR (height of reflector) ft or m (this can be made equal to HI when adjustable prism poles are used).
  4. Temperature (°F or °C).
  5. Atmospheric pressure (inches Hg or mm Hg).
  6. Prism constant (this does not have to be changed unless a prism with a different constant is brought in).
  7. Degrees or gon (grad) selection.

B. Data collection:
  1. The telescope is aimed to the prism optically. Coaxial instruments can be set optically for both optical and electronic alignment.
  2. Noncoaxial instruments must now be "pointed" electronically by adjusting the vertical and horizontal tangent screws until a signal peaking is displayed.
  3. When maximum signal peaking has been achieved, the measuring process can begin. Slope distance, horizontal distance, vertical distance, azimuth angle, and vertical angle buttons are individually engaged until all these parameters have been measured and displayed (often, only the slope distance and the vertical and horizontal angles are recorded).

4. An initial azimuth can be assigned to the first target direction or the coordinates of the initial target station can be entered, if applicable.

5. When the data collection for the initial target station has been finalized, the target station is assigned an identification code, and all the data are stored in the microprocessor memory simply by pushing the Store button.

6. The ETI can then be turned to a second target station and the data collection procedure repeated. At any time, if needed, the initial data entry item can be revised (e.g., HR). The coordinates of the second target station (and all subsequent target stations) can be called for, displayed, and then stored.

7. When used for layout purposes, the ETI can be set to the fast tracking mode, which allows for a very fast measurement update display, usually stated in a less precise manner (e.g., one fewer decimals than is displayed for the normal measure mode). When the prism is very close to its desired location (on-line and marginally too close or too far away), that intermediate point can be marked and the distance then can be completed with the aid of a pocket tape; this manual "touching-up" of the layout measurement may be a more efficient procedure than continued tracking for some of the ETIs.

C. Data transfer:

A very important characteristic of the ETI is that the data can be automatically transferred to a computer. All ETIs can be interfaced to micro, mini, or mainframe computers. Some ETIs' storage can be transferred directly to the computer (e.g., an RS 232 C or IEEE interface with appropriate "handshake" software or even by telephone modems). Other ETIs' storage must first be transferred to data convertors and then to the computer. Once in the computer, the data can be adjusted and processed with output via a printer, hard-copy graphics, or scale drawings via digital plotters.

Table 8.1 summarizes the characteristics of typical ETIs. Generally, they all can do the job. Direct comparison is difficult (it is not intended) because of the different terminology used by the various manufacturers. For example, the data collector storage appears to vary by a wide degree between various models. The authors suspect that the variance is as much due to reporting terminology as it is to hardware capability. In any event, all data collectors listed can store a normal day's work, and an additional storage collectors can be taken on extended field trips.

Reports on these recently introduced instruments have been

TABLE 8.1 *Summary of Typical ETI Manufacturers Specifications*

	ELTA-3, 1 Zeiss, Oberkochen	Wild T2-E (T-2000) with Distomat D14 and GRE 3 Recorder, Wild Heerbrugg, Switzerland	Vectron with Autoranger and Vectron Field Computers, Keuffel and Esser, Morristown, N.J., U.S.A.	E 1 with DM502, Kern, Switzerland	Omni 1, Alpha Electronics, Colorado, U.S.A.	Geodimeter 140, Agatronics, Sueden
Approximate range (km) 1 prism 3 prisms	1.2	1 1.6	2	1.8 3.6	2 3	2.5 3.6
Distance accuracy	±(5mm + 5ppm)	±(5mm + 5ppm)	±(6mm + 6ppm)	±(5mm + 5ppm)	±(5mm + 2ppm)	±(5mm + 3ppm)
Angle accuracy	±.06"	±0.5"	±3.24"	±3.24"	±6 sec.	±2"
Display (LED = light emitting diode) (LCD = liquid crystal display)	9-digit LED	8-digit LED both sides	9-digit LCD	6-digit LCD	9-digit LCD	8-digit LED
Coaxial	Yes	No (only one pointing req'd.)	No	No (only one pointing req'd.)	No	Yes
Measuring time (s)	5	8–15	6	8	8	5–10
Tracking update (s)	6	4	6	4	6	0.04
Telescope magnification	30×	32×	30×	32×	24×	30×
Data collector storage	200 or 400 lines, built in	2000 lines, built in	1600 items, connected	800 points, connected	700 points built in	500–1000 points, connected

	DAC100 RS232C	RS232C	IEEE RS232C	RS232C (V-24)	RS232C	RS232C
Transfer to computer						
Temperature Range	−20°C to +60°C	−20°C to +50°C	−20°C to +54°C		−20°C to +50°C	−20°C to +50°C
Weight	13.5 kg 29.7 lb	11.6 kg 24.9 lb	8.2 kg 20.2 lb	10.3 kg 22.7 lb	7.7 kg 17 lb	8.5 kg 18.7 lb
Carrier wavelength (1 micrometer, μm = 10^{-6} m)	0.910 μm	0.885 μm	0.865 μm	0.860 μm		
Dial in corrections ft/m conversion Automatic corrections (c + r) etc.	Yes	Yes	Yes	Yes	Yes	Yes
Other	ELTA-2 can be programmed	Keyboard and display on both sides of instrument; GRE-3 has programming cababilities	Field computer is programmable for traverse and curve problems	Can be interfaced to HP41C as collector, remote receiver available	Can easily be updated with new computer chip; display "flips" when ETI is transited	One-way voice communication
Approximate cost, 1983 Canadian $	$34,000	$32,000	$21,000	$32,000	$17,750	$32,000

largely positive. Negative reports have been mostly associated with cold weather performance and unreliable battery operation. Since all ETIs can be powered by 12-V batteries, many surveyors now carry inexpensive motorcycle batteries to act as spares in case the EDM battery fails during the day's operations.

A recently introduced ETI (see Figure 8.9 and Table 8.1) is the Alpha Electronics Omni I (Colorado). The price of $17750 (1983 Canadian dollars) is about two-thirds *less* than the cost of the earlier ETIs. Whereas before only very large volume surveying companies could justify the high capital cost of this equipment, now moderate-sized concerns are using ETIs. As electronics technology and computer chip technology continue to advance, the cost of these incredibly useful instruments will continue to decrease until they will be within reach of most surveyors. The advantages of using ETIs are twofold:

1. The data can be collected quickly, accurately, and precisely. (Ground surveys are in some cases, once again, competitive with aerial surveys.)

2. The data can be automatically transferred to a computer, where it can be further processed and finally displayed.

The last point emphasizes the new importance in surveying of the systematic collection (layout), processing, and display of ground data.

9 Survey Drafting and Computations

9.1 GENERAL

Survey drafting is a term that covers a broad spectrum of scale graphics and related computations. Generally, survey data are graphically displayed in the form of maps or plans. Maps are normally small scale, whereas plans are drawn to a much larger scale. The essential difference between maps and plans is characterized by their use. Maps portray, as in an inventory, the detail (e.g., topographic) for which they were designed. Maps can be of a general nature, such as the topographic maps compiled and published by the U.S. Geological Survey (scales normally ranging from 1:24000 down to 1:1000000), or maps can be specific in nature showing only those data (e.g., cropland inventory) for which they were designed.

Plans, on the other hand, not only show existing terrain conditions, but they also depict proposed alterations (i.e., *designs*) to the existing landscape. Most plans are drawn to a large scale, although comprehensive route design plans for state and provincial highways can be drawn to a small scale (e.g., 1:50000) to give a bird's-eye view of a large study area.

Tables 7.1 and 9.1 summarize typical scales along with appropriate contour intervals. Map scales and precision are discussed in Section 7.2.

297

TABLE 9.1 Summary of Map Scales and Contour Intervals

	Metric Scale	Foot/Inch Scale Equivalents	Contour Interval for Average Terrain[a]	Typical Uses
Large scale	1:10	1″ = 1′		Detail
	1:50	¼″ = 1′, 1″ = 5′		Detail
	1:100	⅛″ = 1′, 1″ = 10′, 1″ = 8′		Detail, profiles
	1:200	1″ = 20′		Profiles
	1:500	1″ = 40′,	0.5 m, 1 ft	Municipal design plans
	1:1000	1″ = 80′, 1″ = 100′	1 m, 2 ft	Municipal services and site engineering
Intermediate scale	1:2000	1″ = 200′	2 m, 5 ft	Engineerings studies and planning (e.g., drainage areas, route planning)
	1:5000	1″ = 400′	5 m, 10 ft	
	1:10 000	1″ = 800′	10 m, 20 ft	
Small scale	1:20 000	1:25000, 2½″ = 1 mi		Topographic maps, Canada and United States
	1:25 000			
	1:50 000	1:63 360, 1″ = 1 mi		
	1:100 000	1:126 720, ½″ = 1 mi		Geological maps, Canada and United States
	1:200 000	1:250 000, ¼″ = 1 mi		Special purpose maps and atlases (e.g., climate, minerals)
	1:250 000	1:625 000, 1/10″ = 1 mi		
	1:500 000	1:1000 000, 1/16″ = 1 mi		
	1:1000 000			

[a]The contour interval chosen must reflect the scale of the plan or map, but, additionally, the terrain (flat or steeply inclined) and intended use of the plan are also factors in choosing the appropriate contour interval.

9.2 MAPS AND PLANS

The topic of maps and plans is included in a survey text to show how survey data, obtained from either ground or aerial surveys, can be graphically portrayed in scale drawings. The reproduction of maps involves photographing the finished inked or scribed map and preparing a printing plate from the negative. Lithographic offset printing is used to create the maps; multicolor maps require a separate plate for each color, although shading can be accomplished using screens.

Scribing is a mapping technique in which the map details are directly cut onto drafting film that has a soft opaque coating. This scribed film takes the place of a photographic negative in the photolithography printing process. Scribing is preferred by many because of the sharp definition made possible by this "cutting" technique.

Plans, on the other hand, are reproduced in an entirely different manner. The completed plan, in ink or pencil, can simply be run through a direct-contact negative printing machine (blueprint) or a direct-contact positive printing machine (whiteprint). The whiteprint machine, which is now in use in most drafting and design offices, utilizes paper sensitized with diazo compounds, which when exposed to light and ammonia vapor produce prints. The quality of whiteprints cannot be compared to map quality reproductions; however, the relatively inexpensive whiteprints are widely used in surveying and engineering offices where they are used as working plans, customer copies, and contract plans.

Although reproduction techniques are vastly different for maps and plans, the basic plotting procedures are identical.

9.3 PLOTTING

The size of drafting paper required can be determined by knowing the scale to be used and the area or length of the survey. Standard paper sizes are shown in Table 9.2. The title block is often a standard size and has a format similar to that shown in Figure 9.1. The title block is usually (depending on the filing system) placed in the lower right corner of the plan. Revisions to the plan are usually referenced immediately above the title block, showing the date and a brief description of the revision.

TABLE 9.2 Standard Drawing Sizes

| | Inch Drawing Sizes | | International Standards Organization (ISO) | | | | ACSM[a] Recommendations | |
| | | | Metric Drawing Sizes (Millimeters) | | | | | |
DRAWING SIZE	BORDER SIZE	OVERALL PAPER SIZE	DRAWING SIZE	BORDER SIZE		OVERALL PAPER SIZE		PAPER SIZE
A	8.00 × 10.50	8.50 × 11.00	A4	195 × 282	210 × 297		—	150 × 200
B	10.50 × 18.50	11.00 × 17.00	A3	277 × 400	297 × 420		A4	200 × 300
C	16.25 × 21.25	17.00 × 22.00	A2	400 × 574	420 × 594		A3	300 × 400
D	21.00 × 33.00	22.00 × 34.00	A1	574 × 821	594 × 841		A2	400 × 600
E	33.00 × 43.00	34.00 × 44.00	A0	811 × 1159	841 × 1189		A1	600 × 800
							A0	800 × 1200

[a]American Congress on Surveying and Mapping Metric Workshop, March 14, 1975. Paper sizes rounded off for simplicity, still have cut-in-half characteristic.

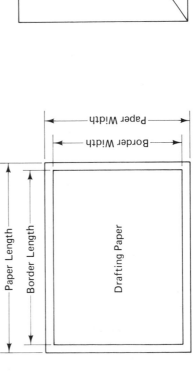

Drafting Paper

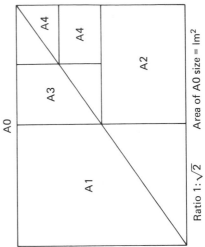

Ratio 1: $\sqrt{2}$ Area of A0 size = $1\mathrm{m}^2$

Metric Drawing Paper

2.			
1.	28·12·1983	RETAINING WALL ADDED @ 8+72–SOUTH SIDE	
No.	DATE	DESCRIPTION	
		REVISIONS	

Owner →

General Title →

Specific Title →

Consultant or Design Department →

CITY OF NORTH YORK

PLAN AND PROFILE
OF
FINCH AVE. RECONSTRUCTION

DON MILLS RD. TO HWY #404

BLACK & ASSOCIATES CONSULTING ENGINEERS 876 BAY ROAD TORONTO	SCALES: 1:500 HORIZONTAL 1:100 VERTICAL
	DATE: DEC. 1. 1983
	DRAWING NO.: D–15–1

DESIGN: E.J.J.	CHECKED: C.C.W.	TRACED:
DRAWN: A.O.B.	APPROVED· *J.J. Smith*	

FIGURE 9.1 Typical title block.

Many consulting firms and engineering departments attempt to limit the variety of their drawing sizes so that plan filing can be standardized. Some vertical hold filing cabinets are designed such that title blocks in the upper right corner are more easily seen.

The actual plotting begins by first plotting the survey control (e.g., ℄, traverse line, coordinate grid, etc.) on the drawing. The control is plotted so that the data plot will be suitably centered on the available paper. Sometimes the data outline is roughly plotted first on tracing paper so that the plan's dimension requirements can be properly oriented on the available drafting paper. It is customary to orient the data plot so that north is toward the top of the plan; a north arrow is included on all property survey plans and many engineering drawings. The north direction on maps is clearly indicated by lines of longitude or by N–S, E–W grid lines. The plan portion of the plan and profile does not usually have a north indication; instead, local practice dictates the direction of increasing chainage (e.g., chainage increasing left to right for west to east and south to north directions).

Once the control has been plotted and checked, the features can be plotted utilizing either rectangular (x, y coordinates) or polar (r, θ coordinates) methods.

Rectangular plots (x, y coordinates) can be laid out using a T-square and set square although the parallel rule has now largely replaced the T-square. When using either the parallel rule or the T-square, the paper is first set square and then taped with masking tape to the drawing board. Once the paper is square and secure, the parallel rule, together with a set square (right-angle triangle) and scale, can be used to lay out and measure rectangular dimensions. Since rectangular plotting is more precise than polar plotting (plotting errors are not accumulated), this technique is used for precise work (e.g., control layout, important details) and for plotting details that have been tied into base lines using right-angle tie-ins (see Section 7.3).

Polar plots are accomplished using a protractor and a scale. The protractor can be a plastic graduated circle or half-circle having various size diameters (the larger, the more precise), a paper full-circle protractor for use under or on the drafting paper, or a flexible-arm drafting machine complete with right-angle-mounted graduated scales. Field data that have been collected using polar techniques (stadia is a good example) can be efficiently plotted utilizing polar techniques. See Figure 9.2 for standard map and plan symbols.

9.4 CONTOURS

Contours are lines drawn on a plan that connect points having the same elevation. Contour lines represent an even value (see Table 9.1), with the contour interval being selected consistent with terrain, scale, and intended use of the plan. It is commonly accepted that elevations can be determined to half the contour interval; this permits, for example, a 10-ft contour interval on a plan where it is required to know elevations to the closest 5 ft.

Contours are plotted by scaling between two adjacent points of known elevation. In Figure 9.3a, the scaled distance (any scale can be used) between points 1 and 2 is 0.75 units, and the difference in elevation is 5.4 ft. The difference in elevation between point 1 and contour line 565 is 2.7 ft; therefore, the distance from point 1 to contour line 565 is

$$\frac{2.7}{5.4} \times 0.75 = 0.38 \text{ units}$$

To verify this computation, the distance from contour line 565 to point 2 is

$$\frac{2.7}{5.4} \times 0.75 = 38 \text{ units.} \quad 0.38 + 0.38 \approx 0.75 \qquad \text{Check}$$

Primary highway, hard surface		Boundaries: National	
Secondary highway, hard surface		State	
Light-duty road, hard or improved surface		County, parish, municipio	
Unimproved road		Civil township, precinct, town, barrio	
Road under construction, alinement known		Incorporated city, village, town, hamlet	
Proposed road		Reservation, National or State	
Dual highway, dividing strip 25 feet or less		Small park, cemetery, airport, etc.	
Dual highway, dividing strip exceeding 25 feet		Land grant	
Trail		Township or range line, United States land survey	

Township or range line, approximate location

Section line, United States land survey

Section line, approximate location

Township line, not United States land survey

Section line, not United States land survey

Railroad: single track and multiple track	Found corner: section and closing
Railroads in juxtaposition	Boundary monument: land grant and other
Narrow gage: single track and multiple track	Fence or field line
Railroad in street and carline	
Bridge: road and railroad	
Drawbridge: road and railroad	
Footbridge	Index contour ... Intermediate contour
Tunnel, road and railroad	Supplementary contour ... Depression contours
Overpass and underpass	Fill ... Cut
Small masonry or concrete dam	Levee ... Levee with road
Dam with lock	Mine dump ... Wash
Dam with road	Tailings ... Tailings pond
Canal with lock	Shifting sand or dunes ... Intricate surface
	Sand area ... Gravel beach

Buildings (dwelling, place of employment, etc.)	
School, church, and cemetery	Cem
Buildings (barn, warehouse, etc.)	
Power transmission line with located metal tower	Perennial streams ... Intermittent streams
Telephone line, pipeline, etc. (labeled as to type)	Elevated aqueduct ... Aqueduct tunnel
Wells other than water (labeled as to type)	o Oil ... o Gas	Water well and spring ... Glacier
Tanks: oil, water, etc. (labeled only if water)	Water	Small rapids ... Small falls
Located or landmark object; windmill	o	Large rapids ... Large falls
Open pit, mine, or quarry; prospect	× ... x	Intermittent lake ... Dry lake bed
Shaft and tunnel entrance		Foreshore flat ... Rock or coral reef
		Sounding, depth curve 10 ... Piling or dolphin
		Exposed wreck ... Sunken wreck

Horizontal and vertical control station:

Rock, bare or awash; dangerous to navigation

Tablet, spirit level elevation	BM △ 5653
Other recoverable mark, spirit level elevation	△ 5455
Horizontal control station: tablet, vertical angle elevation	VABM △ 95I9	Marsh (swamp) ... Submerged marsh
Any recoverable mark, vertical angle or checked elevation	△3775	Wooded marsh ... Mangrove
Vertical control station: tablet, spirit level elevation	BM × 957	Woods or brushwood ... Orchard
Other recoverable mark, spirit level elevation	× 954	Vineyard ... Scrub
Spot elevation	× 7369 × 7369	Land subject to controlled inundation ... Urban area
Water elevation	670 670	

FIGURE 9.2 (a) Topographic map symbols. (U.S. Department of Interior, Geological Survey); (b) municipal works plan symbols, including typical title block. (Municipal Engineers Association—Ontario) (*Continued*)

303

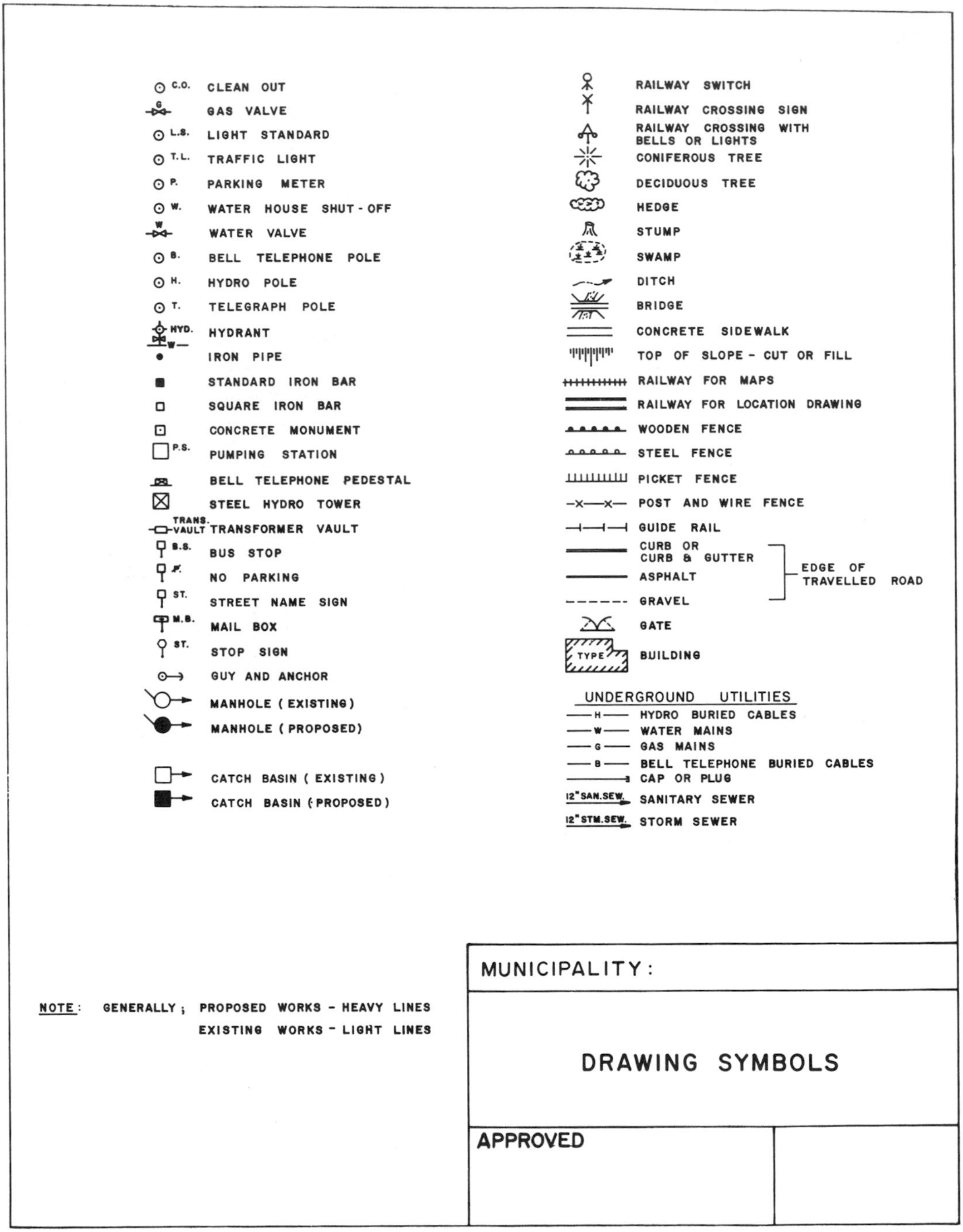

⊙ C.O.	CLEAN OUT		RAILWAY SWITCH
	GAS VALVE		RAILWAY CROSSING SIGN
⊙ L.S.	LIGHT STANDARD		RAILWAY CROSSING WITH BELLS OR LIGHTS
⊙ T.L.	TRAFFIC LIGHT		CONIFEROUS TREE
⊙ P.	PARKING METER		DECIDUOUS TREE
⊙ W.	WATER HOUSE SHUT - OFF		HEDGE
	WATER VALVE		STUMP
⊙ B.	BELL TELEPHONE POLE		SWAMP
⊙ H.	HYDRO POLE		DITCH
⊙ T.	TELEGRAPH POLE		BRIDGE
HYD.	HYDRANT		CONCRETE SIDEWALK
•	IRON PIPE		TOP OF SLOPE - CUT OR FILL
■	STANDARD IRON BAR		RAILWAY FOR MAPS
□	SQUARE IRON BAR		RAILWAY FOR LOCATION DRAWING
⊡	CONCRETE MONUMENT		WOODEN FENCE
□ P.S.	PUMPING STATION		STEEL FENCE
	BELL TELEPHONE PEDESTAL		PICKET FENCE
⊠	STEEL HYDRO TOWER	—x—x—	POST AND WIRE FENCE
TRANS. VAULT	TRANSFORMER VAULT		GUIDE RAIL

NOTE: GENERALLY; PROPOSED WORKS - HEAVY LINES
EXISTING WORKS - LIGHT LINES

B.S.	BUS STOP	
⚐.	NO PARKING	
ST.	STREET NAME SIGN	
M.B.	MAIL BOX	
ST.	STOP SIGN	
⊙→	GUY AND ANCHOR	

MANHOLE (EXISTING)
MANHOLE (PROPOSED)

CATCH BASIN (EXISTING)
CATCH BASIN (PROPOSED)

CURB OR CURB & GUTTER } EDGE OF TRAVELLED ROAD
ASPHALT
GRAVEL

GATE

TYPE   BUILDING

UNDERGROUND   UTILITIES
— H — HYDRO BURIED CABLES
— W — WATER MAINS
— G — GAS MAINS
— B — BELL TELEPHONE BURIED CABLES
CAP OR PLUG
12"SAN.SEW. SANITARY SEWER
12"STM.SEW. STORM SEWER

MUNICIPALITY:

DRAWING SYMBOLS

APPROVED

FIGURE 9.2 (Continued)

304

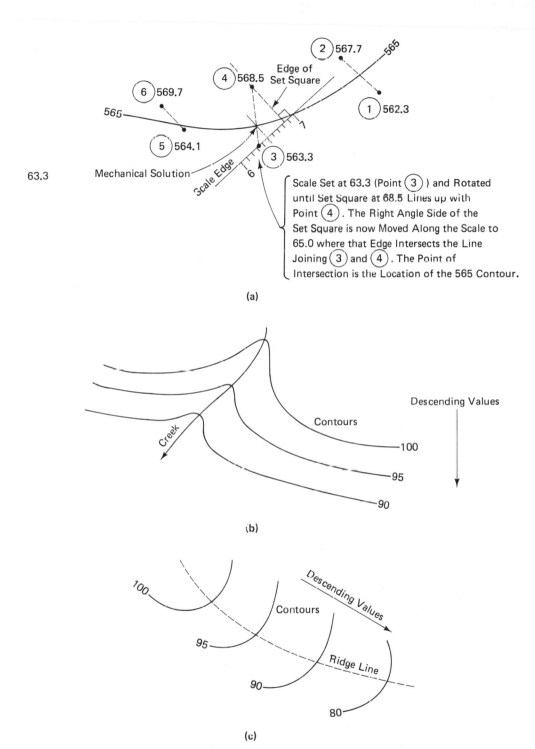

FIGURE 9.3 Contours: (a) plotting contours by interpolation; (b) valley line; (c) ridge line.

The scaled distance between points 3 and 4 is 0.86 units, and their difference in elevation is 5.2 ft. The difference in elevation between point 3 and contour line 565 is 1.7 ft; therefore, the distance from point 3 to contour line 565 is

$$\frac{1.7}{5.2} \times 0.86 = 0.28 \text{ units}$$

This can be verified by computing the distance from contour line 565 to point 4:

$$\frac{3.5}{5.2} \times 0.86 = 0.58 \text{ units}. \quad 0.58 + 0.28 = 0.86 \qquad \text{Check}$$

The scaled distance between points 5 and 6 is 0.49 units, and the difference in elevation is 5.6 ft. The difference in elevation between point 5 and contour line 565 is 0.9 ft; therefore, the distance from point 5 to contour line 565 is

$$\frac{0.9}{5.6} \times 0.49 = 0.08 \text{ units}$$

and from line 565 to point 6 the distance is

$$\frac{4.7}{5.6} \times 0.49 = 0.41 \text{ units}$$

In addition to the foregoing arithmetic solution, contours can be interpolated using mechanical techniques. It is possible to scale off units on a barely taut elastic band and then stretch the elastic so that the marked-off units fit the interval being analyzed. Alternately, the problem can be solved by rotating a scale while using a set square to line up the appropriate divisions with the field points. In Figure 9.3a, a scale is set at 63.3 on point 3 and then rotated until the 68.5 mark lines up with point 4 using a set square on the scale. The set square is then slid along the scale until it lines up with 65.0; the intersection of the set square edge (90° to the scale) with the straight line joining points 3 and 4 yields the solution (i.e., the location of elevation at 565 ft). This latter technique is faster than the arithmetic technique.

Since contours are plotted by analyzing adjacent field points, it is essential that the ground slope be uniform between those points. An experienced survey crew will ensure that enough rod readings are taken to suitably define the ground surface. The survey crew can further define the terrain if care is taken in identifying and tying in valley lines and ridge lines. Figure 9.3b shows how contour lines bend uphill as they cross a valley; the steeper the valley, the more the line diverges uphill. Figure 9.3c shows how contour lines bend downhill as they cross ridge lines. Figure 9.4 shows the plot of control, elevations, and valley and ridge lines. Figure 9.5 shows contours interpolated from the data in Figure 9.4. Figure 9.6b shows the completed plan, with additional detail (roads and buildings) also shown. Figure 9.6a shows typical field notes for a stadia reduction tacheometer.

9.5 SUMMARY OF CONTOUR CHARACTERISTICS

1. Closely spaced contours indicate steep slopes.
2. Widely spaced contours indicate moderate slopes (spacing here is a relative relationship).
3. Contours must be labeled to give the elevation value. Either each line is labeled or every fifth line is drawn darker (wider) and it is labeled.
4. Contours are not shown going through buildings.

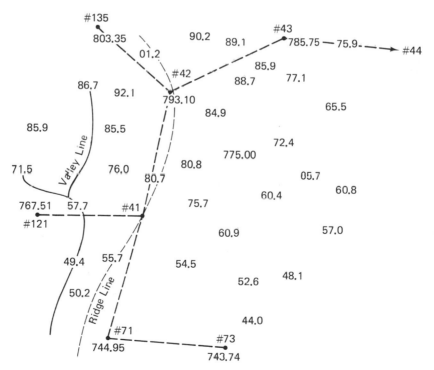

FIGURE 9.4 Plot of survey control, ridge and valley lines, and spot elevations.

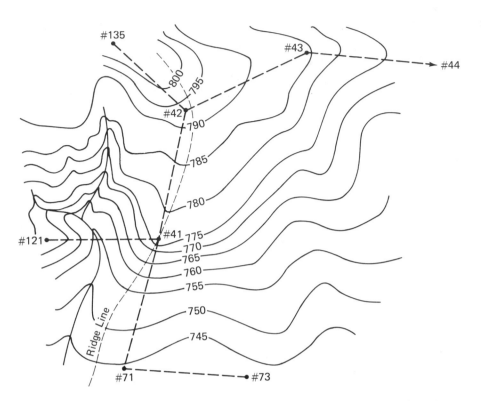

FIGURE 9.5 Contours plotted by interpolating between spot elevations, with additional plotting information given when the location of ridge and valley lines is also known.

5. Contours crossing a man-made horizontal surface (roads, railroads) will be straight parallel lines as they cross the facility.

6. Since contours join points of equal elevation, contour lines cannot cross. (Caves present an exception.)

7. Contour lines cannot begin or end on the plan.

8. Depressions and hills look the same; one must note the contour value to distinguish the terrain (some agencies use hachures or shading to identify depressions).

9. Contours deflect uphill at valley lines and downhill at ridge lines; line crossings are perpendicular: U-shaped for ridge crossings; V-shaped for valley crossings.

10. Contour lines must close on themselves, either on the plan or in locations off the plan.

11. The ground slope between contour lines is uniform. Had the ground slope not been uniform between the points, addition-

al readings (stadia or level) would have been taken at the time of the survey.

12. Important points can be further defined by including a "spot" elevation (height elevation). See Figure 9.7.

13. Contour lines tend to parallel each other on uniform slopes.

Example: Station 42   Elevation 793.10 m   Instr. RDS

Point	Horizontal Angle	Horizontal Distance	Difference in Elevation	Elevation
42				793.10
41	00.0	197.80	− 17.30	775.80
43	232 25.2	199.10	− 7.35	785.75
135	120 05.2	145.20	+ 10.25	803.35
a	234 50	76.10	− 2.60	790.50
b	247 10	76.30	—	—
c	277 22	85.30	− 8.20	784.90
d	322 10	100.50	− 7.60	785.50
etc.				

(a)

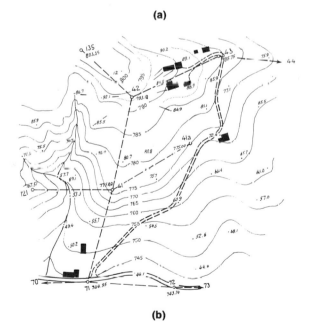

(b)

FIGURE 9.6 (a) Stadia reduction tacheometer (Wild RDS) survey notes; (b) plan plotted from notes shown in Figures 9.4, 9.5, and 9.6(a). (Courtesy of Wild Heerbrugg Leitz, Canada)

9.6 PROFILES

Profiles establish ground elevations along a defined route (see Figure 9.7). Profile data can be directly surveyed, as when a road ℄ has rod readings taken at specific intervals (e.g., 100 ft), as well as at all significant changes in slope; or profile data can be taken from contour drawings.

For example, referring to Figure 9.6, the profile of line 71–41–42 can be determined as follows: The original scale was given as $1'' = 150$ ft, and this is used to scale the distance 71 to 41 (198 ft) and 41 to 42 (198 ft). The distances to the various contour crossings are as follows:

$$
\begin{aligned}
71 \text{ to } 745 &= 16' \\
71 \text{ to } 750 &= 63' \\
71 \text{ to } 755 &= 115' \\
71 \text{ to } 760 &= 138' \\
71 \text{ to } 765 &= 155' \\
71 \text{ to } 770 &= 172' \\
71 \text{ to } 775 &= 188' \\
41 \text{ to } 780 &= 51' \\
41 \text{ to } 785 &= 118' \\
41 \text{ to } 790 &= 165' \\
41 \text{ to } 795 &= 225'
\end{aligned}
$$

CONTOUR LINES

These are drawn through points having the same elevation. They show the height of ground above sea level (M.S.L.) in either feet or metres and can be drawn at any desired interval.

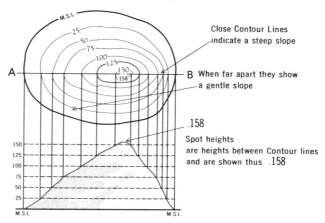

FIGURE 9.7 Contour plan with derived profile (line *AB*). (Courtesy Energy, Mines and Resources—Canada, Surveys and Mapping Branch)

Figure 9.8 shows line 71–41–42 and the contour crossings replotted at $1'' - 100'$ (for clarity). Directly below, the elevations are plotted vertically at $1'' = 30'$. The horizontal scale for both plan and profile is always identical; the vertical scale of the profile is usually exaggerated to properly display the terrain line. In this example, a convenient datum line (740) was suitably placed on the available paper so that the profile would be centered.

9.7 CROSS SECTIONS, END AREAS, AND VOLUMES

Cross sections establish ground elevations at right angles to a proposed route. Cross sections can be developed from a contour plan as were profiles in the previous section, although it is common to have cross sections taken by field surveys [see Section 7.4 and Figure 7.3(1)]. Chapter 16 introduces a computerized system that permits generation of profiles and cross sections from a general data base.

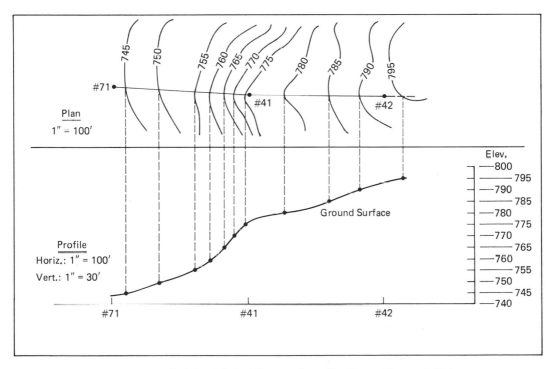

FIGURE 9.8 Plan and profile (from Figure 9.6b).

Cross sections are useful in determining quantities of cut and fill in construction design. If the original ground cross section is plotted, and then the as-constructed cross section is also plotted, the *end area* at that particular station can be computed. In Figure 9.9a, the proposed road at STA 7 + 00 is at an elevation below existing ground. This indicates a *cut* situation (i.e., the contractor will cut out that amount of soil shown between the proposed section and the original section). In Figure 9.9b, the proposed road elevation at station 3 + 00 is above the existing ground, indicating a *fill* situation (i.e., the contractor will bring in or fill in that amount of soil shown). Figure 9.9c shows a transition section between cut and fill sections.

When the end areas of cut or fill have been computed for adjacent stations, the volume of cut or fill between those stations can be computed by simply averaging the end areas and multiplying the average end area by the distance between the end area stations; Figure 9.10 illustrates this concept.

$$V = \left(\frac{A_1 + A_2}{2} \right) L \qquad (9.1)$$

Formula (9.1) gives the general case for volume computation, where A_1 and A_2 are the end areas of two adjacent stations, and L is the distance (feet or meters) between the stations. The answer in cubic feet is divided by 27 to give the answer in cubic yards; when metric units are used, the answer is left in cubic meters.

The average end area method of computing volumes is entirely valid only when the area of the midsection is, in fact, the average of the two end areas. This is seldom the case in actual earthwork computations; however, the error in volume resulting from this assumption is insignificant for the usual earthwork quantities of cut and fill. For special earthwork quantities (e.g., expensive structures excavation) or for higher-priced materials (e.g., concrete in place), a more precise method of volume computation, the prismoidal formula, must be used.

Example 9.1 *Volume by End Areas*

Figure 9.11a shows a pavement cross section for a proposed four-lane curbed road. As shown, the total pavement depth is 605 mm, the total width is 16.30 m, the subgrade is sloping to the side at 2%, and the top of the curb is 20 mm below the elevation of ℄. This proposed cross section is shown in Figure 9.11b along with the existing ground cross section at station 0 + 340. It can be seen that all subgrade elevations were derived from the proposed

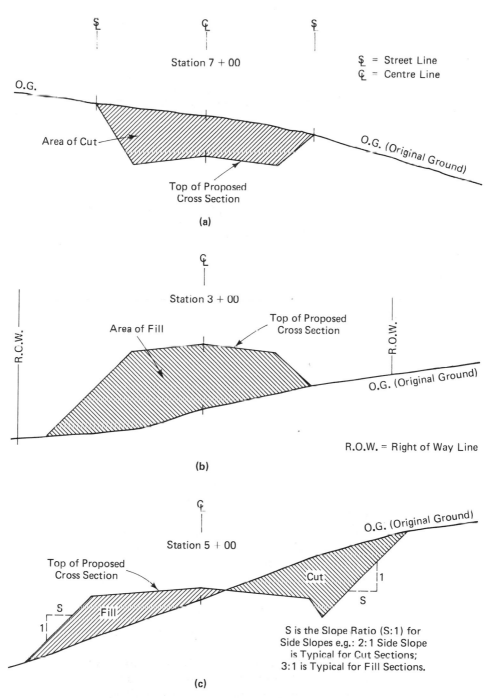

FIGURE 9.9 End areas: (a) cut section; (b) fill section; (c) transition section (i.e., both cut and fill).

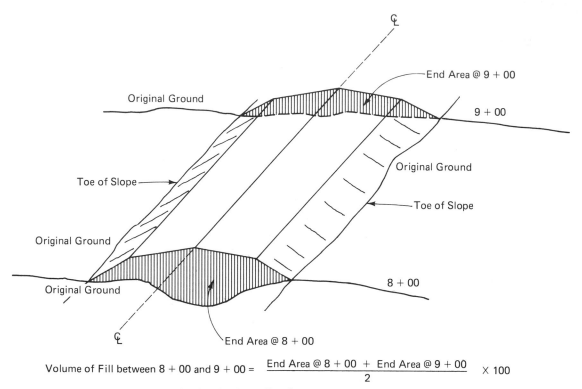

Volume of Fill between 8 + 00 and 9 + 00 = $\dfrac{\text{End Area @ 8 + 00 + End Area @ 9 + 00}}{2}$ × 100

FIGURE 9.10   Fill volume computations using end areas.

cross section, together with the $\mathcal{L}$ design elevation of 221.43. The desired end area is the area shown below the original ground plot and above the subgrade plot.

At this point, an elevation datum line is arbitrarily chosen (220.00). The datum line chosen can be any elevation value rounded to the closest foot, meter, or 5-ft value that is lower than the lowest elevation in the plotted cross section.

Figure 9.12 illustrates that end area computations involve the computation of two areas:

1. Area between the ground cross section and the datum line.
2. Area between the subgrade cross section and the datum line.

The desired end area (cut) is area 1 minus area 2. For fill situations, the desired end area would be area 2 minus area 1.

The end area computation can be determined as follows:

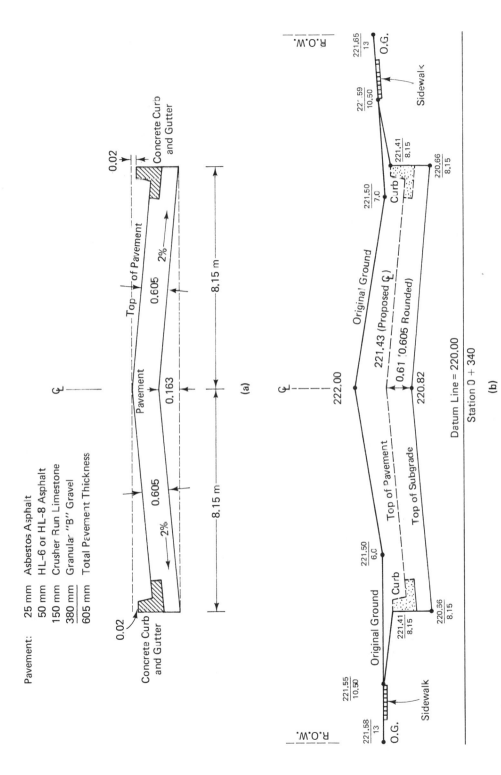

Pavement:
25 mm   Asbestos Asphalt
50 mm   HL-6 or HL-8 Asphalt
150 mm Crusher Run Limestone
380 mm Granular "B" Gravel
605 mm Total Pavement Thickness

(a)

(b)

Datum Line = 220.00

Station 0 + 340

FIGURE 9.11 End area computation, general: (a) typical arterial street cross section; (b) survey and design data plotted to show both the original ground and the design cross sections.

315

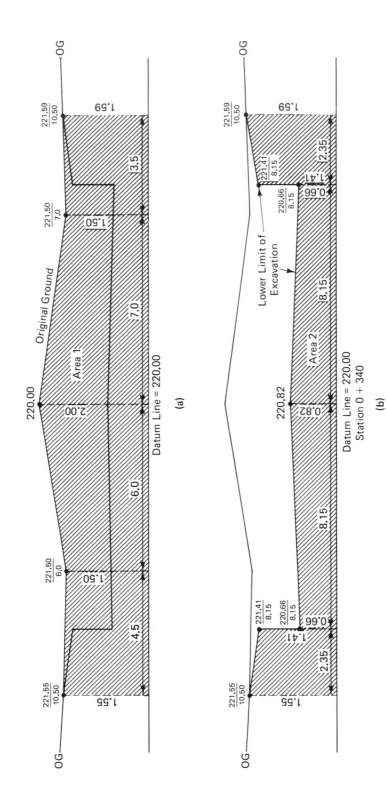

FIGURE 9.12 End area computations for cut areas: (a) area between ground cross section and datum line; (b) area between subgrade cross section and datum line.

316

Station	Plus		Sub area	Minus		Sub area
0 + 340	$\dfrac{1.55 + 1.50}{2} \times 4.5 =$		6.86	$\dfrac{1.55 + 1.41}{2} \times 2.35 =$		3.48
	$\dfrac{1.50 + 2.00}{2} \times 6.0 =$		10.50	$\dfrac{0.66 + 0.82}{2} \times 8.15 =$		6.03
	$\dfrac{2.00 + 1.50}{2} \times 7.0 =$		12.25	$\dfrac{0.82 + 0.66}{2} \times 8.15 =$		6.03
	$\dfrac{1.50 + 1.59}{2} \times 3.5 =$		5.41	$\dfrac{1.41 + 1.59}{2} \times 2.35 =$		3.53
	Check: 21 m		35.02 m²	Check: 21 m		19.07 m²

End area = 35.02 − 19.07 = 15.95 m²

Assuming that the end area at 0 + 300 has been computed to be 18.05 m², the volume of cut between 0 + 300 and 0 + 340 can now be computed:

$$V = \frac{18.05 + 15.95}{2} \times 40 = 680 \text{ m}^3 \qquad (9.1)$$

9.8 PRISMOIDAL FORMULA

If values more precise than end area volumes are required, the prismoidal formula can be used. A prismoid is a solid with parallel ends joined by plane or continuously warped surfaces. The prismoidal formula is

$$V = L\frac{(A_1 + 4_{Am} + A_2)}{6} \qquad \text{ft}^3 \text{ or m}^3 \qquad (9.2)$$

where A_1 and A_2 are the two end areas, Am is the area of a section midway between A_1 and A_2, and L is the distance from A_1 to A_2. Am is not the average of A_1 and A_2, but Am is derived from distances that are the average of corresponding distances required for A_1 and A_2 computations.

This formula is also used for other geometric solids (e.g., truncated prisms, cylinders, and cones). To justify its use, the surveyor must refine the field measurements to reflect the increase in precision being sought. A typical application of the prismoidal formula would be the computation of in-place volumes of concrete.

The difference in cost between a cubic yard or meter of concrete and a cubic yard or meter of earth cut or fill is sufficient reason for the increased precision.

9.9 CONSTRUCTION VOLUMES

In highway construction, for economic reasons, the designers try to optimally balance cut and fill volumes. Cut and fill cannot be precisely balanced because of geometric and esthetic design considerations, and because of the unpredictable effects of shrinkage and swell. *Shrinkage* occurs when a cubic yard (meter) is excavated and then placed while being compacted. The same material formerly occupying 1-yd^3 (m^3) volume now occupies a smaller volume. Shrinkage reflects an increase in density of the material and is obviously greater for silts, clays, and loams than it is for granular materials such as sand and gravel. *Swell* is a term used to describe the placing of shattered (blasted) rock. Obviously 1 yd^3 (m^3) of solid rock will expand significantly when shattered. Swell is usually in the 15% to 20% range, whereas shrinkage is in the 10% to 15% range, although values as high as 40% are possible with organic material in wet areas.

To keep track of cumulative cuts and fills as the profile design proceeds, the cumulative cuts (plus) and fills (minus) are shown graphically in a mass diagram. The total cut-minus-fill is plotted at each station directly below the profile plot. The mass diagram is an excellent method of determining *waste* or borrow volumes and can be adapted to show haul (transportation) considerations (Figure 9.13).

Large fills require *borrow* material usually taken from a nearby *borrow pit.* Borrow pit leveling procedures are described in Section 3.18 and Figure 3.24. The borrow pit in Figure 3.24 was laid out on a 50-ft grid. The volume of a grid square is the average height $(a + b + c + d)/4$ times the area of the base (50^2). The partial grid volumes (along the perimeter of the borrow pit) can be computed by forcing the perimeter volumes into regular geometric shapes (wedge shapes or quarter-cones).

When high precision is less important, volumes can be determined by analysis of contour plans; the smaller the contour interval, the more precise the result. The areas enclosed by a contour line can be taken off by planimeter; electronic planimeters are very useful for this purpose:

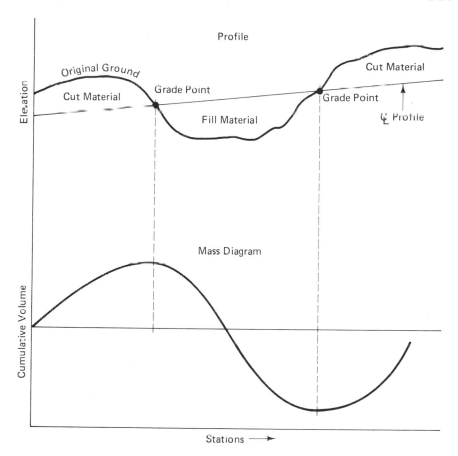

FIGURE 9.13 Mass diagram.

$$V = I\left(\frac{C_1 + C_2}{2}\right) \tag{9.3}$$

V is the volume (ft^3 or m^3) of earth or water; C_1 and C_2 are areas of adjacent contours; and I is the contour interval. The prismoidal formula can be used if m is an intervening contour (C_2) between C_1 and C_3. This method is well suited for many water-storage volume computations.

Finally, perhaps the most popular present-day volume computation technique involves the use of computers utilizing any one of a large number of available software programs. The computer programmer uses techniques similar to those described here, but the surveyor's duties may end with proper data entry to the computer (see Chapter 16).

9.10 AREA COMPUTATIONS

Areas enclosed by closed traverses can be computed using the double meridian distance method (Section 6.16) or the coordinate method (Section 6.15). Reference to Figure 9.14 will illustrate two additional area computation techniques.

9.10.1 Trapezoidal Technique

The area in Figure 9.14 was measured using a cloth tape for the offset distances. A common interval of 15 ft was chosen to suitably delineate the riverbank. Had the riverbank been more uniform, a larger interval could have been used, and had the riverbank been even more irregular, a smaller interval would have been appropriate. The trapezoidal technique assumes that the lines joining the ends of each offset line are straight lines (the smaller the common interval, the more valid this assumption).

The end sections can be treated as triangles:

$$A = \frac{8.1 \times 26.1}{2} = 106 \text{ ft}^2$$

and

$$A = \frac{11.1 \times 20.0}{2} = \underline{111} \ \text{ft}^2$$

$$217 \ \text{ft}^2 = \text{subtotal}$$

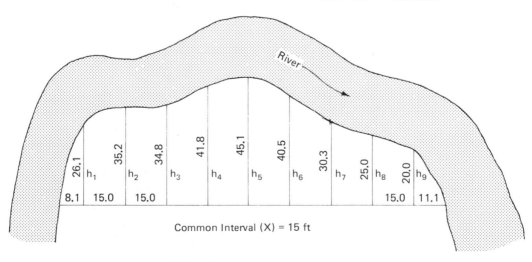

FIGURE 9.14 Irregular area computation.

The remaining areas can be treated as trapezoids. The trapezoidal rule is stated as follows:

$$\text{Area} = X \left(\frac{h_1 + h_n}{2} + h_1 + h_2 + \cdots + h_{n-1} \right) \tag{9.4}$$

where X = common interval between the offset lines
$\quad\quad\ h$ = offset measurement
$\quad\quad\ n$ = number of offset measurements

$$A = 15 \left(\frac{26.1 + 20.0}{2} + 35.2 + 34.8 + 41.8 + 45.1 + \right.$$
$$\left. 40.5 + 30.3 + 25.0 \right)$$

$$= 4136 \text{ ft}^2$$

Total area $= 4136 + 217 = 4353 \text{ ft}^2$

9.10.2 Simpsons One-Third Rule

This technique gives more precise results than the trapezoidal technique and is used where one boundary is irregular in the manner shown in Figure 9.14. The rule assumes that an odd number of offsets are involved and that the lines joining the ends of three successive offset lines are parabolic in configuration.

Simpsons one-third rule is stated as follows:

$$A = \frac{\text{interval}}{3} (h_1 + h_n + 2 \Sigma h \text{ odd} + 4 \Sigma h \text{ even}) \tag{9.5}$$

That is, one-third of the common interval times the sum of the first and last offsets ($h_1 + h_n$) plus twice the sum of the other odd offsets (Σh odd) plus four times the sum of the even-numbered offsets (Σh even). From Figure 9.14

$$A = \frac{15}{3} [26.1 + 20.0 + 2(34.8 + 45.1 + 30.3) +$$
$$4(35.2 + 41.8 + 40.5 + 25.0)]$$

$$= 4183 \text{ ft}^2$$

Total area $= 4183 + 217 \quad$ (from preceding example)
$$= 4400 \text{ ft}^2$$

If a problem is encounted with an even number of offsets, the area between the odd number of offsets is determined by Simpsons one-third rule with the remaining area being determined using the trapezoidal technique.

The discrepancy between the trapezoidal technique and

Simpson's one-third rule is 47 ft^2 in a total of 4400 ft^2 (about 1% in this case).

9.11 AREA BY GRAPHICAL ANALYSIS

We have seen that areas can be determined very precisely by using coordinates or DMDs (Chapter 6), and less precisely using the somewhat approximate methods illustrated by the trapezoidal rule and Simpson's one-third rule. Areas can also be determined by analyzing plotted data on plans and maps. For example, if a transparent sheet is marked off in grid squares to some known scale, an area outlined on a map can be determined by placing the squared paper over (sometimes under) the map and counting the number of squares and partial squares within the boundary limits shown on the map. The smaller the squares are, the more precise will be the result.

Another method of graphic analysis involves the use of a planimeter (Figures 9.15 and 9.16). A planimeter consists of a gradu-

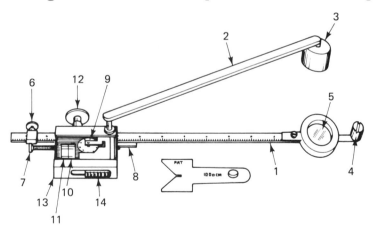

Part 1 Tracer Arm	Part 8 Tracer Arm Vernier
Part 2 Pole Arm	Part 9 Revolution Recording Dial
Part 3 Pole Weight	Part 10 Measuring Wheel
Part 4 Hand Grip	Part 11 Measuring Wheel Vernier
Part 5 Tracing Magnifier	Part 12 Idler Wheel
Part 6 Clamp Screw	Part 13 Carriage
Part 7 Fine Movement Screw	Part 14 Zero Setting Slide Bar

FIGURE 9.15 Polar planimeter.

FIGURE 9.16 Area take-off by polar planimeter.

ated measuring drum attached to an adjustable or fixed tracing arm, which itself is attached to a pole arm, one end of which is anchored to the working surface by a needle. The graduated measuring drum gives partial revolution readings, while a disc keeps count of the number of full revolutions.

Areas are determined by placing the pole-arm needle in a convenient location, setting the measuring drum and revolution counter to zero (some planimeters require recording an initial reading), and then tracing (using the tracing pin) the outline of the area being measured. As the tracing proceeds, the drum, which is also in contact with the working surface, revolves, measuring a value that is proportional to the area being measured. Some planimeters measure directly in square inches, while others can be set to map scales. When in doubt, or as a check on planimeter operation, the surveyor can measure out a scaled figure [e.g., 4-inch (100-mm) square] and then measure the area (16 in.^{2}) with a planimeter so that the planimeter area can be compared with the actual area laid off by scale. If the planimeter gives a result in square inches, say 51.2 in.^{2} , and the map is at a scale of $1'' = 100$ ft., the actual ground area portrayed by 51.2 in.^{2} would be $51.2 \times 100^2 = 512000$ ft^2 = 11.8 acres.

The planimeter is normally used with the pole-arm anchor point outside the area being traced. If it is necessary to locate the pole-arm anchor point inside the area being measured, as in the case of a relatively large area, the area of the zero circle of the planimeter must be added to the planimeter readings.

This constant is supplied by the manufacturer or can be deduced by simply measuring a large area twice, once with the anchor point outside the area and once with the anchor point inside the area.

Planimeters are particularly useful in measuring end areas (Section 9.7) used in volume computations. Planimeters are also effectively used in measuring watershed areas, as a check on various contruction quantities (e.g., areas of sod, asphalt), and as a check on areas determined by DMDs or coordinates.

Electronic planimeters (Figure 9.17) measure larger areas in less time than traditional polar planimeters. Computer software is available for highways and other earthworks applications (e.g., cross sections) and for drainage basin areas. The planimeter shown in Figure 9.17 has a 36×30 in. working area capability with a measuring resolution of 0.01 in.^{3} (0.02-in.^{2} accuracy).

FIGURE 9.17   Area take-off by a Numonics electronic planimeter.

PROBLEMS

A topographic survey was performed on a tract of land using leveling techniques to obtain elevations, and using stadia techniques to locate the topographic detail. The accompanying sketch shows the traverse (A to G) used for stadia control, and the grid baseline (0+00 @A) used to control the leveling survey—offset distances are @90° to the baseline. Also given are bearings and distances of the traverse sides; grid elevations; and angle and distance ties for the topographic detail. Problems 9-1 to 9-8 combine to form a comprehensive engineering project; the project can be completely covered by solving all the problems of this section, or individual problems can be selected to illustrate specific topics. Numeric values can be chosen as meters or feet. All field and design data for these problems are shown on this page and on pages 326, 327 and 328.

9-1   Establish the grid, plot the elevations (the decimal point is the plot point), and interpolate the data to establish contours at 1 m (ft) intervals. Scale @ 1:500 for metric units, or 1" = 10 ft or 15 ft for foot units. Use pencil.

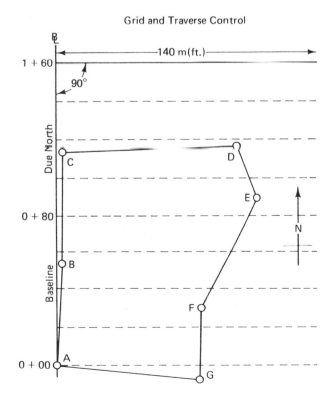

Grid and Traverse Control

9-2   Compute the interior angles of the traverse and check for geometric closure [i.e., n-2(180)].

9-3   Plot the traverse, using the interior angles and the given distances (scale as in 9-1).

9-4   Plot the stadia detail using the plotted traverse as control (scale as in 9-1).

9-5   Determine the area enclosed by the traverse in m^2 (ft^2), by using one or more of the following methods.

   a) Use grid paper as an overlay or underlay; count the squares and partial squares enclosed by the traverse; determine the area represented by one square at the chosen scale, and from that relationship determine the area enclosed by the traverse.

   b) Use a planimeter to determine the area.

   c) Divide the traverse into regular shaped figures (squares, rectangles, trapezoids, triangles) using a scale to determine the figure dimensions. Calculate the areas of the individual figures and sum them to produce the overall traverse area.

   d) Use the given balanced traverse data and the technique of coordinates, or double meridian distances (DMD's) to compute the traverse area.

A highway is to be constructed to pass through points *A* and *E* of the traverse. The proposed highway ℄ grade is + 2.30% rising from *A* to *E*. (℄ elevation at *A* = 68.95). The proposed cut and fill sections are shown in Figures A and B.

Balanced Traverse Data

Course	Bearing	Distance, m(ft.)
AB	N 3°30′ E	56.05
BC	N 0°30′ W	61.92
CD	N 88°40′ E	100.02
DE	S 23°30′ E	31.78
EF	S 28°53′ W	69.11
FG	South	39.73
GA	N 83°37′ W	82.67

Stadia Notes (Problem 9-4)

	Horizontal Angle	Distance m(ft)	Description
Sta. 兀 @ Sta. 'B' (sight 'C', 0°00')			
1	8°15'	45.5	S. limit of treed area
2	17°00'	57.5	"
3	33°30'	66.0	"
4	37°20'	93.5	"
5	45°35'	93.0	"
6	49°30'	114.0	"
兀 @ Sta. A (sight B, 0°00')			
7	50°10'	73.5	℄ gravel road (8m± width)
8	50°10'	86.0	"
9	51°30'	97.5	"
10	53°50'	94.5	N. limit of treed area
11	53°50'	109.0	"
12	55° 0'	58.0	℄ gravel road
13	66°15'	32.0	N. limit of treed area
14	86°30'	19.0	"
兀 @ Sta. D (sight E, 0°00')			
15	0°00'	69.5	℄ gravel road
16	7°30'	90.0	N. limit of treed area
17	64°45'	38.8	N. E. corner of building
18	13°30'	75.0	N. limit of treed area
19	88°00'	39.4	N. W. corner of building
20	46°00'	85.0	N. limit of treed area

9-6   Draw profile A-E showing both the existing ground and the proposed ℄ of highway. Use the following scales—metric: horizontal, 1:500; vertical, 1:100. foot: horizontal, 1"=10 ft or 15 ft; vertical, 1"=2 ft or 3 ft

9-7   Plot the highway ℄, and 16m (ft) width on the plan (see Problems 9-1, 9-3 & 9-4). Show the limits of cut and fill on the plan.

9-8   Combine Problems 9-1, 9-3, 9-4, 9-6 and 9-7 on one sheet of drafting paper. Suitably arrange the plan and profile together with the balanced traverse data and a suitable title block. All line work is to be in ink, and all text to be performed by mechanical lettering (e.g., Leroy) techniques. Use A2 size paper or equivalent.

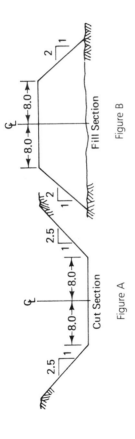

Figure A — Cut Section

Figure B — Fill Section

FIELD DATA   *Surveying Grid Elevations (Problem 9-1)*

Sta.	₵	20 m (ft)E	40 m (ft)E	60 m (ft)E	80 m (ft)E	100 m (ft)E	120 m (ft)E	140 m (ft)E
1+60	68.97	69.51	70.05	70.53	70.32			
1+40	69.34	69.82	71.12	71.00	71.26	71.99		
1+20	69.29	70.75	69.98	71.24	72.07	72.53	72.61	
1+00	69.05	71.02	70.51	69.91	72.02	73.85	74.00	75.18
0+80	69.09	71.90	74.13	71.81	69.87	71.21	74.37	74.69
0+60	69.12	70.82	72.79	72.81	71.33	70.97	72.51	73.40
0+40	68.90	69.66	70.75	72.00	72.05	69.80	71.33	72.42
0+20	68.02	68.98	69.53	70.09	71.11	70.48	69.93	71.51
0+00	67.15	68.11	68.55	69.55	69.92	71.02		

@STA.A

2
SURVEYING APPLICATIONS

10 *Highway Curves*

10.1 ROUTE SURVEYS

Highway and railroad routes are chosen only after a complete and detailed study of all possible locations. Route selection usually involves the use of air photos and ground surveys and the analysis of existing plans and maps. The route selected is chosen because it satisfies all design requirements with minimal social, environmental, and financial impact.

The proposed center line ($\mathcal{L}$) is laid out in a series of straight lines (tangents) beginning at 0 + 00 (0 + 000 metric) and continuing to the route terminal point. Each time the route changes direction, the deflection angle between the back tangent and forward tangent is measured and recorded. Existing detail that could have an effect on the highway design is tied in either by conventional ground surveys, by aerial surveys, or by a combination of the two methods; typical detail would include lakes, streams, trees, structures, existing roads and railroads, and so on. In addition to the detail location, the surveyor will run levels along the proposed route, with rod shots being taken across the route width at right angles to the $\mathcal{L}$ at regular intervals (full stations, half stations, etc.) and at locations dictated by changes in the topography.

The elevations thus determined will be used to aid in the design of horizontal and vertical alignments; in addition, these elevations will form the basis for the calculation of construction cut and fill quantities (see Chapters 9 and 12).

The location of detail and the determination of elevations are normally confined to that relatively narrow strip of land representing the highway right-of-way (ROW). Exceptions would include potential river, highway, and railroad crossings, where approach profiles and sight lines (railroads) must be established.

10.2 CIRCULAR CURVES: GENERAL

It was noted in the previous section that a highway route survey was initially laid out as a series of straight lines (tangents). Once the ℄ location alignment has been confirmed, the tangents are joined by circular curves that allow for smooth vehicle operation at the speeds for which the highway was designed. Figure 10.1 illustrates how two tangents are joined by a circular curve and shows some related circular curve terminology. The point at

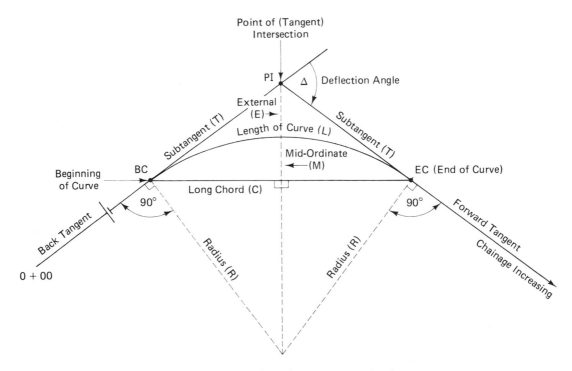

FIGURE 10.1 Circular curve terminology.

which the alignment changes from straight to circular is known as the BC (beginning of curve). The BC is located distance T (subtangent) from the PI (point of tangent intersection). The length of circular curve (L) is dependent on the central angle and the value of R (radius). The point at which the alignment changes from circular back to tangent is known as the EC (end of curve). Since the curve is symmetrical about the PI, the EC is also located distance T from the PI.

The terms BC and EC are referred to by some agencies as PC (point of curve) and PT (point of tangency).

10.3 CIRCULAR CURVE GEOMETRY

Most curve problems are calculated from field measurements (Δ and the chainage of PI) and from design parameters (R). Given R (which is dependent on the design speed) and Δ, all other curve components can be computed.

Analysis of Figure 10.2 will show that the curve deflection angle (PI, BC, EC) is $\Delta/2$, and that the central angle (0) is equal to Δ, the tangent deflection.

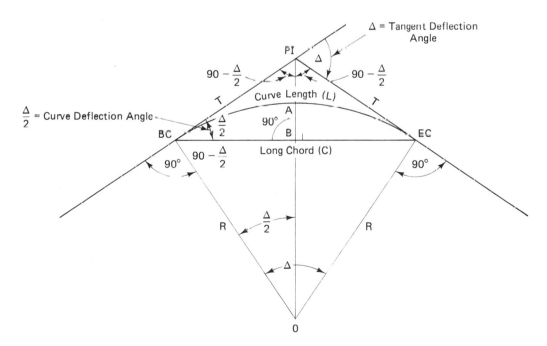

FIGURE 10.2   Geometry of the circle.

The line (0 - PI), joining the center of the curve to the PI, effectively bisects all related lines and angles.

$$\text{In } \Delta \text{ BC, 0, PI,} \qquad \frac{T}{R} = \text{Tan } \frac{\Delta}{2}$$

$$T = R \text{ Tan } \frac{\Delta}{2} \qquad (10.1)$$

$$\text{In } \Delta \text{ BC, 0, } B, \qquad \frac{1/2 \, C}{R} = \text{Sin } \frac{\Delta}{2}$$

$$C = 2R \text{ Sin } \frac{\Delta}{2} \qquad (10.2)$$

Also,

$$\frac{OB}{R} = \text{Cos } \frac{\Delta}{2}$$

$$OB = R \text{ Cos } \frac{\Delta}{2}$$

But

$$OB = R - M$$

$$R - M = R \text{ Cos } \frac{\Delta}{2}$$

$$M = R \left(1 - \text{Cos } \frac{\Delta}{2} \right) \qquad (10.3)$$

$$\text{In } \Delta \text{ BC, 0, PI,} \qquad 0 \text{ to PI} = R + E$$

$$\frac{R}{R + E} = \text{Cos } \frac{\Delta}{2}$$

$$E = R \left(\frac{1}{\text{Cos}(\Delta/2)} - 1 \right) \qquad (10.4)$$

$$= R \left(\text{Sec } \frac{\Delta}{2} - 1 \right) \qquad \text{(alternate)}$$

From Figure 10.3,

$$\frac{L}{2\pi R} = \frac{\Delta}{360}, \qquad L = 2\pi R \frac{\Delta}{360} \qquad (10.5)$$

where Δ is expressed in degrees and decimals of a degree.

The sharpness of the curve is determined by the choice of the radius (R); large radius curves are relatively flat, whereas small radius curves are relatively sharp.

Many highways agencies use the concept of degree of curve (D) to define the sharpness of the curve. Degree of curve D is defined to be that central angle subtended by 100 ft of arc. (In rail-

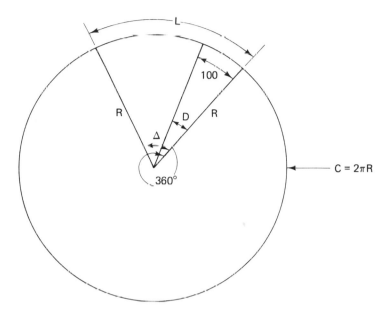

FIGURE 10.3 Relationship between the degree of curve (D) and the circle.

way design, D is defined to be the central angle subtended by 100 ft of chord.)

From Figure 10.3,

$$\frac{D}{360} = \frac{100}{2\pi R}, \qquad D = \frac{5729.58}{R} \qquad (10.6)$$

$$\frac{L}{100} = \frac{\Delta}{D}, \qquad L = 100\frac{\Delta}{D} \qquad (10.7)$$

Example 10.1

Refer to Figure 10.4. Given

$$\Delta = 16°38'$$
$$R = 1000 \text{ ft}$$
$$\text{PI at } 6 + 26.57$$

Calculate the station of the BC and EC; also calculate lengths C, M, and E.

$$T = R \operatorname{Tan} \frac{\Delta}{2} \quad (10.1) \qquad L = 2\pi R \frac{\Delta}{360} \quad (10.5)$$

$$= 1000 \operatorname{Tan} 8°19' \qquad = 2\pi \times 1000 \times \frac{16.6333°}{360}$$

$$= 146.18 \text{ ft} \qquad = 290.31 \text{ ft}$$

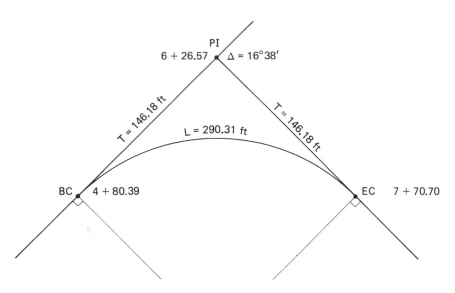

FIGURE 10.4 Sketch for Example 10.1.

$$
\begin{array}{lll}
\text{PI at} & 6\ + & 26.57 \\
-T & 1 & 46.18 \\ \hline
\text{BC} = & 4\ + & 80.39 \\
+L & 2 & 90.31 \\ \hline
\text{EC} = & 7\ + & 70.70
\end{array}
$$

$$C = 2R \operatorname{Sin} \frac{\Delta}{2} \tag{10.2}$$
$$= 2 \times 1000 \times \operatorname{Sin} 8°18'$$
$$= 289.29 \text{ ft}$$

$$M = R\left(1 - \operatorname{Cos} \frac{\Delta}{2}\right) \tag{10.3}$$
$$= 1000(1 - \operatorname{Cos} 8°19')$$
$$= 10.52'$$

$$E = R\left(\operatorname{Sec} \frac{\Delta}{2} - 1\right) \tag{10.4}$$
$$= 1000(\operatorname{Sec} 8°19' - 1)$$
$$= 10.63 \text{ ft}$$

Note: A common mistake made by students first studying circular curves is to determine the station of the EC by adding the T distance to the PI. Although the EC is physically a distance of T from the PI, the stationing (chainage) must reflect the fact that the ₵ no longer goes through the PI. The ₵ now takes the shorter distance (L) from the BC to the EC.

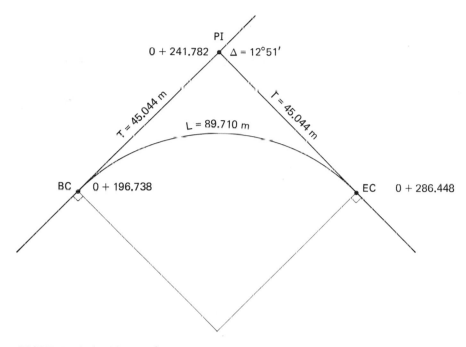

FIGURE 10.5 Sketch for Example 10.2. In this section, the magnitude of the Δ angle has been shown exaggerated.

Example 10.2

Refer to Figure 10.5. Given

$$\Delta = 12°51'$$
$$R = 100 \text{ m}$$
$$\text{PI at } 0 + 241.782$$

Calculate the station of the BC and EC.

$$T = R \operatorname{Tan} \frac{\Delta}{2} \quad (10.1)$$
$$= 400 \operatorname{Tan} 6°25'30''$$
$$= 45.044 \text{ m}$$

$$L = 2\pi R \frac{\Delta}{360} \quad (10.5)$$
$$= 2\pi \times 400 \times \frac{12.850}{360}$$
$$= 89.710 \text{ m}$$

PI at	0 +	241.782	
−T		45.044	
BC =	0 +	196.738	
+L		89.710	
EC =	0 +	286.448	

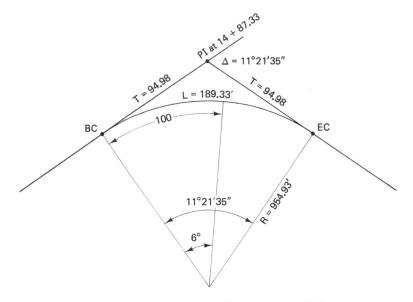

FIGURE 10.6   Sketch for Example 10.3.

Example 10.3

Refer to Figure 10.6. Given

$$\Delta = 11°21'35''$$
$$\text{PI at } 14 + 87.33$$
$$D = 6°$$

Calculate the station of the BC and EC.

$$R = \frac{5729.58}{D} = 954.93 \text{ ft} \qquad (10.6)$$

$$T = R \operatorname{Tan} \frac{\Delta}{2} = 954.93 \operatorname{Tan} 5.679861° \qquad (10.1)$$
$$= 94.98 \text{ ft}$$

$$L = 100 \frac{\Delta}{D} = \frac{100 \times 11.359722}{6} \qquad (10.7)$$
$$= 189.33 \text{ ft}$$

or,
$$L = \frac{2\pi R\Delta}{360} = 2\pi \times \frac{954.93 \times 11.359722}{360} \qquad (10.5)$$
$$= 189.33 \text{ ft}$$

$$
\begin{array}{rrcr}
\text{PI at} & 14 & + & 87.33 \\
-T & & & 94.98 \\
\hline
\text{BC} = & 13 & + & 92.35 \\
+L & 1 & & 89.33 \\
\hline
\text{EC} = & 15 & + & 81.68 \\
\end{array}
$$

10.4 CIRCULAR CURVE DEFLECTIONS

The most used method of locating a curve in the field is by deflection angles. Typically, the theodolite is set up at the BC and the deflection angles are turned from the tangent line (see Figure 10.7).

If we use the data from Example 10.2,

$$BC \text{ at } 0 + 196.738$$
$$EC \text{ at } 0 + 286.448$$
$$\frac{\Delta}{2} = 6°25'30''$$
$$L = 89.710$$
$$T = 45.044$$

And if the layout is to proceed at 20-m intervals, the procedure would be as follows. First, compute the deflection angle for 1 m of arc:

$$\frac{\text{Total deflection angle}}{\text{Length of arc}} = \frac{\Delta/2}{L}$$
$$\frac{\Delta/2}{L} = \frac{6°25'30''}{89.710} = \frac{385.50'}{89.710}$$
$$= 4.2971798'$$
$$= 4.2972 \text{ minutes per meter}$$

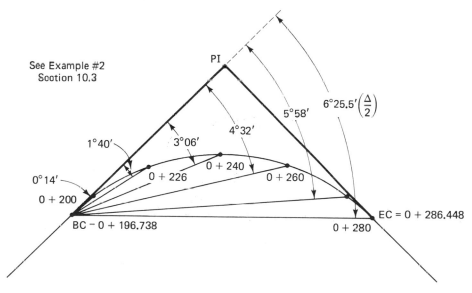

FIGURE 10.7 Field location for deflection angles.

Second, compute the deflection angles for the three required arc distances.

1. BC to first even station (0 + 200):
   $3.262 \times 4.2972 = 14.017' = 0°14.02'$
   (= 14.017401'; see calculation of subchord, Section 10.5)
2. Even station interval:
   $20 \times 4.2972 = 85.944' = 1°25.94'$
3. Last even station (0 + 280) to EC:
   $6.448 \times 4.2972 = 27.708' = 0°27.71'$

Third, prepare a list of appropriate stations together with *cumulative* deflection angles.

Stations	Deflection Angle	
BC 0 + 196.738	0°00.00'	
0 + 200	0°14.02'	+1°25.94'
0 + 220	1°39.96'	+1°25.94'
0 + 240	3°05.90'	+1°25.94'
0 + 260	4°31.84'	+1°25.94'
0 + 280	5°57.78'	+0°27.71'
EC 0 + 286.448	6°25.49'	$\approx 6°25.5' = \frac{\Delta}{2}$

For most engineering layouts, the deflection angles are rounded to the closest minute or half-minute. The two decimals are shown in the example only for bookkeeping purposes.

10.5 CHORD CALCULATIONS

In the previous example, it was determined that the deflection angle for station 0 + 200 was 0°14.02'; it follows that 0 + 200 could be located by placing a stake on the transit line at 0°14' and at a distance of 3.262 m (200 − 196.738) from the BC.

Furthermore, station 0 + 220 could be located by placing a stake on the transit line at 1°40' and at a distance of 20 m from the stake locating 0 + 200. The remaining stations could be located in a similar manner. However, it must be noted that the distances measured with a steel tape are not arc distances; they are straight lines known as *subchords*.

To calculate the subchord, Eq. (10.2), $C = 2R \ Sin(\Delta/2)$ may be used. This equation, derived from Figure 10.2, is the special

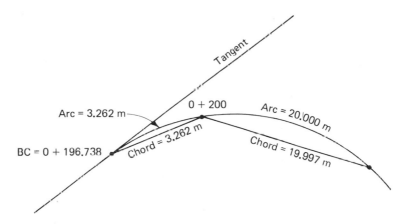

FIGURE 10.8   Curve arcs and chords.

case of the long chord and the total deflection angle. The general case can be stated as follows:

$$C = 2R \text{ Sin deflection angle} \qquad (10.8)$$

and any subchord can be computed if its deflection angle is known.

Relevant chords for the previous example can be computed as follows (see Figure 10.8):

First chord: $C = 2 \times 400 \times \text{Sin } 0°14.02$
                           $= 3.262$ m (at three decimals, chord $=$ arc)
                           ($C = 3.261988$ if deflection/m $=$ 4.2971798; see below)

Even station chord: $C = 2 \times 400 \times \text{Sin } 1°25.94'$
                           $- 19.997$ m

Last chord: $C = 2 \times 400 \times \text{Sin } 0°27.71'$
                           $- 6.448$

If these chord distances were used, the curve layout could proceed without error.

Note: Although the calculation of the first and last subchord shows the chord and arc to be equal (i.e., 3.262 m and 6.448 m), the chords are always marginally shorter than the arcs. In the cases of short distances (above) and in the case of flat (large radius) curves, the arcs and chords can often appear to be equal. If more decimal places are introduced into the computation, the marginal difference between arc and chord will become evident.

10.6 METRIC CONSIDERATIONS

Countries that have switched from English to metric (SI) units (e.g., Canada) have adopted for highway use a reference station of 1 km (e.g., 1 + 000), cross sections at 50-, 20-, and 10-m intervals, and a curvature design parameter based on a rational (even foot or meter) value radius, as opposed to a rational value degree (even degree) of curve (D).

The degree of curve found favor with most highway agencies because of the somewhat simpler calculations associated with its use, a factor that was significant in the pre-electronics age when most calculations were performed using logarithms. A comparison of techniques involving both D and R (radius) shows that the only computation in which the rational aspect of D is carried through is that for the arc length, that is, $L = 100\Delta/D$, Eq. (10.7), and even in that one case, the ease of calculation depends on delta (Δ) also being a rational number. In all other formulas, the inclusion of trigonometric functions or pi (π) ensures a more complex computation requiring the use of a calculator.

In fieldwork, the use of D (as opposed to R) permits quick determination of the deflection angle for even stations. For example, in English units, if the degree of curve were 2°, the deflection angle for a full station (100 ft) would be $D/2$ or 1°; for 50 ft, the deflection would be 0°30′; and so on.

In metric units, the degree of curve would be the central angle subtended by 100 m of arc, and the deflections would be similarly computed. That is, for a metric D of 6°, the deflections would be as follows: for 100 m, 3°; for 50 m, 1°30′; for 20 m, 0°36′; and for 10 m, 0°18′. The metric curve deflections here are not quite as simple as in the English system, but they are still uncomplicated and rational. However, curve stake outs require more stations than just those on the even chainages. For example, the BC and EC, catchbasins or culverts, vertical curve stations, and the like, usually occur on odd chainages, and the deflection angles for those odd chainages involve irrational number calculations requiring the use of a calculator.

The advent and proliferation of hand-held calculators and office computers have greatly reduced the importance of techniques that permit only marginal reductions in computations. Surveyors are now routinely solving their problems with calculators rather than utilizing the endless array of tables that once characterized the back section of survey texts.

An additional reason for the lessening importance of D in computing deflection angles is that many curves (particularly at interchanges) are now being laid out by control-point-based polar or intersection techniques (i.e., angle/distance or angle/angle) instead of deflection angles (see Chapter 11).

Those countries using the metric system, almost without exception, use a rational value for the radius (R) as a design parameter and have stopped using the degree of curve (D) for new designs.

10.7 FIELD PROCEDURE

With the PI location and Δ angle measured in the field, and with the radius or degree of curve (D) chosen consistent with the design speed, all curve computations can be completed. The surveyor then goes back out to the field and measures off the tangent (T) distance from the PI to locate the BC and EC on the appropriate tangent lines. The transit is then set up at the BC and zeroed and sighted in on the PI. The Δ/2 angle (6°25′30″ in Example 10.2) is then turned off in the direction of the EC mark (wood stake, nail, etc.). If the computations for T and the field measurements of T have been performed correctly, the line of sight at the Δ/2 angle will fall over the EC mark. If this does not occur, the T computations and then the field measurements are repeated.

Note: The Δ/2 line of sight over the EC mark will, of necessity, contain some error. In each case, the surveyor will have to decide if the resultant alignment error is acceptable for the type of survey in question. For example, if the Δ/2 line of sight misses the EC mark by 0.10 ft (30 mm) in a ditched highway ℄ survey, the surveyor would probably find the error acceptable and then proceed with the deflections. However, a similar error in the Δ/2 line of sight in a survey to lay out an elevated portion of urban freeway would not be acceptable; in that case an acceptable error would be roughly one-third of the preceding error (0.03 ft or 10 mm).

After the Δ/2 check has been satisfactorily completed, the curve stakes are set by turning off the deflection angle and measuring the chord distance for the appropriate stations. The theodolite is, if possible, left at the BC (see next section) for the entire curve stake out whereas the distance measuring moves continually forward from station to station. The rear tapeman keeps his body to the outside side of the curve to avoid blocking the line of sight from the instrument.

A final verification of the work is available after the last even station has been set; the chord distance from the last even station to the EC stake is measured and compared to the theoretical value; if the check indicates an unacceptable discrepancy, the work is checked.

Finally, after the curve has been deflected in, the party chief usually walks the curve, looking for any abnormalities; if a mistake has been made (e.g., putting in two stations at the same deflection angle), it will probably be very evident. The circular curve's symmetry is such that even minor mistakes are obvious in a visual check.

10.8 MOVING UP ON THE CURVE

The curve deflections shown in Section 10.4 are presented in a form suitable for deflecting in while set up at the BC, with a zero setting at the PI. However, it often occurs that the entire curve cannot be deflected in from the BC, and possibly two or more instrument setups may be required before the entire curve has been located. The reasons for this include a loss of line of sight due to intervening obstacles (i.e., detail or elevation rises).

In Figure 10.9, the data of Example 10.2 are used to illustrate the geometric considerations in moving up on the curve. In this case, station 0 + 260 cannot be established with the theodo-

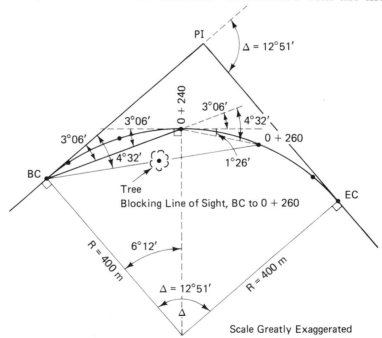

FIGURE 10.9 Moving up on the curve.

lite at the BC (as were the previous stations). The line of sight from the BC to 0 | 260 is obscured by a large tree. To establish station 0 + 260, the instrument is moved forward to the last station (0 +240) established from the BC. The horizontal circle is zeroed, and the BC is then sighted with the telescope in its inverted position. When the telescope is transited, the theodolite is once again oriented to the curve; that is, to set off the next (0 + 260) deflection, the surveyor refers to the previously prepared list of deflections and sets the appropriate deflection (4°32′) for the desired station location, and then for all subsequent stations.

Figure 10.9 shows the geometry involved in this technique. A tangent to the curve is shown by a dashed line through station 0 + 240 (the proposed setup location). The angle from that tangent line to a line joining 0 + 240 to the BC is the deflection angle 3°06′. When the line from the BC is produced through station 0 + 240, the same angle (3°06′) occurs between that line and the tangent line through 0 + 240 (opposite angles). It was determined that the deflection angle for 20 m was 1°26′ (Section 10.4). When 1°26′ is added to 3°06′, the angle of 4°32′ for station 0 + 260 results, the same angle previously calculated for that station.

This discussion has limited the move up to one station; in fact, the move up can be repeated as often as is necessary to complete the curve layout. The technique can generally be stated as follows: *When the instrument is moved up on the curve and the instrument is sighted, with the telescope inverted, at any other station, the theodolite will be "oriented to the curve," if the horizontal circle is first set to the value of the deflection angle for the sighted station;* i.e., in the case of a BC sight, the deflection angle to be set is obviously zero; if the instrument were set on 0 + 260 and sighting 0 + 240, a deflection angle of 3°06′ would first be set on the scale.

When the inverted telescope is transited to its normal position, all subsequent stations can then be sighted using the original list of deflections; this is the meaning of "theodolite oriented to the curve," and this is why the list of deflections can be made first, before the instrument setup stations have been determined and (as we shall see in the next section) even before it has been decided whether to run in the curve center line (₵) or whether it would be more appropriate to run in the curve on some offset line.

10.9 OFFSET CURVES

Curves being laid out for construction purposes must be established on offsets so that the survey stakes are not disturbed by construction activities. Many highways agencies prefer to lay out the curve on ℄ (center line) and then offset each ℄ stake a set distance left and right (left and right are oriented by facing to forward chainage).

The stakes can be offset to one side using the arm-swing technique described in Section 7.3, with the hands pointing to the two adjacent stations. If this is done with care, the offsets on that one side can be established on radial lines without too much error (see Figure 10.11 or 10.12). After one side has been offset in this manner, the other side is then offset by lining up the established offset stake with the ℄ stake and measuring out the offset distance, ensuring that all three stakes are visually in a straight line. Keeping the three stakes in a straight line will ensure that any alignment error existing at the offset stakes will steadily diminish as one moves toward the ℄ and the construction works.

In the construction of most municipal roads, particularly curbed roads, the center line may not be established; instead, the road alignment will be established directly on offset lines that are located a safe distance from the construction works. To illustrate, consider the curve in Example 10.2 being used to construct a curbed road as shown in Figure 10.10. The face of the curb is to be 4.00 m left and right of the center line. Assume that the curbs can be offset 2 m (each side) without interfering with construction (generally, the less cut or fill required, the smaller can be the offset distance).

Figure 10.11 shows that if the layout is to be kept on radial

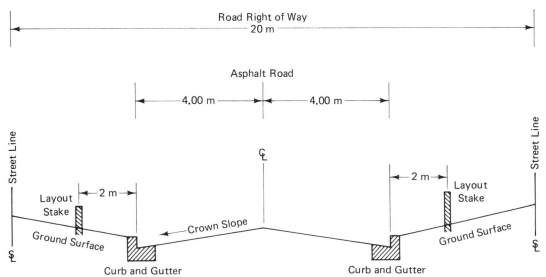

FIGURE 10.10 Municipal road cross section.

lines through the ℄ stations, the station arc distances on the left side (outside) curve will be longer than the corresponding ℄ arc distances, whereas the station arc distances on the right side (inside) curve will be shorter than the corresponding ℄ arc distances. The figure also clearly shows that the ratio of the outside arc to the ℄ arc is identical to the ratio of the ℄ arc to the inside arc. (See arc computations in the next section.) *By keeping the offset stations on radial lines, the surveyor is able to use the ℄ deflections previously computed.*

10.10 ILLUSTRATIVE PROBLEM FOR OFFSET CURVES (METRIC UNITS)

Consider the problem of a construction offset layout utilizing the data of Example 10.2, the deflections developed in Section 10.4, and the offset of 2 m introduced in the previous section.

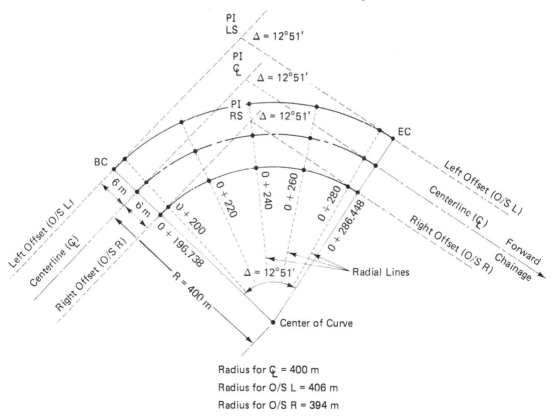

Radius for ℄ = 400 m
Radius for O/S L = 406 m
Radius for O/S R = 394 m

FIGURE 10.11   Offset curves.

Given data: $\Delta = 12°51'$
$R = 400$ m
PI at $0 + 241.782$
Calculated data: $T = 45.044$ m
$L = 89.710$ m
BC at $0 + 196.738$
EC at $0 + 286.448$
Required: Curbs to be laid out on 2-m offsets
at 20-m stations

Calculated Deflections

Station	Computed Deflection	Field Deflection
BC 0 + 196.738	0°00.00'	0°00'
0 + 200	0°14.02'	0°14'
0 + 220	1°39.96'	1°40'
0 + 240	3°05.90'	3°06'
0 + 260	4°31.84'	4°32'
0 + 280	5°57.78'	5°58'
EC 0 + 286.448	6°25.49'	6°25.5' $= \frac{\Delta}{2}$, Check

Reference to Figures 10.10 and 10.11 will show that the left side (outside) curb face will have a radius of 404 m. A 2-m offset for that curb will result in an offset radius of 406 m. Similarly, the offset radius for the right side (inside) curb will be $400 - 6 = 394$ m.

Since we are going to use the deflections already computed, it only remains to calculate the corresponding left-side arc or chord distances and the corresponding right-side arc or chord distances. Although layout procedure (angle and distance) indicates that chord distances will be required, for illustrative purposes we will compute both the arc and chord distances on offset.

Arc Distance Computations

Reference to Figure 10.12 will show that the offset (o/s) arcs can be computed by direct ratio.

$$\frac{\text{o/s arc}}{\text{₡ arc}} = \frac{\text{o/s radius}}{\text{₡ radius}}$$

For the first arc (BC to $0 + 200$),

Left side: o/s arc $= 3.262 \times \dfrac{406}{400} = 3.311$ m

Right side: o/s arc $= 3.262 \times \dfrac{394}{400} = 3.213$ m

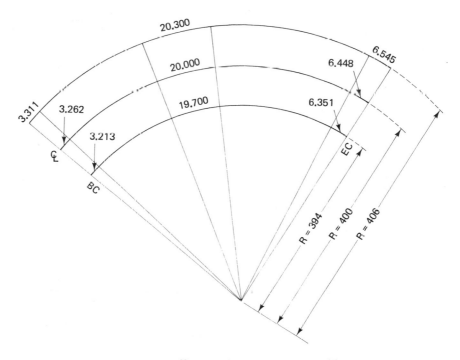

FIGURE 10.12 Offset arc lengths computed by ratios.

For the even station arcs,

$$\text{Left side:} \quad \text{o/s arc} = 20 \times \frac{406}{400} = 20.300 \text{ m}$$

$$\text{Right side:} \quad \text{o/s arc} = 20 \times \frac{394}{400} = 19.700 \text{ m}$$

For the last arc (0 + 280 to EC),

$$\text{Left side:} \quad \text{o/s arc} = 6.448 \times \frac{406}{400} = 6.545 \text{ m}$$

$$\text{Right side:} \quad \text{o/s arc} = 6.448 \times \frac{394}{400} = 6.351 \text{ m}$$

Summary of Arc Distances(m)

Interval	Left Side		℄		Right Side
		Difference		Difference	
BC to 0 + 200	3.311 ◄— 0.049 —►		3.262 ◄— 0.049 —►		3.213
Even stations	20.300 ◄— 0.300 —►		20.000 ◄— 0.300 —►		19.700
0 + 280 to EC	6.545 ◄— 0.097 —►		6.448 ◄— 0.097 —►		6.351

Arithmetic check: $LS - ℄ = ℄ - RS$

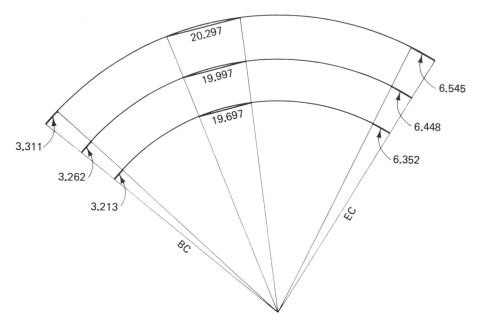

FIGURE 10.13 Offset chords computed from deflection angles and offset radii.

Chord Distance Computations

Refer to Figure 10.13. For any deflection angle, the equation for chord length is (see Section 10.5)

$$C = 2R \text{ Sin deflection} \qquad (10.8)$$

In this problem, the deflection angles have been calculated previously, and it is the radius (R) that is the variable.

For the first chord (BC to 0 + 200),

Left side: $C = 2 \times 406 \times \text{Sin } 0°14.017 = 3.312 \text{ m}$
Right side: $C = 2 \times 394 \times \text{Sin } 0°14.017 = 3.214 \text{ m}$

(If the angle here is rounded to 0°14.02', the check shown below is marginally off.)

For the even station chords,

Left side: $C = 2 \times 406 \times \text{Sin } 1°25.94' = 20.297 \text{ m}$
Right side: $C = 2 \times 394 \times \text{Sin } 1°25.94' = 19.697 \text{ m}$

See Section 10.4.

For the last chord,

Left side: $C = 2 \times 406 \times \text{Sin } 0°27.71' = 6.545 \text{ m}$
Right side: $C2 = 394 \times \text{Sin } 0°27.71'$ $= 6.352 \text{ m}$

Summary of Chord Distances(m)

Interval	Left Side		℄ (See Section 10.5)		Right Side
		Difference		Difference	
BC to 0 + 200	3.311	◄— 0.049 —►	3.262	◄— 0.049 —►	3.213
Even stations	20.297	◄— 0.300 —►	19.997	◄— 0.300 —►	19.697
0 + 280 to EC	6.545	◄— 0.097 —►	6.448	◄— 0.096 —►	6.351

10.11 CURVE PROBLEM (FOOT UNITS)

Given the following ℄ data,

$$D = 5°$$
$$\Delta = 16°28'30''$$
$$\text{PI at } 31 + 30.62$$

it is required to furnish stake out information for the curve on 50-ft offsets left and right of ℄ at 50-ft stations.

$$R = \frac{5729.58}{D} \tag{10.6}$$
$$= 1145.92 \text{ ft}$$

$$T = R \text{ Tan} \frac{\Delta}{2} \tag{10.1}$$
$$= 1145.92 \text{ Tan } 8°14'15'' = 165.90 \text{ ft}$$

$$L = 100 \frac{\Delta}{D} \tag{10.7}$$
$$= \frac{100 \times 16.475}{5} = 329.50 \text{ ft}$$

Alternately,

$$L = 2\pi R \frac{\Delta}{360} \quad (10.5) \quad = 329.50 \text{ ft}$$

$$
\begin{array}{lrr}
\text{PI at} & 31 + & 30.62 \\
-T & 1 & 65.90 \\
\hline
\text{BC} = & 29 + & 64.72 \\
+L & 3 & 29.50 \\
\hline
\text{EC} = & 32 + & 94.22 \\
\end{array}
$$

Computation of Deflections

$$\text{Total deflection for curve} = \frac{\Delta}{2} = 8°14'15'' = 494.25'$$

$$\text{Deflection per foot} = \frac{494.25}{329.50} = 1.5'/\text{ft}$$

Alternately, since $D = 5°$, the deflection for 100 ft is $D/2$ or $2°30'$ $= 150'$. The deflection, therefore, for 1 ft is $150/100$ or $1.5'/\text{ft}$.

Deflection for first station:

$$35.28 \times 1.5 = 52.92' = 0°52.9'$$

Deflection for even 50-ft stations:

$$50 \times 1.5 = 75' = 1°15'$$

Deflection for last station:

$$44.22 \times 1.5 = 66.33' = 1°06.3'$$

	Deflections (cumulative)	
Stations	Office	Field (closest minute)
BC 29 + 64.72	0°00.0'	0°00'
30 + 00	0°52.9'	0°53'
30 + 50	2°07.9'	2°08'
31 + 00	3°22.9'	3°23'
31 + 50	4°37.9'	4°38'
32 + 00	5°52.9'	5°53'
32 + 50	7°07.9'	7°08'
EC 32 + 94.22	8°14.2'	8°14'
	8°14.25'	
	$= \frac{\Delta}{2}$, Check	

Chord calculations for left- and right-side curves on 50-ft (from ℄) offsets (see Figure 10.14):

$$\text{Radius for } ℄ = 1145.92 \text{ ft}$$
$$\text{Radius for LS} = 1195.92 \text{ ft}$$
$$\text{Radius for RS} = 1095.92 \text{ ft}$$

10.12 COMPOUND CIRCULAR CURVES

A compound curve consists of two (usually) or more circular arcs between two main tangents turning in the same direction and joining at common tangent points. Figure 10.15 shows a com-

Chord Calculations

Interval	Left Side	℄	Right Side
BC to 30 + 00	$C = 2 \times 1195.92 \times \text{Sin } 0°52.9'$ $= 36.80$ ft	$C = 2 \times 1145.92 \times \text{Sin } 0°52.9'$ $= 35.27$ ft	$C = 2 \times 1095.97 \times \text{Sin } 0°52.9'$ $= 33.73$ ft
	Diff. $= 1.53$	Diff. $= 1.54$	
50-ft stations	$C = 2 \times 1195.92 \times \text{Sin } 1°15'$ $= 52.18$ ft	$C = 2 \times 1145.92 \times \text{Sin } 1°15'$ $= 50.00$ (to 2 decimals)	$C = 2 \times 1095.92 \times \text{Sin } 1°15'$ $= 47.81$ ft
	Diff. $= 2.18$	Diff. $= 2.19$	
32 + 50 to EC	$C = 2 \times 1195.92 \times \text{Sin } 1°06.3'$ $= 46.13$ ft	$C = 2 \times 1145.92 \times \text{Sin } 1°06.3'$ $= 44.20$ ft	$C = 2 \times 1095 \times \text{Sin } 1°05.3'$ $= 42.27$ ft
	Diff. $= 0.93$	Diff. $= 0.93$	

Differences are equal or nearly so, giving a check on the work.

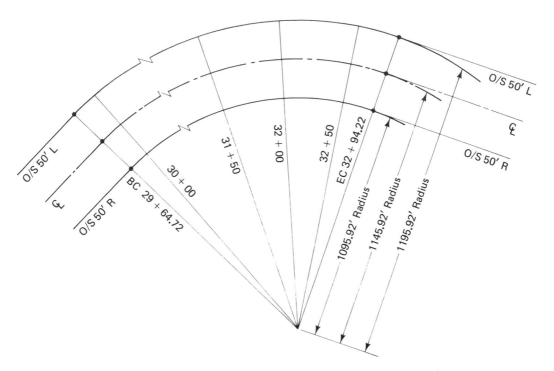

FIGURE 10.14   Sketch for the curve problem of Section 10.11.

pound curve consisting of two circular arcs joined at a point of compound curve (PCC). The lower chainage curve is number 1, whereas the higher chainage curve is number 2.

The parameters are R_1, R_2, Δ_1, Δ_2 ($\Delta_1 + \Delta_2 = \Delta$), T_1, and T_2. If four of these six or seven parameters are known, the others can be solved. Under normal circumstances, Δ_1 and Δ_2, or Δ, are measured in the field, and R_1 and R_2 are given by design considerations with minimum values governed by design speed.

Although compound curves can be manipulated to provide practically any vehicle path desired by the designer, they are not employed where simple curves or spiral curves can be utilized to achieve the same desired effect. Practically, compound curves are reserved for those applications where design constraints (topographic or cost of land) preclude the use of simple or spiral curves, and they are now usually found chiefly in the design of interchange loops and ramps. Smooth driving characteristics require that the larger radius be no more than $1\frac{1}{3}$ times larger than the smaller radius (this ratio increases to $1\frac{1}{2}$ when dealing with interchange curves).

Solutions to compound curve problems vary, as several possi-

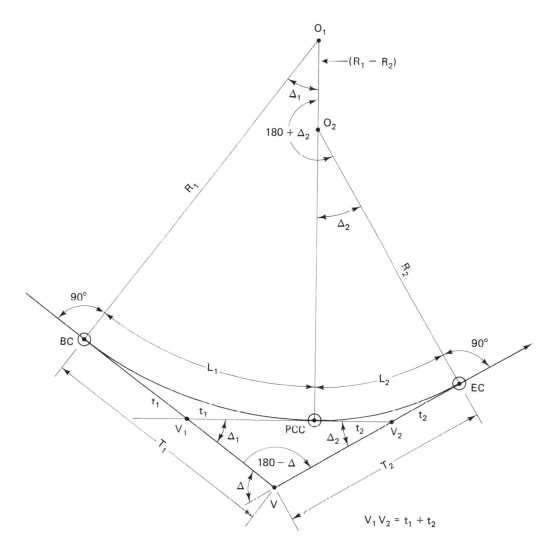

FIGURE 10.15 Compound circular curve.

bilities exist as to which of the data are known in any one given problem. All problems can be solved by use of the sine law or cosine law or by the omitted measurement traverse technique illustrated in Example 6.3 of Section 6.11.

If the omitted measurement traverse technique is used, the problem becomes a five-sided traverse (Figure 10.15) with sides R_1, T_1, T_2, R_2, and $(R_1 - R_2)$, and with angles 90°, $180 - \Delta°$, 90°, $180 + \Delta°_2$, and $\Delta°_1$. An assumed azimuth can be chosen that will simplify the computations (i.e., set direction of R_1 to be 0°00′00″).

10.13 REVERSE CURVES

Reverse curves (see Figure 10.16a and b) are seldom used in highway or railway alignment. The instantaneous change in direction occurring at the PRC would cause discomfort and safety problems for all but the slowest of speeds. Additionally, since the change in curvature is instantaneous, there is no room to provide superelevation transition from cross-slope right to cross-slope left (see Sections 10.26 and 10.27). However, reverse curves can be used to advantage where the instantaneous change in direction poses no threat to safety or comfort.

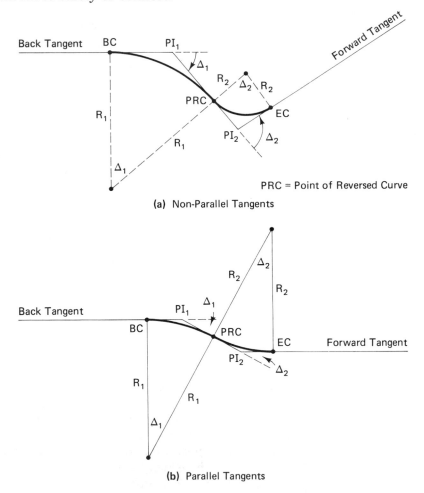

FIGURE 10.16 Reverse curves: (a) non-parallel tangents; (b) parallel tangents.

The reverse curve is particularly pleasing to the eye and is used with great success on park roads, formal paths, waterway channels, and the like. This curve can be encountered in both situations illustrated in Figure 10.16a and b; the parallel tangent application is particularly common (R_1 is often equal to R_2). As with compound curves, reverse curves have six independent parameters (R_1, Δ_1, T_1, R_2, Δ_2, T_2); the solution technique depends on which parameters are unknown, and the techniques noted for compound curves will also provide the solution to reverse curve problems.

10.14 VERTICAL CURVES: GENERAL

Vertical curves are used in highway and street vertical alignment to provide a gradual change between two adjacent grade lines. Some highways and municipal agencies introduce vertical curves at every change in grade-line slope, whereas other agencies introduce vertical curves into the alignment only when the net change in slope direction exceeds a specific value (e.g., 1.5% or 2%).

In Figure 10.17, vertical curve terminology is introduced: g_1 is the slope (%) of the lower chainage grade line, g_2 is the slope of the higher chainage grade line, BVC is the beginning of the vertical curve, EVC is the end of the vertical curve, and PVI is the point of intersection of the two adjacent grade lines. The length of vertical curve (L) is the projection of the curve onto a horizontal surface and as such corresponds to plan distance.

The algebraic change in slope direction is A, where $A = g_2 - g_1$. For example, if $g_1 = +1.5\%$ and $g_2 = -3.2\%$, A would be equal to $(-3.2 - 1.5) = -4.7$.

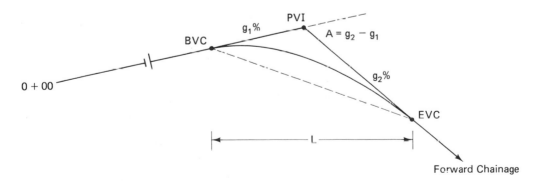

FIGURE 10.17   Vertical curve terminology. (Profile view shown)

The geometric curve used in vertical alignment design is the vertical axis parabola. The parabola has the desirable characteristics of (1) a constant rate of change of slope, which contributes to smooth alignment transition, and (2) ease of computation of vertical offsets, which permits easily computed curve elevations.

The general equation of the parabola is

$$y = ax^2 + bx + c \qquad (10.9)$$

The slope of this curve at any point is given by the first derivative,

$$\frac{dy}{dx} = 2ax + b \qquad (10.10)$$

and the rate of change of slope is given by the second derivative,

$$\frac{d^2y}{dx^2} = 2a \qquad (10.11)$$

which, as was previously noted, is a constant. The rate of change of slope (2a) can also be written as A/L.

If, for convenience, the origin of the axes is placed at the BVC (Figure 10.18), the general equation becomes

$$y = ax^2 + bx$$

and because the slope at the origin is g_1, the expression for slope of the curve at any point becomes

$$\frac{dy}{dx} = \text{slope} = 2ax + g_1 \qquad (10.12)$$

The general equation can finally be written as

$$y = ax^2 + g_1 x \qquad (10.13)$$

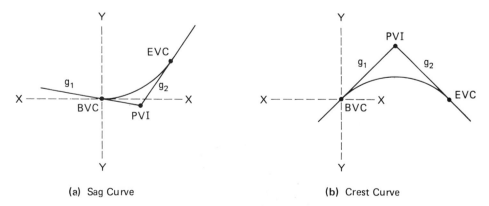

(a) Sag Curve   (b) Crest Curve

FIGURE 10.18   Types of vertical curves: (a) sag curve; (b) crest curve.

10.15 GEOMETRIC PROPERTIES OF THE PARABOLA

Figure 10.19 illustrates the following relationships:

1. The difference in elevation between the BVC and a point on the g_1 grade-line at a distance X units (feet or meters) is g_1X (g_1 is expressed as a decimal).

2. The tangent offset between the grade line and the curve is given by ax^2, where x is the horizontal distance from the BVC; that is, tangent offsets are proportional to the squares of the horizontal distances.

3. The elevation of the curve at distance x from the BVC is given by BVC $+ g_1x - ax^2 = $ curve elevation (the signs would be reversed in a sag curve).

4. The grade lines (g_1 and g_2) intersect midway between the BVC and the EVC; that is, BVC to $V = \frac{1}{2}L = V$ to EVC.

5. Offsets from the two grade lines are symmetrical with respect to the PVI (V).

6. The curve lies midway between the PVI and the mid-point of the chord; that is, CM = MV.

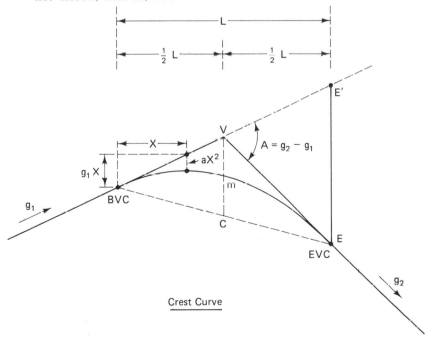

FIGURE 10.19 Geometric properties of the parabola.

10.16 COMPUTATION OF THE HIGH
OR LOW POINT ON A VERTICAL CURVE

The locations of curve high and low points (if applicable) are important for drainage considerations; for example, on curbed streets catch basins must be installed precisely at the drainage low point.

It was noted earlier that the slope was given by the expression

$$\text{Slope} = 2aX + g_1 \tag{10.12}$$

Figure 10.20 shows a sag vertical curve with a tangent drawn through the low point; it is obvious that the tangent line is horizontal with a slope of zero; that is,

$$2ax + g_1 = 0 \tag{10.14}$$

Had a crest curve been drawn, the tangent through the high point would have exhibited the same characteristics.

Since
$$2a = \frac{A}{L}$$

expression (10.14) can be rewritten as

$$x = \frac{-g_1 L}{A} \tag{10.15}$$

where x is the distance from the BVC to the high or low point.

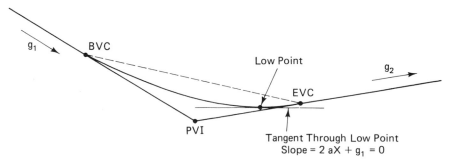

FIGURE 10.20 Tangent at curve low point.

10.17 PROCEDURE FOR COMPUTING A VERTICAL CURVE

1. Compute the algebraic difference in grades: $A = g_2 - g_1$.
2. Compute the distance from the BVC to the high or low point (if applicable):

$$x = \frac{-g_1 L}{A} \qquad (10.15)$$

and determine the station of the high/low point.

3. Compute the chainage of the BVC and EVC. If the chainage of the PVI is known, $\frac{1}{2}L$ is simply subtracted and added to the PVI chainage.
4. Compute the tangent grade-line elevation of the BVC and the EVC.
5. Compute the tangent grade-line elevation for each required station.
6. Compute the mid-point of chord elevation:

$$\frac{\text{Elevation of BVC + elevation of EVC}}{2}$$

7. Compute the tangent offset (d) at the PVI (i.e., distance VM in Figure 10.19):

$$d = \frac{\text{Difference in elevation of PVI and midpoint of curve}}{2}$$

8. Compute the tangent offset for each individual station (see line ax^2 in Figure 10.19):

$$\text{tangent offset} = d\left(\frac{x}{L/2}\right)^2 \qquad (10.16)$$

where x is the distance from the BVC or EVC (whichever is closer) to the required station.

9. Compute the elevation on the curve at each required station by combining the tangent offsets with the appropriate tangent grade-line elevations—add for sag curves and subtract for crest curves.

10.18 ILLUSTRATIVE EXAMPLE

The ease with which vertical curves can be computed is shown in the following example. Given that $L = 300$ ft, $g_1 = -3.2\%$, $g_2 = +1.8\%$, PVI at 30 + 30, and elevation = 465.92, determine the location of the low point and elevations on the curve at even stations, as well as at the low point.

Solution

1. $A = 1.8 - (-3.2) = 5.0$.
2. PVI $- \frac{1}{2}L$ = BVC

   BVC at $(30 + 30) - 150$
   $= 28 + 80.00$

   PVI $+ \frac{1}{2}L$ = EVC
   EVC $(30 + 30) + 150$
   $= 31 + 80.00$
   EVC $-$ BVC $= L$
   $(31 + 80) - (28 + 80) = 300$ Check

3. Elevation of PVI $= 465.92$
   150 ft at 3.2% $= 4.80$ (see Figure 10.21)
   Elev. BVC $= 470.72$

   Elevation PVI $= 465.92$
   150 ft at 1.8% $= 2.70$
   Elevation EVC $= 468.62$

4. Location of low point

   $$x = \frac{-g_1 L}{A} \qquad (10.15)$$

   $$= \frac{3.2 \times 300}{5} = 192.00 \text{ ft} \qquad \text{(from the BVC)}$$

5. Tangent grade-line computations are entered in Table 10.1.
   Example: Elevation at $29 + 00 = 470.72 - (0.032 \times 20)$
   $= 470.72 - 0.64$
   $= 470.08$

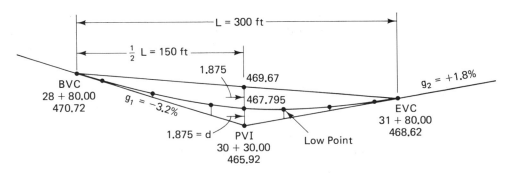

FIGURE 10.21   Sketch for Example 10.18.

6. Mid-chord elevation:

$$\frac{470.72 \ (BVC) + 468.62 \ (EVC)}{2} = 469.67$$

7. Tangent offset at PVI (d):

$$d = \frac{\text{Difference in elevation of PVI and mid-chord}}{2}$$

$$= \frac{469.67 - 465.92}{2}$$

$$= \frac{3.75}{2}$$

$$= 1.875 \ \text{ft}$$

8. Tangent offsets are computed by multiplying the distance ratio squared, $(x/L/2)^2$, by the maximum tangent offset (d). See Table 10.1.

9. The computed tangent offsets are added (in this example) to the tangent elevation in order to determine the curve elevation.

TABLE 10.1 *Parabolic Curve Elevations by Tangent Offsets*

Station	*Tangent Elevation* +	*Tangent Offset* $\left(\frac{x}{\frac{1}{2}l}\right)^2 d*$	= *Curve Elevation*
BVC 28 + 80	470.72	$(0/150)^2 \times 1.875 = 0$	470.72
29 + 00	470.08	$(20/150)^2 \times 1.875 = .03$	470.11
30 + 00	466.88	$(120/150)^2 \times 1.875 = 1.20$	468.08
PVI 30 + 30	465.92	$(150/150)^2 \times 1.875 = 1.875$	467.80
Low			
Point 30 + 72	466.68	$(108/150)^2 \times 1.875 = .97$	467.65
31 + 00	467.18	$(80/150)^2 \times 1.875 = .53$	467.71
EVC 31 + 80	468.62	$(0/150)^2 \times 1.875 = 0$	468.62
See Section 10.19			
30 + 62	466.50	$(118/150)^2 \times 1.875 = 1.16$	467.66
30 + 72	466.68	$(108/150)^2 \times 1.875 = 0.97$	467.65
30 + 82	466.86	$(98/150)^2 \times 1.875 = 0.80$	467.66

*Where x is distance from BVC or EVC, whichever is closer.

10.19 DESIGN CONSIDERATIONS

From Section 10.14, $2a = A/L$ is an expression giving the constant rate of change of slope for the parabola. Another useful relationship is the inverse, or

$$K = \frac{L}{A} \tag{10.17}$$

where K is the horizontal distance required to effect a 1% change in slope on the vertical curve.

Substituting for L/A in formula (10.15) yields

$$x = -g_1 K \qquad (10.18)$$

(the result is always positive), where x is the distance to the low point from the BVC, or

$$X = +g_2 K \qquad (10.19)$$

where x is the distance to the low point from the EVC.

In Figure 10.19, it is seen that EE' is the distance generated by the divergence of g_1 and g_2 over distance $L/2$:

$$EE' = \frac{g_2 - g_1}{100} \times \frac{L}{2}$$

Also in Figure 10.19, it can be seen that $VC = \frac{1}{2} EE'$ (similar triangles) and that $VM = d = \frac{1}{4} EE'$; thus

$$d = \frac{1}{4} \left(\frac{g_2 - g_1}{100} \right) \frac{L}{2}$$

$$= \frac{AL}{800} \qquad (10.20)$$

or, from Eq. (10.17),

$$d = \frac{KA^2}{800} \qquad (10.21)$$

Equations (10.18), (10.19), and (10.21) are useful when design criteria are defined in terms of K.

Table 10.2 shows values of K for minimum stopping sight distances. On crest curves it is assumed that the driver's eye height is at 1.05 m and the object that the driver must see is at least 0.38 m. The defining conditions for sag curves would be night-time restrictions and relate to the field of view given by headlights with an angular beam divergence of 1°.

In practice, the length of vertical curve is rounded to the nearest even meter and, if possible, the PVI is located at an even station so that the symmetrical characteristics of the curve can be fully utilized. To avoid the esthetically unpleasing appearance of very short vertical curves, as a rule of thumb some agencies insist that the length of vertical curve (L) be at least as long in meters as the design velocity is in kilometers per hour.

Closer analysis of the data shown in Table 10.1 indicates a possible concern. Unfortunately, the vertical curve at the low or high point has a relatively small change of slope. This is not a

TABLE 10.2 *Design Controls for Crest and Sag Vertical Curves Based on Minimum Stopping Sight Distances for Various Design Speeds*

Design Speed, v (km/h)	Min. Stopping Sight Distance, s (m)	K Factor Crest (m)	K Factor Sag (m)
40	45	4	8
50	65	8	12
60	85	15	18
70	110	25	25
80	135	35	30
90	160	50	40
100	185	70	45
110	215	90	50
120	245	120	60
130	275	150	70
140	300	180	80

$$K_{crest} = \frac{S^2}{200 h_1 (1 + \sqrt{h_2/h_1})^2}$$

$$K_{sag} = \frac{S^2}{200 (h + S \tan\alpha)}$$

$h_1 = 1.05$ m, height of driver's eye

$h = 0.60$ m, height of headlights

$h_2 = 0.38$ m, height of object

$\alpha = 1°$, angular spread of light beam

$L = KA$, from Eq. (10.15), where L (meters) cannot be less than V (km/h). When $KA < V$, use $L = V$. (See section 10.29 for illustration.)

Source: "Vertical Curve Tables," Ministry of Transportation and Communications, Ontario.

problem for crest curves or for sag curves being used for ditched roads and highways (ditches can have grade lines independent of the ₵ grade line). However, when sag curves are used for curbed municipal street design, a drainage problem is introduced. The curve elevations in brackets in Table 10.1 cover 10 ft either side of the low point. These data illustrate that for a distance of 20 ft there is virtually no change in elevation (i.e., only 0.01 ft). If the road were built according to the design, chances are that the low point catch basin would not completely drain the extended low-point area. The solution to this problem is for the surveyor to arbitrarily lower the catch-basin grate (1 in. is often used) to ensure proper drainage.

10.20 SPIRALS: GENERAL

Spirals are used in highway and railroad alignment to overcome the abrupt change in direction that occurs when the alignment changes from tangent to circular curve, and vice versa. The length of the spiral curve is also utilized for the transition from normally crowned pavement to fully superelevated (banked) pavement.

Figure 10.22 illustrates how the spiral curve is inserted between tangent and circular curve alignment. It can be seen that at the beginning of the spiral (TS = tangent to spiral) the radius of the spiral is the radius of the tangent line (infinitely large), and that the radius of the spiral curve decreases at a uniform rate until, at the point where the circular curve begins (SC = spiral to curve), the radius of the spiral equals the radius of the circular curve. In the previous section, we noted that the parabola, which is used in vertical alignment, had the important property of having a uniform rate of change of slope. Here we find that the spiral, used in horizontal alignment, has a uniform rate of change of radius (curvature). This property permits the driver to leave a tangent section of highway at relatively high rates of speed without experiencing problems with safety or comfort.

Figure 10.23 illustrates how the circular curve is moved inward (toward the center of the curve), leaving room for the inser-

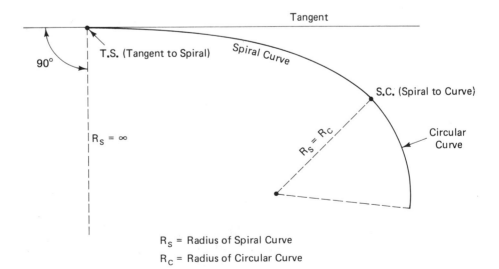

R_S = Radius of Spiral Curve

R_C = Radius of Circular Curve

FIGURE 10.22   Spiral curves.

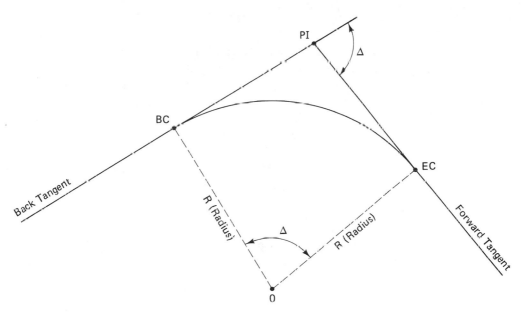

(a) Circular Curve Joining Two Tangents

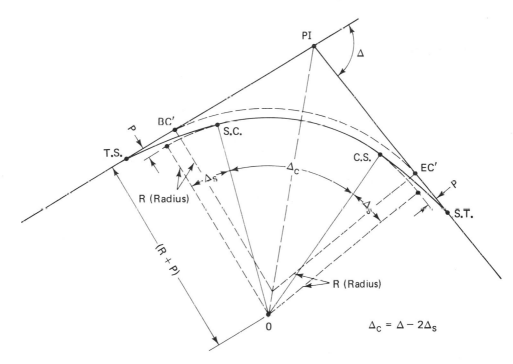

$$\Delta_c = \Delta - 2\Delta_s$$

(b) Circular Curve Shifted Inwards (Toward Curve Centre) to make Room for the Insertion of Spiral Curves at Either End of the Circular Curve

FIGURE 10.23 Shifting the circular curve to make room for the insertion of spirals.

tion of a spiral at either end of the shortened circular curve. The amount that the circular curve is shifted in from the main tangent line is known as P. This shift results in the curve center (0) being at the distance $(R + P)$ from the main tangent lines.

The spirals illustrated in this text reflect the common practice of using equal spirals to join the ends of a circular or compound curve to the main tangents. For more complex spiral applications, such as unequal spirals and spirals joining circular arcs, the reader is referred to a text on route surveying.

This text shows excerpts from spiral tables (Tables 10.3, 10.4 and 10.5). Each state and province prepare and publish similar tables for use by their personnel. A wide variety of spirals, both geometric and empirical, have been used to develop spiral tables. Geometric spirals include the cubic parabola and the clothoid curve, and empirical spirals include the A.R.E.A. 10-chord spiral, used by many railroads. Generally, the use of tables is giving way to computer programs for spiral solutions. All spirals give essentially the same appearance when staked out in the field.

10.21 SPIRAL CURVE COMPUTATIONS
(Refer to Figure 10.24)

Usually, data for a spiral computation are obtained as follows:

1. Δ is determined in the field.
2. R or D (degree of curve) is given by design considerations (limited by design speed).
3. Chainage of PI is determined in the field.
4. L_s is chosen with respect to design speed and the number of traffic lanes.

All other spiral parameters can be determined by computation and/or by use of spiral tables.

$$T_s = (R + P) \operatorname{Tan} \frac{\Delta}{2} + q \qquad (10.22)$$

(see Figure 10.24).

$$\Delta s = \frac{L_s D}{200} \qquad (10.23)$$

In circular curves

$$\Delta - \frac{LD}{100}$$

(see Eq. 10.7).

Since the spiral has a uniformly changing D, the spiral angle $(\Delta s) = $ length of spiral (L_s) in stations times the average degree of curve $(D/2)$,

$$L = L_c + 2L_s \qquad (10.24)$$

(see Figure 10.24, equal spirals) where L is the total length of the curve system.

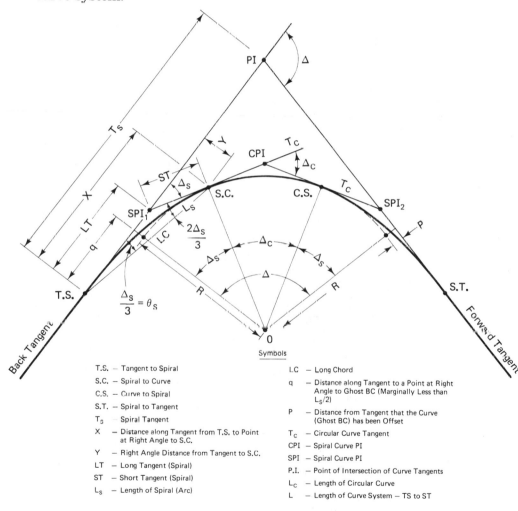

Symbols

T.S.	— Tangent to Spiral		LC	— Long Chord
S.C.	— Spiral to Curve		q	— Distance along Tangent to a Point at Right Angle to Ghost BC (Marginally Less than $L_s/2$)
C.S.	— Curve to Spiral			
S.T.	— Spiral to Tangent		P	— Distance from Tangent that the Curve (Ghost BC) has been Offset
T_s	— Spiral Tangent			
X	— Distance along Tangent from T.S. to Point at Right Angle to S.C.		T_c	— Circular Curve Tangent
			CPI	— Spiral Curve PI
Y	— Right Angle Distance from Tangent to S.C.		SPI	— Spiral Curve PI
			P.I.	— Point of Intersection of Curve Tangents
LT	— Long Tangent (Spiral)		L_c	— Length of Circular Curve
ST	— Short Tangent (Spiral)			
L_s	— Length of Spiral (Arc)		L	— Length of Curve System — TS to ST

FIGURE 10.24 Summary of spiral geometry and spiral symbols.

TABLE 10.3 Spiral tables for $L_s = 150$ feet

D	$\triangle_s$	R	p	$R + p$	q	LT	ST
7°30′	5°37′30″	763.9437	1.2268	765.1705	74.9759	100.0305	50.0459
8°00′	6°00′00″	716.1972	1.3085	717.5057	74.9726	100.0575	50.0523
30′	6°22′30″	674.0680	1.3902	675.4582	74.9691	100.0649	50.0590
9°00′	6°45′00″	636.6198	1.4719	638.0917	74.9653	100.0728	50.0662
30′	7°07′30″	603.1135	1.5536	604.6671	74.9614	100.0811	50.0738
10°00′	7°30′00″	572.9578	1.6352	574.5930	74.9572	100.0899	50.0817
30′	7°52′30″	545.6741	1.7169	547.3910	74.9528	100.0991	50.0901
11°00′	8°15′00″	520.8707	1.7985	522.6692	74.9482	100.1088	50.0989
30′	8°37′30″	498.2242	1.8802	500.1044	74.9434	100.1190	50.1082
12°00′	9°00′00″	477.4648	1.9618	479.4266	74.9384	100.1295	50.1178
13°00′	9°45′00″	440.7368	2.1249	442.8617	74.9277	100.1521	50.1383
14°00′	10°30′00″	409.2556	2.2880	411.5436	74.9161	100.1765	50.1605
15°00′	11°15′00″	381.9719	2.4510	384.4229	74.9037	100.2027	50.1843
16°00′	12°00′00″	358.0986	2.6139	360.7125	74.8905	100.2307	50.2098
17°00′	12°45′00″	337.0340	2.7767	339.8107	74.8764	100.2606	50.2370
18°00′	13°30′00″	318.3099	2.9394	321.2493	74.8614	100.2924	50.2659
19°00′	14°15′00″	301.5567	3.1020	304.6587	74.8456	100.3259	50.2964
20°00′	15°00′00″	286.4789	3.2645	289.7434	74.8290	100.3614	50.3287
21°00′	15°45′00″	272.8370	3.4269	276.2639	74.8115	100.3987	50.3627
22°00′	16°30′00″	260.4354	3.5891	264.0245	74.7932	100.4379	50.3983
23°00′	17°15′00″	249.1121	3.7512	252.8633	74.7740	100.4790	50.4357
24°00′	18°00′00″	238.7324	3.9132	242.6456	74.7539	100.5219	50.4748
25°00′	18°45′00″	229.1831	4.0750	233.2581	74.7331	100.5668	50.5157
26°00′	19°30′00″	220.3684	4.2367	224.6051	74.7114	100.6135	50.5582

Source: "Spiral Tables," foot units, courtesy Ministry of Transportation and Communication, Ontario.
$L_s = 150$

TABLE 10.3 *(Continued)*

D	$\triangle_s$	X_c	Y_c	$\dfrac{D}{10\,L_s}$
7°30′	5.62500°	149.86	4.91	0.00500
8°00′	6.00000°	149.84	5.23	0.00533
30′	6.37500°	149.81	5.56	0.00567
9°00′	6.75000°	149.79	5.88	0.00600
30′	7.12500 °	149.77	6.21	0.00633
10°00′	7.50000°	149.74	6.54	0.00667
30′	7.87500°	149.72	6.86	0.00700
11°00′	8.25000°	149.69	7.19	0.00733
30′	8.62500°	149.66	7.51	0.00767
12°00′	9.00000°	149.63	7.84	0.00800
13°00′	9.75000°	149.57	8.49	0.00867
14°00′	10.50000°	149.50	9.14	0.00933
15°00′	11.25000°	149.42	9.79	0.01000
16°00′	12.00000°	149.34	10.44	0.01067
17°00′	12.75000°	149.26	11.09	0.01133
18°00′	13.50000°	149.17	11.73	0.01200
19°00′	14.25000°	149.07	12.38	0.01267
20°00′	15.00000°	148.98	13.03	0.01333
21°00′	15.75000°	148.87	13.67	0.01400
22°00′	16.50000°	148.76	14.31	0.01467
23°00′	17.25000°	148.65	14.96	0.01533
24°00′	18.00000°	148.53	15.60	0.01600
25°00′	18.75000°	148.40	16.26	0.01667
26°00′	19.50000°	148.27	16.88	0.01733

TABLE 10.4(a) *Spiral Curve Lengths and Superelevation Rates: (a) Superelevation (e) Maximum of 0.06, Typical for Northern Climate.*

(a)

D	V = 30 e	L, ft 2 Lane	L, ft 4 Lane	V = 40 e	L, ft 2 Lane	L, ft 4 Lane	V = 50 e	L, ft 2 Lane	L, ft 4 Lane	V = 60 e	L, ft 2 Lane	L, ft 4 Lane	V = 70 e	L, ft 2 Lane	L, ft 4 Lane	V = 80 e	L, ft 2 Lane	L, ft 4 Lane
0°15′	NC	0	0	NC	0	0	NC	0	0	NC	0	0	NC	0	0	RC	250	250
0°30′	NC	0	0	NC	0	0	NC	0	0	RC	200	200	RC	200	200	.023	250	250
0°45′	NC	0	0	NC	0	0	RC	150	150	.021	200	200	.026	200	200	.033	250	250
1°00′	NC	0	0	RC	150	150	.020	150	150	.027	200	200	.033	200	200	.041	250	250
1°30′	RC	100	100	.020	150	150	.028	150	150	.036	200	200	.044	200	200	.053	250	300
2°00′	RC	100	100	.026	150	150	.035	150	150	.044	200	200	.052	200	250	.059	250	300
2°30′	.020	100	100	.031	150	150	.040	150	150	.050	200	200	.057	200	300	.060	250	300
3°00′	.023	100	100	.035	150	150	.044	150	200	.054	200	250	.060	200	300	$D_{max} = 2°30′$		
3°30′	.026	100	100	.038	150	150	.048	150	200	.057	200	250	$D_{max} = 3°00′$					
4°00′	.029	100	100	.041	150	150	.051	150	200	.059	200	250						
5°00′	.034	100	100	.046	150	150	.056	150	200	.060	200	250						
6°00′	.038	100	100	.050	150	200	.059	150	250	$D_{max} = 4°30′$								
7°00′	.041	100	150	.054	150	200	.060	150	250									
8°00′	.043	100	150	.056	150	200	$D_{max} = 7°00′$											
9°00′	.046	100	150	.058	150	200												
10°00′	.048	100	150	.059	150	200												
11°00′	.050	100	150	.060	150	200												
12°00′	.052	100	150	$D_{max} = 11°00′$														
13°00′	.053	100	150															
14°00′	.055	100	150															
16°00′	.058	100	200															
18°00′	.059	150	200															
20°00′	.060	150	200															
21°00′	.060	150	200															
$D_{max} = 21°00′$																		

Legend:
V, design speed, mph
e, rate of superelevation, feet per foot of pavement width
L, length of superelevation runoff or spiral curve
NC, normal crown section
RC, remove adverse crown, superelevate at normal crown slope
D, degree of circular curve
   Above the heavy line spirals are not required but superelevation is to be run off in distances shown.

Source: Ministry of Transportation and Communications, Ontario.

372

TABLE 10.4(b) *Spiral Curve Lengths and Superelevation Rates: (b) Superelevation (e) Maximum of 0.100, Typical for Southern Climate.*

(b)

D	R	e	V = 30 L 2 Lane	4 Lane	e	V = 40 L 2 Lane	4 Lane	e	V = 50 L 2 Lane	4 Lane	e	V = 60 L 2 Lane	4 Lane	e	V – 70 L 2 Lane	4 Lane
0°15'	22918'	NC	0	0	NC	0	0	NC	0	0	NC	0	0	RC	200	200
0°30'	11459'	NC	0	0	NC	0	0	RC	150	150	RC	175	175	RC	200	200
0°45'	7639'	NC	0	0	RC	125	125	RC	150	150	0.018	175	175	0.020	200	200
1°00'	5730'	NC	0	0	RC	125	125	0.018	150	150	0.022	175	175	0.028	200	200
1°30'	3820'	RC	100	100	0.020	125	125	0.027	150	150	0.034	175	175	0.042	200	200
2°00'	2865'	RC	100	100	0.027	125	125	0.036	150	150	0.046	175	190	0.055	200	250
2°30'	2292'	0.020	100	100	0.033	125	125	0.045	150	160	0.059	175	240	0.069	210	310
3°00'	1910'	0.024	100	100	0.038	125	125	0.054	150	190	0.070	190	280	0.083	250	370
3°30'	1637'	0.027	100	100	0.045	125	140	0.063	150	230	0.081	220	330	0.096	290	430
4°00'	1432'	0.030	100	100	0.050	125	160	0.070	170	250	0.090	240	360	0.100	300	450
5°00'	1146'	0.038	100	100	0.060	130	190	0.083	200	300	0.099	270	400	$D_{max} = 3.9°$		
6°00'	955'	0.044	100	120	0.068	140	210	0.093	220	330	0.100	270	400			
7°00'	819'	0.050	100	140	0.076	160	240	0.097	230	350	$D_{max} = 5.5°$					
8°00'	716'	0.055	100	150	0.084	180	260	0.100	240	360						
9°00'	637'	0.061	110	160	0.089	190	280	0.100	240	360						
10°00'	573'	0.065	120	180	0.093	200	290	$D_{max} = 8.3°$								
11°00'	521'	0.070	130	190	0.096	200	300									
12°00'	477'	0.074	130	200	0.098	210	310									
13°00'	441'	0.078	140	210	0.099	210	310									
14°00'	409'	0.082	150	220	0.100	210	320									
16°00'	358'	0.087	160	240	$D_{max} = 13.4°$											
18°00'	318'	0.093	170	250												
20°00'	286'	0.096	170	260												
22°00'	260'	0.099	180	270												
24.8°	231'	0.100	180	270												
		$D_{max} = 24.8°$														

Source: American Association of State Highway Officials (A.A.S.H.O.)

NOTES. NC = normal crown section. RC = remove adverse crown, superelevate at normal crown slope. Spirals desirable but not as essential above heavy line. Lengths rounded in multiples of 25 or 50 ft permit simpler calculations.
The higher e value (0.100) in (b) permits a sharper maximum curvature.

373

TABLE 10.5 *Spiral Curve Lengths and Superelevation Rates (Metric)*

Design speed, km/h	40			50			60			70			80		
Radius, m	e	A 2 lane	A 3 & 4 lane	e	A 2 lane	A 3 & 4 lane	e	A 2 lane	A 3 & 4 lane	e	A 2 lane	A 3 & 4 lane	e	A 2 lane	A 3 & 4 lane
7000	NC			NC			NC			NC			NC		
5000	NC			NC			NC			NC			NC		
4000	NC			NC			NC			NC			NC		
3000	NC			NC			NC			NC			NC		
2000	NC			NC			NC			RC	275	275	RC	300	300
1500	NC			NC			RC	225	225	RC	250	250	0.024	250	250
1200	NC			NC			RC	200	200	0.023	225	225	0.028	225	225
1000	NC			RC	170	170	0.021	175	175	0.027	200	200	0.032	200	200
900	NC			RC	150	150	0.023	175	175	0.029	180	180	0.034	200	200
800	NC			RC	150	150	0.025	160	160	0.031	175	175	0.036	175	175
700	NC			0.021	140	140	0.027	150	150	0.034	175	175	0.039	175	175
600	NC	120	120	0.024	125	125	0.030	140	140	0.037	150	150	0.042	175	175
500	RC	100	100	0.027	120	120	0.034	125	125	0.041	140	150	0.046	150	160
400	0.023	90	90	0.031	100	100	0.038	115	120	0.045	125	135	0.051	135	150
350	0.025	90	90	0.034	100	100	0.041	110	115	0.048	120	125	0.054	125	140
300	0.028	80	80	0.037	90	100	0.044	100	110	0.051	120	125	0.057	125	135
250	0.031	75	80	0.040	85	90	0.048	90	100	0.055	110	120	0.060	125	125
220	0.034	70	80	0.043	80	90	0.050	90	100	0.057	110	110	0.060	125	125
200	0.036	70	75	0.045	75	90	0.052	85	100	0.059	110	110	Minimum R = 250		
180	0.038	60	75	0.047	70	90	0.054	85	95	0.060	110	110			
160	0.040	60	75	0.049	70	85	0.056	85	90	Minimum R = 190					
140	0.043	60	70	0.052	65	80	0.059	85	90						
120	0.046	60	65	0.055	65	75	0.060	85	90						
100	0.049	50	65	0.058	65	70	Minimum R = 130								
90	0.051	50	60	0.060	65	70									
80	0.054	50	60	0.060	65	70									
70	0.058	50	60	Minimum R = 90											
60	0.059	50	60												
	0.059	50	60												
	Minimum R = 55														

374

TABLE 10.5 *(Continued)*

90 e	90 A 2 lane	90 A 3 & 4 lane	100 e	100 A 2 lane	100 A 3 & 4 lane	110 e	110 A 2 lane	110 A 3 & 4 lane	120 e	120 A 2 lane	120 A 3 & 4 lane	130 e	130 A 2 lane	130 A 3 & 4 lane	140 e	140 A 2 lane	140 A 3 & 4 lane
NC			NC			NC			NC			NC	700	700	RC	700	700
NC			NC			NC	500	500	RC	600	600	0.021	600	600	0.024	625	625
NC			RC	480	480	RC	500	500	0.022	500	500	0.025	500	500	0.028	560	560
RC	390	400	0.025	400	400	0.023	450	450	0.027	450	450	0.030	450	450	0.034	495	495
0.023	300	350	0.027	340	340	0.031	350	350	0.035	350	350	0.039	400	400	0.043	400	400
0.029	270	275	0.033	300	300	0.037	300	300	0.041	300	300	0.046	330	330	0.050	345	340
0.033	240	240	0.038	250	250	0.042	275	275	0.047	285	285	0.051	300	300	0.056	330	330
0.037	225	225	0.042	240	240	0.046	250	260	0.051	250	275	0.055	300	300	0.060	325	325
0.039	200	200	0.044	225	225	0.049	230	250	0.053	250	270	0.057	300	300	0.060	325	325
0.042	200	200	0.047	200	225	0.051	225	250	0.056	250	260	0.060	275	275	Minimum R=1000		
0.045	185	195	0.049	200	220	0.054	225	235	0.059	250	250	0.060	275	275			
0.048	175	185	0.053	200	200	0.057	220	220	0.060	250	250	Minimum R=800					
0.052	160	175	0.057	200	200	0.060	220	220	Minimum R=650								
0.057	160	165	0.060	200	200	Minimum R=525											
0.059	160	160	Minimum R=420														
0.060	160	160															
Minimum R=340																	

Source: Roads and Transportation Association of Canada (RTAC).

NOTES:
e_{max} = 0.06
e is super elevation.
A is spiral parameter in meters.
NC is normal cross section.
RC is remove adverse crown and superelevate at normal rate.
Spiral length, $L = A^2 \div$ radius.
Spiral parameters are minimum and higher values should be used where possible.
Spirals are desirable but not essential above the heavy line.
For 6-lane pavement: above the dashed line use 4-lane values, below the dashed line use 4-lane values $\times$ 1.15.
A divided road having a median less than 7 m may be treated as a single pavement.

375

$$\Delta = \Delta_c + 2\Delta_s \qquad\qquad (10.25)$$

(see Figure 10.24).

$$\theta_s = \frac{\Delta_s}{3} \text{ (approx)} \qquad\qquad (10.26)$$

θ_s is the total spiral deflection angle; compare to circular curves where the deflection angle is $1/2 \Delta$).

This approximate formula gives realistic results for the vast majority of spiral problems. When, for example, Δ_s is as large as 21°—which is seldom the case—the correction to $\Delta_s/3$ is approximately $+30''$.

$$L_c = \frac{2\pi R \Delta_c}{360}, \qquad \text{foot or meter units} \qquad (10.27)$$

$$L_c = \frac{100\Delta_c}{D}, \qquad \text{foot units} \qquad (10.28)$$

$$\Delta_s = \frac{90}{\pi} \times \frac{L_s}{R}, \qquad\qquad (10.29)$$

$$\phi = \left(\frac{l}{L_s}\right)^2 \theta_s \qquad\qquad (10.30)$$

from spiral definition, where ϕ is the deflection angle for any distance l, and ϕ and θ_s are in the same units. Practically,

$$\phi' = l^2 \left(\frac{\theta_s \times 60}{L_s^2}\right)$$

where ϕ' is the deflection angle in minutes for any distance l measured from the TS or the ST.

Other values such as x, y, P, q, ST, and LT are routinely found in spiral tables issued by state and provincial highway agencies and can also be found in route surveying texts. For the past few years, solutions to these problems have almost exclusively been achieved by the use of computers and appropriate computer software such as COGO.

With the switchover to metric that took place in the 1970s in Canada, a decision was reached to work exclusively with the radius when defining horizontal curves. It was decided that new spiral tables, based solely on R definition, would be appropriate. The Roads and Transportation Association of Canada (RTAC) prepared spiral tables based on the spiral property defined as follows:

> The product of any instantaneous radius r and the corresponding spiral length λ (i.e., l) from the beginning of the spiral to that point, is equal to the product of the spiral end radius R and the entire length (L_s) of that spiral, which means it is a constant.

Thus

$$r\lambda = RL_s = A^2 \qquad (10.31)$$

The constant is denoted as A^2 to retain dimensional consistency since it represents a product of two lengths. It follows that

$$\frac{A}{R} = \frac{L_s}{A} \qquad (10.32)$$

The RTAC tables (see Table 10.5) are based on design speed and the number of traffic lanes, together with a design radius. The constant value A is taken from the table and used in conjunction with R to find all the spiral table curve parameters noted earlier (see Table 10.7).

It should be noted that, concurrent with the changeover to metric units, a major study in Ontario of highway geometrics resulted in spiral curve lengths that reflected design speed, attainment of superelevation, driver comfort, and esthetics. Accordingly, the spiral lengths vary somewhat from the foot unit system to the metric system.

10.22 SPIRAL LAYOUT PROCEDURE SUMMARY

Refer to Figure 10.24:

1. Select L_s (foot units) or A (metric units) in conjunction with the design speed, number of traffic lanes, and sharpness of the circular curve (radius or D).
2. From the spiral tables, determine P, q, x, y, and so on.
3. Compute the spiral tangent (T_s), Eq. (10.22), and the circular tangent (T_c), Eq. (10.1).
4. Compute the spiral angle, Δ_s. Use Eq. (10.23) for foot units or Eq. (10.29) for metric units (or use tables).
5. Prepare a list of relevant layout stations. This list will include all horizontal alignment key points (e.g., TS, SC, CS, and ST), as well as all vertical alignment key points, such as grade points, BVC, low point, and EVC.
6. Calculate the deflection angles. See Eq. (10.30).
7. From the established PI, measure out the T_s distance to locate the TS and ST.
8a. From the TS, turn off the spiral deflection ($\theta_s = 1/3\ \Delta_s$ ap-

proximately), measure out the long chord (LC), and thus locate the SC.

8b. Or, from the TS, measure out the LT distance along the main tangent and locate the spiral PI (SPI); the spiral angle (Δ_s) can now be turned, and the ST distance measured out to locate the SC.

9. From the SC, measure out the circular tangent (T_c) along the line SPI_1 - SC to establish the CPI.

10. The procedure to this point is repeated, starting at the ST.

11. The key points are verified by checking all angles and redundant distances. Some surveyors prefer to locate the CPI by intersecting the two tangent lines (i.e., lines through SPI_1 and SC and through SPI_2 and CS). The location can be verified by checking the angle Δ_c and by checking the two tangents (T_c).

12. Only after all key control points have been verified can the deflection angle stake-out commence. The lower chainage spiral is run in from the TS, whereas the higher chainage spiral is run in from the ST. The circular curve can be run in from the SC, although it is common practice to run half the curve from the SC and the other half from the CS so that any acceptable errors that accumulate can be isolated in the middle of the circular arc, where, relatively speaking, they will be less troublesome.

10.23 ILLUSTRATIVE SPIRAL PROBLEM: FOOT UNITS

Given the following data:

$$\Delta = 25°45' \; RT$$
$$V = 40 \text{ mph}$$
$$D = 9°$$
$$\text{PI at } 36 + 17.42$$
$$\text{Two-lane highway, 24 ft wide}$$

Solution

1. From Table 10.4(a)

$$L_s = 150 \text{ ft}$$
$$e = 0.058$$

See Section 10.26 for superelevation.

2. From Table 10.3,

$$
\begin{aligned}
\Delta_s &= 6.75000° \\
R &= 636.6198 \text{ ft} \\
P &= 1.4719 \text{ ft} \\
R + P &= 638.0917 \text{ ft} \\
q &= 74.9653 \text{ ft} \\
LT &= 100.0728 \text{ ft} \\
ST &= 50.0662 \text{ ft} \\
X &= 149.79 \text{ ft} \\
Y &= 5.88 \text{ ft}
\end{aligned}
$$

$$
\Delta_s = \frac{L_s D}{200} \quad (10.23)
$$

$$
= \frac{150}{200} \times 9 = 6.75000°
$$

3. $T_s = (R + p) \operatorname{Tan} \dfrac{\Delta}{2} + q$ (10.22)

$$
= 638.0917 \operatorname{Tan} 12°52.5' + 74.9653
$$
$$
= 220.82 \text{ ft}
$$

4. $\Delta_c = \Delta - 2\Delta_s$
$$
= 25°45' - 2(6°45')
$$
$$
= 12°15'
$$

5. $L_c = \dfrac{100\Delta_c}{D}$ (10.7)

$$
= \frac{100 \times 12.25}{9}
$$
$$
= 136.11 \text{ ft}
$$

6. Key station computation:

PI at	36 +	17.42
$-T_s$	2	20.82
TS =	33 +	96.60
$+L_s$	1	50.00
SC =	35 +	46.60
$+L_c$	1	36.11
CS =	36 +	82.71
$+L_s$	1	50.00
ST	38 +	32.71

7. $\theta_s = \dfrac{\Delta_s}{3}$

$\quad = \dfrac{6.75000°}{3}$

$\quad = 2.25°$

$\quad = 2°15'00''$

Circular Curve Deflections

See Table 10.6

$$\Delta_c = 12°15'$$

$$\frac{\Delta_c}{2} = 6°07.5'$$

$$= 367.5'$$

From Section 10.4, the deflection angle for one unit of distance is $(\Delta/2)/L$. Here the deflection angle for one foot of arc (minutes) is

$$\frac{(\Delta c/2)}{L_c} = \frac{367.5}{136.11} = 2.700'$$

TABLE 10.6   Curve System Deflection Angles

Station	Distance from TS (or ST) l (ft)	l^2	$\dfrac{\theta_s° \times 60}{L_s^2}$	$\dfrac{l^2(\theta_s \times 60)}{L_s^2}$ Deflection Angle (minutes)	Deflection
TS 33 + 96.60	0	0	0.006	0	0°00'00''
34 + 00	3.4	11.6	0.006	0.070	0°00'04''
34 + 50	53.4	2851.4	0.006	17.108	0°17'06''
35 + 00	103.4	10691.6	0.006	64.149	1°04'09''
SC 35 + 46.60	150	22500	0.006	135	$\theta_s = 2°15'00''$

Station	Circular Curve Data			Deflection Angle	Deflection
SC 35 + 46.60	$\Delta_c = 12°15'$, $\frac{\Delta_c}{2} = 6°07'30''$			0°00.0	0°00'00''
35 + 50	Deflection for 3.40' = 9.18'			0°09.18	0°09'11''
36 + 00	Deflection for 50' = 135'			2°24.18	2°24'11''
36 + 50	Deflection for 32.71' = 88.32'			4°39.18	4°39'11''
CS 36 + 82.71				6°07.50	6°07'30''

Station	l (ft)	l^2		Deflection Angle (minutes)	Deflection
CS 36 + 82.71	150	22500	0.006	135	$\theta_s = 2°15'00''$
37 + 00	132.71	17611.9	0.006	105.672	1°45'40''
37 + 50	82.71	6840.9	0.006	41.046	0°41'03''
38 + 00	32.71	1069.9	0.006	6.420	0°06'25''
ST 38 + 32.71	0*	0	0.006	0	0°00'00''

*Note that l is measured from the ST.

Alternately, since $D = 9°$, then the deflection for 100 ft is $D/2$ or 4°30', which is 270'. The deflection angle for 1 ft $= 270/100 = 2.700'$ (as previously).

The required distances (from Table 10.6) are

$(35 + 50) - (35 + 46.60) = 3.4'$; deflection angle $= 3.4 \times 2.7 = 9.18'$
Even interval $= 50'$; deflection angle $= 50 \times 2.7 = 135'$
$(36 + 82.71) - (36 + 50) = 32.71$; deflection angle $= 32.71 \times 2.7 = 88.32'$

These values are now entered cumulatively in Table 10.6.

10.24 ILLUSTRATIVE SPIRAL PROBLEM: METRIC UNITS

Given the following data:

PI at $1 + 086.271$
$V = 80$ kmh
$R = 300$ m
$\Delta = 16°00'$ RT
Two-lane road (7.5 m wide)

From Table 10.5, $A = 125$ and $e = 0.057$. See Section 10.26 for a discussion of superelevation. From Table 10.7, we obtain the following

Steps	*For A = 125*	*and R = 300*
1 and 2	$L_s = 52.083$ m	LT $= 34.736$ m
	$P = 0.377$ m	ST $= 17.374$ m
	$X = 52.044$ m	$\Delta_0 = 4°58'24.9''$
	$Y = 1.506$ m	$\theta_s = 1/3\,\Delta_s = 1°39'27.9''$
	$q = 26.035$ m	LC $= 52.066$ m long chord

Also, $\Delta_s = \dfrac{90}{\pi} \times \dfrac{L_s}{R}$ Eq. (10.29)

$= \dfrac{90}{\pi} \times \dfrac{52.083}{300} = 4.9735601°$

$= 4°58'24.8''$

3     $T_s = (R + p)\,\text{Tan}\,\dfrac{\Delta}{2} + q$ (10.22)

$= 300.377\,\text{Tan}\,8° + 26.035$

$= 68.250$ m

4     $\Delta_c = \Delta - 2\Delta_s$

$= 16° - 2(4°58'24.8'')$

$= 6°03'10.2''$

TABLE 10.7 Functions of the Standard Spiral* for A = 125m

R (m)	A/R	L_s	X	Y	q	P	LT	ST	L_c	Δ_s Deg.	Min.	Sec.	θ_s** Deg.	Min.	Sec.
					Meters										
115	1.0870	135.870	131.204	26.095	67.152	6.606	92.293	46.851	133.774	33	50	48.3	11	14	55.1
120	1.0417	130.208	126.428	23.057	64.471	5.825	88.183	44.658	128.513	31	05	05.8	10	20	08.3
125	1.0000	125.000	121.911	20.464	61.983	5.162	84.451	42.685	123.617	28	38	52.4	9	31	44.3
130	0.9615	120.192	117.649	18.240	59.671	4.595	81.044	40.898	119.055	26	29	11.7	8	48	46.1
140	0.8929	111.607	109.847	14.661	55.509	3.686	75.034	37.755	110.821	22	50	16.5	7	36	08.5
150	0.8333	104.167	102.918	11.953	51.875	3.001	69.888	35.126	103.610	19	53	39.7	6	37	28.8
160	0.7813	97.656	96.751	9.868	48.677	2.475	65.425	32.844	97.253	17	29	07.0	5	49	25.8
170	0.7353	91.912	91.242	8.239	45.844	2.065	61.511	30.852	91.614	15	29	19.3	5	09	34.9
180	0.6944	86.086	86.302	6.948	43.319	1.741	58.048	29.096	86.581	13	48	55.9	4	36	10.5
190	0.6579	82.237	81.853	5.913	41.054	1.481	54.960	27.535	82.066	12	23	58.3	4	07	53.5
200	0.6250	78.125	77.828	5.072	39.013	1.270	52.188	26.137	77.993	11	11	26.1	3	43	44.4
210	0.5952	74.405	74.172	4.384	37.163	1.097	49.685	24.876	74.301	10	09	00.7	3	22	57.0
220	0.5682	71.023	70.838	3.814	35.481	0.954	47.413	23.733	70.941	9	14	54.3	3	04	55.6
230	0.5435	67.935	67.787	3.339	33.943	0.835	45.342	22.692	67.869	8	27	42.1	2	49	12.1
240	0.5208	65.104	64.984	2.940	32.532	0.735	43.455	21.739	65.051	7	46	16.5	2	35	24.0
250	0.5000	62.500	62.402	2.601	31.234	0.651	41.701	20.864	62.457	7	09	43.1	2	23	13.2
280	0.4464	55.804	55.748	1.852	27.983	0.463	37.222	18.619	55.779	5	24	34.1	1	54	10.8
300	0.4167	52.083	52.044	1.506	26.035	0.377	34.736	17.374	52.066	4	58	24.9	1	39	27.9
320	0.3906	48.828	48.800	1.241	24.409	0.310	32.562	16.285	48.815	4	22	16.8	1	27	25.3
340	0.3676	45.956	45.935	1.035	22.974	0.259	30.645	15.325	45.947	3	52	19.8	1	17	26.4
350	0.3571	44.643	44.625	0.949	22.318	0.237	29.768	14.887	44.635	3	39	14.6	1	13	04.7
380	0.3289	41.118	41.106	0.741	20.557	0.185	27.416	13.710	41.113	3	05	59.6	1	01	59.8
400	0.3125	39.063	39.063	0.636	19.530	0.159	26.045	13.024	39.058	2	47	51.5	0	55	57.1
420	0.2976	37.202	37.195	0.549	18.600	0.137	24.804	12.403	37.195	2	32	15.2	0	50	45.0
450	0.2778	34.722	34.717	0.446	17.360	0.112	23.150	11.576	34.720	2	12	37.7	0	44	12.5
475	0.2632	32.895	32.891	0.380	16.447	0.095	21.931	10.966	32.893	1	59	02.1	0	39	40.7
500	0.2500	31.250	31.247	0.325	15.624	0.081	20.834	10.418	31.249	1	47	25.8	0	35	48.6
525	0.2381	29.762	29.760	0.281	14.881	0.070	19.842	9.921	29.761	1	37	26.5	0	32	20.8
550	0.2273	28.409	28.407	0.245	14.204	0.061	18.940	9.470	28.408	1	28	47.1	0	29	35.7
575	0.2174	27.174	27.172	0.214	13.587	0.054	18.116	9.058	27.173	1	21	13.9	0	27	04.6

Source: "Metric Curve Tables," Table IV, Roads and Transportation Association of Canada (RTAC).
A = 125m

* See Table 4 for a complete listing.
**Short radius (i.e., < 150m) may require a correction to $\Delta_s/3$ in order to determine precise value of θ_s.

$$\frac{\triangle_c}{2} = 3°01'35.1''$$

5

$$L_c = \frac{2\pi \, R\triangle_c}{360} \ (10.5)$$

$$= \frac{2\pi \times 300 \times 6.052833}{360}$$

$$= 31.693 \text{ m}$$

6   Key station computation:

PI at		1 + 086.271
$- T_s$		68.250
TS	=	1 + 018.021
$+ L_s$		52.083
SC	=	1 + 070.104
$+ L_c$		31.693
CS		1 + 101.797
$+ L_s$		52.083
ST		1 + 153.880

TABLE 10.8   *Curve System Deflection Angles*

Station	Distance from TS (or ST) l (m)	l_2	$\dfrac{\theta_s \times 60}{L_s^2}$	$\dfrac{l^2(\theta_s \times 60)}{L_s^2}$ Deflection Angle (minutes)	Deflection Angle
TS 1 + 018.021	0	0			0°00'00"
1 + 020	1.979	3.9	0.036667	0.1436	0°00'09"
1 + 040	21.979	483.1	0.036667	17.71	0°17'43"
1 + 060	41.979	1762.2	0.036667	64.62	1°04'37"
SC 1 + 070.104	52.083	2712.6	0.036667	99.464	1°39'28"

	Circular Curve Data	Deflection Angle (cumulative)	Deflection Angle
SC 1 + 070.104	$\triangle_c = 6°03'10''$, $\frac{\triangle_c}{2} = 3°01'35''$	0°00'00"	0°00'00"
1 + 080	Deflection for 9.896 m = 56.70'	0°56.70'	0°56'42"
1 + 100	Deflection for 20 m = 114.59'	2°51.29'	2°51'17"
CS 1 + 101.797	Deflection for 1.797 m = 10.30'	3°01.59'	3°01'35"

Station	l (m)	l_2		Deflection Angle (minutes)	Deflection Angle (cumulative)
CS 1 + 101.797	52.083	2712.6	0.036667	99.464	1°39'28"
1 + 120	33.880	1147.9	0.036667	42.09	0°42'05"
1 + 140	13.880	192.7	0.036667	7.06	0°07'04"
ST 1 + 153.880	0*	0	0.036667	0	0°00'00"

*Note that l is measured from the ST.

Circular Curve Deflections

See Table 10.8.

$$\Delta_c = 6°03'10''$$

$$\frac{\Delta_c}{2} = 3°01'35''$$

$$= 181.58'$$

From Section 10.4, the deflection angle for one unit of distance is $(\Delta/2)/L$. Here the deflection angle for 1 m of arc is

$$\frac{\Delta_c/2}{L_c} = \frac{181.58}{31.693} = 5.7293'$$

The required distances (deduced from Table 10.8) are

$1 + 080 - 1 + 070.104 = 9.896;$ deflection angle $= 5.7293 \times 9.896 = 56.70'$
Even interval $= 20.000;$ deflection angle $= 5.7293 \times 20 = 114.59'$
$1 + 101.797 - 1 + 100 = 1.797;$ deflection angle $= 5.7293 \times 1.787 = 10.30'$

These values are now entered cumulatively in Table 10.8.

10.25 APPROXIMATE SOLUTION FOR SPIRAL PROBLEMS

It is possible to lay out spirals using the approximate relationships illustrated in Figure 10.25. Since $L_s \approx LC$ (long chord), the following can be assumed:

$$\frac{Y}{L_s} = \operatorname{Sin} \theta_s, \qquad Y = L_s \operatorname{Sin} \theta_s \qquad (10.33)$$

$$X^2 = L_s^2 - Y^2$$
$$X = \sqrt{L_s^2 - Y^2} \qquad (10.34)$$
$$q = \tfrac{1}{2}X \qquad (10.35)$$
$$p = \tfrac{1}{4}Y \qquad (10.36)$$

Using the sine law,

$$LT = \operatorname{Sin} \tfrac{2}{3} \Delta_s \times \frac{L_s}{\operatorname{Sin} \Delta_s} \qquad (10.37)$$

Using the sine law,

$$ST = \operatorname{Sin} \tfrac{1}{3}\Delta_s \times \frac{L_s}{\operatorname{Sin} \Delta_s} \qquad (10.38)$$

For comparison, the values in the illustrative examples in Sec-

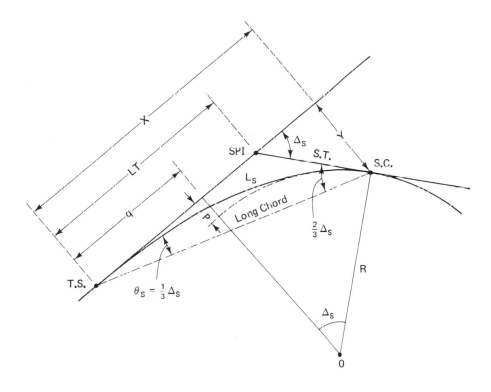

Basic Assumption: $L_s \approx$ Long Chord

FIGURE 10.25 Sketch for approximate formulas.

tions 10.23 and 10.24 are compared to the values obtained by the approximate methods.

Data Summary

Parameter	Precise Methods		Approximate Methods	
	Example 10.23 (ft)	Example 10.24 (m)	Example 10.23	Example 10.24
Y	5.88	1.506	5.89	1.507
X	149.79	52.044	149.88	52.061
q	74.97	26.035	74.94	26.031
P	1.47	0.377	1.47	0.377
LT	100.07	34.736	100.13	34.744
ST	50.07	17.374	50.10	17.379

It can be seen from the data summary that the precise and approximate values for Y, X, q, P, LT, and ST are quite similar. The largest discrepancy shows up in the X value, which is not required for spiral layout. The larger the Δ_s value, the larger will be the discrepancy between the precise and approximate values. For the normal range of spirals in use, the approximate method could be adequate for the layout of an asphalt-surfaced ditched highway. For curbed highways or elevated highways, precise methods should be employed. See Section 11.11 for further consideration of these techniques.

10.26 SUPERELEVATION: GENERAL

If a vehicle travels too fast on a horizontal curve, the vehicle may either skid off the road or overturn. The factors that cause this phenomenon are based on the radius of curvature and the velocity of the vehicle: the sharper the curve and the faster the velocity, the larger will be the centrifugal force requirement.

Two factors can be called on to help stabilize the radius and velocity factors: (1) side friction, which is always present to some degree between the vehicle tires and the pavement, and (2) superelevation, which is a banking of the pavement toward the center of the curve.

The side friction factor (f) has been found to vary linearly with velocity. Design values for f range from 0.16 at 30 mph (50 km/h) to 0.11 at 80 mph (130 km/h).

Superelevation must satisfy normal driving practices and climatic conditions. In practice, values for superelevation range from 0.125 (i.e., 0.125 ft/ft or 12.5% cross slope) in ice-free southern states to 0.06 in the northern states and Canadian provinces. Typical values for superelevation can be found in Table 10.4.

10.27 SUPERELEVATION DESIGN

Figure 10.26 illustrates how the length of spiral (L_s) is utilized to change the pavement cross slope from normal crown to full superelevation. Figure 10.26b illustrates that the pavement can be revolved about the center line, the usual case, or the pavement can be revolved about the inside or outside edges, a technique that is often encountered on divided four-lane highways where a narrow median restricts drainage profile manipulation.

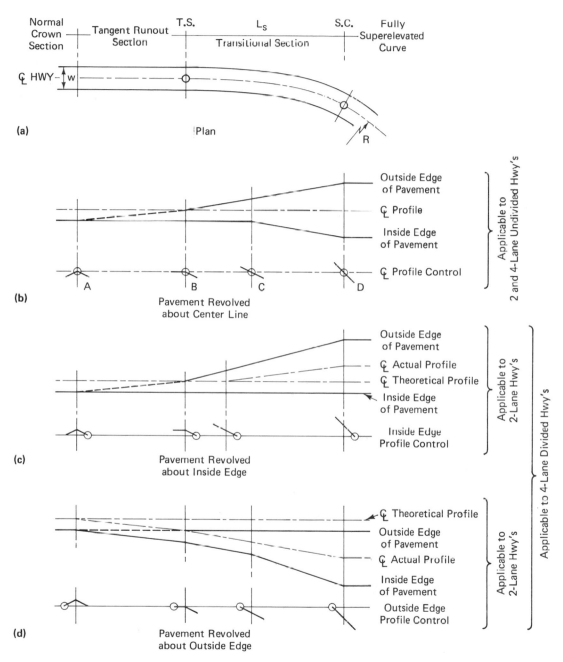

FIGURE 10.26 Methods of attaining superelevation for spiraled curves. (From *Geometric Design Standards for Ontario Highways*, Ministry of Transportation and Communications)

Figures 10.28b and 10.29 clearly show the technique used to achieve pavement superelevation when revolving the pavement edges about the centerline (℄) profile. At points A and A' the pavement is at normal crown cross section—with both edges (inside and outside) of pavement a set distance below the ℄ elevation. At points SC (D) and CS (D') the pavement is at full superelevation—with the outside edge a set distance above the ℄ elevation and the inside edge the same set distance below the ℄ elevation. The transition from normal crown cross-section to full superelevation cross section proceeds as follows.

Outside Edge:

- From A to the TS (B) the outside edge rises at a ratio of 400:1 —relative to the ℄ profile—and becomes equal in elevation to the ℄.
- From the TS (B) the outside edge rises (relative to the ℄ profile) at a uniform rate from being equal to the ℄ elevation at the TS (B) until it is at full superelevation above the ℄ elevation at the SC (D.)

Inside Edge:

- From A, through the TS (B), to point C, the inside edge remains below the ℄ profile at normal crown depth.
- From C to the SC (D) the inside edge drops at a uniform rate from being at normal crown depth below the ℄ profile to being at full superelevation depth below the ℄ profile.

The transition from full superelevation at the CS (D') to normal cross section at A' proceeds in a reverse manner to that just described.

10.28 ILLUSTRATIVE SUPERELEVATION
PROBLEM: FOOT UNITS

This problem utilizes the horizontal curve data of Section 10.23. Given data:

$V = 40$ mph
$\Delta = 25°45'\ RT$
$D = 9°$
PI at 36 + 17.42
Two-lane highway, 24 ft wide, each lane 12 ft wide

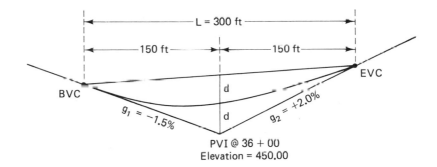

BVC = (36 + 00) − 150 = 34 + 50

EVC = (36 + 00) + 150 = 37 + 50

Elevation BVC = 450.00 + (150 × 0.015) = 452.25

Elevation EVC = 450.00 + (150 × 0.020) = 453.00

Mid-Chord Elevation = $\dfrac{452.25 + 453.00}{2}$ = 452.63

Tangent Offset: d = $\dfrac{452.63 - 450.00}{2}$ = 1.315 ft.

FIGURE 10.27   Vertical curve solution for the problem of Section 10.28.

Chainage of low point:

$$X = \frac{-g_1 L}{A} \tag{10.15}$$

$$= \frac{1.5 \times 300}{2 - (-1.5)}$$

$$= 128.57$$

BVC at   34 + 50

$X =$ 1   28.57

Low point =   35 + 78.57

Additional data are as follows:

PVI at 36 + 00
Elevation PVI = 450.00
$g_1 = -1.5\%$
$g_2 = +2\%$
$L = 300$ ft
Tangent runout at 400 : 1
Normal crown at 2%
Pavement revolved about the ₵

From Table 10.4,

$$L_s = 150 \text{ ft, e} = 0.058$$

It is required to determine the ℄ and edge of pavement elevations for even 50-ft stations and all other key stations.

Solution

1. Compute the key horizontal alignment stations. (These stations have already been computed in Section 10.23.)
2. Solve the vertical curve for ℄ elevations at 50-ft stations, the low point, *plus any horizontal alignment key stations that may fall between the BVC and EVC.*
3. Begin preparation of Table 10.9, showing all key stations for both horizontal and vertical alignment. List the ℄ grade elevation for each station.
4. Compute station A; see Figure 10.28c.

$$\text{Cross fall at 2\% for 12'} = 0.02 \times 12 = 0.24 \text{ ft}$$

$$\text{Tangent runout} = \frac{400}{1} \times 0.24 = 96 \text{ ft}$$

$$
\begin{aligned}
\text{Tangent runout} &= 96.00 \text{ ft}\\
\text{TS} &= 33 + 96.60\\
X_1 \text{ (Tangent runout)} &= \underline{- 96.00}\\
A &= 33 + 00.60
\end{aligned}
$$

TABLE 10.9   ℄ *Pavement Elevations*

Station		Tangent Elevation	Tangent Offset $\left(\dfrac{X}{L/2}\right)^2 d$		℄ Elevation
A	33 + 00.60	454.49			454.49
	33 + 50	453.75			453.75
TS(B)	33 + 96.60	453.05			453.05
	34 + 00	453.00			453.00
C	34 + 48.03	452.28			452.28
BVC	34 + 50	452.25	$(0/150)^2 \times 1.32 =$	0	452.25
	35 + 00	451.50	$(50/150)^2 \times 1.32 =$	0.15	451.65
SC(D)	35 + 46.60	450.79	$(96.6/150)^2 \times 1.32 =$	0.55	451.35
	35 + 50	450.75	$(100/150)^2 \times 1.32 =$	0.58	451.33
Low point	35 + 78.57	450.32	$(128.6/150)^2 \times 1.32 =$	0.97	451.29
PVI	36 + 00	450.00	$(150/150)^2 \times 1.32 =$	1.32	451.32
	36 + 50	451.00	$(100/150)^2 \times 1.32 =$	0.58	451.58
CS(D')	36 + 82.71	451.65	$(67.3/150)^2 \times 1.32 =$	0.27	451.92
	37 + 00	452.00	$(50/150)^2 \times 1.32 =$	0.15	452.15
EVC	37 + 50	453.00	$(0/150)^2 \times 1.32 =$	0	453.00
C'	37 + 81.28	453.63			453.63
	38 + 00	454.00			454.00
ST(B')	38 + 32.71	454.65			454.65
	39 + 00	456.00			456.00
A'	39 + 28.71	456.57			456.57

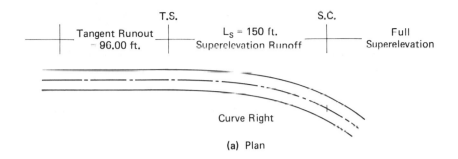

Curve Right

(a) Plan

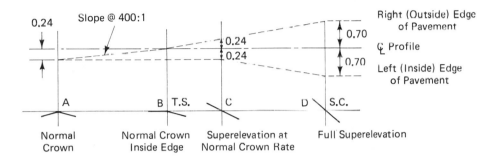

(b) Profile and Cross Sections

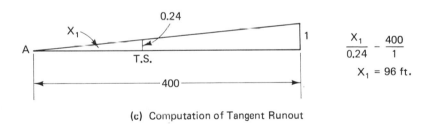

(c) Computation of Tangent Runout

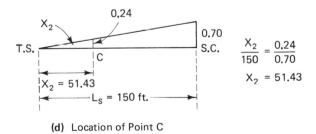

(d) Location of Point C

FIGURE 10.28 Sketches for superelevation example of Sections 10.27 and 10.28: (a) plan; (b) profile and cross sections; (c) computation of tangent runout; (d) location of point C.

391

Compute station A' (i.e., tangent runout at higher chainage spiral).

$$
\begin{aligned}
\text{ST} &= 38 \ + \ 32.71 \\
\text{Tangent runout} &= + \ 96.00 \\
\hline
A' &= 39 \ + \ 28.71
\end{aligned}
$$

5. Compute station C; see Figure 10.28d.

$$
\begin{aligned}
\text{Cross fall at 5.8\% for 12-ft lane} &= 0.058 \times 12 = 0.70 \text{ ft} \\
\text{Distance from TS to C} &= 150 \times \frac{0.24}{0.70} = 51.43 \text{ ft} \\
\text{TS} &= 33 \ + \ 96.60 \\
X \text{ distance} &= 51.43 \\
\hline
C &= 34 \ + \ 48.03
\end{aligned}
$$

Compute station C' (i.e., at higher chainage spiral).

$$
\begin{aligned}
\text{ST} &= 38 \ + \ 32.71 \\
X_2 \text{ distance} &= - \ 51.43 \\
\hline
C' &= 37 \ + \ 81.28
\end{aligned}
$$

6. Reference to Figure 10.29 will show that right-side pavement elevations are 0.24 ft below ₵ elevation from A to C and from C' to A', and 0.70 ft below ₵ from SC to CS. Right-side pavement elevations between C and SC and CS and C' must be interpolated. Reference to Figure 10.29 will also show that left-side pavement elevations must be interpolated between A and TS, between TS and SC, between CS and ST, and between ST and A'. Between SC and CS the left-side pavement elevation is 0.70 higher than the corresponding ₵ elevations.

7. Fill in the left- and right-edge pavement elevations (Table 10.10), where the computation simply involves adding or subtracting normal crown (0.24) or full superelevation (0.70).

8. Perform the computations necessary to interpolate for the missing pavement-edge elevations in Table 10.10 (values are underlined). See Figures 10.30 and 10.31.

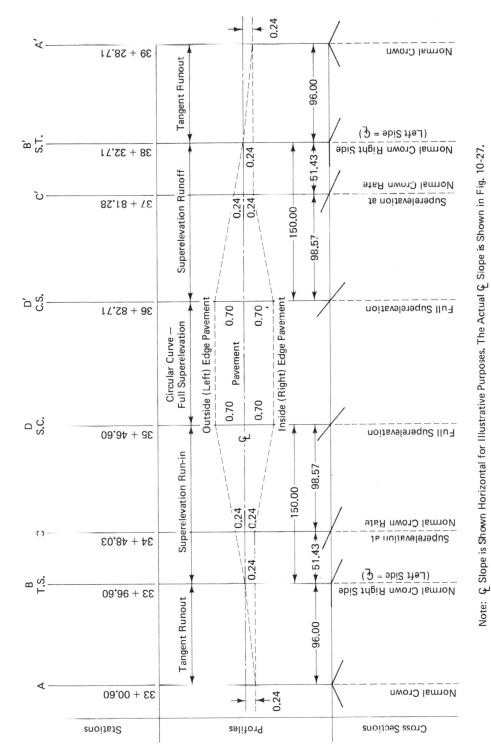

FIGURE 10.29 Superelevated pavement profiles and cross sections for the problem of Section 10.28.

Note: ₵ Slope is Shown Horizontal for Illustrative Purposes. The Actual ₵ Slope is Shown in Fig. 10-27.

393

TABLE 10.10 Pavement Elevations for the Problem of Section 10.28

Station		℄ Grade	Left-Edge Pavement		Right-Edge Pavement	
			Above/Below ℄	Elevation	Below ℄	Elevation
A	33 + 00.60	454.49	−0.24	454.25	−0.24	454.25
	33 + 50	453.75	−0.12	453.63	−0.24	453.51
TS(B)	33 + 96.60	453.05	0.00	453.05	−0.24	452.81
	34 + 00	453.00	+0.02	453.02	−0.24	452.76
C	34 + 48.03	452.28	+0.24	452.52	−0.24	452.04
BVC	34 + 50	452.25	+0.25	452.50	−0.25	452.00
	35 + 00	451.65	+0.48	452.13	−0.48	451.17
SC(D)	35 + 46.60	451.35	+0.70	452.05	−0.70	450.65
	35 + 50	451.33	+0.70	452.02	−0.70	450.62
Low pt.	35 + 78.57	451.29	+0.70	451.99	−0.70	450.59
PVI	36 + 00	451.32	+0.70	452.02	−0.70	450.62
	36 + 50	451.58	+0.70	452.28	−0.70	450.88
CS(D')	36 + 82.71	451.92	+0.70	452.62	−0.70	451.22
	37 + 00	452.15	+0.62	452.77	−0.62	451.53
EVC	37 + 50	453.00	+0.39	453.39	−0.39	452.61
C'	37 + 81.28	453.63	+0.24	453.87	−0.24	453.39
	38 + 00	454.00	+0.15	454.15	−0.24	453.76
ST(B')	38 + 32.71	454.65	0.00	454.65	−0.24	454.41
	38 + 50	455.00	−0.04	454.96	−0.24	454.76
	39 + 00	456.00	−0.17	455.83	−0.24	455.76
A'	39 + 28.71	456.57	−0.24	456.33	−0.24	456.33

Interpolated values are shown underlined.
Pavement revolved about the center line (℄).

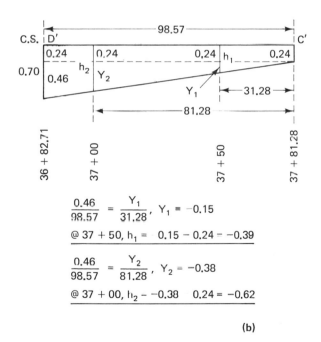

$$\frac{0.46}{98.57} = \frac{Y_1}{1.97}, \quad Y_1 - -0.01$$

@ 34 + 50, $h_1 = -0.01 - 0.24 = -0.25$

$$\frac{0.46}{98.57} = \frac{Y_2}{51.97}, \quad Y_2 = -0.24$$

@ 35 + 00, $h_2 = -0.24 - 0.24 = -0.48$

(a)

$$\frac{0.46}{98.57} = \frac{Y_1}{31.28}, \quad Y_1 = -0.15$$

@ 37 + 50, $h_1 = \ 0.15 - 0.24 = -0.39$

$$\frac{0.46}{98.57} = \frac{Y_2}{81.28}, \quad Y_2 = -0.38$$

@ 37 + 00, $h_2 - -0.38 \quad 0.24 = -0.62$

(b)

Note: Pavement Edge Differentials (with ₵) Between C and D or D' and C' are Identical at Each Station.

FIGURE 10.30 Right-edge pavement elevation interpolation for the problem of Section 10.28.

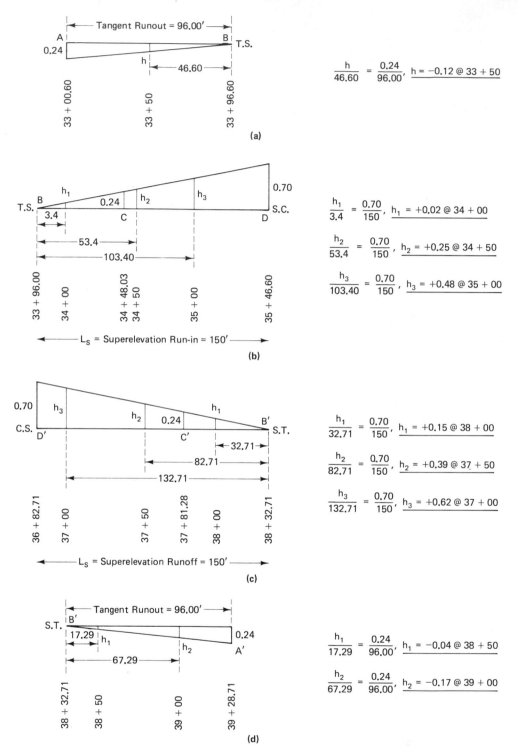

$$\frac{h}{46.60} = \frac{0.24}{96.00}, \quad \underline{h = -0.12 \; @ \; 33 + 50}$$

$$\frac{h_1}{3.4} = \frac{0.70}{150}, \quad \underline{h_1 = +0.02 \; @ \; 34 + 00}$$

$$\frac{h_2}{53.4} = \frac{0.70}{150}, \quad \underline{h_2 = +0.25 \; @ \; 34 + 50}$$

$$\frac{h_3}{103.40} = \frac{0.70}{150}, \quad \underline{h_3 = +0.48 \; @ \; 35 + 00}$$

$$\frac{h_1}{32.71} = \frac{0.70}{150}, \quad \underline{h_1 = +0.15 \; @ \; 38 + 00}$$

$$\frac{h_2}{82.71} = \frac{0.70}{150}, \quad \underline{h_2 = +0.39 \; @ \; 37 + 50}$$

$$\frac{h_3}{132.71} = \frac{0.70}{150}, \quad \underline{h_3 = +0.62 \; @ \; 37 + 00}$$

$$\frac{h_1}{17.29} = \frac{0.24}{96.00}, \quad \underline{h_1 = -0.04 \; @ \; 38 + 50}$$

$$\frac{h_2}{67.29} = \frac{0.24}{96.00}, \quad \underline{h_2 = -0.17 \; @ \; 39 + 00}$$

FIGURE 10.31 Left-edge pavement elevation interpolation for the problem of Section 10.28.

396

10.29 ILLUSTRATIVE SUPERELEVATION PROBLEM: METRIC UNITS

This problem utilizes the horizontal curve data of Section 10.24.

Given Data	Additional Data
$V = 80$ km/h	Lane width = 3.75 m
$\Delta = 16°00'RT$	Normal crown (NC) at 2%
$R = 300$ m	Tangent runout at 400:1
PT at 1 + 086.271	PVI at 1 + 140
Two-lane road	Elevation at 1 + 000 = 186.261
	$g_1 = +1\%$, $g_2 = 0.5\%$
	$(A = 0.5 - 1 = -0.5)$

From Table 10.2, for $V = 80$ km/h and crest curve,

$$K = 35, \qquad L = KA = 35 \times 0.5 = 17.5 \text{ m}$$

Since $KA < V$, use $L = V$; thus, $L = 80$ m.

From Table 10.5, for $R = 300$ m, $V - 80$ km/h, and a two-lane road,

$$A = 125$$
$$e = 0.057$$

where e = superelevation rate.

From Table 10.7,

$$L_s - 52.083$$

That is, $L_s = A^2/R$, from Eq. (10.31).

It is required to determine the ℄ and edge of pavement elevations for even 20-m stations and all other key stations.

Solution

1. Compute the key horizontal alignment stations. (These stations have already been computed in Section 10.24.)

2. Solve the vertical curve for ℄ elevations at even 20-m stations, plus any horizontal alignment key stations that may fall between the BVC and the EVC (see Table 10.11).

3. Begin preparation of Table 10.12, showing all key stations for both horizontal and vertical alignment. List the ℄ grade for each station as computations progress.

4. Compute cross fall for normal crown and for full superelevation:

   Normal crown (NC) = 2%

      Pavement lane width = 3.75 m

      Cross fall = $3.75 \times 0.02 = 0.075$ m

TABLE 10.11 ℄ *Pavement Elevations for the Problem of Section 10.29*

	Station	Tangent Elevation	Tangent Offset (−)	℄ Elevation
BVC	1 + 100	187.261	$(0/40)^2 \times 0.050 = 0$	187.261
CS	1 + 101.797	187.279	$(1.797/40)^2 \times 0.050 = 0$	187.279
	1 + 120	187.461	$(20/40)^2 \times 0.050 = 0.013$	187.448
C_1	1 + 135.605	187.617	$(35.605/40)^2 \times 0.050 = 0.040$	187.577
PVI	1 + 140	187.661	$(40/40)^2 \times 0.050 = 0.050$	187.611
ST	1 + 153.880	187.730	$(26.120/40)^2 \times 0.050 = 0.021$	187.709
	1 + 160	187.761	$(20/40)^2 \times 0.050 = 0.013$	187.748
EVC	1 + 180	187.861	$(0/40)^2 \times 0.050 = 0$	187.861

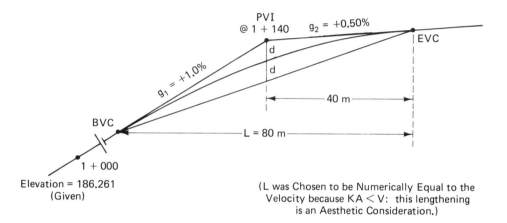

(L was Chosen to be Numerically Equal to the Velocity because KA < V: this lengthening is an Aesthetic Consideration.)

BVC = (1 + 140) − 40 = 1 + 100

EVC = (1 + 140) + 40 = 1 + 180

Elevation BVC = 186.261 + (140 × 0.01) = 187.261

Elevation PVI = 187.261 + (40 × 0.01) = 187.661

Elevation EVC = 187.661 + (40 × 0.005) = 187.861

Mid-Chord = $\dfrac{187.261 + 187.861}{2} = 187.561$

Tangent Offset: d = $\dfrac{187.661 - 187.561}{2} = 0.050$

or

Since L = 80, K = L/A, therefore K = 160

$d = \dfrac{KA^2}{800}$ (equation 10-21), d = $\dfrac{160 \times 0.5^2}{800} = \underline{0.050 \text{ m}}$

FIGURE 10.32 Vertical curve solution for the problem of Section 10.29.

Full superelevation (e) at 0.057
Pavement width = 3.75 m
Cross fall = $3.75 \times 0.057 = 0.214$ m (0.21375)

5. Compute tangent runout and stations for A and A' (Figure 10.33a).

$$\frac{X_1}{0.075} = \frac{400}{1}, \qquad X_1 = 30.000 \text{ m}$$

$$
\begin{array}{lll}
\text{TS at} & 1 + 018.021 \\
-X_1 & \underline{30.000} \\
A \text{ at} & 0 + 988.021
\end{array}
$$

$$
\begin{array}{lll}
\text{ST at} & 1 + 153.880 \\
+X_1 & \underline{30.000} \\
A' \text{ at} & 1 + 183.880
\end{array}
$$

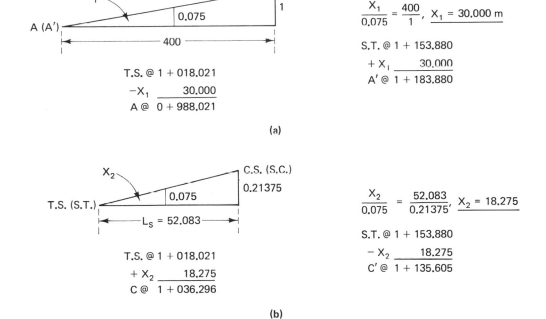

(a)

$$\frac{X_1}{0.075} = \frac{400}{1}, \; X_1 = 30.000 \text{ m}$$

S.T. @ 1 + 153.880
+ X_1 _____ 30.000
A' @ 1 + 183.880

T.S. @ 1 + 018.021
−X_1 _____ 30.000
A @ 0 + 988.021

$$\frac{X_2}{0.075} = \frac{52.083}{0.21375}, \; X_2 = 18.275$$

S.T. @ 1 + 153.880
− X_2 _____ 18.275
C' @ 1 + 135.605

T.S. @ 1 + 018.021
+ X_2 _____ 18.275
C @ 1 + 036.296

(b)

FIGURE 10.33 Sketch for tangent runout computations: (a) point A; (b) point C for problem in Section 10.29.

6. Compute location of C and C', at the points where superelevation is at normal crown rate (i.e., at 0.075 m). See Figure 10.33b.

$$\frac{X_2}{0.075} = \frac{52.083}{0.21375}, \qquad X_2 = 18.275 \text{ m}$$

$$
\begin{array}{rl}
\text{TS at} & 1 + 018.021 \\
+X_2 & \underline{18.275} \\
C \text{ at} & 1 + 036.296
\end{array}
$$

$$
\begin{array}{rl}
\text{ST at} & 1 + 153.880 \\
-X_2 & \underline{18.275} \\
C' \text{ at} & 1 + 135.605
\end{array}
$$

7. Reference to Figure 10.34 will show that right-side pavement elevations are 0.075 m below ℄ from A to C' and from C' to A', and 0.214 m below ℄ from SC to CS. Right-side pavement elevations between C and SC and between CS and C' must be interpolated. Reference to the same figure will show that left-side pavement elevations must be interpolated between A and TS, between TS and SC, between CS and ST, and between ST and A'. Between the SC and the CS the left-side pavement elevation is 0.214 higher than the corresponding ℄ elevations.

8. Fill in the left- and right-side pavement elevation columns in Table 10.12 where the computation simply involves adding or subtracting normal crown (0.075) or full superelevation (0.214).

9. Perform the computations necessary to interpolate for the missing pavement-edge elevations in Table 10.12 (values are underlined). See Figures 10.35 and 10.36.

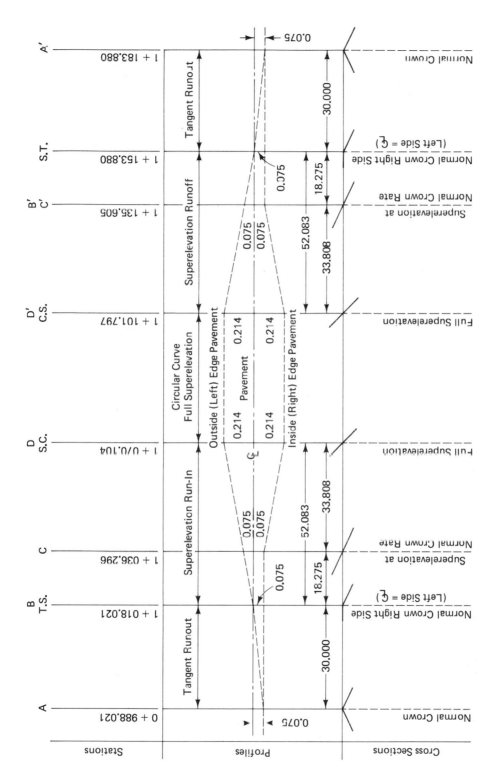

FIGURE 10.34 Superelevated pavement profiles and cross sections for the problem of Section 10.29.

Note: ₵ Slope is Shown Horizontal for Illustrative Purposes. The Actual ₵ Slope is Shown in Fig. 10-32.

401

TABLE 10.12 Pavement Elevations for the Problem of Section 10.29

Station		₵ Grade	Left-Edge Pavement		Right-Edge Pavement	
			Above/Below ₵	Elevation	Below ₵	Elevation
	0 + 980	186.061	−0.075	185.986	−0.075	185.986
A	0 + 988.021	186.141	−0.075	186.066	−0.075	186.066
	1 + 000	186.261	−0.045	186.216	−0.075	186.186
TS(B)	1 + 018.021	186.441	−0.000	186.441	−0.075	186.366
	1 + 020	186.461	+0.008	186.469	−0.075	186.386
C	1 + 036.296	186.624	+0.075	186.699	−0.075	186.549
	1 + 040	186.661	+0.090	186.751	−0.090	186.571
	1 + 060	186.861	+0.172	187.033	−0.172	186.689
SC(D)	1 + 070.104	186.962	+0.214	187.176	−0.214	186.748
	1 + 080	187.061	+0.214	187.275	−0.214	186.847
BVC	1 + 100	187.261	+0.214	187.475	−0.214	187.047
CS(D')	1 + 101.797	187.279	+0.214	187.493	−0.214	187.065
	1 + 120	187.448	+0.139	187.587	−0.139	187.309
C'	1 + 135.605	187.577	+0.075	187.652	−0.075	187.502
PVI	1 + 140	187.611	+0.057	187.668	−0.075	187.536
ST(B')	1 + 153.880	187.709	+0.000	187.709	−0.075	187.634
	1 + 160	187.748	−0.015	187.733	−0.075	187.673
EVC	1 + 180	187.861	−0.065	187.796	−0.075	187.786
A'	1 + 183.880	187.880	−0.075	187.805	−0.075	187.805
	1 + 200	187.961	−0.075	187.886	−0.075	187.886

Pavement revolved about center line (₵).
Interpolated values are underlined.

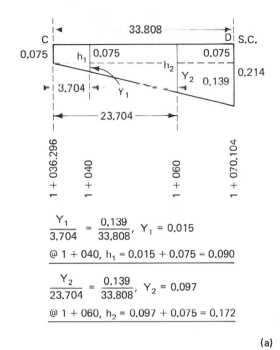

$$\frac{Y_1}{3.704} = \frac{0.139}{33.808}, \quad Y_1 = 0.015$$

@ 1 + 040, h_1 = 0.015 + 0.075 = 0.090

$$\frac{Y_2}{23.704} = \frac{0.139}{33.808}, \quad Y_2 = 0.097$$

@ 1 + 060, h_2 = 0.097 + 0.075 = 0.172

(a)

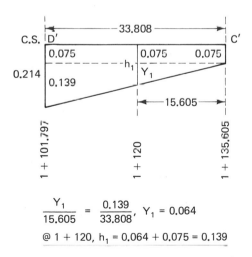

$$\frac{Y_1}{15.605} = \frac{0.139}{33.808}, \quad Y_1 = 0.064$$

@ 1 + 120, h_1 = 0.064 + 0.075 = 0.139

(b)

FIGURE 10.35 Right-edge pavement elevation interpolation for the problem of Section 10.29.

403

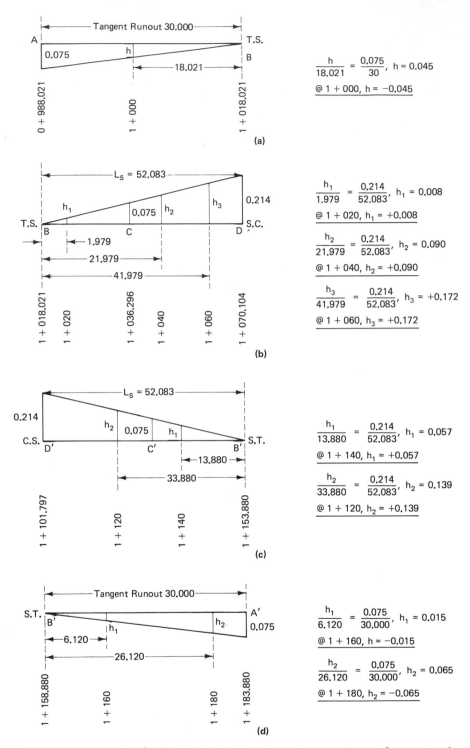

FIGURE 10.36 Left-edge pavement elevation interpolation for the problem of Section 10.29.

404

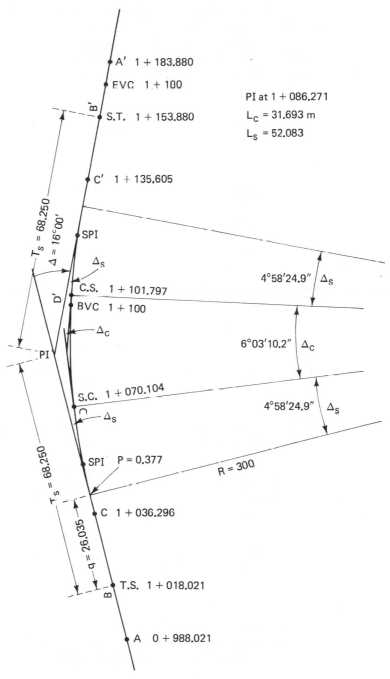

A′ 1 + 183.880

EVC 1 + 100

S.T. 1 + 153.880

PI at 1 + 086.271

L_C = 31.693 m

L_S = 52.083

C′ 1 + 135.605

T_s = 68.250

Δ = 16°00′

SPI

Δ_s

4°58′24.9″ Δ_s

C.S. 1 + 101.797

BVC 1 + 100

Δ_c

6°03′10.2″ Δ_c

S.C. 1 + 070.104

Δ_s

4°58′24.9″ Δ_s

SPI P = 0.377

R = 300

C 1 + 036.296

T_s = 68.250

q = 26.035

T.S. 1 + 018.021

A 0 + 988.021

FIGURE 10.37 Summary of chainage data for the problem of Section 10.29.

All these values are entered in Table 10.12. Figure 10.37 provides a summary for the problem of Section 10.29.

Note: Pavement edge differentials at each station between C and D (D' and C') are identical—that is, the outside edge is higher than ℄ by the same amount that the inside edge is lower than ℄.

PROBLEMS

10–1 Given PI @ 11 + 41.67, $\triangle$ = 31°32′, R = 600 ft, compute tangent (T) and length of arc (L).

10–2 Given PI @ 6 + 27.98, $\triangle$ = 6°30′, D = 7°, compute tangent (T) and length of arc (L).

10–3 From the data in Problem 10–1, compute the stationing of the BC and EC.

10–4 From the data in Problem 10–2, compute the stationing of the BC and EC.

10–5 Given PI @ 3 + 162.789, $\triangle$ = 12°47′, and R = 300 m, compute the deflections for even 20-m stations.

10–6 Given PI @ 8 + 272.311, $\triangle$ = 28°28′20″, R = 500 m, compute E (external), M (mid-ordinate) and the stations of the BC and EC.

10–7 Given PI @ 12 + 32.33, $\triangle$ = 40°43′30″ RT, R = 1500 ft, compute the deflections for even 100-ft stations.

10–8 From the distances and deflections computed in Problem 10–5, compute the three key ℄ chord layout lengths, i.e., (1) BC to first 20-m station, (2) chord distance for 20-m (arc) stations and (3) from the last even 20-m station to the EC.

10–9 From the distances and deflections computed in Problem 10–7, compute the three key ℄ chord layout lengths, i.e., (1) BC to first 100-ft station, (2) chord distance for 100-ft (arc) stations, and (3) from the last even 100-ft station to the EC.

10–10 From the distances and deflections computed in Problem 10–7, compute the chords (6) required for layout directly on offsets 50-ft right and 50-ft left of ℄.

10–11 It is required to join two highway ℄ tangents with a circular curve of radius 1000 ft. The PI is inaccessible as it's lo-

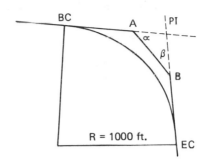

cation falls in a river. Point A is established near the river on the back tangent and point B is established near the river on the forward tangent. Distance AB is measured to be 615.27 ft. Angle $\alpha - 51°31'20''$, and angle $\beta = 32°02'45''$. Perform the calculations required to locate the BC and the EC in the field.

10–12 Two street curb lines intersect with $\triangle = 71°36'$. A curb radius must be selected so that an existing catch basin (CB) will abut the future curb. The curb-side of the catch basin ₵ is located from point V: V to CB = 8.713 m and angle E, V, CB = 21°41'. Compute the radius which will permit the curb to abut the existing catch basin.

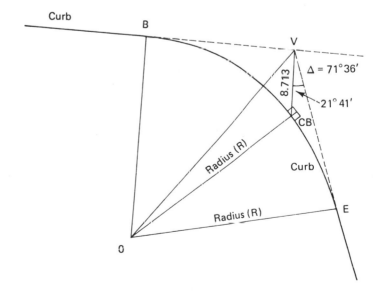

10-13 Given the following compound curve data: R_1 = 200 ft, R_2 = 300 ft, $\triangle_1$ = 44°26' and $\triangle_2$ = 45°18', compute T_1 and T_2 (see Figure 10-15).

10-14 Given the following vertical curve data: PVI @ 7 + 25.712; L = 100 m; g_1 = -3.2%; g_2 = +1.8%; elevation of PVI = 210.440, compute the elevations of the curve low point and even 20-m stations.

10-15 Given the following vertical curve data: PVI @ 19 + 00; L = 500 ft; g_1 = +2.5%; g_2 = -1%; elevation @ PVI = 723.86 ft., compute the elevations of the curve summit and even full stations (i.e., 100-ft even stations).

10-16 Given the following vertical curve data: g_1 = +3%; g_2 = -1%; design speed = 100 Km/h (from Table 10-2, K = 90 m); PVI @ 0 + 360.100, with an elevation of 156.663, compute the elevations of the high point and even 50-m stations.

10-17 Given the following spiral curve data: D = 8°, V = 40 mph, $\triangle$ = 16°44', PI @ 11 + 66.18, determine the value of each key spiral and circular curve component (Ls, R, P, q, LT, ST, X, Y, $\triangle$s, $\triangle$c, Tc, and Lc) and determine the stationing (chainage) of the TS, SC, CS and ST.

10-18 Given the same data as in Problem 10-17, use the approximate formulae (#10-33 to #10-38) to compute X, Y, q, P, LT, and ST. Enter these values in a table together with equivalent values as determined in Problem 10-17 in order to compare the results.

10-19 Use the data from Problem 10-17 to compute the deflections for the curve system (spirals and circular curves) at even 50-ft stations.

10-20 Given the following spiral curve data: R = 220 m; V = 80 Km/hr; $\triangle$ = 28°08' RT; two-lane road; PI @ 1 + 286.441: Determine the value of each key spiral and horizontal curve component (Ls, X, Y, q, P, LT, ST, Lc, $\triangle$s, $\ominus$s, $\triangle$c, Tc, and Lc) and determine the chainage of the TS, SC, CS, and ST.

10-21 Given the same data as in Problem 10-20, use the approximate formulae (#10-33 to #10-38) to compute X, Y, q, P, LT, and ST. Enter these values in a table together with equivalent values as determined in Problem 10-20 in order to compare the results.

10–22 Use the data from Problem 10–20 to compute the deflections for the curve system (spirals and circular curve) at even 20-m stations.

10–23 Given the following vertical curve data: $g_1 = -1\%$; $g_2 = +3\%$, design speed (V) = 80 Km/h, PVI @ 1 + 300, elevation 210.400, compute the elevations of the low point and the even 20-m stations (use K minimum).

10–24 Combine the data from Problems 10–20 and 10–23 and add data for crown slope, tangent runout, and lane width:

V = 80 Km/h	Lane width – 4 m
R = 220 m	Normal crown @ 2%
$\triangle$ = 28°08′RT	Tangent runout @ 400:1
PI @ 1 + 286.441	PVI @ 1 + 300
Two-lane road	$g_1 = -1\%$, $g_2 = +3\%$

Compute the elevations of the $\underline{\mathcal{C}}$ and edges of pavement at 20-m stations and at all key stations between 1 + 160 and 1 + 420. From Problem 10–20, the superelevation rate (e) – 0.060 (Table 10–5).

11 *Control Surveys*

11.1 GENERAL

We have seen in several prior sections the need to tie our survey data to some control fabric. Characteristically, the survey control fabric should be relatively permanent and established with a higher level of precision than will be the various surveys that will use it as a reference. When one considers nationwide or continental control, one thinks in terms of the highest possible precision combined with guaranteed permanence or scheduled replacement. The control net of the United States has been tied into the control nets of Canada and Mexico, giving a consistent continental net. The specifications and accuracy levels for both horizontal and vertical control are discussed in Appendix A.

The first major adjustment in control data was made in 1927, resulting in the North American Datum (NAD). Since that time a great deal more has been learned about the shape and mass of the earth; these new and expanded data come to us from releveling surveys, precise traverses, satellite positioning surveys, earth movement studies, and gravity surveys. The mass of data thus accumulated has been utilized to update and expand existing control data, and, as well, the new geodetic data have provided scientists with the means to more precisely define the actual geometric shape of the earth. The reference ellipsoid previously used for this purpose (the Clarke spheroid of 1866) has been modified to reflect our current knowledge of the earth. Accordingly, a World Geodetic System, first proposed in 1972 (WGS '72) and later endorsed in 1979 by the International Association of Geodesy (IAG), included proposals for an earth-mass-centered ellipsoid that would more closely represent the planet on which we live. Proposed parameters include a semimajor axis (equatorial radius) of 6378135 m (the Clark spheroid has a semimajor axis of 6378206.4 m).

The new adjustment, NAD 83, covers the North American continent, including Greenland and parts of Central America. All individual control nets are included in a weighted simultaneous computation. A good tie to the global system is given by Doppler satellite positioning.

The federal agencies responsible (National Ocean Survey— U.S.A., and Energy, Mines, and Resources—Canada) will publish revised coordinates (in meters) for all existing control stations. The new ellipsoid gives relatively large changes in meridian arc values, and this results in somewhat larger discrepancies in the northings as opposed to the eastings. It should be noted that although position values will change somewhat in the new system, distances between positions will change only very slightly (e.g., an east–west continental adjustment along the total U.S.–Canada border of about 25 to 30 m). The NAD 83 will include about 250 000 control points all reduced to the new geodetic ellipsoid.

Work is also progressing on a vertical control net with revised values for about 600 000 bench marks in the United States and Canada. It is expected that this work will be largely completed by 1988, resulting in a new North American Vertical Datum ("NAVD-88"); the original adjustment of continental vertical values was performed in 1929.

First-order control monuments in North America were established using very precise triangulation techniques (see Appendix A). Triangulation was favored because the basic measurement, angles, could be more quickly and precisely taken than could distances. However, with the advent of EDM instruments, traditional triangulation techniques are giving way to (1) combined triangulation and trilateration techniques, (2) simple trilateration, and (3) precise traverses. One instrument setup can now give both angle and distances, ensuring a more accurate result. See Figures 11.1 to 11.5 for a selection of precise theodolites, levels, and EDM.

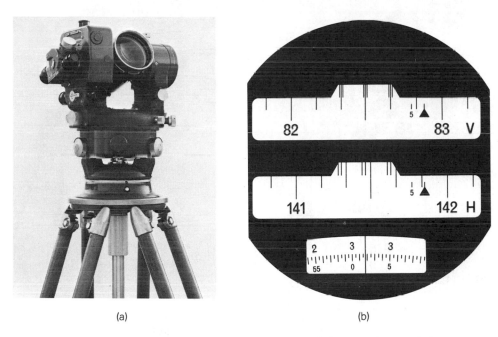

(a)   (b)

FIGURE 11.1 (a) Kern DKM 3 precise theodolite; angles read directly to 0.5 second; used in first-order surveys; (b) Kern DKM 3 scale reading (vertical angle = 82°53′01.8″).

FIGURE 11.2 Wild T-4 precise astronomic theodolite.

FIGURE 11.3 Wild N3 precise level (spirit level). SE for 1 mile (double run) = ±0.0008 ft; 1 km (double run) = ±0.2 mm. Used with a parallel plate micrometer, the N3 is employed on National Geodetic surveys. Split bubble and micrometer settings are viewed simultaneously, ensuring precise readings. (Courtesy Wild Heerbrugg Co. Ltd.)

FIGURE 11.4 Zeiss Ni 1 precise automatic level featuring 40× magnification with a nominal accuracy of ± 0.2 mm √distance in kilometers. (Courtesy Carl Zeiss—Oberkochen)

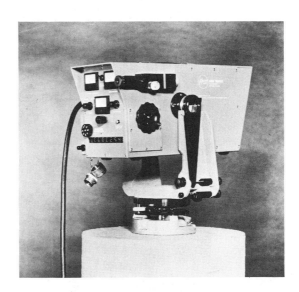

FIGURE 11.5 Kern Mekometer, ME 3000, a high-precision EDM [SE = ±(0.2 mm + 1 ppm)] with a triple prism distance range of 2.5 km. Used wherever first-order results are required, for example, deformation studies, network surveys, plant engineering, and base-line calibration. (Courtesy Kern Instruments, Inc.)

414

In addition to the advent of EDMs, precise positioning is also being achieved using Doppler satellite positioning techniques, the Navstar Global Positioning System, inertial surveying techniques, and very long base line interferometry (VLBI). The latter technique utilizes radio waves emitted by quasars to provide precise results over very long distances.

When the relative positions of control stations have been determined, it must be decided how best to tie in all these widely spaced points to a reference fabric. In Section 1, it was noted that a basic reference fabric for the earth is the geodetic coordinate system employing latitude and longitude (ϕ, λ). Although this system is widely used for other purposes (e.g., navigation), it has been found too cumbersome for use in surveying; for instance, the latitude and longitude angles must be expressed to four decimals of a second (01.0000″) to give position to the closest 0.01 ft. (At latitude 44°, 1″ latitude = 101 ft and 1″ longitude = 73 ft.)

To allow the surveyor to perform reasonably precise surveys and still use plane geometry and trigonometry for problem solutions, several forms of plane coordinate grids have been utilized.

11.2 UNIVERSAL TRANSVERSE MERCATOR GRID SYSTEM

The universal transverse Mercator (UTM) projection (Figure 11.6) is developed by placing a cylinder around the earth with its circumference tangent to the earth along a meridian (central meridian). See Figure 11.7. When the cylinder is developed, a plane is established that can be used for grid purposes. It can be noted (Figures 11.7 and 11.9a) that at the central meridian the scale is exact, and the scale becomes progressively more distorted as the distance from the central meridian increases. The distortion (which is always present when a spherical surface is projected onto a plane) can be minimized in two ways. First, the distortion can be minimized by keeping the zone width relatively narrow; and, second, the distortion can be lessened by reducing the radius of the projection cylinder so that, instead of being tangent to the earth's surface, the cylinder cuts through the earth's surface at an optimal distance on either side of the central meridian (see Figures 11.8 and 11.9b). This means that the scale factor at the central meridian is less than unity; it is unity at the line of intersection and more than unity between the lines of intersection and the zone limit meridians. The plane surfaces developed from these tangent and secant cylinders are illustrated in Figures 11.9a

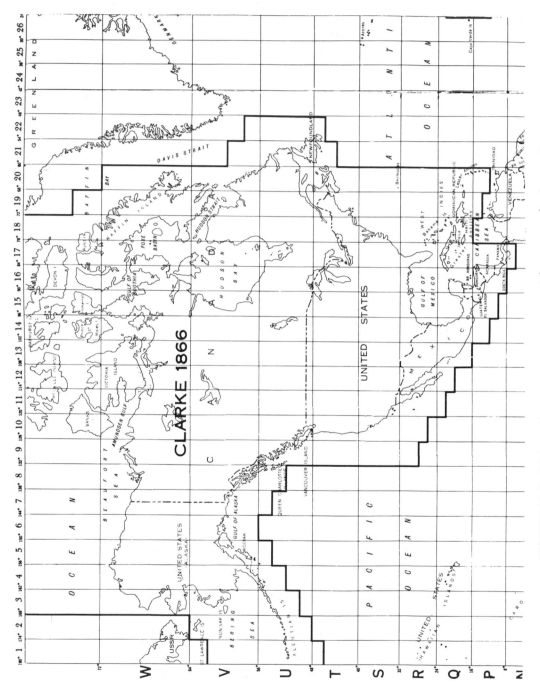

FIGURE 11.6 Universal transverse mercator grid zone numbering system.

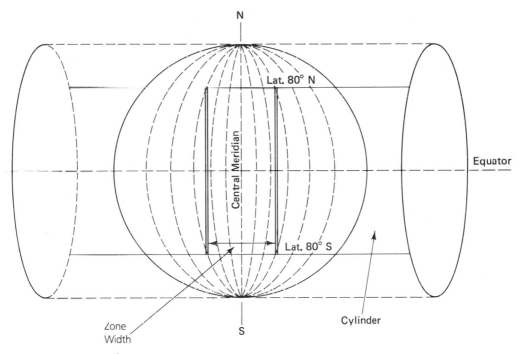

1. 6° for UTM
2. 3° for MTM (some regions in Canada)
3. About 158 Miles for State Plane Coordinate Grid (U.S.A.)

FIGURE 11.7 Transverse mercator projection cylinder *tangent* to earth's surface at central meridan (CM) (see Figure 11.6 for zone numbers).

and 11.9b. Figure 11.10 shows a cross section of a UTM 6° zone. For full treatment of this topic, the reader is referred to texts on Geodesy and Cartography.

Characteristics of the Universal Transverse Mercator Grid System

1. Zone is 6° wide. Zone overlap of 0°30'. See Table 11.1.
2. Latitude of the origin is the equator.
3. Easting value of each central meridian = 500 000.000 m.
4. Northing value of the equator = 0.000 m (10 000 000.000 m in the southern hemisphere).
5. Scale factor at the central meridian is 0.9996 (i.e., $^1/_{2500}$).
6. Zone numbering commences with 1 in the zone 180° W to

TABLE 11.1   *UTM Zone Width in Kilometers*

North Latitude	Width in Kilometers
42°00'	497.11827
43°00'	489.25961
44°00'	481.25105
45°00'	473.09497
46°00'	464.79382
47°00'	456.35005
48°00'	447.76621
49°00'	439.04485
50°00'	430.18862

Source: Ontario Geographical Referencing Grid. Ministry of Natural Resources, Ontario

174° W and increases eastward to zone 60 at the zone 174° E to 180° E (see Figure 11.6 for zone numbering in North America).

8. Projection limits of latitude 80° S to 80° N.

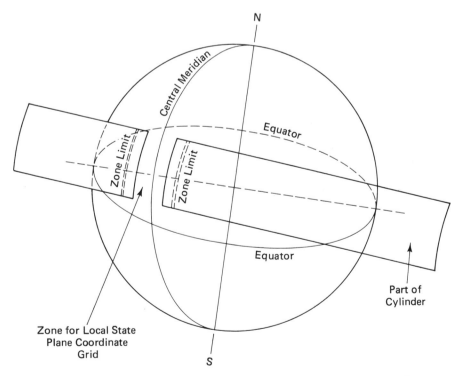

FIGURE 11.8   Transverse mercator projection. *Secant* cylinder for state plane coordinate grids.

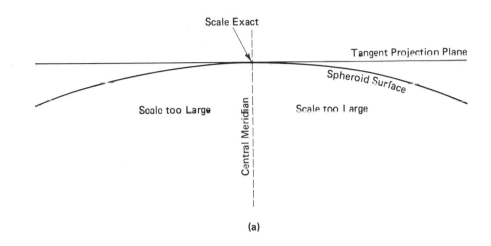

Scale Exact

Tangent Projection Plane

Spheroid Surface

Scale too Large

Central Meridian

Scale too Large

(a)

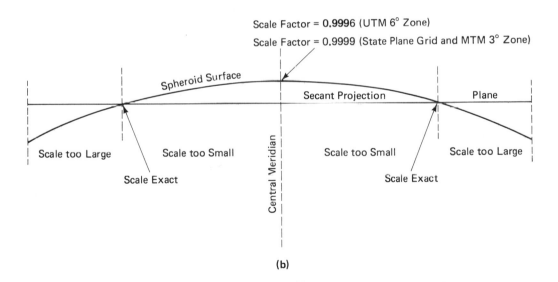

Scale Factor = **0.9996** (UTM 6° Zone)

Scale Factor = **0.9999** (State Plane Grid and MTM 3° Zone)

Spheroid Surface

Secant Projection

Plane

Scale too Large

Scale too Small

Central Meridian

Scale too Small

Scale too Large

Scale Exact

Scale Exact

(b)

FIGURE 11.9 (a) A section view of the projection plane and earth's surface (*tangent projection*); (b) a section view of the projection plane and earth's surface (*secant projection*).

419

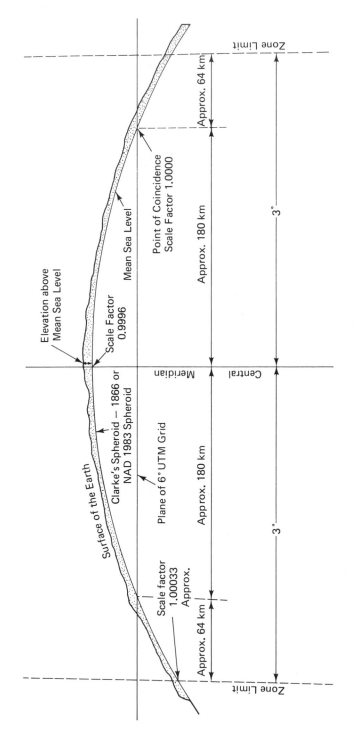

FIGURE 11.10 Cross section of a 6° zone (UTM).

420

11.3 MODIFIED TRANSVERSE MERCATOR GRID SYSTEM

Some regions and agencies in Canada have adopted a modified transverse Mercator system (MTM). The modified projection is based on 3° wide zones instead of 6° wide zones. By narrowing the zone width, the scale factor is improved from 0.9996 ($^1/_{2500}$) to 0.9999 ($^1/_{10000}$). The improved scale factor permits surveyors to work at moderate levels of accuracy (e.g., $^1/_{5000}$) without having to account for projection corrections. The zone width of 3° (about 152 miles wide at latitude 43°) compares very closely with the 158-mile-wide zones used in the United States for transverse Mercator and Lambert projections in the state plane coordinate system.

The characteristics of the 3° zone are as follows:

1. Zone is 3° wide.
2. Latitude of origin is the equator, 0°.
3. Easting value of the central meridian is 1 000 000.000 ft.
4. Northing value of the equator is 0.000 ft.
5. Scale factor at the central meridian is 0.9999 (i.e., $^1/_{10000}$).
6. Zone numbering commences with Newfoundland and continues west across Canada.

Although the dimensions for the MTM were defined in feet, many agencies have made a direct conversion to meters (e.g., easting at CM = 304 800.000 m) and have been working in metric units for the past few years.

As noted above, the 3° zone scale factor at the central meridian of .9999 ($^1/_{10.000}$) permits the surveyor to work at moderate levels of precision without correcting for scale error. However, the introduction of new instrumentation (e.g., electronic tacheometers) has resulted in new or up-graded surveying techniques (e.g., layout by polar co-ordinates)—techniques that require control surveys at higher orders of precision (see Appendix A, equations A-17 and A-18).

This requirement for higher precision has resulted in proposals for even narrower grid zones (e.g., 2° and even 1° zones) so that field work can still proceed without the need for scale corrections.

It should be kept in mind that narrow grid zones only permit the surveyor to ignore corrections for scale, and that other corrections to field measurements such as for elevation, temperature, sag, etc., and the balancing of errors are still routinely required.

Further, the irreversible trend to computer-use for field data processing means that scale errors can be easily rectified—along with all other data processing requirements.

Proponents of the UTM (6° zone) grid cite the advantages to a uniform, world-wide system:

- eliminates the confusion resulting from different grid coordinate axes in the same area
- permits quick ground data correlation between neighboring (or distant) government agencies
- permits a uniformity in maps and map referencing systems
- facilitates worldwide or continental data-base sharing (see Chapter 16)

11.4 STATE PLANE COORDINATE GRID SYSTEMS

In 1933, the U.S. Coast and Geodetic Survey devised a system of coordinate grids for state coverage. Each state was covered with one or more zones, utilizing either the transverse Mercator projection (cylindrical) or the Lambert projection (conical). Both projections are conformal, with the transverse Mercator projection being relatively distortion-free in a north–south direction, and the Lambert projection being relatively distortion-free in an east–west direction.

To minimize distortion, the transverse Mercator projection was limited to about 158 miles in east–west width, and the Lambert projection was limited to about 158 miles in a north–south direction.

11.5 LAMBERT PROJECTION

The Lambert projection is a conical conformal projection. The apex of the cone is on the earth's axis of rotation above the north pole for Northern Hemisphere projections and below the south pole for Southern Hemisphere projections. The location of the apex (and the θ angle) depends on the area of the ellipsoid that is being projected. Reference to Figure 11.11 will confirm that although the east–west direction is relatively distortion-free, the north–south coverage must be restrained (e.g., 158 miles) to maintain the integrity of the projection; therefore, the Lambert projection is used for states having a greater east west dimension. Table 11.2 gives a list of all the states indicating the type of pro jection being used; New York, Florida, and Alaska utilize both the transverse Mercator and Lambert projections. The U.S. National Ocean Survey publishes tables for each state plane projection so that computations can be easily accomplished. These data are presently (1983) being revised to reflect the new reference spheroid (NAD 1983) and new computer-based techniques for data manipulation. The National Ocean Survey will publish data (in meters) for both the state plane coordinate grid systems and the universal transverse Mercator grid system.

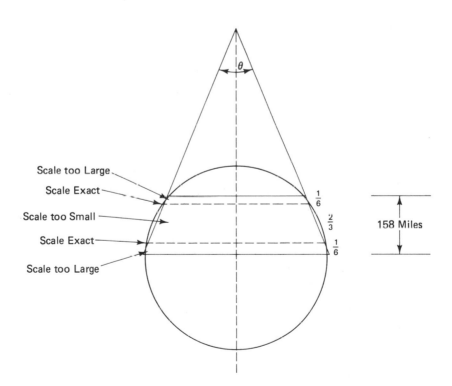

FIGURE 11.11   Lambert secant projection.

TABLE 11.2 *State Plane Coordinate Grid Systems*

Transverse Mercator System		Lambert System		Both Systems
Alabama	Mississippi	Arkansas	North Dakota	Alaska
Arizona	Missouri	California	Ohio	Florida
Delaware	Nevada	Colorado	Oklahoma	New York
Georgia	New Hampshire	Connecticut	Oregon	
Hawaii	New Jersey	Iowa	Pennsylvania	
Idaho	New Mexico	Kansas	South Carolina	
Illinois	Rhode Island	Kentucky	South Dakota	
Indiana	Vermont	Louisiana	Tennessee	
Maine	Wyoming	Maryland	Texas	
		Massachusetts	Utah	
		Michigan	Virginia	
		Minnesota	Washington	
		Montana	West Virginia	
		Nebraska	Wisconsin	
		North Carolina		

11.6 COMPUTATIONS FOR THE LAMBERT PROJECTION

Computations for the Lambert secant projection grid are illustrated in Figure 11.12. Most Lambert projection grids in the United States assign an X value at the central meridian (Y axis) of 2 000 000 ft, and a Y value at the X axis of 0 ft.

The geographic coordinates (ϕ, λ; latitude, longitude) of point P are also shown. The distance R_b is the largest latitude radius of the zone and is obtained from the state plane tables. The angle θ is the angle between the meridian through point P and the central meridian. X_c is the X value assigned to the meridian (usually 2 000 000 ft) so that all the state X values will remain positive. The values of R are found in the state plane tables for the range of values for latitude, as are the values for θ for the range of values for longitude.

Using the relationships shown in Figure 11.12 and data from the appropriate state plan tables, the grid coordinates can be computed:

$$X_p = X_c + R \sin \theta \tag{11.1}$$
$$Y_p = R_b - R \cos \theta \tag{11.2}$$

The direction of a line can be analyzed referring to Figure 11.13. It can be seen that the grid and geodetic meridians coincide at the central meridian, and that the difference in azimuth between the

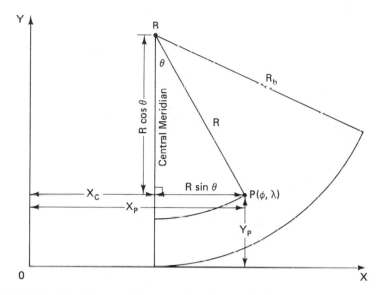

FIGURE 11.12   Rectangular coordinates on the Lambert projection.

grid meridian and the geodetic meridian is angle θ. Angle θ becomes progressively larger as the distance from the central meridian increases (i.e., as the difference in longitude increases).

Angle θ is considered positive ($+$) if point P is east of the central meridian and negative if point P is west of the central meridian.

$$\text{Grid azimuth} = \text{geodetic azimuth} - \theta \qquad (11.3)$$

For distances of sights beyond 5 miles, a second term is added to θ is improve precision; the value of the second term is also available in the state plane tables.

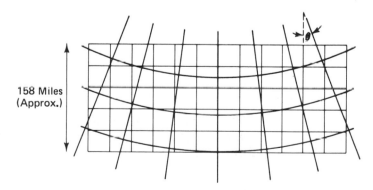

FIGURE 11.13   Lines of latitude (parallels) and lines of longitude (meridians) on the Lambert projection grid.

If, in addition to limiting the north–south zone width to 158 miles, two-thirds of the zone width is between the secant lines (Figure 11.11), the distortion effect is kept to a level of $1/10\ 000$ or better.

11.7 COMPUTATIONS FOR THE TRANSVERSE MERCATOR SECANT PROJECTION

Most Mercator projection grids in the United States assign an X value to the central meridian (Y axis) of 500 000 ft and a Y value to the X axis of 0 ft. The X axis is chosen so that all Y values for a specific state will be positive.

Figure 11.14 illustrates the features of the transverse Mercator projection. The difference between grid north and geodetic north, also called convergence, is given by $\Delta\alpha$:

$$\Delta\alpha'' = \Delta\lambda''\ \text{Sin}\ \phi p \qquad (11.4a)$$

Where $\Delta\lambda''$ is the difference in longitude in seconds between the central meridian and point P, and ϕp is the latitude of point P. When long sights (>5 miles) are taken, a second term (given in the state plane tables) is required to maintain directional accuracy. When the direction of a line from P_1 to P_2 is being considered, the expression becomes

$$\Delta\alpha'' = \Delta\lambda''\ \text{Sin}\ \left(\frac{\phi P_1 + \phi P_2}{2}\right)$$

Alternately, if the distance from the central meridian is known, the expression becomes

$$\theta'' = 52.13d\ \text{Tan}\ \phi \qquad (11.4b)$$

where $\theta =$ convergence angle in seconds
$\quad\ \ d =$ departure distance from the central meridian (CM) in miles
$\quad\ \ \phi =$ average latitude of the line.

See Section 13.2 for further discussion of convergence.

Using the relationships shown in Figure 11-14 and data from the appropriate state plane tables, the grid coordinates can be computed:

$$X'_p = H \cdot \Delta\lambda'' \pm ab \qquad (11.5)$$
$$X_p = X'_p + X_c \qquad (11.6)$$

(X_c is the X value assigned to the Y axis for a specific state grid, usually 500 000 ft)

$$YP = Y_0 + V \left(\frac{\Delta\lambda''}{100}\right)^2 \pm c \tag{11.7}$$

Y_0, H, V, and a are based on the latitude and are found in the state zone tables. λ'' is the longitude of the central meridian minus the longitude of point P given in seconds of arc; b and c are related to λ'' and are also found in the appropriate zone tables.

Since the transverse Mercator projection is relatively distortion-free in the N–S direction, it is utilized for states having a predominantly N–S orientation (e.g., Indiana).

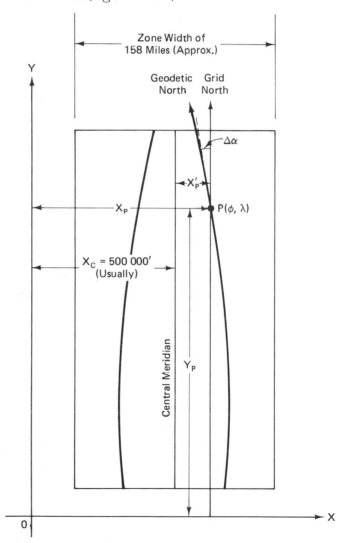

FIGURE 11.14 Transverse mercator secant projection.

11.8 UTILIZATION OF GRID COORDINATES

When local surveys (traverse or trilateration) are tied into coordinate grids, corrections must be provided so that (1) grid and ground distances can be reconciled, and (2) grid and geodetic directions can be reconciled.

Figures 11.15 and 11.16 give a section view of MTM and state plane grids. It can be seen that a distance measured on the earth's surface must first be reduced for equivalency on the spheroid, and then it must be further reduced (in this illustration) for equivalency on the projection plane. The first reduction involves multiplication by an elevation factor (sea-level factor); the second reduction (adjustment) involves multiplication by the scale factor.

The elevation (sea level) factor can be determined by establishing a ratio as is illustrated in Figure 11.17:

$$\frac{\text{Sea level distance}}{\text{Ground distance}} = \frac{R}{R + H} \tag{11.8}$$

where R is the average radius of the earth, and H is the elevation above mean sea level.

For example, at 500 ft the elevation factor would be

$$\frac{20\ 906\ 000}{20\ 906\ 500} = 0.999976$$

and a ground distance of 800.00 ft at an average elevation of 500 ft would become $800 \times 0.999976 = 799.98$ at sea level. If this ground distance were being measured at the central meridian of a MTM zone with a central scale factor of 0.9999, the ground distance converted for both elevation and scale factors would be $800.00 \times 0.999976 \times 0.9999 = 799.90$ ft. This solution is illustrated in Figure 11.18a.

If the elevation factor is multiplied by the scale factor ($0.9999 \times 0.999976 = 0.999876$), the resultant (0.999876) is known as the *grid factor*:

$$\text{Ground distance} \times \text{grid factor} = \text{grid distance}$$

Thus

$$800.00 \times 0.999876 = 799.90 \text{ ft}$$

Figure 11.18b illustrates a similar conversion in a UTM zone.

Alternatively, $\dfrac{\text{Grid distance}}{\text{Grid factor}} = \text{ground distance}$

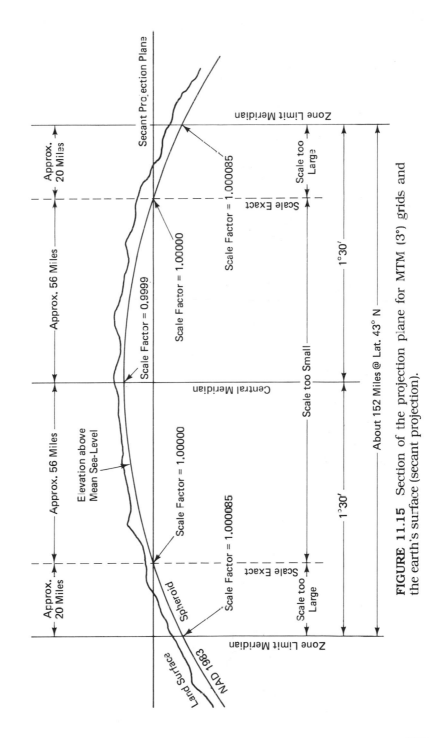

FIGURE 11.15 Section of the projection plane for MTM (3°) grids and the earth's surface (secant projection).

429

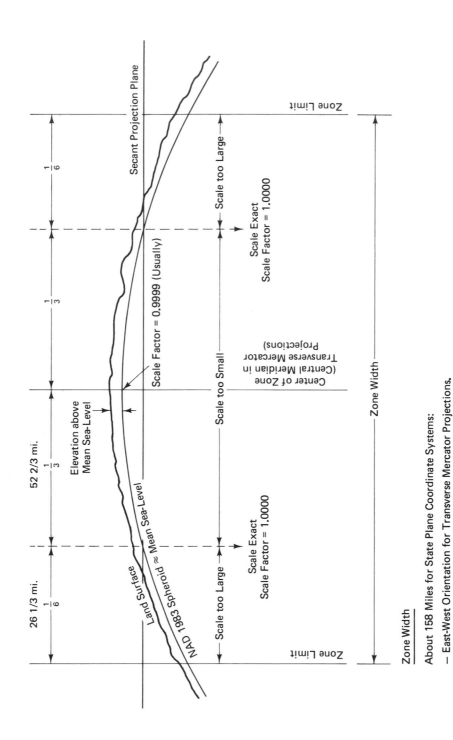

FIGURE 11.16 Section of the projection plane and the earth's surface for State Plane grids (secant projection).

Zone Width

About 158 Miles for State Plane Coordinate Systems:

– East-West Orientation for Transverse Mercator Projections,
– North-South Orientation for Lambert Projections,

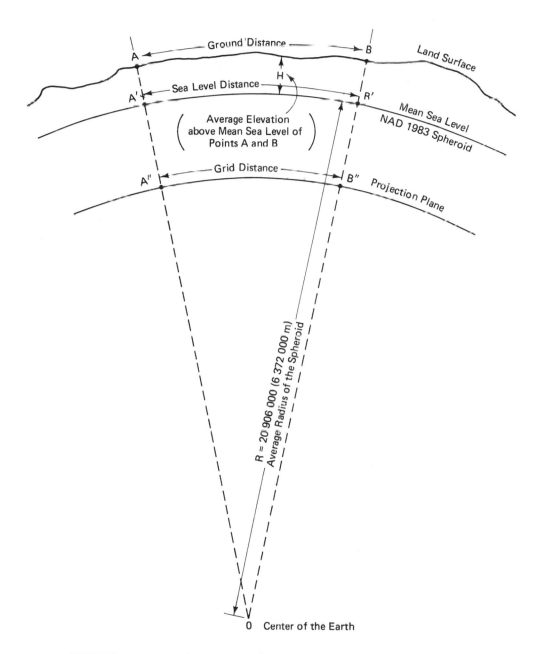

FIGURE 11.17 Relationship of ground distances to sea level distances and grid distances.

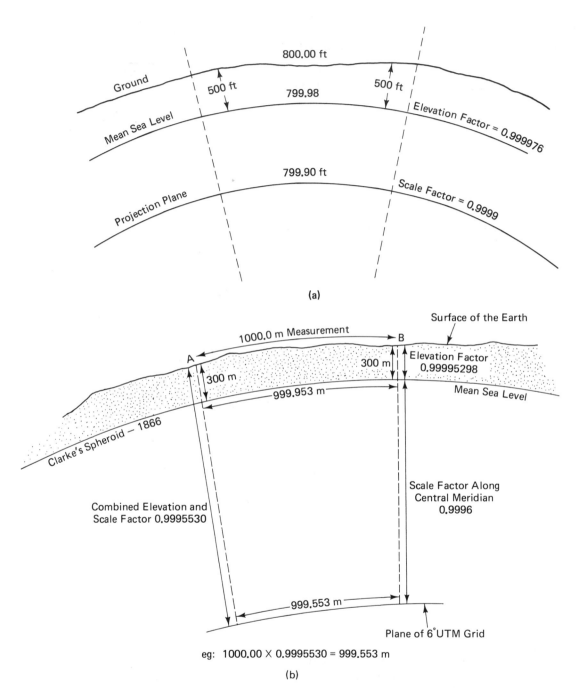

FIGURE 11.18 Conversion of a ground distance to a grid distance using the elevation factor and the scale factor: (a) modified transverse mercator (MTM) grid, 3° zone; (b) universal transverse mercator (UTM) grid, 6° zone.

For state plane projections, the published tables will give scale factors for positions of latitude difference (Lambert projection) or for distances east or west of the central meridian (transverse Mercator projections).

Scale factors for the transverse Mercator projections can also be computed by the formula

$$M_p = M_0 \left(1 + \frac{X^2}{2R^2}\right) \tag{11.9}$$

where M_p = scale factor at the survey station, M_o = scale factor at the central meridian, X = east–west distance of the survey station from the central meridian, R average radius of the sphroid. ($X^2/2R^2$ can be expressed in feet, meters, miles, or kilometers.)

For example, survey stations 12 000 ft from a central meridian having a scale factor of 0.9999 would have a scale factor determined as follows:

$$M_p = 0.9999 \left(1 + \frac{12\ 000^2}{2 \times 20\ 906\ 000^2}\right)$$
$$= 0.9999002$$

In practice, it is seldom necessary to use Formula (11.9) as tables are available for all state plane grids, the universal transverse Mercator grid (see Figure 11.19, and Table 11.4) and the modified transverse Mercator grid (see Table 11.3). Grid factors can easily be interpolated (double interpolation) from these tables.

Elev.	Distance from CM (1000s of Feet)		
	100	*150*	*125 (interpolated)*
500	999888	999902	999895
750	999876	999890	999883

Diff. = 12

For example, using Table 11.3 (MTM projection), determine the combined scale and elevation factor (grid factor) of a point 125 000 ft from the central meridian (scale factor = 0.9999) and at an elevation of 600 ft above mean sea level:

Solution: From Table 11.3

For 600,

$$\frac{100}{250} \times 12 = 5$$

SCALE FACTOR

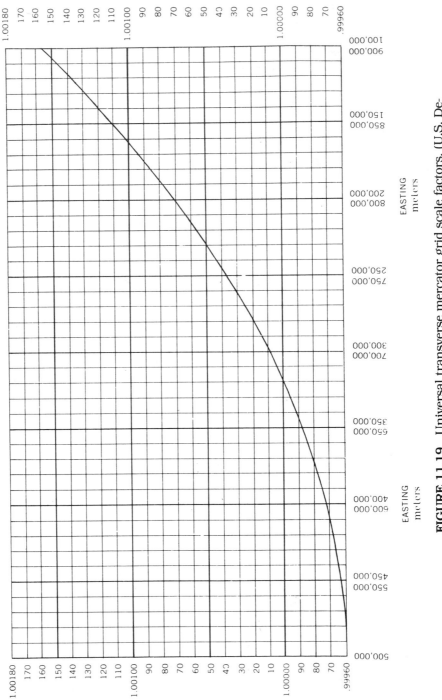

FIGURE 11.19 Universal transverse mercator grid scale factors. (U.S. Department of the Army, TM5-241-4/1)

$$\text{Grid factor} = \quad 0.999895$$
$$\underline{0.000005}$$
$$= \quad 0.999890$$

For important survey lines, grid factors can be determined for both ends and then averaged. For lines longer than 5 miles, intermediate computations will be required to maintain high precision.

11.9 EXAMPLES

Example 11.1

Given the coordinates on the MTM coordinate grid of two horizontal control monuments and their elevations (see Figure 11.20), compute the ground distance and geodetic direction between them:

Station	Elev.	Northing	Easting
Monument 870	595 ft	15912 789.795 ft	1039 448.609 ft
Monument 854	590 ft	15913 044.956 ft	1040 367.657 ft

Scale factor at CM = 0.9999.

By subtraction, coordinate distances, $\Delta N = 255.161$, $\Delta E = 919.048$

The solution is obtained as follows:

1. Distance $870 - 854 = \sqrt{255.161^2 + 919.048^2} = 953.811$ ft

2. Tan bearing $= \dfrac{\Delta E}{\Delta N} = \dfrac{919.048}{255.161}$

   Grid bearing $= 74.48\ 342°$

   $= N\ 74°29'00''\ E$

3. Convergence: Use method (a) or (b). Average latitude $-$ $43°47'31''$; average longitude $= 79°20'54''$ (Figure 11.20).

   (a) $\theta'' = 52.13D\ \text{Tan}\ \phi \qquad (11.4B)$

   $= 52.13 \times \dfrac{40\ 367.657 + 39\ 448.609}{2 \times 5280}\ \text{Tan}\ 43°47'31''$

   $= 377.74''$

   $\theta = 0°06'18''$

   (b) $\Delta\alpha'' = \Delta\lambda''\ \text{Sin}\ \phi p \qquad (11.4a)$

   $= (79°30 - 79°20'54'')\ \text{Sin}\ 43°47'31''$

   $- 546'' \times \text{Sin}\ 43°47'31''$

   $= 377.85''$

   $\Delta\alpha = 0°06'18''$

TABLE 11.3 *Grid Factors for the Modified Transverse Mercator Projection*

3° MTM Combined Grid Factor Based on Central Scale Factor of 0.9999

Elevation in Feet	Distance from Central Meridian in Thousands of Feet												
	0	50	100	150	200	250	300	350	400	450	500	550	600
0	0.999900	0.999903	0.999911	0.999926	0.999946	0.999971	1.000003	1.000040	1.000083	1.000131	1.000186	1.000245	1.000311
250	0.999888	0.999891	0.999899	0.999914	0.999934	0.999959	0.999991	1.000028	1.000071	1.000119	1.000174	1.000234	1.000299
500	0.999876	0.999879	0.999888	0.999902	0.999922	0.999947	0.999979	1.000016	1.000059	1.000107	1.000162	1.000222	1.000287
750	0.999864	0.999867	0.999876	0.999890	0.999910	0.999936	0.999967	1.000004	1.000047	1.000095	1.000150	1.000210	1.000275
1000	0.999852	0.999855	0.999864	0.999878	0.999898	0.999924	0.999955	0.999992	1.000035	1.000083	1.000138	1.000198	1.000263
1250	0.999840	0.999843	0.999852	0.999864	0.999886	0.999912	0.999943	0.999980	1.000023	1.000072	1.000126	1.000186	1.000251
1500	0.999828	0.999831	0.999840	0.999854	0.999874	0.999900	0.999931	0.999968	1.000011	1.000060	1.000114	1.000174	1.000239
1750	0.999816	0.999819	0.999828	0.999842	0.999862	0.999888	0.999919	0.999956	0.999999	1.000048	1.000102	1.000162	1.000227
2000	0.999804	0.999807	0.999816	0.999830	0.999850	0.999876	0.999907	0.999944	0.999987	1.000036	1.000090	1.000150	1.000216
2250	0.999792	0.999795	0.999804	0.999818	0.999838	0.999864	0.999895	0.999932	0.999975	1.000024	1.000078	1.000138	1.000204
2500	0.999781	0.999784	0.999792	0.999806	0.999826	0.999852	0.999883	0.999920	0.999963	1.000012	1.000066	1.000126	1.000192
2750	0.999796 ·	0.999771	0.999780	0.999794	0.999814	0.999840	0.999871	0.999908	0.999951	1.000000	1.000054	1.000114	1.000180

Ground distance × grid factor = grid distance.
Adapted from "Horizontal Control Surveys Precis." Ministry of Transportation and Communications, Ontario, 1974.

TABLE 11.4 Combined Scale and Elevation Factors (Grid Factors) – UTM

	Elevation above Mean Sea Level in Meters										
	0	100	200	300	400	500	600	700	800	900	1000
0	0.999600	0.999584	0.999569	0.999553	0.999537	0.999522	0.999506	0.999490	0.999475	0.999459	0.999443
10	0.999601	0.999586	0.999570	0.999554	0.999539	0.999523	0.999507	0.999492	0.999476	0.999460	0.999445
20	0.999605	0.999589	0.999574	0.999558	0.999542	0.999527	0.999511	0.999495	0.999480	0.999464	0.999448
30	0.999611	0.999595	0.999580	0.999564	0.999548	0.999553	0.999517	0.999501	0.999486	0.999470	0.999454
40	0.999620	0.999604	0.999588	0.999573	0.999557	0.999541	0.999526	0.999510	0.999494	0.999479	0.999463
50	0.999631	0.999615	0.999599	0.999584	0.999568	0.999552	0.999537	0.999521	0.999505	0.999490	0.999474
60	0.999644	0.999629	0.999613	0.999597	0.999582	0.999566	0.999550	0.999535	0.999519	0.999503	0.999488
70	0.999660	0.999645	0.999629	0.999613	0.999598	0.999582	0.999566	0.999551	0.999535	0.999519	0.999504
80	0.999679	0.999663	0.999647	0.999632	0.999616	0.999600	0.999585	0.999569	0.999553	0.999538	0.999522
90	0.999700	0.999684	0.999668	0.999653	0.999637	0.999621	0.999606	0.999590	0.999574	0.999559	0.999543
100	0.999723	0.999707	0.999692	0.999676	0.999660	0.999645	0.999629	0.999613	0.999598	0.999582	0.999566
110	0.999749	0.999733	0.999717	0.999702	0.999686	0.999670	0.999655	0.999639	0.999623	0.999608	0.999592
120	0.999777	0.999761	0.999746	0.999730	0.999714	0.999699	0.999683	0.999667	0.999652	0.999636	0.999620
130	0.999808	0.999792	0.999777	0.999761	0.999745	0.999730	0.999714	0.999698	0.999683	0.999667	0.999651
140	0.999841	0.999825	0.999810	0.999794	0.999778	0.999763	0.999747	0.999731	0.999716	0.999700	0.999684
150	0.999877	0.999861	0.999845	0.999830	0.999814	0.999798	0.999783	0.999767	0.999751	0.999736	0.999720
160	0.999915	0.999899	0.999884	0.999868	0.999852	0.999837	0.999821	0.999805	0.999830	0.999774	0.999758
170	0.999955	0.999940	0.999924	0.999908	0.999893	0.999877	0.999861	0.999846	0.999830	0.999814	0.999799
180	0.999999	0.999983	0.999967	0.999952	0.999936	0.999920	0.999905	0.999889	0.999873	0.999858	0.999842
190	1.000044	1.000028	1.000013	0.999997	0.999981	0.999966	0.999950	0.999934	0.999919	0.999903	0.999887
200	1.000092	1.000076	1.000061	1.000045	1.000029	1.000014	0.999998	0.999982	0.999967	0.999951	0.999935
210	1.000142	1.000127	1.000111	1.000095	1.000080	1.000064	1.000048	1.000033	1.000017	1.000001	0.999986
220	1.000195	1.000180	1.000164	1.000148	1.000133	1.000117	1.000101	1.000086	1.000070	1.000054	1.000039
230	1.000251	1.000235	1.000219	1.000204	1.000188	1.000172	1.000157	1.000141	1.000125	1.000110	1.000094
240	1.000309	1.000293	1.000277	1.000262	1.000246	1.000230	1.000215	1.000199	1.000183	1.000168	1.000152
250	1.000369	1.000353	1.000338	1.000322	1.000306	1.000290	1.000275	1.000259	1.000243	1.000228	1.000212

Distance from Central Meridian in Kilometers

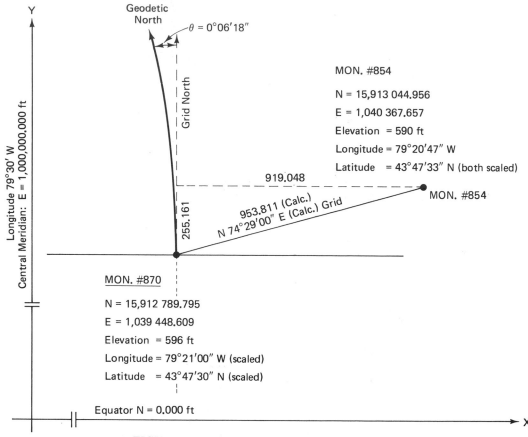

FIGURE 11.20   Illustration for Example 11.1.

Convergence is in this case computed to the closest second of arc. (*Note*: The average latitude need not have been computed since the latitude range, 03″, was in this case insignificant.) Referring to Figure 11.21,

$$\begin{aligned} \text{Grid bearing} &= \text{N } 74°29'00'' \text{ E} \\ + \text{ Convergence} &= 0°06'18'' \\ \text{Geodetic bearing} &= \text{N } 74°35'18'' \text{ E} \end{aligned}$$

4. Scale factor:

$$\text{Scale factor at CM, } M_0 = 0.9999$$

$$\text{Distance } (X) \text{ from CM} = \frac{40\ 368 + 39\ 449}{2}$$

$$= 39\ 908 \text{ ft}$$

Scale factor at midpoint 870–854:

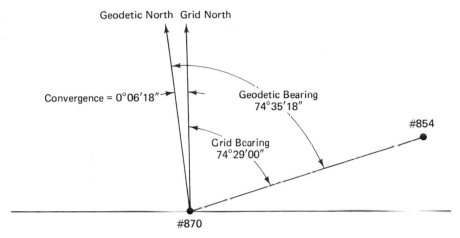

FIGURE 11.21   Illustration for Example 11.1.

$$M_p = M_0 \left(1 + \frac{X^2}{2R^2}\right) \tag{11.9}$$

$$= 0.9999 \left(1 + \frac{39\ 908^2}{2 \times 20\ 906\ 000^2}\right)$$

$$= 0.9999018$$

5. Elevation factor:

$$\text{Elevation factor} = \frac{\text{sea level distance}}{\text{ground distance}} = \frac{R}{R + H} \tag{11.8}$$

$$= \frac{20\ 906\ 000}{20\ 906\ 000 + 593} = 0.9999716$$

6. Grid factor: Use method (a) or (b).

(a) Grid factor = scale factor × elevation factor
 = 0.9999018 × 0.9999716
 = 0.9998734

(b) Alternately, the grid factor can be determined through double interpolation of Table 11.3, as follows: The following values are taken from Table 11.3; required elevation is 593 ft at 39900 ft from the CM.

	Distance from Central Meridian (1000s of feet)		
Elev.	*0*	*50*	*39.9 (interpolated)*
500	0.999876	0.999879	0.999878
593			
750	0.999864	0.999867	0.999866

After first interpolating the values at 0 and 50 for 39.9 thousand feet from the CM, it is a simple matter to interpolate for the elevation of 593 ft.

$$0.999878 - (.000012) \times \frac{93}{250} = .999873$$

Thus, the grid factor at 39 900 ft from the CM at an elevation of 593 ft is 0.999873.

7. Ground distance:

Given the relationship discussed previously,

$$\text{Ground distance} \times \text{grid factor} = \text{grid distance}$$

it can be seen that

$$\text{Ground distance} = \frac{\text{grid distance}}{\text{grid factor}}$$

In this example,

$$\text{Ground distance (870–854)} = \frac{953.811}{0.9998734} = 953.93 \text{ ft}$$

Example 11.2

Given the coordinates on the UTM coordinate grid of two horizontal control monuments (Mon. 113 and Mon. 115), and their elevations, compute the ground distance and geodetic direction between them. (See Figure 11.22).

Station	Elevation	Northing	Easting
#113	181.926	4,849,872.066	632,885.760
#115	178.444	4,849,988.216	632,971.593

Zone 17 UTM; CM at 81° longitude West:

Scale factor at CM = 0.9996
ϕ(lat.) = 43°47′31″) (scaled from topographic map)
λ (long.) = 79°20′35″) for mid-point of line
                                          joining Mon. 113 and Mon. 115

By subtraction, coordinate distances are

$$\Delta N = 116.150 \text{ m., } \Delta E = 85.833 \text{ m}$$

1. Grid Distance   Mon 113 to Mon 115
$$= \sqrt{116.150^2 + 85.833^2} = 144.423 \text{ m}$$

2. Grid Bearing:

$$\text{Tan Brg.} = \frac{\triangle E}{\triangle N} = \frac{85.833}{116.150}$$

Brg. $= 36.463811°$
Grid Bearing $= \text{N}.36°27'50''\text{E}$

3. Convergence

$$\Delta \, \alpha'' = \Delta\lambda'' \, \text{Sin} \, \phi \quad (11\text{-}4a)$$
$$= (81° - 79°20'35'')'' \, \text{Sin} \, 43°47'31''$$
$$= 4128''$$
$$= 1°08'48''$$

See Figure 11.22 for application of convergence.

4. Scale Factor:

Scale factor at CM $= .9996$

$$\text{Distance from CM} = \frac{132885.760 + 132971.593}{2}$$
$$= 132{,}928.677 \text{ m} = 132.929 \text{ Km}$$

Scale factor at mid-point line Mon. 113 − Mon. 115:

$$Mp = Mo\left(1 + \frac{X^2}{2R^2}\right) \quad (11\text{-}9)$$

$$Mp = .9996\left(1 + \frac{132.929^2}{2 \times 6372^{2*}}\right) = .999818$$

* average radius of sea level surface
$= 20{,}906{,}000$ ft or $6{,}372{,}000$ m

5. Elevation Factor:

$$\text{Elevation Factor} = \frac{\text{sea level distance}}{\text{ground distance}} = \frac{R}{R+H} \quad (11\text{-}8)$$
$$= \frac{6372}{6372 + 0.180} = .999972$$

6. Grid Factor:

a)

Grid Factor $=$ Elevation factor $\times$ scale factor
$$= .999972 \times .999818 = .999790$$

or

b) using Table 11-4 "Combined Scale and Elevation Factors for UTM"

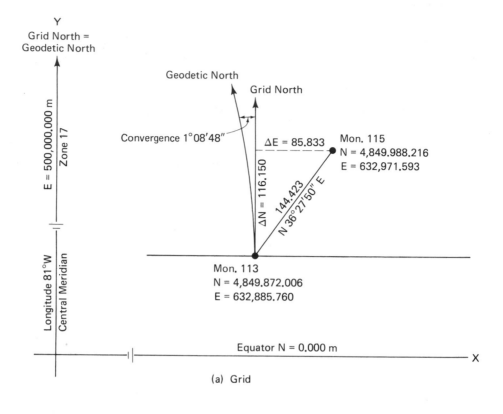

(a) Grid

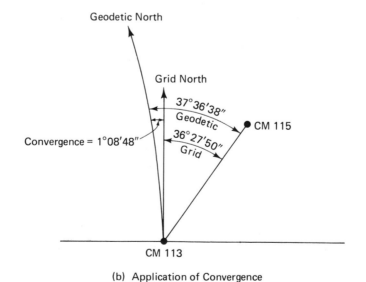

(b) Application of Convergence

FIGURE 11.22 Illustration for Example 11.2.

	Distance from CM (Km)		
Elev.	130	140	132.9 (Interpolated)
100	.999792	.999825	999802
200	.999777	.999810	999787

It now only remains to interpolate for the elevation value of

$$180 \text{ m, i.e., } 999802 - \left(\frac{80}{100} \times .000015\right) = .999790$$

7. Ground Distance:

$$\text{Ground distance} = \frac{\text{Grid distance}}{\text{Grid factor}}$$

In this example:

$$\text{ground distance Mon. 113 to Mon. 115}$$
$$= \frac{144.423}{.999790} = 144.453 \text{ m}$$

11.10 HORIZONTAL CONTROL TECHNIQUES

The original control nets in North America were established using the technique of *triangulation*. This technique involved (1) a precisely measured base line as a starting side for a series of triangles or chains of triangles (see Section A.11.1), (2) the determination of each angle in the triangle using a precise theodolite, and (3) a check on the work made possible by precisely measuring a subsequent side of a triangle (the spacing of check lines depended on the desired accuracy level). Table A.15 gives characteristics and specifications for pure triangulation surveys.

The advent of EDM instruments changed the approach to control surveys. It became possible to measure precisely the length of a triangle side in about the same length of time as was required for angle determination. The solution of triangles using only side lengths is known as *trilateration*. Table A.16 gives the characteristics and specifications for trilateration surveys. Trilateration on its own has not been widely used as there is no obvious check on the accuracy of the work as it proceeds, without resorting to computations. (Triangulation allows the quick check of simply summing the triangle angles to 180°.)

It is expected that specifications will soon be published to reflect the ever increasing practice of measuring both the angles and the sides in a triangle net (a combination of triangulation and trilateration). Once a station is occupied, it takes little extra time or effort to read and record both the angles and the distances.

Whereas triangular-type control surveys are used for basic state or provincial controls, precise traverses are now used to densify the basic control net. The advent of reliable and precise EDM instruments has elevated the traverse to a valuable role, both in strengthening a triangulation net and in providing its own stand-alone control figure. To provide reliability, traverses must close on themselves or on previously coordinated points. Table A.17 gives characteristics and specifications for traverses (Federal Geodetic Control Committee, U.S.A., 1974).

Figure 11.23 shows some typical control survey figures. Part (a) depicts a simple chain of single triangles. In triangulation (angles only) this figure suffers from the weakness that essentially only one route can be followed to solve for side *KL*. Figure 11.23b shows a chain of double triangles or quadrilaterals. This configuration is preferred for triangulation as side *KL* can be solved using different routes (many more redundant measurements). Modern survey practice favors a combination of triangulation and trilateration (i.e., measure both the angles and the distances), thus ensuring many redundant measurements even for the simple chain of triangles shown in Figure 11.23a. Figure 11.23c shows the typical control configuration for an urban area. Typically, the highest-order control is established by federal agencies, the secondary control is established by state or provincial agencies, and the lower-order control is established by municipal agencies.

It is sometimes the case that the federal agency will establish all three orders of control when requested to do so by the state, province, or municipality.

In triangulation surveys, a great deal of attention was paid to the geometric *strength of figure* of each control configuration (see Table A.15). Generally, an equilateral triangle is considered *strong*, whereas triangles with small (less than 10°) angles are considered relatively weak. Trigonometric functions vary in precision as the angle varies in magnitude. The sines of small angles (near zero), the cosines of large angles (near 90°), and the tangents of both small (zero) and large (90°) angles are all relatively imprecise. That is, there are relatively large changes in the values of the trigonometric functions that result from relatively small changes in angular values.

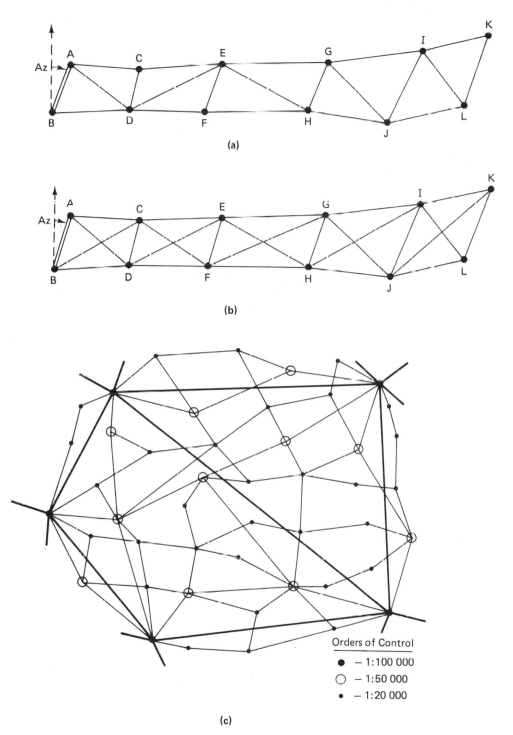

FIGURE 11.23 Control survey configurations: (a) chain of single triangles; (b) chain of double triangles (quadrilaterals); (c) urban control figure.

445

For example, the angular error in 5 seconds for the sine of 10° is 1/7300, whereas the angular error in 5 seconds for the sine of 20° is 1/15000, and the angular error in 5 seconds for the sine of 80° is 1/234000 (see Example 11.3).

One can see that if sine or cosine functions are used in triangulation to calculate the triangle side distances, care must be exercised to ensure that the trigonometric function itself is *not* contributing errors to the solution more significant than the specified surveying error limits.

When all angles *and* distances are measured for each triangle, the redundant measurements ensure an accurate solution, and the configuration strength of figure becomes somewhat less important. However, given the opportunity, most surveyors still prefer to use well-balanced triangles and to avoid using the sine and tangent of small angles and the cosine and tangent of large angles to compute control distances.

Example 11.3 *Effect of the Angle Magnitude on the Accuracy of Computed Distances*

(a) Consider a right-angle triangle with a hypotenuse 1000.00 ft long.

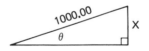

$$X = 1000 \, \text{Sin} \, \theta$$

Use various values for θ to investigate the affect of 05″ errors.

1. $\theta = 10°$ $X = 173.64818$ ft
   $\theta = 10°00'05''$ $X = 173.67205$ ft
                          diff. = 0.02387 in 173.65 ft, an
                                   accuracy of 1/7300

2. $\theta = 20°$ $X = 342.02014$ ft
   $\theta = 20°00'05''$ $X = 342.04292$ ft
                          diff. = 0.022782 in 342.02 ft, an
                                   accuracy of 1/15000

3. $\theta = 80°$ $X = 984.80775$ ft
   $\theta = 80°00'05''$ $X = 984.81196$ ft
                          diff. = 0.00421 in 984.81 ft, an
                                   accuracy of 1/234000

(b) Consider a right triangle with the adjacent side 1 000.00 ft long.

$$X = 1000 \text{ Tan } \theta$$

Use various values for θ to investigate the affect of 05″ errors.

1. $\theta = 10°$ $X = 176.32698$ ft
    $\theta = 10°00'05''$ $X = 176.35198$
                   diff. = 0.025, an
                            accuracy of 1/7100

2. $\theta = 45°$ $X = 1000.00$
    $\theta = 45°00'05''$ $X = 1000.0485$
                   diff. = 0.0485, an
                            accuracy of 1/20600

3. $\theta = 80°$ $X = 5671.2818$
    $\theta = 80°00'05''$ $X = 5672.0858$
                   diff. = 0.804, an
                            accuracy of 1/7100

4. In the preceding example, if the angle can be determined to the closest second, the accuracy would be as follows:
    $\theta = 80°$ $X = 5671.2818$
    $\theta = 80°00'01''$ $X = 5671.4426$
                   diff. = 0.1608, an
                            accuracy of 1/35270

The foregoing illustrates that the surveyor should either avoid using weak angles in distance computations, or if weak angles must be used, they should be measured more precisely than would normally be required. Also illustrated here is the need for the surveyor to preanalyze the proposed control survey configuration to determine optimal field techniques and attendant precisions.

11.11 PROJECT CONTROL

Project control begins with either a boundary survey (e.g., subdivisions) or an all-inclusive peripheral survey (e.g., construction sites). The boundary or site peripheral survey will, if possible, be tied into state or provincial grid control monuments so that reference can be made to the state or provincial coordinate grid system. The peripheral survey is densified with judiciously placed control stations over the entire site. The survey data for all control points are entered into the computer for accuracy verification and error adjustment, and finally for coordinate determination of all control points. All key layout points (e.g., lot corners, radius points, ₵ stations, curve points, construction points) are also coordinated on the computer using coordinate geometry computer programs (e.g., COGO). Printout sheets as shown in Figure 16.9a are used by the surveyor to lay out the proposed facility from coordinated control stations. The computer will give the surveyor the azimuth and distance from one, two, or perhaps three different control points to one layout point. Positioning a layout point from more than one control station provides an exceptional check on the accuracy of the work.

Generally, a layout point can be positioned by simultaneous angle sightings from two control points, with the distance being established by EDM from one of those stations, or a layout point can be positioned by simultaneous angle sightings from three control points. Both techniques provide a redundancy in measurement that permits positional accuracy determination.

To ensure that the layout points have been accurately located (e.g., with an accuracy level of between $1/5000$ and $1/10000$), the control points themselves must be located to an even higher level of accuracy (i.e., typically better than $1/20000$). These accuracies can be achieved if EDMs are used for distances and one- or two-second theodolites are used for angle measurement.

As noted earlier, in addition to quality instrumentation, the surveyor must use "quality" geometrics in designing the shape of the control net; a series of interconnected equilateral triangles provides the strongest control net.

Control points are positioned keeping in mind the following:

1. Good visibility to other control points and an optimal number of layout points.
2. The visibility factor is considered not only for existing ground conditions, but consideration is also given to visibility lines during all stages of construction.
3. A minimum of two (three is preferred) reference ties is required for each control point so that it can be reestablished if destroyed. Consideration must be given to the availability of

features suitable for referencing (i.e., features into which nails can be driven or cut-crosses chiseled, etc.). Ideally, the three ties would be each 120° apart.

4. Control points should be placed in locations that will not be affected by primary or secondary construction activity. In addition to keeping clear of the actual construction site positions, the surveyor must anticipate temporary disruptions to the terrain resulting from access roads, materials stockpiling, and so on. If possible, control points are safely located adjacent to features that will *not* be moved (e.g., hydro towers, concrete walls, large valuable trees).

5. Control points must be established on solid ground (or rock). Swampy areas or loose fill areas must be avoided (see Section 11.12 for various types of markers).

Once the control point locations have been tentatively chosen, they are plotted so that the quality of the control net geometrics can be considered. At this stage it may be necessary to go back into the field and locate additional control points to strengthen weak geometric figures. When the locations have been finalized on paper, each station is given a unique identification code number, and then the control points are set in the field. Field notes, showing reference ties to each point, are carefully taken and then filed. Now the actual measurements of the distances and angles of the control net are taken. When all the field data have been collected, the closures and adjustments are computed. The coordinates of all layout points are then computed, with polar ties being generated for each layout point, from two or possibly three control stations.

Figure 11.24a shows a single layout point being positioned by angle only from three control sights. The three control sights could simply be referenced to the farthest away of the control points themselves (e.g., angles *A*, *B*, and *C*), or if a reference azimuth point (RAP) has been identified and coordinated in the locality, it would be preferred as it no doubt would be farther away and thus capable of providing more precise sightings (e.g., angles 1, 2, and 3). RAPs are typically communications towers, church spires, or other identifiable points that can be seen from widely scattered control stations. Coordinates of RAPs are computed by turning angles to the RAP from project control monuments or preferably from state or provincial control grid monuments.

Figure 11.24b shows a bridge layout involving azimuth and distance ties for abutment and pier locations. It will be noted that,

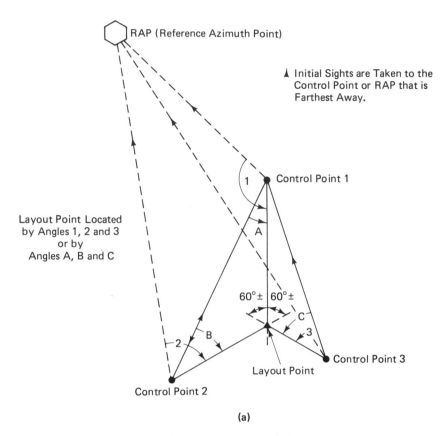

(a)

FIGURE 11.24 Examples of coordinate control for polar layout: (a) single point layout, located by three angles; (b) bridge layout, located by angle and distance. (Adapted from *Construction Manual,* Ministry of Transportation and Communications—Ontario) *(Continued)*

although the perfect case of equilateral triangles is not always present, the figures are quite "strong," with redundant measurements providing accuracy verification. (Angles of less than 20° are not used in computations.)

Figure 11.25 illustrates a method of recording angle directions and distances to control stations with a listing of derived azimuths. Station 17 can be quickly found by the surveyor from the distance and alignment ties to the hydrant, cutcross on curb, and nail in pole. Had station 17 been destroyed, it could have been reestablished from these and other reference ties.

The bottom row marked "check" indicates that the surveyor has "closed the horizon" by continuing to revolve the theodolite back to the initial target point (100 in this example) and then

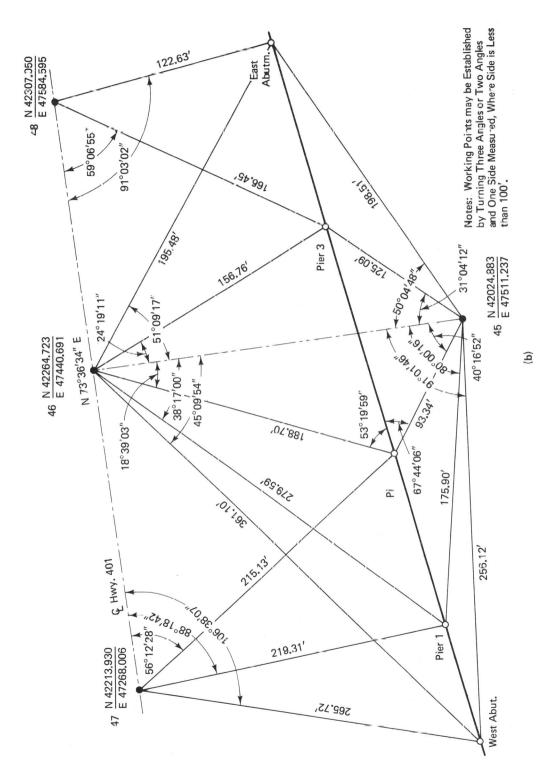

FIGURE 11.24 (Continued)

(b)

451

FAIRVIEW SUBDIVISION
CONTROL NET

CLEAR 23°C
RICHARDSON—NOTES
DAWES—π(WILD T-2 #08)
BALOG—PRISM,

INST. @ STA. 17

DIRECTION 17-100 KNOWN—119° 39' 06"

POINT	DIRECT	REVERSE	DIRECT	REVERSE	DISTANCE DIRECTION
100°	0°00'07"	180°00'17"	90°05'02"	270°04'55"	0°00'00"
REDUCED	0°00'00"	180°00'00"	90°00'00"	270°00'00"	0°00'00"
18.	41°45'44"	221°45'47"	131°50'28"	311°50'22"	142-188 m.
REDUCED	41°45'37"	221°45'30"	45'26"	45'27"	41°45'30"
HYD.	113°47'41"	293°48'08"			2.906 m
REDUCED	113°47'34"	47'51"			113°47'42"
13.	232°50'25"	52°50'30"	332°55'13"	142°55'11"	250.113 m.
REDUCED	232°50'18"	50'13"	50'11"	50'16"	232°50'14"
11.	268°27'56"	88°28'02"	358°32'41"	178°32'37"	179.375 m.
REDUCED	268°27'49"	27'45"	27'39"	27'42"	268°27'44"
8.	334°36'22"	154°36'27"	64°41'15"	244°41'07"	138.125 m.
REDUCED	334°36'15"	36'10"	36'13"	36'12"	334°36'12"
CHECK	0°00'08"	180°00'15"	90°05'05"	270°04'52"	90°05'05"

LINE	AZIMUTH
STA. 17—100	119° 39' 06" (Given)
STA. 17—STA. 18	161° 24' 36"
STA. 17—HYD.	233° 26' 48"
STA. 17—STA. 13	352° 29' 20"
STA. 17—STA. 11	28° 06' 50"
STA. 17—STA. 8	94° 15' 18"

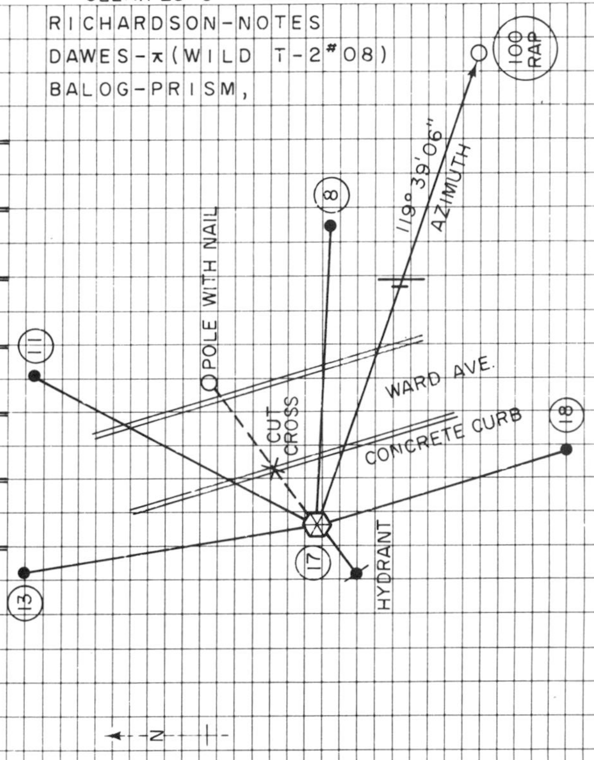

DISTANCE BY WILD DI 4
SEE PAGE #51

100 RAP   81   8   17   11   13

WARD AVE.   CONCRETE CURB   HYDRANT   CUT CROSS   POLE WITH NAIL

AZIMUTH "119°39'06""

N

FIGURE 11.25 Field notes for control point directions and distances.

reading the horizontal circle. A difference of more than 5″ between the initial reading and the check reading usually means that the series of angles in that column must be repeated.

After the design of a facility has been coordinated, polar layout coordinates can be generated for layout points from selected stations. The surveyor can copy the computer data directly into the field book (see Figure 11.26) for later use in the field. On large projects (expressways, dams, etc.) it is common practice to have bound volumes printed that include polar coordinate data for all control stations and all layout points (see Figure 16-4a). Many surveyors now predict that most electronic tacheometer instruments (ETI), discussed in Chapter 8, will soon have automatic computer-to-ETI layout data transfer capabilities. (The Zeiss Elta 3 now has programmable features.) This means that the surveyor will be able to store all the layout data for a project in the ETI itself and will be able to access specific data simply by entering the appropriate point identification code. The automatic transfer of polar layout data to the ETI will eliminate transcription errors that can occur when the technique illustrated in Figure 11.26 is used.

In addition to providing rectangular coordinates (based on the coordinate grid) and polar coordinates based on specific control stations, the computer program will also generate direct distance and bearing data between layout points. These point-to-point data are valuable for checking the location of points already established by polar layout. These checks are usually accomplished by measuring the point-to-point distances, although angular checks are also possible.

Figure 11.27 shows a primary control net established to provide control for a construction site. The primary control stations are tied into a national, state or provincial coordinate grid by a series of precise traverses or triangle networks. Points on base lines (secondary points) can be tied into the primary control net by polar ties, intersection, or resection. The actual layout points of the structure (columns, walls, footings, etc.) are established from these secondary points. International standard 4463 (International Organization for Standardization) points out that the accuracy of key building or structural layout points should not be influenced by possible discrepancies in the state or provincial coordinate grid. For that reason, the primary project control net is analyzed and adjusted independently of the state or provincial coordinate grid. This "free net" is tied to the state or provincial coordinate grid without becoming an integrated adjusted component of that grid.

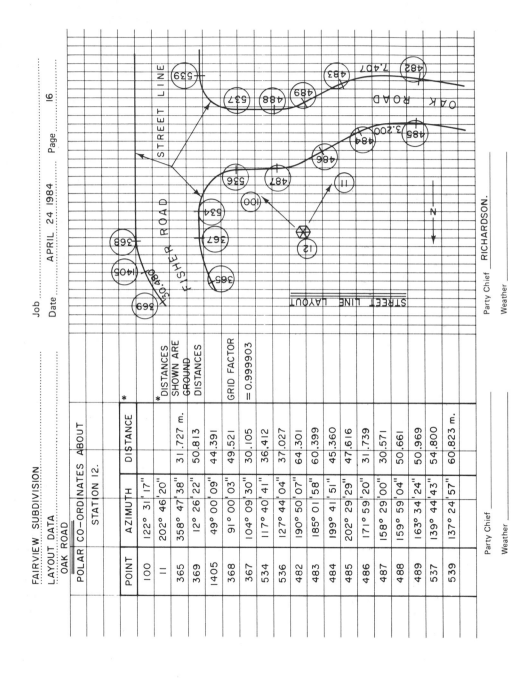

POLAR CO-ORDINATES ABOUT STATION 12.

POINT	AZIMUTH	DISTANCE	*
100	122° 31' 17"		
11	202° 46' 20"		
365	358° 47' 38"	31.727 m.	*DISTANCES SHOWN ARE GROUND DISTANCES
369	12° 26' 22"	50.813	
1405	49° 00' 09"	44.391	
368	91° 00' 03"	49.521	GRID FACTOR = 0.999903
367	104° 09' 30"	30.105	
534	117° 40' 41"	36.412	
536	127° 44' 04"	37.027	
482	190° 50' 07"	64.301	
483	185° 01' 58"	60.399	
484	199° 41' 51"	45.360	
485	202° 29' 29"	47.616	
486	171° 59' 20"	31.739	
487	158° 29' 00"	30.571	
488	159° 59' 04"	50.661	
489	163° 34' 24"	50.969	
537	139° 44' 43"	54.800	
539	137° 24' 57"	60.823 m.	

Party Chief _____

Weather _____

Party Chief ___ RICHARDSON.

Weather _____

FIGURE 11.26 Prepared polar coordinate layout notes.

454

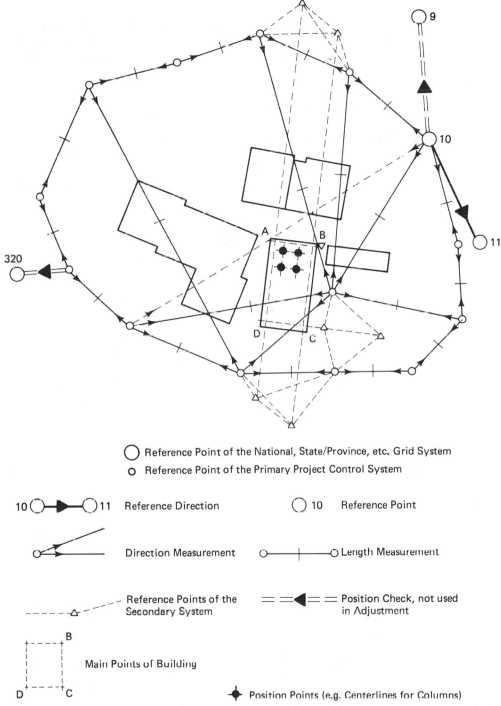

○ Reference Point of the National, State/Province, etc. Grid System

○ Reference Point of the Primary Project Control System

10 ○━▶━○ 11 Reference Direction ○ 10 Reference Point

○━▶ Direction Measurement ○━━┼━━○ Length Measurement

- - - -△- - - Reference Points of the Secondary System

══◀══ Position Check, not used in Adjustment

Main Points of Building

Position Points (e.g. Centerlines for Columns)

FIGURE 11.27 Project control net. (Adapted from International Organization for Standardization, ISO Standards 4463, 1979)

455

The relative positional accuracy of project layout points with each other is more important than the positional accuracy of these layout points relative to a state or provincial coordinate grid.

11.11.1 Positional Accuracies (ISO 4463)

Primary System Control Stations

1. Permissible deviations of the distances and angles obtained when *measuring the positions of primary points, and those calculated from the adjusted coordinates of these points* shall not exceed:

$$\text{Distances:} \quad \pm 0.75\sqrt{L} \text{ mm}$$
$$\text{Angles:} \quad \pm \frac{0.045}{\sqrt{L}} \text{ degrees, or}$$
$$\pm \frac{0.05}{\sqrt{L}} \text{ gon*}$$

where L is the distance in meters between primary stations; in the case of angles, L is the shorter side of the angle.

2. Permissible deviations of the distance and angles obtained when *checking the positions of primary points* shall not exceed

$$\text{Distances:} \quad \pm 2\sqrt{L} \text{ mm}$$
$$\text{Angles:} \quad \pm \frac{0.135}{\sqrt{L}} \text{ degree}$$
$$\pm \frac{0.15}{\sqrt{L}} \text{ gon}$$

where L is the distance in meters between primary stations; in the case of angles, L is the length of the shorter side of the angle.

Angles are measured with a one-second theodolite, with the measurements being made in two sets (each set is formed by two observations, one on each face of the instrument). See Figure 11-25.

Distances can be measured with steel tapes or EDMs, and

---

*1 revolution $= 360° = 400$ gon   (formerly grad— a European angle unit.)
    1 gon $= 0.9$ degrees   (exactly)

will be measured at least twice by either method. Steel tape measurements will be corrected for temperature, sag, slope, and tension; a tension device will be used while taping. *EDMs should be checked against a range of known distances regularly.*

Secondary System Control Stations

1. Secondary control stations and main layout points (e.g., *ABCD*, Figure 11.27) constitute the secondary system. The permissible deviations for *a checked distance from a given or calculated distance between a primary control station and a secondary point* shall not exceed

$$\text{Distances: } \pm 2 \sqrt{L} \text{ mm}$$

2. The permissible deviations for *a checked distance from the given or calculated distance between two secondary points in the same system* shall not exceed

$$\text{Distances: } \pm 2 \sqrt{L} \text{ mm}$$

where L is distance in meters. For L less than 10 m, permissible deviations are ± 6 mm.

$$\text{Angles: } \pm \frac{0.135}{\sqrt{L}} \text{ degree } \text{ or}$$
$$\pm \frac{0.15}{\sqrt{L}} \text{ gon}$$

where L is the length in meters of the shorter side of the angle.

3. The permissible deviations for *a checked distance from the given or calculated distance between two points in different secondary systems for the same project* should not exceed

$$\pm K \sqrt{L} \text{ mm}$$

where L is the distance in meters and K is a constant derived as shown in Table 11.5.

Angles are measured with a transit or theodolite reading to at least one minute. The measurement shall be made in at least one set (i.e., two observations, one on each face of the instrument).

Distances can be measured using steel tapes or EDMs and will be measured at least twice by either method. Distances will be corrected for temperature, sag, slope, and tension; a tension device is to be used with the tape. EDMs should be checked against a range of known distances regularly.

TABLE 11.5   *Accuracy Requirement Constants for Layout Surveys*

K	Application
10	Earthwork without any particular accuracy requirement (e.g., rough excavation, embankments)
5	Earthwork subject to accuracy requirements (e.g., roads, pipelines, structures)
2	Poured concrete structures (e.g., curbs, abutments)
1	Precast concrete structures, steel structures (e.g., bridges, buildings)

Adapted from Table 8-1, ISO 4463.

Layout Points

The permissible deviations of *a checked distance between a secondary point and a layout point, or between two layout points,* are

$$\pm K \sqrt{L}\ mm$$

where L is the specified distance in meters, and K is a constant taken from Table 11.5. For L less than 5 m, permissible deviation is $\pm 2K$ mm.

The permissible deviations for *a checked angle between two lines, dependent upon each other, through adjacent layout points* are

$$\pm \frac{0.0675}{\sqrt{L}}\ K \text{ degrees}\quad \text{or}\quad \pm \frac{0.075}{\sqrt{L}}\ K \text{ gon}$$

where L is the length in meters of the shorter side of the angle and K is a constant from Table 11.5.

Figure 11.28 illustrates the foregoing specifications for the case involving a curved stake out for a concrete curb. The layout point on the curve has a permissible area of uncertainty generated by ± 0.015/m due to angle uncertainties and by ± 0.013/m due to distance uncertainties. See also Section A.10.

11.12 CONTROL SURVEY MARKERS

Generally, *horizontal control survey markers* are used for (1) state or provincial coordinate grids, (2) property and boundary delineation, and (3) project control. The type of marker used varies with the following factors:

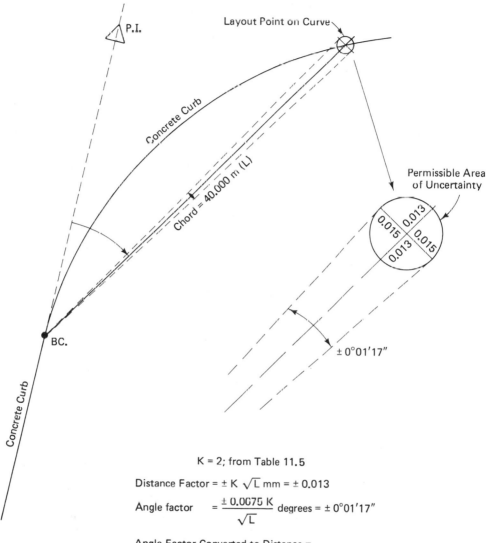

K = 2; from Table 11.5

Distance Factor = $\pm K \sqrt{L}$ mm = ± 0.013

Angle factor $= \dfrac{\pm 0.0075 K}{\sqrt{L}}$ degrees = $\pm 0°01'17''$

Angle Factor Converted to Distance =

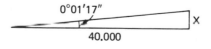

X = 40 tan 0°01'17" = 0.015 m

FIGURE 11.28 Accuracy analysis for a concrete curb layout point. (See ISO Standards, 4463, 1979)

1. Type of soil or material at the marker site.

2. Degree of permanence required.

3. Cost of replacement.

4. Precision requirements.

Early North American surveyors marked important points with suitably inscribed 4 × 4 in. cedar posts, sometimes embedded in rock cairns. Adjacent trees were used for reference ties, with the initials BT (bearing tree) carved into a blazed portion of the trunk. As time went on, surveyors used a wide assortment of markers (e.g., glass bottles, gun barrels, iron tubes, iron bars, concrete monuments with brass top plates, tablet and bolt markers for embedding into rock and concrete drill holes, and aluminum break-off markers with built-in magnets to facilitate relocation).

The most popular markers in use today are as follows:

1. For property markers: iron tubes, square iron bars (1 or $1/2$ in. square), and round iron bars.

2. For construction control: reinforcing steel bars (with and without aluminum caps), concrete monuments with brass caps.

3. For control surveys: bronze tablet markers, bronze bolt markers, post markers, sleeve-type survey markers, and aluminum break-off markers.

The latter type of aluminum marker (see Figures 11.29 and 11.30) has become quite popular for both control and construction monuments over the past few years. The chief features of these markers are their light weight (compared to concrete or steel) and the break-off feature, which ensures that, if disturbed, the monument will not bend (as iron bars will) and thus give erroneous location. Instead, these monuments will break cleanly off, leaving the lower portion (including the base) in its correct location. The base can itself be used as a monument or it can be utilized in the relocation of its replacement. The base of these monuments is equipped with a magnet (as is the top portion) to facilitate relocation when magnetic locating instruments are used by the surveyor.

Whichever type of monument is being considered for use, the key characteristic must be that of horizontal directional stability. Some municipalities establish secondary and tertiary coordinate

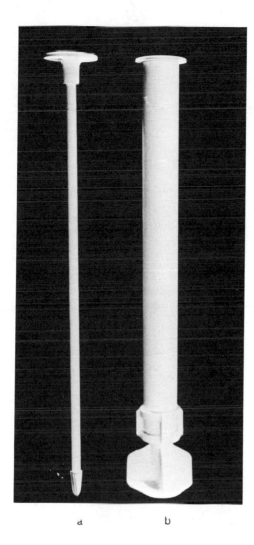

a   b

FIGURE 11.29 Lightweight aluminum monuments: (a) rod with aluminum cap: (b) break-off pipe monument. (Courtesy Berntsen Inc.)

grid monuments (survey tablet markers, Figure 11.32) in concrete curbs and sidewalks. In areas that experience frost penetration, vertical movement will take place as the frost enters and then leaves the soil supporting the curb or walk. Normally, the movement is restricted to the vertical direction only, leaving the monument's horizontal coordinates valid.

Vertical control survey markers (bench marks) are established on structures that restrict vertical movement. It is seldom the case that the same survey marker is used for both horizontal and vertical control.

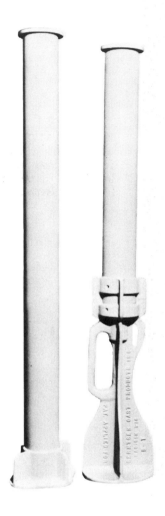

FIGURE 11.30 Lightweight aluminum monuments: (a) federal monument with ceramic magnets in cap and base; (b) highway monument, aluminum cap on aluminum pipe attached to a $13\frac{1}{4}$-in. cast aluminum base, magnets included. (Courtesy Berntsen Inc.)

a   b

Markers such as those illustrated in Figures 11.31 to 11.33 can be unreliable for elevation reference unless they have been installed on concrete structures that are not affected by frost or loading movement. The marker shown in Figure 11.34 was designed specifically for use as both a horizontal and vertical marker; however this monument can only be used for vertical control if the base plate is set below the level to which frost penetrates.

Most agencies prefer to place vertical control markers on ver-

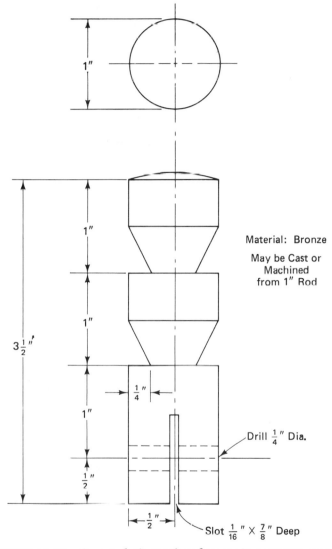

FIGURE 11.31 Survey bolt marker for use in concrete or rock.

tical structural members (e.g., masonry building walls, concrete piers, abutments, walls), that is, on any structural component whose footing is well below the level of frost penetration. Markers such as those shown in Figure 11.32 can be placed in drilled holes and secured with fast bonding grout. Tablet markers used for bench marks are sometimes manufactured with a protruding ledge so that the rod can be supported on the mark.

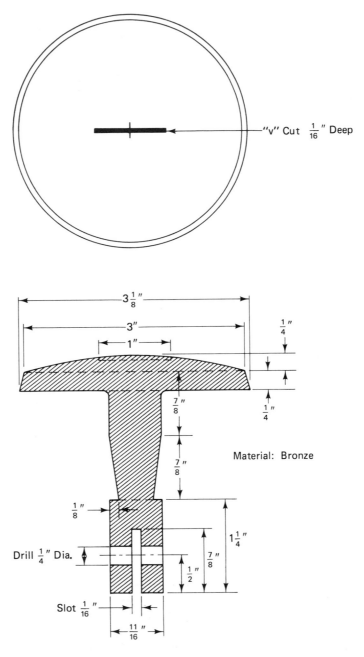

FIGURE 11.32 Survey tablet marker for use in concrete, masonry, or rock.

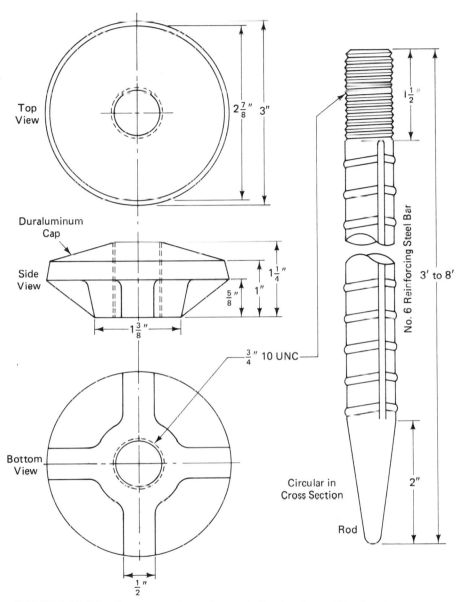

FIGURE 11.33 Survey post marker, reinforcing bar with aluminum cap.

Temporary bench marks (TBMs) can be chiseled marks on concrete, rock, or steel or spikes in hydro poles, tree roots, and the like. The unambiguous description of TBM locations along with their elevations are not published (as are regular bench marks) but are kept on file for same-agency use.

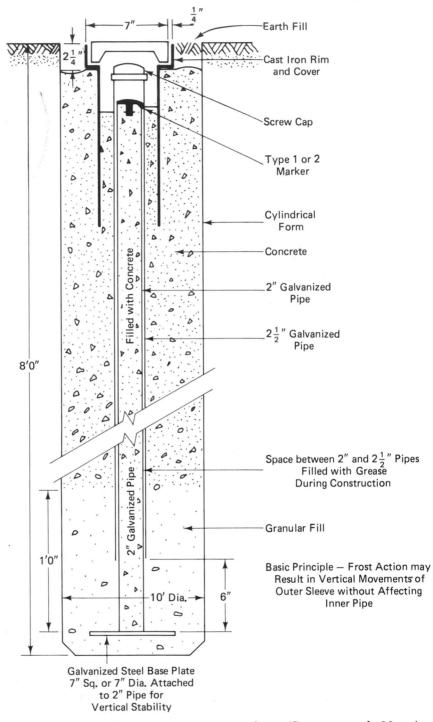

FIGURE 11.34 Sleeve-type survey marker. (Surveys and Mapping Branch, Department of Energy, Mines and Resources, Ottawa)

11.13 GLOBAL POSITIONING SYSTEMS (GPS)

The U.S. Navy Navigation Satellite System encircling the globe is called the Transit System. The five transit satellites are in polar orbit at an altitude of about 1000 km. This system, originally designed for military guidance purposes, has been adopted by civil authorities for positioning applications. This system is presently being replaced by the Navstar (Navigation Satellite Timing and Ranging) System. The Navstar System will comprise 18 satellites at an altitude of 20000 km providing 24-hour service anywhere on earth. Initially, six satellites were placed in orbit, and it is planned to have all 18 satellites fully operational by 1987. The satellites will provide guidance and position information to the military, as well as atomic time and global geodetic referencing on a worldwide basis for civil agencies.

Global positioning by satellite originally utilized a ground receiver capable of noting the change in satellite transmission as the satellite first approached and then receded from the observer. The velocity of sound of the transmission was, of course, affected by the velocity of the satellite itself. The change in velocity of the radio transmissions from the approaching and then receding satellite, known as the Doppler shift, is directly proportional to the amount of shift in frequency of the transmitted signal. When the satellite's orbit is known precisely and when the location of the satellite in its orbit is also given precisely by its atomic clock, the position of the receiving station on earth can be computed.

Doppler receivers can be used for positioning in several different ways. A relatively precise technique, known as *field translocation* (see Figure 11.35), involves simultaneous range measurement from two or more stations with the coordinates of one station being known; this technique can provide position to within 0.2 or 0.3 m if multiple passes are utilized. Doppler receivers can be left unattended for days in all types of weather with satellite multiple-pass data being recorded on cassette tapes. Table 11.6 summarizes accuracy requirements together with Doppler positioning techniques.

More recent Navstar satellite receivers utilize ultrahigh-precision atomic clocks on board the satellites to give precise times of satellite transmissions and the times of reception of these signals at control stations. Each geodetic receiver can decode the satellite transmissions and determine precise geodetic position (X, Y, Z co-

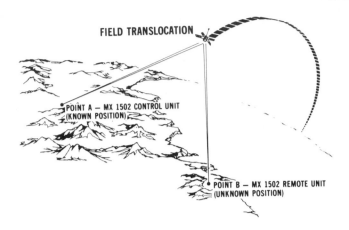

FIELD TRANSLOCATION PROCESSING

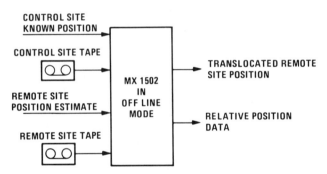

FIGURE 11.35   Field translocation. (Courtesy Magnavox Advanced Products and Systems Company)

ordinates) by computing the range from the control station to each of the four satellites that are always within communications reach. It is expected that these new geodetic Geostar receivers, manufactured by Texas Instruments Co. for the U.S. Department of Defense, will have accuracies better than the present-day Doppler satellite receivers.

Another recently designed geodetic receiver is the Macrometer* Interferometric Surveyor, manufactured by Macrometrics, Inc.

This instrument uses a variation of the very long base line interferometry (VLBI) technique mentioned earlier. Whereas the

_____

*Macrometer is a registered trademark (pending) of Macrometrics, Inc.

TABLE 11.6 *Doppler Positioning Techniques Applicable to Different Position Accuracy Requirements*

Horizontal Position Accuracy Requirement, 1 sigma (m)	Doppler Positioning Technique	Minimum Number of MX 1502 Units Required	Minimum Number of Passes Required
≥ 37	Single-pass point positioning	1	1 acceptable pass (height must be known to $\pm$ 20 m)
≥ 10	Point positioning	1	15 acceptable passes
≥ 1	Field or real-time translocation	2	4 acceptable passes common to each station
≥ 0.5 ≤ 1.0	Field or real-time translocation	2	17 acceptable passes common to each station
≥ 0.3	Network adjustment	3	20 acceptable passes common to at least two stations

Courtesy Magnavox Advanced Products and Systems Co.

VLBI technique utilizes large parabolic dish antennas to receive radio waves from distant stars, the macrometer receives radio waves (19-cm wavelength) from the Navstar satellites using small portable instrument antennas (see Figure 11.36). The manufacturers of the macrometer expect to have an instrument on the market in 1983 capable of a positional accuracy of 1 cm in X, Y, and Z, coordinates after 1-hour observation of all five satellites (90-minute observation of four satellites).

The measuring technique is illustrated in Figure 11.37. Macrometer units at points A and B receive signals from a GPS satellite at different orbit locations (positions S_1 and S_2 in Figure 11.37. The signal from location S_1 will be received at A shortly before it is received at B, permitting the determination of an equivalent phase difference (phase difference 1); a second phase difference is determined when the satellite is at position S_2. The phase differences can be developed into sides AC and BC of the triangle ABC. Angle θ can also be determined by analyzing multipositional data. Desired distance AB (chord length) can now be readily computed using the cosine law.

The preceding simple explanation fairly describes the process for two-dimensional positioning. In reality, when several satellites are observed in many different positions, systems ambiguities are resolved and three-dimensional vectors are produced. A present

FIGURE 11.36 Macrometer$^{(TM)}$. (Courtesy Macrometrics, Inc., Woburn, Mass.)

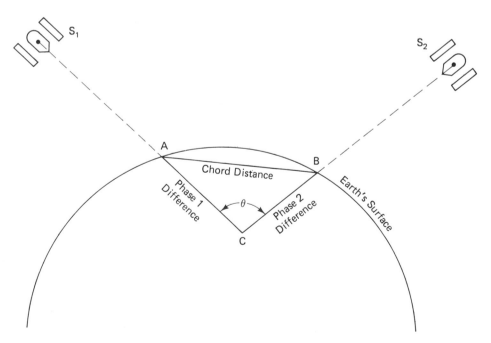

FIGURE 11.37 Cross section of the earth in the plane of the satellite orbit.

disadvantage to the system is that satellite coverage of an area is only for 6 to 8 hours a day and that coverage could well occur during normal sleeping hours. As the full 18-satellite system becomes operational (1988), this problem will disappear. A marked advantage of this system is that relative positions between ground stations (X, Y, Z coordinates) can be determined simply by observing the Navstar (or other satellite) signals without actually decoding those signals. Precise positioning now requires the surveyor to wait, sometimes many months, for declassification and decoding of the satellite signals.

The manufacturers of this remarkable instrument envisage a local system operating with a master Macrometer antenna mounted on a public building; both private or government surveyors, each with their own Macrometer units, could take appropriate field station readings. The collected data would be transmitted (mobile phone) back to the central office, where coordinates of the new point would be computed within a few minutes. The manufacturers see the day coming when some permanent control monuments may not be needed, as the GPS satellites will permit immediate determination of positional control to within a "few centimeters."

It is anticipated that the receiver will require updated satellite orbital information every two weeks. These data can be manually entered or entered through a computer interface. The relatively high cost ($200 000 approximately, including computer) is expected to decrease as production increases.

11.14 INTERTIAL SURVEYING SYSTEMS (ISS)

Research and development into military inertial guidance systems has resulted in civilian use of this high technology for coordinating control point positions. The system, which is presently manufactured by three or four companies (e.g., Honeywell, Ferranti, Litton), consists of three computer-controlled accelerometers, each aligned to a north–south, east–west, and vertical axis. Accelerometers are essentially pendulums. When acceleration occurs, the pendulum swings accordingly, and with the pendulum displacement and time of acceleration being constantly monitored, velocities and directions of the ISS vehicle are continuously updated. The platform carrying the accelerometers is also oriented N–S,

E–W, and plumb by means of three two-motion, computer-controlled gyroscopes, each of which is aligned to one of the three axes (X is aligned north; Y axis is aligned east; Z axis is aligned vertically up). See Figure 11.38. Analysis of acceleration data gives rectangular (latitude and departure) displacement factors for horizontal movement in addition to vertical displacement.

The typical inertial surveying system, mounted in a truck or helicopter, consists of (1) the inertial measuring unit, (2) onboard computer, (3) cassette recorders, and (4) display and command unit. Measurement commences at a point of known coordinates and elevation. The X, Y, Z (latitude, longitude, and elevation) values of the point are fed into the computer, and the inertial measuring unit is allowed (automatically) to orient itself to that position. One part of the truck or helicopter is plumbed over the control mark, and this offset point is tied to the measuring unit by known-length offset arms or lever arms. (Values for the offset arms or lever arms are also entered into the computer.) As the vehicle moves, the gyroscopes keep the platform properly aligned while the accelerometers measure the three-dimensional displacement. At frequent intervals (5 minutes), the vehicle is stopped to permit zero calibration of the accelerometers, that is, zero-velocity

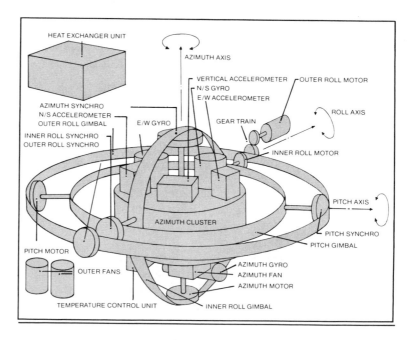

FIGURE 11.38 Inertial platform schematic. (Courtesy Nortech Canada)

update (ZUPT); usually the more ZUPTs taken, the more accurate the final results. Additionally, when possible, stations having known coordinates are occupied to provide a check on equipment operation and as a basis for error adjustment. The latitude, longitude, and elevation of all survey stations can be determined simply by stopping the vehicle, on station, for about 30 seconds.

Some systems provide EDM capabilities so that stations which cannot be directly occupied may yet be tied in; this latter capability is very useful for truck-mounted inertial surveying systems.

Accuracies have steadily improved over the few years that the system has been operating. In 1980 accuracies were reported as follows:

$$\text{Truck-mounted system: Lat., long.} = 20 \text{ cm}$$
$$\text{Elevation} = 15 \text{ cm}$$

$$\text{Helicopter-mounted} \quad \text{Lat., long.} = 60 \text{ cm}$$
$$\text{system:} \quad \text{Elevation} = 50 \text{ cm}$$

And in 1981, positional accuracies as high as 10 cm were reported. Over long distances, these accuracies are acceptable for high-order control surveys, and the system has been used to good effect on many of these control surveys. The high cost of the equipment (over $1 million not including the vehicle, or $4000 per day rental) restricts use of this technique to well-planned, large-scale surveys.

The system has been used effectively for establishing control for large-scale engineering projects, for aerial surveying and mapping, and for control monument densification.

ISS traverses begin and end at stations of known coordinate position. These traverses are normally run a minimum of two times (i.e., in opposite directions with resultant relative accuracy ratios of 1/20000). Higher orders of accuracy (including first order) can be achieved by utilizing multiple double-traverse runs.

11.15 DIRECTION OF A LINE
BY OBSERVATION ON POLARIS

Polaris, also called the North Star, has been used for centuries by sailors navigating in the northern hemispheres of our globe. Polaris is particularly useful as its apparent path of rotation keeps the star very close to the extension of the earth's polar axis through the North Pole. The actual position of Polaris along with the position of the sun and many other stars is computed and published annually in a nautical almanac or ephemeris.

Polaris is also used by surveyors to establish astronomic directions on survey control lines. In built-up areas, surveyors can establish control line directions by tying in to existing surveys having known azimuths or bearings, or they can tie into horizontal control monuments whose grid coordinates are known; but in more remote areas often the easiest method of establishing direction is to take an observation on the sun or a star.

Observations on the sun are usually more conveniently taken than are observations on the stars (they can be accomplished during normal working hours), but due to the size of the sun and the speed at which it appears to move, directions that are based on solar observations are generally less precise than those taken on the stars (i.e., to the closest 1 to 2 min for solar observations and to the closest 0.1 min for observations on Polaris, using normal procedures).

In this text we will limit discussion to observations on Polaris and we will use the ephemeris tables published each year by the Keuffel & Esser (K&E) Co. of Morristown, New Jersey. This ephemeris, which also includes data for observations on the sun and 26 other stars, is available from all local K&E sales offices for a nominal charge. The tables shown in this text are for the year 1983, and although the data changes slightly from year to year, the techniques outlined in the text are time valid.

Since many astronomical concepts are abstract in nature, it simplifies overall comprehension and positional computations to consider the earth as being stationary at the center of a celestial sphere of infinitely large radius. The stars then appear to be fixed on the surface of this infinitely large sphere, which itself appears to be rotating from east to west (see Figure 11.39).

To determine the bearing of Polaris at any time, the position of Polaris on the celestial sphere must first be determined. As can be seen in Figure 11.39, the position of Polaris (S) is given by the declination (d), which is equivalent to the latitude on earth, and by the Greenwich hour angle (GHA), which is measured westward from the Greenwich meridian (0°) and which for the first 180° is equivalent to west longitude on earth. (Longitudes are measured east and west from the Greenwich meridian for 180°, whereas the GHA is simply measured west a full 360°.)

Reference to Figure 11.39 will also show the spherical triangle PZS, which can be solved to provide the bearing of Polaris.

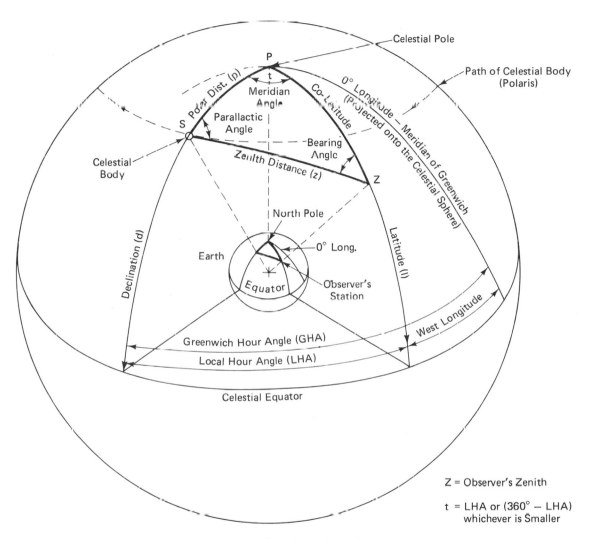

FIGURE 11.39   Celestial sphere.

Point Z is the zenith of the observer (i.e., a point made on the celestial sphere when the line of gravity through the theodolite is produced straight upward). S is the location of the celestial body (Polaris in this case), and P is the celestial pole (i.e., the extension of the earth's polar axis upward through the North Pole.

Angle t, the meridian angle, is given by the local hour angle (LHA) or by $360° -$ LHA, whichever is smaller. As can be seen in Figure 11.39, the LHA is an angle measured westward from the meridian through the observer's station to the meridian occupied by Polaris. If the LHA is less than 180°, Polaris is west of the North Pole; and if the LHA is more than 180°, Polaris is east of the North Pole.

It can also be seen in Figure 11.39 that the LHA is determined by subtracting the west longitude from the GHA (western hemispheres):

$$LHA = GHA - \text{west longitude} \qquad (11.10)$$

In eastern hemispheres,

$$LHA = GHA + \text{east longitude} \qquad (11.11)$$

11.16 TIME

We generally consider one day to be one complete revolution of the earth on its axis. We can see from Figure 11.40 that the earth, while rotating on its axis, is also traveling in an elliptical orbit around the sun. If we reference the earth to a point (a star) that is an infinite distance away, one complete revolution of the earth, which is called a sidereal (star) day, is exactly 360°. In Figure 11.40, it can be seen that the earth while in position 2 has turned 360° to describe a sidereal day, but that the earth now has to revolve an additional angle (K) in order to complete its revolution with respect to the sun.

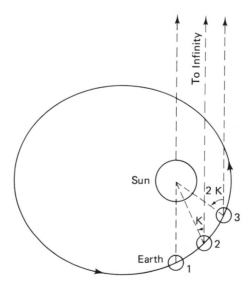

FIGURE 11.40   Elliptical path of the earth around the sun.

A solar day is, therefore, 360° plus the partial revolution of angle K. Since the earth travels about the sun once in one year, all the partial revolutions of angle K must sum to one complete revolution of the earth.

The earth makes 366.2422 revolutions on its axis (i.e., 366.2422 sidereal days) while completing the annual solar orbit; since a solar day is longer than a sidereal day by the partial revolution angle K and since all partial revolutions sum to one day, there must therefore be 365.2422 solar days in one year.

The additional partial revolutions (angle K) are not all equal because the earth's orbit about the sun is not uniform; the average partial revolution is 0°59.15′.

Since it is not possible to have days of different lengths, we keep time on a 24-hour basis and call that *mean solar* or *civil* time. The difference between mean solar time and the time required for a complete solar revolution on any one given day of the year is called the equation of time. Values for this time equation, and for solar day–sidereal day relationships, are given in the ephemeris.

In 24 hours the earth revolves 360° of longitude and in 1 hour the earth revolves through 15° of longitude. To facilitate time keeping, local time zones are established at 15° intervals as follows:

		At	Add Hours to Convert to GCT
GCT	Greenwich Civil Time	0° long. W.	—
AST	Atlantic Standard Time	60° long. W.	4
EST	Eastern Standard Time	75° long. W.	5
CST	Central Standard Time	90° long. W.	6
MST	Mountain Standard Time	105° long. W.	7
PST	Pacific Standard Time	120° long. W.	8
YST	Yukon Standard Time	135° long. W.	9
AHST	Alaska/Hawaii Standard Time	150° long. W.	10

DST—Daylight saving time. DST in a zone is equal to standard time in the next zone to the east.

The relationship between time and longitude is summarized as follows:

$$24^h = 360° \qquad\qquad 360° = 24^h$$
$$1^h = 15° \qquad\qquad 1° = 4^m$$
$$1^m = 15′ \qquad\qquad 1′ = 4^s$$
$$1^s = 15″ \qquad\qquad 1″ = 0.067^s$$

From these relationships, it can be seen that the solar day,

which included an additional partial revolution on average of
0°59.15′, is an average $3^m56.6^s$ longer than a sidereal day. This relationship is used in ephemeris conversions from Greenwich civil
time to Greenwhich sidereal time.

11.17 POLARIS

Reference to Figure 11.41 will show that because of the direction
of the earth's revolution the star appears to trace a counterclockwise path around the North Celestial Pole. The star makes one
complete revolution in one sidereal day. The star is at upper culmination at 2 hours sidereal every day, at western elongation at 8
hours sidereal every day, at lower culmination at 14 hours sidereal every day, and at eastern elongation at 20 hours sidereal every
day.

At the equator, the star appears to describe a circular orbit
about the pole, with a horizontal elongation angle of about ±0°50′
and a vertical culmination angle also of about ±0°50′. As the observer moves northward, the orbit becomes elliptical in appearance, with the horizontal elongation angle varying with latitude;
the range in elongation angles varies from about 0°50′ at the
equator to about 2°20′ at latitude 70°. See Figure 11.42 and
Table 11.11.

With respect to the altitude of Polaris, Figure 11.43 demonstrates that the altitude of Polaris is directly related to the latitude
of the observer's station. A correction in altitude ($\triangle h$), found in Table 11.10, is required to account for the culmination movement of
the star. It follows that at eastern and western elongation the altitude of Polaris is, in fact, the latitude of the observer, and at all other points in the star's orbit the altitude is equal to latitude $\pm\triangle h$.

11.18 PROCEDURE FOR OBSERVING
POLARIS

The success of a star observation depends a great deal on being
prepared well in advance of the actual sighting. The ideal time for
observations is just before nightfall. At that time, although Polaris
is visible only through the telescope, it is the only star in the tele-

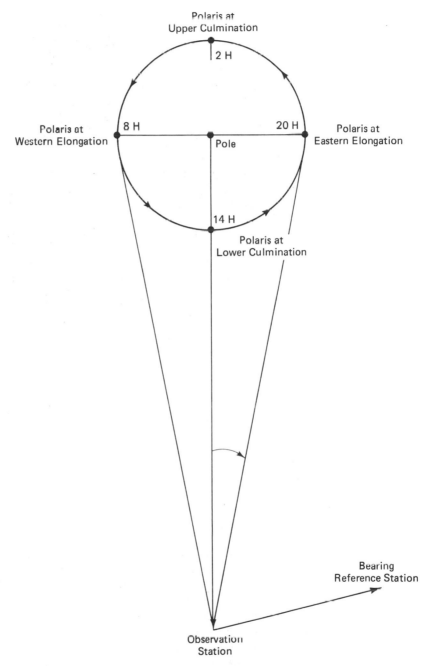

FIGURE 11.41 The apparent path taken by Polaris, along with the four major positions.

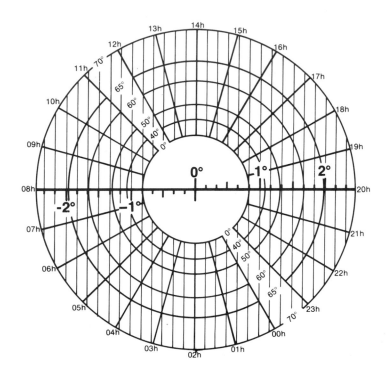

Find the point on the diagram that corresponds to the latitude and local
sidereal time of observation. The projection of this point on the horizontal
scale shows the azimuth of Polaris. To find Δh, the correction to the
latitude, for the altitude of Polaris, subtract 6 hours from the local sidereal
time of observation and read the corresponding azimuth for 0° latitude.

FIGURE 11.42   The effect of latitude on Polaris bearings and altitudes.

scopic field of view. Later, after nightfall, Polaris is easily seen
with the naked eye, but when a sight on Polaris is taken, confusion could result from the large number of stars now visible in
the magnified field of view. An additional advantage to taking the
observation prior to nightfall is that illumination is not required
for targets, scales, or note keeping, and the work can proceed
quickly. The wise surveyor will, however, be prepared with well-
charged batteries and illuminators just in case the observation
cannot be completed prior to nightfall.

To sight Polaris at dusk, north must be approximated; a
compass reading corrected for magnetic declination (see Section
4.12) is used by many surveyors.

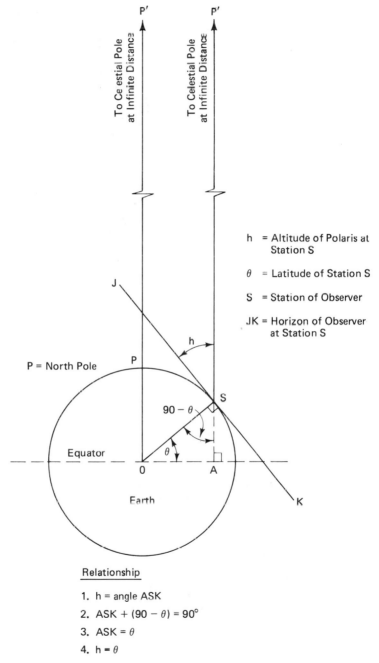

FIGURE 11.43 Relationship between the latitude of the observer station and the altitude of Polaris when viewed at that station.

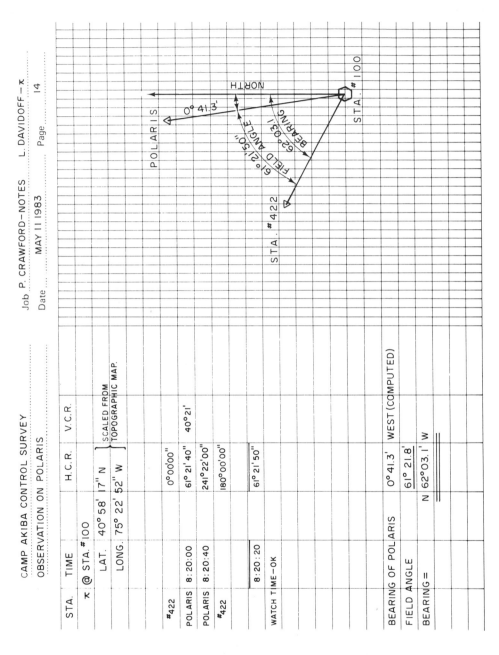

FIGURE 11.44 Field notes for Polaris observation.

The following is a procedure for observation on Polaris; see Figure 11.44 and Section 11.19.

1. Prepare note forms in advance.

2. Check to see that equipment is working properly and that a good supply of batteries (including spares) is available.

3. Predetermine the necessary correction to the latitude to give the altitude of Polaris (see Section 11.19).

4. After carefully setting up the instrument (STA 100), approximately determine the direction of north and establish a target.

5. With the horizontal scales zeroed, sight at the reference station (STA 422).

6. Sight an object about 250 m (800 ft) away and focus carefully; this is the instrument's infinite focus, which must be set when sighting the star. Some surveyors mark this point on the focus ring so that the infinite focus can be reestablished after dark when there may be no suitable long-range sight available. Proper identification of the infinite focus position on the focusing ring is emphasized, because if the telescope is only slightly off focus, the star will not even appear in the telescopic field of view and much time will be wasted.

7. If the telescope has been properly directed toward north, and if the correct altitude has been set on the vertical circle, and if the focus adjustment has been properly set (infinite focus), Polaris should appear in the telescope at least 15 minutes prior to nightfall. It may be necessary to move the telescope through slight horizontal and vertical arcs to find the star.

8. At the instant the star has been carefully centered on the cross hairs, the time is recorded.

9. The telescope is transited (plunged), and with the upper motion free (lower motion still clamped) the star is resighted. The time is once again recorded.

10. The telescope is finally sighted back on the original reference station (422) and the angle (it should be 180°00′) is noted. This procedure may be repeated if higher accuracies are required.

11. The average time is used for the bearing calculation for Polaris (see Section 11.19), and the average angle is used to determine the bearing of line 100–422 (see Figure 11.44).

Note: When Polaris is to be observed during darkness, Polaris can be located by using the two Big Dipper (Ursa Major) stars as pointers—as shown in Figure 11.45.

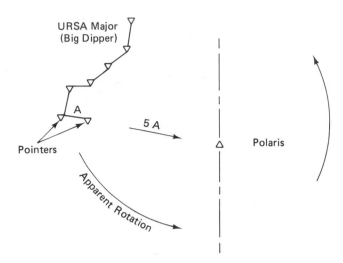

FIGURE 11.45   Aid for locating Polaris.

11.19 ILLUSTRATION 11.1

The techniques used in determining the direction of Polaris are demonstrated in the following example.

Instrument at Control Station 100
Camp Akiba, Reeders, Pa.
Latitude   = 40°58'17" N ⎫
Longitude = 75°22'52" W ⎭ Scaled from a topo. map
Date: May 11, 1983
Time: 8:20:20 P.M. EST

$\quad\quad$ +12

$\quad\quad$ 20:20:20 Local mean time

$\quad\quad$ +5$\quad\quad$ Correction at long. 75°W (i.e., 1 h for each 15°)

$\quad\quad$ 25:20:20

$\quad\quad\quad$ 1:20:20 May 12, 1983, Greenwich Civil Time

$\quad$ 195°56.6' = GHA for May 12/83 (Table 11.7)

$\quad\quad$ 20°03.3' = correction for 1^h20^m (Table 11.8)

$\quad\quad\quad\quad$ 5.0 = correction for 20^s (Table 11.8)

$\quad$ 216°04.9 = GHA (Greenwich hour angle)

$\quad$ −75°22.9$\quad$ minus west longitude (see Figure 11.39)

$\quad$ 140°42.0 = **LHA** (local hour angle) (see Figure 11.39)

For LHA < 180°, Polaris is west of north
t = LHA or 360 − LHA, whichever is smaller (Figure 11.39)
t = 140°42.0'

Long Method

$$Z = \frac{\text{Sin } t\, p}{\text{Cos } h}$$

where Z = bearing of Polaris (minutes)
t = meridian angle
h = altitude
P = polar distance in minutes

From Table 11.9, p = 48.86′ (May 12/83).
From Table 11.10, $\triangle h = -37.65'$:

Latitude = 40°58.28′
h (altitude) = 40°58.28′ −37.65′
h = 40°20.6′

$$Z = \frac{\text{Sin } 140°42.0'}{\text{Cos } 40°20.6'} \times 48.86$$
$$= 40.6'$$

The bearing of Polaris is N 0°40.6′ W.

Short Method

LHA = 140°42.0′ (from previous computation)
Latitude = 40°58.3′

From Table 11.11, by double interpolation:

Z (Bearing of Polaris) for p = 48.5

LHA	θ = 40°	0 − 40°58.3′	θ = 42°
140°	0°40.3′	0°40.9′	0°41.5′
140°42′		0°40.3′	
145°	0°36.0′	0°36.5′	0°37.0′

From Table 11.12

Correction Z = 0°40.3, for P = 48.86
= + 0.3

Therefore,

$$Z = 0°40.3 + 0.3 = 0°40.6'$$

The bearing of Polaris is N 0°40.6 W.

TABLE 11.7 GHA of Polaris (0ʰ Greenwich Civil Time), 1983

January		February		March		April		May		June	
Day of Month and Week	GHA of Polaris °	Day of Month and Week	GHA of Polaris °	Day of Month and Week	GHA of Polaris °	Day of Month and Week	GHA of Polaris °	Day of Month and Week	GHA of Polaris °	Day of Month and Week	GHA of Polaris °
1 SA	066 20.4	1 TU	097 05.4	1 TU	124 51.5	1 FR	155 32.2	1 SU	185 07.2	1 W	215 35.0
2 SU	067 19.9	2 W	098 04.9	2 W	125 51.0	2 SA	156 31.4	2 M	186 06.2	2 TH	216 34.0
3 M	068 19.4	3 TH	099 04.5	3 TH	126 50.5	3 SU	157 30.6	3 TU	187 05.2	3 FR	217 32.9
4 TU	069 18.9	4 FR	100 04.4	4 FR	127 49.9	4 M	158 29.8	4 W	188 04.3	4 SA	218 31.8
5 W	070 18.5	5 SA	101 03.5	5 SA	128 49.3	5 TU	159 29.1	5 TH	189 03.4	5 SU	219 30.7
6 TH	071 18.0	6 SU	102 03.0	6 SU	129 48.7	6 W	160 28.3	6 FR	190 02.4	6 M	220 29.6
7 FR	072 17.5	7 M	103 02.5	7 M	130 48.1	7 TH	161 27.6	7 SA	191 01.5	7 TU	221 28.4
8 SA	073 16.9	8 TU	104 02.0	8 TU	131 47.5	8 FR	162 26.8	8 SU	192 00.6	8 W	222 27.3
9 SU	074 16.4	9 W	105 01.5	9 W	132 46.9	9 SA	163 26.1	9 M	192 59.6	9 TH	223 26.1
10 M	075 15.9	10 TH	106 01.0	10 TH	133 46.3	10 SU	164 25.4	10 TU	193 58.6	10 FR	224 24.9
11 TU	076 15.3	11 FR	107 00.5	11 FR	134 45.8	11 M	165 24.6	11 W	194 57.6	11 SA	225 23.6
12 W	077 14.8	12 SA	108 00.0	12 SA	135 45.2	12 TU	166 23.9	12 TH	195 56.6	12 SU	226 22.4
13 TH	078 14.3	13 SU	108 59.6	13 SU	136 44.7	13 W	167 23.1	13 FR	196 55.6	13 M	227 21.2
14 FR	079 13.8	14 M	109 59.1	14 M	137 44.1	14 TH	168 22.3	14 SA	197 54.5	14 TU	228 20.0
15 SA	080 13.3	15 TU	110 58.7	15 TU	138 43.5	15 FR	169 21.4	15 SU	198 53.4	15 W	229 18.9
16 SU	081 12.9	16 W	111 58.2	16 W	139 42.9	16 SA	170 20.5	16 M	199 52.3	16 TH	230 17.7
17 M	082 12.4	17 TH	112 57.8	17 TH	140 42.3	17 SU	171 19.6	17 TU	200 51.3	17 FR	231 16.6
18 TU	083 12.0	18 FR	113 57.3	18 FR	141 41.7	18 M	172 18.8	18 W	201 50.2	18 SA	232 15.5
19 W	084 11.5	19 SA	114 56.8	19 SA	142 41.0	19 TU	173 17.9	19 TH	202 49.2	19 SU	233 14.3
20 TH	085 11.1	20 SU	115 56.2	20 SU	143 40.3	20 W	174 17.0	20 FR	203 48.2	20 M	234 13.1
21 FR	086 10.6	21 M	116 55.7	21 M	144 39.6	21 TH	175 16.2	21 SA	204 47.2	21 TU	235 11.9
22 SA	087 10.2	22 TU	117 55.1	22 TU	145 38.9	22 FR	176 15.3	22 SU	205 46.2	22 W	236 10.6
23 SU	088 09.7	23 W	118 54.5	23 W	146 38.3	23 SA	177 14.5	23 M	206 45.1	23 TH	237 09.4
24 M	089 09.2	24 TH	119 54.0	24 TH	147 37.6	24 SU	178 13.7	24 TU	207 44.0	24 FR	238 08.1
25 TU	090 08.6	25 FR	120 53.5	25 FR	148 36.9	25 M	179 12.8	25 W	208 42.9	25 SA	239 06.9
26 W	091 08.1	26 SA	121 53.0	26 SA	149 36.3	26 TU	180 11.9	26 TH	209 41.8	26 SU	240 05.6
27 TH	092 07.6	27 SU	122 52.5	27 SU	150 35.7	27 W	181 11.0	27 FR	210 40.7	27 M	241 04.4
28 FR	093 07.1	28 M	123 52.0	28 M	151 35.0	28 TH	182 10.1	28 SA	211 39.5	28 TU	242 03.2
29 SA	094 06.6	29 TU	124 51.5	29 TU	152 34.4	29 FR	183 09.1	29 SU	212 38.4	29 W	243 02.0
30 SU	095 06.2			30 W	153 33.7	30 SA	184 08.1	30 M	213 37.3	30 TH	244 00.8
31 M	096 05.8			31 TH	154 32.9	31 SU	185 07.2	31 TU	214 36.1	31 FR	244 59.7
32 TU	097 05.4			32 FR	155 32.2			32 W	215 35.0		

TABLE 11.7 (Continued)

July Day of Month and Week	July GHA of Polaris	August Day of Month and Week	August GHA of Polaris	September Day of Month and Week	September GHA of Polaris	October Day of Month and Week	October GHA of Polaris	November Day of Month and Week	November GHA of Polaris	December Day of Month and Week	December GHA of Polaris
1 FR	244 59.7	1 M	275 21.0	1 TH	305 42.7	1 SA	335 08.4	1 TU	005 37.9	1 TH	035 13.9
2 SA	245 58.5	2 TU	276 19.8	2 FR	306 41.5	2 SU	336 07.4	2 W	006 37.0	2 FR	036 13.2
3 SU	246 57.3	3 W	277 18.5	3 SA	307 40.3	3 M	337 06.3	3 TH	007 36.1	3 SA	037 12.5
4 M	247 56.1	4 TH	278 17.2	4 SU	308 39.1	4 TU	338 05.3	4 FR	008 35.3	4 SU	038 11.8
5 TU	248 54.8	5 FR	279 15.9	5 M	309 37.9	5 W	339 04.3	5 SA	009 34.4	5 M	039 11.0
6 W	249 53.5	6 SA	280 14.6	6 TU	310 36.8	6 TH	340 03.3	6 SU	010 33.4	6 TU	040 10.3
7 TH	250 52.3	7 SU	281 13.3	7 W	311 35.6	7 FR	341 02.3	7 M	011 32.5	7 W	041 09.6
8 FR	251 51.0	8 M	282 12.1	8 TH	312 34.5	8 SA	342 01.2	8 TU	012 31.6	8 TH	042 09.0
9 SA	252 49.7	9 TU	283 10.8	9 FR	313 33.4	9 SU	343 00.1	9 W	013 30.7	9 FR	043 08.3
10 SU	253 48.4	10 W	284 09.6	10 SA	314 32.2	10 M	343 59.0	10 TH	014 29.9	10 SA	044 07.7
11 M	254 47.1	11 TH	285 08.5	11 SU	315 31.0	11 TU	344 58.0	11 FR	015 29.1	11 SU	045 07.1
12 TU	255 45.9	12 FR	286 07.3	12 M	316 29.8	12 W	345 56.9	12 SA	016 28.3	12 M	046 06.5
13 W	256 44.7	13 SA	287 06.0	13 TU	317 28.6	13 TH	346 55.9	13 SU	017 27.5	13 TU	047 05.9
14 TH	257 43.5	14 SU	288 04.8	14 W	318 27.4	14 FR	347 54.9	14 M	018 26.7	14 W	048 05.4
15 FR	258 42.3	15 M	289 03.5	15 TH	319 26.2	15 SA	348 53.9	15 TU	019 25.9	15 TH	049 04.8
16 SA	259 41.1	16 TU	290 02.2	16 FR	320 25.1	16 SU	349 52.9	16 W	020 25.2	16 FR	050 04.1
17 SU	260 39.8	17 W	291 01.0	17 SA	321 23.9	17 M	350 52.0	17 TH	021 24.4	17 SA	051 03.5
18 M	261 38.6	18 TH	291 59.7	18 SU	322 22.8	18 TU	351 51.0	18 FR	022 23.6	18 SU	052 02.8
19 TU	262 37.3	19 FR	292 58.4	19 M	323 21.7	19 W	352 50.1	19 SA	023 22.8	19 M	053 02.2
20 W	263 36.0	20 SA	293 57.2	20 TU	324 20.6	20 TH	353 49.1	20 SU	024 22.0	20 TU	054 01.5
21 TH	264 34.7	21 SU	294 56.0	21 W	325 19.6	21 FR	354 48.2	21 M	025 21.2	21 W	055 00.9
22 FR	265 33.4	22 M	295 54.8	22 TH	326 18.5	22 SA	355 47.2	22 TU	026 20.4	22 TH	056 00.3
23 SA	266 32.1	23 TU	296 53.6	23 FR	327 17.4	23 SU	356 46.3	23 W	027 19.6	23 FR	056 59.8
24 SU	267 30.8	24 W	297 52.4	24 SA	328 16.3	24 M	357 45.3	24 TH	028 18.8	24 SA	057 59.2

TABLE 11.7 *(Continued)*

	July		August		September		October		November		December	
	Day of Month and Week	GHA of Polaris ° ′	Day of Month and Week	GHA of Polaris ° ′	Day of Month and Week	GHA of Polaris ° ′	Day of Month and Week	GHA of Polaris ° ′	Day of Month and Week	GHA of Polaris ° ′	Day of Month and Week	GHA of Polaris ° ′
	25 M	268 29.6	25 TH	298 51.2	25 SU	329 15.2	25 TU	358 44.3	25 FR	029 18.1	25 SU	058 58.7
	26 TU	269 28.4	26 FR	299 50.1	26 M	330 14.1	26 W	359 43.3	26 SA	030 17.3	26 M	059 58.2
	27 W	270 27.2	27 SA	300 48.9	27 TU	331 12.9	27 TH	000 42.3	27 SU	031 16.7	27 TU	060 57.7
	28 TH	271 25.9	28 SU	301 47.7	28 W	332 11.8	28 FR	001 41.4	28 M	032 16.0	28 W	061 57.2
	29 FR	272 24.7	29 M	302 46.5	29 TH	333 10.6	29 SA	002 40.4	29 TU	033 15.3	29 TH	062 56.6
	30 SA	273 23.5	30 TU	303 45.2	30 FR	334 09.5	30 SU	003 39.6	30 W	034 14.6	30 FR	063 56.0
	31 SU	274 22.3	31 W	304 44.0	31 SA	335 08.4	31 M	004 38.7	31 TH	035 13.9	31 SA	064 55.4
	32 M	275 21.0	32 TH	305 42.7			32 TU	005 37.9			32 SU	065 54.9

Courtesy of Keuffel and Esser Co., Morristown, NJ.

11.19.1 Azimuth of Polaris Summaries

Short Method Summary

$$Z = \text{Bearing of Polaris}$$

1. Find GHA for day of observation (Table 11.7).
2. Find adjustment to GHA for partial day (Table 11.8).
3. Find LHA; LHA = GHA − west longitude.
4. Find Z value for LHA when $P = 0°48.50$ (Table 11.11).
5. Find p value for day (Table 11.9).
6. Find correction to Z value (4) for amount by which $P \neq 0°48.50'$ (Table 11.12).

Long Method Summary

$$Z = \frac{\text{Sin } t}{\text{Cos } h} \, p$$

where Z = bearing of Polaris in minutes
   t = meridian angle (t = LHA or (360 − LHA), whichever is smaller)
   h = true altitude
   p = polar distance in minutes

1. Find GHA for day (Table 11.7).
2. Find adjustment to GHA for hours, minutes, seconds (Table 11.8).
3. Find LHA (LHA = GHA − west longitude).
4. Find t (t = LHA or (360 − LHA), whichever is smaller).
5. Find h (h = latitude $\pm \triangle h$) (Table 11.10).
6. Find P (Table 11.9).
7. Solve for Z

11.20 DIRECTION OF A LINE BY GYROTHEODOLITE

In Section 11.14, gyroscopes were introduced as the platform-stabilizing agents used with accelerometers in the inertial survey system technique of position determination. Several surveying equipment manufacturers produce gyro attachments for use with repeating

TABLE 11.8   *Increase in GHA for Elapsed Time*

Min.	0ʰ		1ʰ		2ʰ		3ʰ		Sec.	Corr.
	°	′	°	′	°	′	°	′	°	′
0	0	0.0	15	2.5	30	4.9	45	7.4	0	0.0
1	0	15.0	15	17.5	30	19.9	45	22.4	1	0.3
2	0	30.1	15	32.5	30	35.0	45	37.5	2	0.5
3	0	45.1	15	47.6	30	50.1	45	52.5	3	0.8
4	1	0.2	16	2.6	31	5.1	46	7.6	4	1.0
5	1	15.2	16	17.7	31	20.1	46	22.6	5	1.3
6	1	30.2	16	32.7	31	35.2	46	37.6	6	1.5
7	1	45.3	16	47.8	31	50.2	46	52.7	7	1.8
8	2	0.3	17	2.8	32	5.3	47	7.7	8	2.0
9	2	15.4	17	17.8	32	20.3	47	22.8	9	2.3
10	2	30.4	17	32.9	32	35.3	47	37.8	10	2.5
11	2	45.5	17	47.9	32	50.4	47	52.8	11	2.8
12	3	0.5	18	3.0	33	5.4	48	7.9	12	3.0
13	3	15.5	18	18.0	33	20.5	48	22.9	13	3.3
14	3	30.6	18	33.0	33	35.5	48	38.0	14	3.5
15	3	45.6	18	48.1	33	50.5	48	53.0	15	3.8
16	4	0.7	19	3.1	34	5.6	49	8.0	16	4.0
17	4	15.7	19	18.2	34	20.6	49	23.1	17	4.3
18	4	30.7	19	33.2	34	35.7	49	38.1	18	4.5
19	4	45.8	19	48.2	34	50.7	49	53.2	19	4.8
20	5	0.8	20	3.3	35	5.7	50	8.2	20	5.0
21	5	15.9	20	18.3	35	20.8	50	23.3	21	5.3
22	5	30.9	20	33.4	35	35.8	50	38.3	22	5.5
23	5	45.9	20	48.4	35	50.9	50	53.3	23	5.8
24	6	1.0	21	3.4	36	5.9	51	8.4	24	6.0
25	6	16.0	21	18.5	36	21.0	51	23.4	25	6.3
26	6	31.1	21	33.5	36	36.0	51	38.5	26	6.5
27	6	46.1	21	48.6	36	51.0	51	53.5	27	6.8
28	7	1.1	22	3.6	37	6.1	52	8.5	28	7.0
29	7	16.2	22	18.7	37	21.1	52	23.6	29	7.3
30	7	31.2	22	33.7	37	36.2	52	38.6	30	7.5
31	7	46.3	22	48.7	37	51.2	52	53.7	31	7.8
32	8	1.3	23	3.8	38	6.2	53	8.7	32	8.0
33	8	16.4	23	18.8	38	21.3	53	23.7	33	8.3
34	8	31.4	23	33.9	38	36.3	53	38.8	34	8.5
35	8	46.4	23	48.9	38	51.4	53	53.8	35	8.8
36	9	1.5	24	3.9	39	6.4	54	8.9	36	9.0
37	9	16.5	24	19.0	39	21.4	54	23.9	37	9.3
38	9	31.6	24	34.0	39	36.5	54	39.0	38	9.5
39	9	46.6	24	49.1	39	51.5	54	54.0	39	9.8
40	10	1.6	25	4.1	40	6.6	55	9.0	40	10.0
41	10	16.7	25	19.1	40	21.6	55	24.1	41	10.3
42	10	31.7	25	34.2	40	36.7	55	39.1	42	10.5
43	10	46.8	25	49.2	40	51.7	55	54.2	43	10.8
44	11	1.8	26	4.3	41	6.7	56	9.2	44	11.0
45	11	16.8	26	19.3	41	21.8	56	24.2	45	11.3
46	11	31.9	26	34.4	41	36.8	56	39.3	46	11.5
47	11	46.9	26	49.4	41	51.9	56	54.3	47	11.8
48	12	2.0	27	4.4	42	6.9	57	9.4	48	12.0
49	12	17.0	27	19.5	42	21.9	57	24.4	49	12.3
50	12	32.1	27	34.5	42	37.0	57	39.4	50	12.5
51	12	47.1	27	49.6	42	52.0	57	54.5	51	12.8
52	13	2.1	28	4.6	43	7.1	58	9.5	52	13.0
53	13	17.2	28	19.6	43	22.1	58	24.6	53	13.3
54	13	32.2	28	34.7	43	37.1	58	39.6	54	13.5
55	13	47.3	28	49.7	43	52.2	58	54.6	55	13.8
56	14	2.3	29	4.8	44	7.2	59	9.7	56	14.0
57	14	17.3	29	19.8	44	22.3	59	24.7	57	14.3
58	14	32.4	29	34.8	44	37.3	59	39.8	58	14.5
59	14	47.4	29	49.9	44	52.4	59	54.8	59	14.8
60	15	2.5	30	4.9	45	7.4	60	9.9	60	15.0

TABLE 11.8 (Continued)

						Hours of Greenwich Civil Time						
Min.	4h		5h		6h		7h		8h		9h	
	°	′	°	′	°	′	°	′	°	′	°	′
0	60	9.9	75	12.3	90	14.8	105	17.2	120	19.7	135	22.2
1	60	24.9	75	27.4	90	29.8	105	32.3	120	34.8	135	37.2
2	60	39.9	75	42.4	90	44.9	105	47.3	120	49.8	135	52.3
3	60	55.0	75	57.4	90	59.9	106	2.4	121	4.8	136	7.3
4	61	10.0	76	12.5	91	14.9	106	17.4	121	19.9	136	22.3
5	61	25.1	76	27.5	91	30.0	106	32.5	121	34.9	136	37.4
6	61	40.1	76	42.6	91	45.0	106	47.5	121	50.0	136	52.4
7	61	55.1	76	57.6	92	0.1	107	2.5	122	5.0	137	7.5
8	62	10.2	77	12.6	92	15.1	107	17.6	122	20.0	137	22.5
9	62	25.2	77	27.7	92	30.2	107	32.6	122	35.1	137	37.5
10	62	40.3	77	42.7	92	45.2	107	47.7	122	50.1	137	52.6
11	62	55.3	77	57.8	93	0.2	108	2.7	123	5.2	138	7.6
12	63	10.3	78	12.8	93	15.3	108	17.7	123	20.2	138	22.7
13	63	25.4	78	27.9	93	30.3	108	32.8	123	35.2	138	37.7
14	63	40.4	78	42.9	93	45.4	108	47.8	123	50.3	138	52.8
15	63	55.5	78	57.9	94	0.4	109	2.9	124	5.3	139	7.8
16	64	10.5	79	13.0	94	15.4	109	17.9	124	20.4	139	22.8
17	64	25.6	79	28.0	94	30.5	109	32.9	124	35.4	139	37.9
18	64	40.6	79	43.1	94	45.5	109	48.0	124	50.4	139	52.9
19	64	55.6	79	58.1	95	0.6	110	3.0	125	5.5	140	8.0
20	65	10.7	80	13.1	95	15.6	110	18.1	125	20.5	140	23.0
21	65	25.7	80	28.2	95	30.6	110	33.1	125	35.6	140	38.0
22	65	40.8	80	43.2	95	45.7	110	48.2	125	50.6	140	53.1
23	65	55.8	80	58.3	96	0.7	111	3.2	126	5.7	141	8.1
24	66	10.8	81	13.3	96	15.8	111	18.2	126	20.7	141	23.2
25	66	25.9	81	28.3	96	30.8	111	33.3	126	35.7	141	38.2
26	66	40.9	81	43.4	96	45.8	111	48.3	126	50.8	141	53.2
27	66	56.0	81	58.4	97	0.9	112	3.4	127	5.8	142	8.3
28	67	11.0	82	13.5	97	15.9	112	18.4	127	20.9	142	23.3
29	67	26.0	82	28.5	97	31.0	112	33.4	127	35.9	142	38.4
30	67	41.1	82	43.6	97	46.0	112	48.5	127	50.9	142	53.4
31	67	56.1	82	58.6	98	1.1	113	3.5	128	6.0	143	8.4
32	68	11.2	83	13.6	98	16.1	113	18.6	128	21.0	143	23.5
33	68	26.2	83	28.7	98	31.1	113	33.6	128	36.1	143	38.5
34	68	41.2	83	43.7	98	46.2	113	48.6	128	51.1	143	53.6
35	68	56.3	83	58.8	99	1.2	114	3.7	129	6.1	144	8.6
36	69	11.3	84	13.8	99	16.3	114	18.7	129	21.2	144	23.7
37	69	26.4	84	28.8	99	31.3	114	33.8	129	36.2	144	38.7
38	69	41.4	84	43.9	99	46.3	114	48.8	129	51.3	144	53.7
39	69	56.5	84	58.9	100	1.4	115	3.9	130	6.3	145	8.8
40	70	11.5	85	14.0	100	16.4	115	18.9	130	21.4	145	23.8
41	70	26.5	85	29.0	100	31.5	115	33.9	130	36.4	145	38.9
42	70	41.6	85	44.0	100	46.5	115	49.0	130	51.4	145	53.9
43	70	56.6	85	59.1	101	1.5	116	4.0	131	6.5	146	8.9
44	71	11.7	86	14.1	101	16.6	116	19.1	131	21.5	146	24.0
45	71	26.7	86	29.2	101	31.6	116	34.1	131	36.6	146	39.0
46	71	41.7	86	44.2	101	46.7	116	49.1	131	51.6	146	54.1
47	71	56.8	86	59.2	102	1.7	117	4.2	132	6.6	147	9.1
48	72	11.8	87	14.3	102	16.8	117	19.2	132	21.7	147	24.1
49	72	26.9	87	29.3	102	31.8	117	34.3	132	36.7	147	39.2
50	72	41.9	87	44.4	102	46.8	117	49.3	132	51.8	147	54.2
51	72	56.9	87	59.4	103	1.9	118	4.3	133	6.8	148	9.3
52	73	12.0	88	14.5	103	16.9	118	19.4	133	21.8	148	24.3
53	73	27.0	88	29.5	103	32.0	118	34.4	133	36.9	148	39.3
54	73	42.1	88	44.5	103	47.0	118	49.5	133	51.9	148	54.4
55	73	57.1	88	59.6	104	2.0	119	4.5	134	7.0	149	9.4
56	74	12.2	89	14.6	104	17.1	119	19.5	134	22.0	149	24.5
57	74	27.2	89	29.7	104	32.1	119	34.6	134	37.0	149	39.5
58	74	42.2	89	44.7	104	47.2	119	49.6	134	52.1	149	54.6
59	74	57.3	89	59.7	105	2.2	120	4.7	135	7.1	150	9.6
60	75	12.3	90	14.8	105	17.2	120	19.7	135	22.2	150	24.6

TABLE 11.8 *(Continued)*

				Hours of Greenwich Civil Time						
Min.	10ʰ		11ʰ		12ʰ		13ʰ		Sec.	Corr.
	°	′	°	′	°	′	°	′	°	′
0	150	24.6	165	27.1	180	29.6	195	32.0	0	0.0
1	150	39.7	165	42.1	180	44.6	195	47.1	1	0.3
2	150	54.7	165	57.2	180	59.7	196	2.1	2	0.5
3	151	9.8	166	12.2	181	14.7	196	17.2	3	0.8
4	151	24.8	166	27.3	181	29.7	196	32.2	4	1.0
5	151	39.8	166	42.3	181	44.8	196	47.2	5	1.3
6	151	54.9	166	57.3	181	59.8	197	2.3	6	1.5
7	152	9.9	167	12.4	182	14.9	197	17.3	7	1.8
8	152	25.0	167	27.4	182	29.9	197	32.4	8	2.0
9	152	40.0	167	42.5	182	44.9	197	47.4	9	2.3
10	152	55.1	167	57.5	183	0.0	198	2.4	10	2.5
11	153	10.1	168	12.6	183	15.0	198	17.5	11	2.8
12	153	25.1	168	27.6	183	30.1	198	32.5	12	3.0
13	153	40.2	168	42.6	183	45.1	198	47.6	13	3.3
14	153	55.2	168	57.7	184	0.1	199	2.6	14	3.5
15	154	10.3	169	12.7	184	15.2	199	17.6	15	3.8
16	154	25.3	169	27.8	184	30.2	199	32.7	16	4.0
17	154	40.3	169	42.8	184	45.3	199	47.7	17	4.3
18	154	55.4	169	57.8	185	0.3	200	2.8	18	4.5
19	155	10.4	170	12.9	185	15.3	200	17.8	19	4.8
20	155	25.5	170	27.9	185	30.4	200	32.9	20	5.0
21	155	40.5	170	43.0	185	45.4	200	47.9	21	5.3
22	155	55.5	170	58.0	186	0.5	201	2.9	22	5.5
23	156	10.6	171	13.0	186	15.5	201	18.0	23	5.8
24	156	25.6	171	28.1	186	30.6	201	33.0	24	6.0
25	156	40.7	171	43.1	186	45.6	201	48.1	25	6.3
26	156	55.7	171	58.2	187	0.6	202	3.1	26	6.5
27	157	10.7	172	13.2	187	15.7	202	18.1	27	6.8
28	157	25.8	172	28.2	187	30.7	202	33.2	28	7.0
29	157	40.8	172	43.3	187	45.8	202	48.2	29	7.3
30	157	55.9	172	58.3	188	0.8	203	3.3	30	7.5
31	158	10.9	173	13.4	188	15.8	203	18.3	31	7.8
32	158	26.0	173	28.4	188	30.9	203	33.3	32	8.0
33	158	41.0	173	43.5	188	45.9	203	48.4	33	8.3
34	158	56.0	173	58.5	189	1.0	204	3.4	34	8.5
35	159	11.1	174	13.5	189	16.0	204	18.5	35	8.8
36	159	26.1	174	28.6	189	31.0	204	33.5	36	9.0
37	159	41.2	174	43.6	189	46.1	204	48.6	37	9.3
38	159	56.2	174	58.7	190	1.1	205	3.6	38	9.5
39	160	11.2	175	13.7	190	16.2	205	18.6	39	9.8
40	160	26.3	175	28.7	190	31.2	205	33.7	40	10.0
41	160	41.3	175	43.8	190	46.2	205	48.7	41	10.3
42	160	56.4	175	58.8	191	1.3	206	3.8	42	10.5
43	161	11.4	176	13.9	191	16.3	206	18.8	43	10.8
44	161	26.4	176	28.9	191	31.4	206	33.8	44	11.0
45	161	41.5	176	43.9	191	46.4	206	48.9	45	11.3
46	161	56.5	176	59.0	192	1.5	207	3.9	46	11.5
47	162	11.6	177	14.0	192	16.5	207	19.0	47	11.8
48	162	26.6	177	29.1	192	31.5	207	34.0	48	12.0
49	162	41.6	177	44.1	192	46.6	207	49.0	49	12.3
50	162	56.7	177	59.2	193	1.6	208	4.1	50	12.5
51	163	11.7	178	14.2	193	16.7	208	19.1	51	12.8
52	163	26.8	178	29.2	193	31.7	208	34.2	52	13.0
53	163	41.8	178	44.3	193	46.7	208	49.2	53	13.3
54	163	56.9	178	59.3	194	1.8	209	4.2	54	13.5
55	164	11.9	179	14.4	194	16.8	209	19.3	55	13.8
56	164	26.9	179	29.4	194	31.9	209	34.3	56	14.0
57	164	42.0	179	44.4	194	46.9	209	49.4	57	14.3
58	164	57.0	179	59.5	195	1.9	210	4.4	58	14.5
59	165	12.1	180	14.5	195	17.0	210	19.5	59	14.8
60	165	27.1	180	29.6	195	32.0	210	34.5	60	15.0

TABLE 11.8 *(Continued)*

	Hours of Greenwich Civil Time											
Min	14h		15h		16h		17h		18h		19h	
	°	′	°	′	°	′	°	′	°	′	°	′
0	210	34.5	225	37.0	240	39.4	255	41.9	270	44.4	285	46.8
1	210	49.5	225	52.0	240	54.5	255	56.9	270	59.4	286	1.9
2	211	4.6	226	7.0	241	9.5	256	12.0	271	14.4	286	16.9
3	211	19.6	226	22.1	241	24.5	256	27.0	271	29.5	286	31.9
4	211	34.7	226	37.1	241	39.6	256	42.1	271	44.5	286	47.0
5	211	49.7	226	52.2	241	54.6	256	57.1	271	59.6	287	2.0
6	212	4.7	227	7.2	242	9.7	257	12.1	272	14.6	287	17.1
7	212	19.8	227	22.2	242	24.7	257	27.2	272	29.6	287	32.1
8	212	34.8	227	37.3	242	39.8	257	42.2	272	44.7	287	47.1
9	212	49.9	227	52.3	242	54.8	257	57.3	272	59.7	288	2.2
10	213	4.9	228	7.4	243	9.8	258	12.3	273	14.8	288	17.2
11	213	19.9	228	22.4	243	24.9	258	27.3	273	29.8	288	32.3
12	213	35.0	228	37.5	243	39.9	258	42.4	273	44.8	288	47.3
13	213	50.0	228	52.5	243	55.0	258	57.4	273	59.9	289	2.4
14	214	5.1	229	7.5	244	10.0	259	12.5	274	14.9	289	17.4
15	214	20.1	229	22.6	244	25.0	259	27.5	274	30.0	289	32.4
16	214	35.2	229	37.6	244	40.1	259	42.5	274	45.0	289	47.5
17	214	50.2	229	52.7	244	55.1	259	57.6	275	0.0	290	2.5
18	215	5.2	230	7.7	245	10.2	260	12.6	275	15.1	290	17.6
19	215	20.3	230	22.7	245	25.2	260	27.7	275	30.1	290	32.6
20	215	35.3	230	37.8	245	40.2	260	42.7	275	45.2	290	47.6
21	215	50.4	230	52.8	245	55.3	260	57.8	276	0.2	291	2.7
22	216	5.4	231	7.9	246	10.3	261	12.8	276	15.3	291	17.7
23	216	20.4	231	22.9	246	25.4	261	27.8	276	30.3	291	32.8
24	216	35.5	231	37.9	246	40.4	261	42.9	276	45.3	291	47.8
25	216	50.5	231	53.0	246	55.4	261	57.9	277	0.4	292	2.8
26	217	5.6	232	8.0	247	10.5	262	13.0	277	15.4	292	17.9
27	217	20.6	232	23.1	247	25.5	262	28.0	277	30.5	292	32.9
28	217	35.6	232	38.1	247	40.6	262	43.0	277	45.5	292	48.0
29	217	50.7	232	53.2	247	55.6	262	58.1	278	0.5	293	3.0
30	218	5.7	233	8.2	248	10.7	263	13.1	278	15.6	293	18.0
31	218	20.8	233	23.2	248	25.7	263	28.2	278	30.6	293	33.1
32	218	35.8	233	38.3	248	40.7	263	43.2	278	45.7	293	48.1
33	218	50.8	233	53.3	248	55.8	263	58.2	279	0.7	294	3.2
34	219	5.9	234	8.4	249	10.8	264	13.3	279	15.7	294	18.2
35	219	20.9	234	23.4	249	25.9	264	28.3	279	30.8	294	33.3
36	219	36.0	234	38.4	249	40.9	264	43.4	279	45.8	294	48.3
37	219	51.0	234	53.5	249	55.9	264	58.4	280	0.9	295	3.3
38	220	6.1	235	8.5	250	11.0	265	13.4	280	15.9	295	18.4
39	220	21.1	235	23.6	250	26.0	265	28.5	280	31.0	295	33.4
40	220	36.1	235	38.6	250	41.1	265	43.5	280	46.0	295	48.5
41	220	51.2	235	53.6	250	56.1	265	58.6	281	1.0	296	3.5
42	221	6.2	236	8.7	251	11.1	266	13.6	281	16.1	296	18.5
43	221	21.3	236	23.7	251	26.2	266	28.7	281	31.1	296	33.6
44	221	36.3	236	38.8	251	41.2	266	43.7	281	46.2	296	48.6
45	221	51.3	236	53.8	251	56.3	266	58.7	282	1.2	297	3.7
46	222	6.4	237	8.8	252	11.3	267	13.8	282	16.2	297	18.7
47	222	21.4	237	23.9	252	26.4	267	28.8	282	31.3	297	33.7
48	222	36.5	237	38.9	252	41.4	267	43.9	282	46.3	297	48.8
49	222	51.5	237	54.0	252	56.4	267	58.9	283	1.4	298	3.8
50	223	6.5	238	9.0	253	11.5	268	13.9	283	16.4	298	18.9
51	223	21.6	238	24.1	253	26.5	268	29.0	283	31.4	298	33.9
52	223	36.6	238	39.1	253	41.6	268	44.0	283	46.5	298	48.9
53	223	51.7	238	54.1	253	56.6	268	59.1	284	1.5	299	4.0
54	224	6.7	239	9.2	254	11.6	269	14.1	284	16.6	299	19.0
55	224	21.8	239	24.2	254	26.7	269	29.1	284	31.6	299	34.1
56	224	36.8	239	39.3	254	41.7	269	44.2	284	46.6	299	49.1
57	224	51.8	239	54.3	254	56.8	269	59.2	285	1.7	300	4.2
58	225	6.9	240	9.3	255	11.8	270	14.3	285	16.7	300	19.2
59	225	21.9	240	24.4	255	26.8	270	29.3	285	31.8	300	34.2
60	225	37.0	240	39.4	255	41.9	270	44.4	285	46.8	300	49.3

493

TABLE 11.8 *(Continued)*

	Hours of Greenwich Civil Time								
Min.	20ʰ °	′	21ʰ °	′	22ʰ °	′	23ʰ °	′	Sec.	Corr. ′
0	300	49.3	315	51.7	330	54.2	345	56.7	0	0.0
1	301	4.3	316	6.8	331	9.3	346	11.7	1	0.3
2	301	19.4	316	21.8	331	24.3	346	26.8	2	0.5
3	301	34.4	316	36.9	331	39.3	346	41.8	3	0.8
4	301	49.4	316	51.9	331	54.4	346	56.8	4	1.0
5	302	4.5	317	7.0	332	9.4	347	11.9	5	1.3
6	302	19.5	317	22.0	332	24.5	347	26.9	6	1.5
7	302	34.6	317	37.0	332	39.5	347	42.0	7	1.8
8	302	49.6	317	52.1	332	54.5	347	57.0	8	2.0
9	303	4.7	318	7.1	333	9.6	348	12.0	9	2.3
10	303	19.7	318	22.2	333	24.6	348	27.1	10	2.5
11	303	34.7	318	37.2	333	39.7	348	42.1	11	2.8
12	303	49.8	318	52.2	333	54.7	348	57.2	12	3.0
13	304	4.8	319	7.3	334	9.7	349	12.2	13	3.3
14	304	19.9	319	22.3	334	24.8	349	27.2	14	3.5
15	304	34.9	319	37.4	334	39.8	349	42.3	15	3.8
16	304	49.9	319	52.4	334	54.9	349	57.3	16	4.0
17	305	5.0	320	7.4	335	9.9	350	12.4	17	4.3
18	305	20.0	320	22.5	335	24.9	350	27.4	18	4.5
19	305	35.1	320	37.5	335	40.0	350	42.5	19	4.8
20	305	50.1	320	52.6	335	55.0	350	57.5	20	5.0
21	306	5.1	321	7.6	336	10.1	351	12.5	21	5.3
22	306	20.2	321	22.6	336	25.1	351	27.6	22	5.5
23	306	35.2	321	37.7	336	40.2	351	42.6	23	5.8
24	306	50.3	321	52.7	336	55.2	351	57.7	24	6.0
25	307	5.3	322	7.8	337	10.2	352	12.7	25	6.3
26	307	20.3	322	22.8	337	25.3	352	27.7	26	6.5
27	307	35.4	322	37.9	337	40.3	352	42.8	27	6.8
28	307	50.4	322	52.9	337	55.4	352	57.8	28	7.0
29	308	5.5	323	7.9	338	10.4	353	12.9	29	7.3
30	308	20.5	323	23.0	338	25.4	353	27.9	30	7.5
31	308	35.6	323	38.0	338	40.5	353	42.9	31	7.8
32	308	50.6	323	53.1	338	55.5	353	58.0	32	8.0
33	309	5.6	324	8.1	339	10.6	354	13.0	33	8.3
34	309	20.7	324	23.1	339	25.6	354	28.1	34	8.5
35	309	35.7	324	38.2	339	40.6	354	43.1	35	8.8
36	309	50.8	324	53.2	339	55.7	354	58.2	36	9.0
37	310	5.3	325	8.3	340	10.7	355	13.2	37	9.3
38	310	20.8	325	23.3	340	25.8	355	28.2	38	9.5
39	310	35.9	325	38.3	340	40.8	355	43.3	39	9.8
40	310	50.9	325	53.4	340	55.8	355	58.3	40	10.0
41	311	6.0	326	8.4	341	10.9	356	13.4	41	10.3
42	311	21.0	326	23.5	341	25.9	356	28.4	42	10.5
43	311	36.0	326	38.5	341	41.0	356	43.4	43	10.8
44	311	51.1	326	53.6	341	56.0	356	58.5	44	11.0
45	312	6.1	327	8.6	342	11.1	357	13.5	45	11.3
46	312	21.2	327	23.6	342	26.1	357	28.6	46	11.5
47	312	36.2	327	38.7	342	41.1	357	43.6	47	11.8
48	312	51.3	327	53.7	342	56.2	367	58.6	48	12.0
49	313	6.3	328	8.8	343	11.2	358	13.7	49	12.3
50	313	21.3	328	23.8	343	26.3	358	28.7	50	12.5
51	313	36.4	328	38.8	343	41.3	358	43.3	51	12.8
52	313	51.4	328	53.9	343	56.3	358	58.8	52	13.0
53	314	6.5	329	8.9	344	11.4	359	13.8	53	13.3
54	314	21.5	329	24.0	344	26.4	359	28.9	54	13.5
55	314	36.5	329	39.0	344	41.5	359	43.9	55	13.8
56	314	51.6	329	54.0	344	56.5	359	59.0	56	14.0
57	315	6.6	330	9.1	345	11.5	0	14.0	57	14.3
58	315	21.7	330	24.1	345	26.6	0	29.1	58	14.5
59	315	36.7	330	39.2	345	41.6	0	44.1	59	14.8
60	315	51.7	330	54.2	345	56.7	0	59.1	60	15.0

Courtesy of Keuffel and Esser Co., Morristown, N.J.

TABLE 11.9 *Polar Distance of Polaris, 1983 for 0ʰ Universal Time or Greenwich Civil Time*

1983		Polar Distance Angle ° ′	Cotan	1983		Polar Distance Angle ° ′	Cotan
Jan.	1	0 48.55	70.80	July	10	0 49.02	70.12
	11	0 48.52	70.85		20	0 49.02	70.12
	21	0 48.50	70.88		30	0 49.00	70.15
	31	0 48.50	70.88				
				Aug.	9	0 48.98	70.18
Feb.	10	0 48.50	70.88		19	0 48.96	70.21
	20	0 48.52	70.85		29	0 48.92	70.27
Mar.	2	0 48.54	70.82	Sep.	8	0 48.88	70.33
	12	0 48.57	70.77		18	0 48.83	70.40
	22	0 48.62	70.70		28	0 48.78	70.47
Apr.	1	0 48.66	70.64	Oct.	8	0 48.72	70.56
	11	0 48.71	70.57		18	0 48.65	70.66
	21	0 48.76	70.50		28	0 48.59	70.75
May	1	0 48.81	70.43	Nov.	7	0 48.53	70.83
	11	0 48.86	70.35		17	0 48.47	70.92
	21	0 48.90	70.30		27	0 48.40	71.02
	31	0 48.94	70.24				
				Dec.	7	0 48.35	71.10
June	10	0 48.98	70.18		17	0 48.30	71.17
	20	0 49.00	70.15		27	0 48.26	71.23
	30	0 49.01	70.14				

Declination = 90° − polar distance
Courtesy of Keuffel and Esser, Morristown, NJ.

TABLE 11.10   *Corrections to Be Applied to Latitude to Obtain the True Altitude of Polaris, 1983*

t	Corr.	t	Corr.	t	Corr.	t	Corr.
°	′	°	′	°	′	°	′
000	+48.5	045	+34.1	090	−00.3	135	−34.5
001	+48.5	046	+33.5	091	−01.2	136	−35.1
002	+48.5	047	+32.9	092	−02.0	137	−35.6
003	+48.4	048	+32.3	093	−02.9	138	−36.2
004	+48.4	049	+31.6	094	−03.7	139	−36.7
005	+48.3	050	+31.0	095	−04.6	140	−37.3
006	+48.2	051	+30.3	096	−05.4	141	−37.8
007	+48.1	052	+29.6	097	−06.2	142	−38.3
008	+48.0	053	+29.0	098	−07.1	143	−38.9
009	+47.9	054	+28.3	099	−07.9	144	−39.4
010	+47.8	055	+27.6	100	−08.8	145	−39.8
011	+47.6	056	+26.9	101	−09.6	146	−40.3
012	+47.4	057	+26.2	102	−10.4	147	−40.8
013	+47.2	058	+25.5	103	−11.2	148	−41.2
014	+47.0	059	+24.7	104	−12.1	149	−41.7
015	+46.8	060	+24.0	105	−12.9	150	−42.1
016	+46.6	061	+23.2	106	−13.7	151	−42.5
017	+46.4	062	+22.5	107	−14.5	152	−42.9
018	+46.1	063	+27.1	108	−15.3	153	−43.3
019	+45.8	064	+21.0	109	−16.1	154	−43.7
020	+45.5	065	+20.2	110	−16.9	155	−44.0
021	+45.2	066	+19.4	111	−17.7	156	−44.4
022	+44.9	067	+18.7	112	−18.5	157	−44.7
023	+44.6	068	+17.9	113	−19.2	158	−45.0
024	+44.2	069	+17.1	114	−20.0	159	−45.3
025	+43.9	070	+16.3	115	−20.8	160	−45.6
026	+43.5	071	+15.5	116	−21.5	161	−45.9
027	+43.1	072	+14.7	117	−22.3	162	−46.2
028	+42.7	073	+13.9	118	−23.0	163	−46.4
029	+42.3	074	+13.1	119	−23.8	164	−46.6
030	+41.9	075	+12.2	120	−24.5	165	−46.9
031	+41.5	076	+11.4	121	−25.2	166	−47.1
032	+41.0	077	+10.6	122	−25.9	167	−47.3
033	+40.6	078	+09.8	123	−26.7	168	−47.5
034	+40.1	079	+08.9	124	−27.4	169	−47.6
035	+39.6	080	+08.1	125	−28.0	170	−47.8
036	+39.1	081	+07.3	126	−28.7	171	−47.9
037	+38.6	082	+06.4	127	−29.4	172	−48.0
038	+38.1	083	+05.6	128	−30.1	173	−48.1
039	+37.6	084	+04.7	129	−30.7	174	−48.2
040	+37.0	085	+03.9	130	−31.4	175	−48.3
041	+36.5	086	+03.0	131	−32.0	176	−48.4
042	+35.9	087	+02.2	132	−32.6	177	−48.4
043	+35.3	088	+01.4	133	−33.3	178	−48.5
044	+34.7	089	+00.5	134	−33.9	179	−48.5
045	+34.1	090	−00.3	135	−34.5	180	−48.5

This table is computed for latitude 45° 00′00″N and declination of 89° 11′30″N.
Courtesy of Keuffel and Esser Co., Morristown, NJ.

TABLE 11.11   *Bearing of Polaris at All Local Hour Angles, 1983*

Lat.	10°	20°	26°	30°	32°	34°	36°	38°	Lat.
LHA	° ′	° ′	° ′	° ′	° ′	° ′	° ′	° ′	LHA
0	0 00.0	0 00.0	0 00.0	0 00.0	0 00.0	0 00.0	0 00.0	0 00.0	360
5	0 04.3	0 04.5	0 04.7	0 04.9	0 05.0	0 05.1	0 05.3	0 05.4	355
10	0 08.6	0 09.0	0 09.4	0 09.8	0 10.0	0 10.3	0 10.5	0 10.8	350
15	0 12.8	0 13.4	0 14.1	0 14.6	0 14.9	0 15.3	0 15.7	0 16.1	345
20	0 16.9	0 17.7	0 18.6	0 19.3	0 19.7	0 20.2	0 20.7	0 21.3	340
25	0 20.9	0 21.9	0 22.9	0 23.8	0 24.4	0 24.9	0 25.6	0 26.3	335
30	0 24.7	0 25.9	0 27.1	0 28.2	0 28.8	0 29.5	0 30.2	0 31.1	330
35	0 28.3	0 29.7	0 31.1	0 32.3	0 33.0	0 33.8	0 34.7	0 35.6	325
40	0 31.7	0 33.3	0 34.9	0 36.2	0 37.0	0 37.9	0 38.8	0 39.9	320
45	0 34.9	0 36.6	0 38.3	0 39.8	0 40.7	0 41.6	0 42.7	0 43.9	315
50	0 37.8	0 39.7	0 41.5	0 43.1	0 44.1	0 45.1	0 46.2	0 47.5	310
55	0 40.4	0 42.4	0 44.4	0 46.1	0 47.1	0 48.2	0 49.4	0 50.7	305
60	0 42.7	0 44.8	0 46.9	0 48.7	0 49.7	0 50.9	0 52.2	0 53.6	300
65	0 44.7	0 46.9	0 49.0	0 50.9	0 52.0	0 53.2	0 54.6	0 56.0	295
70	0 46.3	0 48.6	0 50.8	0 52.8	0 53.9	0 55.2	0 56.5	0 58.1	290
75	0 47.6	0 49.9	0 52.2	0 54.2	0 55.4	0 56.6	0 58.1	0 59.6	285
80	0 48.5	0 50.9	0 53.2	0 55.2	0 56.4	0 57.7	0 59.1	1 00.7	280
85	0 49.1	0 51.4	0 53.8	0 55.8	0 57.0	0 58.3	0 59.8	1 01.4	275
90	0 49.2	0 51.6	0 54.0	0 56.0	0 57.2	0 58.5	0 59.9	1 01.5	270
95	0 49.1	0 51.4	0 53.7	0 55.7	0 56.9	0 58.2	0 59.7	1 01.3	265
100	0 48.5	0 50.8	0 53.1	0 55.1	0 56.2	0 57.5	0 58.9	1 00.5	260
105	0 47.5	0 49.8	0 52.0	0 54.0	0 55.1	0 56.4	0 57.8	0 59.3	255
110	0 46.2	0 48.4	0 50.6	0 52.5	0 53.6	0 54.8	0 56.1	0 57.6	250
115	0 44.6	0 46.7	0 48.8	0 50.6	0 51.6	0 52.8	0 54.1	0 55.5	245
120	0 42.6	0 44.6	0 46.6	0 48.3	0 49.3	0 50.4	0 51.7	0 53.0	240
125	0 40.3	0 42.2	0 44.0	0 45.7	0 46.6	0 47.7	0 48.8	0 50.1	235
130	0 37.7	0 39.4	0 41.2	0 42.7	0 43.6	0 44.5	0 45.6	0 46.8	230
135	0 34.8	0 36.4	0 38.0	0 39.4	0 40.2	0 41.1	0 42.1	0 43.2	225
140	0 31.6	0 33.0	0 34.5	0 35.8	0 36.5	0 37.3	0 38.2	0 39.2	220
145	0 28.2	0 29.5	0 30.8	0 31.9	0 32.6	0 33.3	0 34.1	0 35.0	215
150	0 24.6	0 25.7	0 26.8	0 27.8	0 28.4	0 29.0	0 29.7	0 30.5	210
155	0 20.8	0 21.7	0 22.7	0 23.5	0 24.0	0 24.5	0 25.1	0 25.8	205
160	0 16.8	0 17.6	0 18.3	0 19.0	0 19.4	0 19.8	0 20.3	0 20.8	200
165	0 12.7	0 13.3	0 13.9	0 14.4	0 14.7	0 15.0	0 15.4	0 15.8	195
170	0 08.5	0 08.9	0 09.3	0 09.6	0 09.8	0 10.1	0 10.3	0 10.6	190
175	0 04.3	0 04.5	0 04.7	0 04.8	0 04.9	0 05.1	0 05.2	0 05.3	185
180	0 00.0	0 00.0	0 00.0	0 00.0	0 00.0	0 00.0	0 00.0	0 00.0	180

TABLE 11.11   *(Continued)*

Lat.	40°	42°	44°	46°	48°	50°	60°	70°	Lat.
LHA	° ′	° ′	° ′	° ′	° ′	° ′	° ′	° ′	LHA
0	0 00.0	0 00.0	0 00.0	0 00.0	0 00.0	0 00.0	0 00.0	0 00.0	360
5	0 05.6	0 05.8	0 06.0	0 06.2	0 06.4	0 06.7	0 08.7	0 12.9	355
10	0 11.1	0 11.5	0 11.9	0 12.3	0 12.8	0 13.3	0 17.3	0 25.6	350
15	0 16.6	0 17.1	0 17.7	0 18.3	0 19.0	0 19.9	0 25.7	0 38.1	345
20	0 21.9	0 22.6	0 23.4	0 24.2	0 25.2	0 26.2	0 34.0	0 50.3	340
25	0 27.0	0 27.9	0 28.9	0 29.9	0 31.1	0 32.4	0 41.9	1 02.1	335
30	0 32.0	0 33.0	0 34.1	0 35.4	0 36.7	0 38.3	0 49.5	1 13.4	330
35	0 36.7	0 37.8	0 39.1	0 40.5	0 42.1	0 43.9	0 56.8	1 24.0	325
40	0 41.1	0 42.4	0 43.8	0 45.4	0 47.2	0 49.1	1 03.5	1 33.9	320
45	0 45.1	0 46.6	0 48.1	0 49.9	0 51.8	0 54.0	1 09.8	1 43.1	315
50	0 48.9	0 50.4	0 52.1	0 54.0	0 56.1	0 58.4	1 15.5	1 51.4	310
55	0 52.2	0 53.9	0 55.7	0 57.7	0 59.9	1 02.4	1 20.6	1 58.8	305
60	0 55.2	0 56.9	0 58.8	1 00.9	1 03.3	1 05.9	1 25.0	2 05.2	300
65	0 57.7	0 59.5	1 01.5	1 03.7	1 06.1	1 08.9	1 28.8	2 10.6	295
70	0 59.7	1 01.6	1 03.7	1 05.9	1 08.5	1 11.3	1 31.9	2 15.0	290
75	1 01.3	1 03.2	1 05.4	1 07.7	1 10.3	1 13.2	1 34.3	2 18.3	285
80	1 02.5	1 04.4	1 06.6	1 08.9	1 11.6	1 14.5	1 35.9	2 20.5	280
85	1 03.1	1 05.1	1 07.2	1 09.6	1 12.3	1 15.3	1 36.8	2 21.7	275
90	1 03.3	1 05.3	1 07.4	1 09.8	1 12.5	1 15.4	1 37.0	2 21.7	270
95	1 03.0	1 04.9	1 07.1	1 09.5	1 12.1	1 15.0	1 36.4	2 20.7	265
100	1 02.2	1 04.1	1 06.2	1 08.6	1 11.2	1 14.1	1 35.1	2 18.7	260
105	1 01.0	1 02.8	1 04.9	1 07.2	1 09.7	1 12.6	1 33.1	2 15.6	255
110	0 59.3	1 01.1	1 03.1	1 05.3	1 07.7	1 10.5	1 30.4	2 11.5	250
115	0 57.1	0 58.8	1 00.8	1 02.9	1 05.3	1 07.9	1 27.0	2 06.4	245
120	0 54.5	0 56.2	0 58.0	1 00.0	1 02.3	1 04.8	1 23.0	2 00.4	240
125	0 51.5	0 53.1	0 54.8	0 56.7	0 58.8	1 01.2	1 18.4	1 53.6	235
130	0 48.1	0 49.6	0 51.2	0 53.0	0 55.0	0 57.2	1 13.2	1 46.0	230
135	0 44.4	0 45.7	0 47.2	0 48.9	0 50.7	0 52.7	1 07.4	1 37.6	225
140	0 40.3	0 41.5	0 42.9	0 44.4	0 46.0	0 47.9	1 01.2	1 28.5	220
145	0 36.0	0 37.0	0 38.2	0 36.9	0 41.0	0 42.7	0 54.5	1 18.8	215
150	0 31.3	0 32.3	0 33.3	0 34.5	0 35.8	0 37.2	0 47.5	1 08.6	210
155	0 26.5	0 27.3	0 28.1	0 29.1	0 30.2	0 31.4	0 40.1	0 57.9	205
160	0 21.4	0 22.1	0 22.8	0 23.6	0 24.4	0 25.4	0 32.4	0 46.8	200
165	0 16.2	0 16.7	0 17.2	0 17.8	0 18.5	0 19.2	0 24.5	0 35.4	195
170	0 10.9	0 11.2	0 11.6	0 12.0	0 12.4	0 12.9	0 16.4	0 23.7	190
175	0 05.5	0 05.6	0 05.8	0 06.0	0 06.2	0 06.5	0 08.3	0 11.9	185
180	0 00.0	0 00.0	0 00.0	0 00.0	0 00.0	0 00.0	0 00.0	0 00.0	180

Courtesy of Keuffel and Esser Co., Morristown, NJ.

498

TABLE 11.12 *Correction to Bearing of Polaris for Other Polar Distances, 1983*

Bearing	0°	0°20'	0°40'	1°00'	1°20'	1°40'	2°00'	2°20'	2°40'
Polar Dist.									
° '	'	'	'	'	'	'	'	'	'
0 49.00	0.0	+0.2	+0.4	+0.6	+0.8	+1.0	+1.2	+1.4	+1.6
0 48.90	0.0	+0.2	+0.3	+0.5	+0.7	+0.8	+1.0	+1.2	+1.3
0 48.80	0.0	+0.1	+0.2	+0.4	+0.5	+0.6	+0.7	+0.9	+1.0
0 48.70	0.0	+0.1	+0.2	+0.2	+0.3	+0.4	+0.5	+0.6	+0.7
0 48.60	0.0	+0.0	+0.1	+0.1	+0.2	+0.2	+0.2	+0.3	+0.3
0 48.50	0.0	0.0	0.0	0.0	0.0	0.0	0.0	0.0	0.0
0 48.40	0.0	−0.0	−0.1	−0.1	−0.2	−0.2	−0.2	−0.3	−0.3
0 48.30	0.0	−0.1	−0.2	−0.2	0.3	−0.4	−0.5	−0.6	−0.7
0 48.20	0.0	−0.1	−0.2	−0.4	−0.5	−0.6	−0.7	−0.9	−1.0
0 48.10	0.0	−0.2	−0.3	−0.5	−0.7	−0.8	−1.0	−1.2	1.3
0 48.00	0.0	−0.2	−0.4	−0.6	0.8	−1.0	−1.2	−1.4	−1.6

Courtesy of Keuffel and Esser Co., Morristown, N.J.

theodolites (usually 6 or 20 second theodolites). Figure 11.46 shows a gyro attachment mounted on a 20-second theodolite.

The gyro attachment (also called a gyrocompass) consists of a wire-hung pendulum supporting a high-speed, perfectly balanced gyro motor capable of attaining the required speed of 12000 rpm

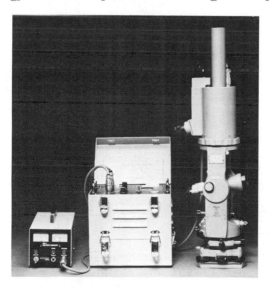

FIGURE 11.46 Gyro attachment, mounted on a 20-second theodolite. Shown with battery charger and control unit. (Courtesy Sokkisha Co. Ltd.)

in 1 min. Basically, the rotation of the earth affects the orientation of the spin axis of the gyroscope such that the gyroscope spin axis orients itself toward the pole in an oscillating motion that is observed and measured in a plane perpendicular to the pendulum. This north-seeking oscillation, which is known as *precession*, is measured on the horizontal circle of the theodolite; extreme left (west) and right (east) readings are averaged to arrive at the meridian direction.

The theodolite with gyro attachment is set up and oriented approximately to north, using a compass; the gyro motor is engaged until the proper angular velocity has been reached (about 12000 rpm for the instrument shown in Figure 11.46, for about 22000 rpm for the instrument shown in Figure 11.47, and then the gyroscope is released. The precession oscillations are observed through the gyro attachment viewing eyepiece, and the theodolite is adjusted closer to the northerly direction if necessary. When the theodolite is pointed to within a few minutes of north, the extreme precession positions (west and east) are noted in the viewing eyepiece and then recorded on the horizontal circle; as noted earlier, the position of the meridian is the value of the averaged precession readings. This technique, which takes about a half-hour to complete, is accurate to within 20 seconds of azimuth.

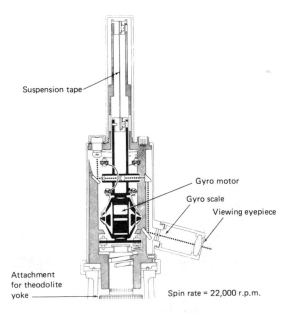

FIGURE 11.47 Cross section of a theodolite gyro attachment, Wild GAK1 gyro attachment. (Courtesy Wild Heerbrugg Co. Ltd.)

These instruments can be well utilized in mining and tunneling surveys for azimuth determination; more precise (3 to 5 seconds) gyro theodolites can be used in the extension of surface control surveys.

PROBLEMS

Problems 11–1 to 11–5 utilize the following control point data. Given the following urban control monument data:

Monument	Elevation	Northing	Easting
A	179.832	4,850 296.103	317 104.062
B	181.356	4,850 218.330	316 823.936
C	188.976	4,850 182.348	316 600.889
D	187.452	4,850 184.986	316 806.910

- Longitude @ monument B = 79° 21′00″ West
- Average latitude @ 43°47′30″
- Central meridian (CM) at longitude 79°30′ west; scale factor = 0.9999; easting at CM = 304 800 m; northing at equator = 0.000 m.

11–1 Determine the grid distances and grid bearings of sides AB, BC, CD and DA.

11–2 From the grid bearings computed in the Problem 11–1, compute the interior angles (and their sum) of the traverse A, B, C, D, A thus verifying the correctness of the grid bearing computations.

11–3 Determine the ground distances for the four traverse sides by applying the scale and elevation factors (i.e., grid factors).

11–4 Determine the convergence correction for each traverse side and determine the geodetic bearings for each traverse side.

11–5 From the geodetic bearings computed in Problem 11–4, compute the interior angles (and their sum) of the traverse A, B, C, D, A–thus verifying the correctness of the geodetic bearing computations.

11–6 A star observation was taken on October 10/83 at 7:20:13 p.m. (mean) in the Toronto, Ontario area; latitude = 43°47′30″N., longitude = 79°21′00″W. An angle of 43°43′38″ (mean) was measured right from Polaris to station #17 while occupying station #16. Compute the azimuth of line #16–#17.

11–7   A star observation was taken on April 4, 1983 at 9:27:30 p.m. (mean) in the Santa Barbara, Ca. area; latitude = 34°29′30″N, longitude = 119°40′00″W. An angle of 61°32′20″ (mean) was turned left from Polaris to monument 331 while occupying monument 332. Compute the azimuth of line 332–331.

12 Construction Surveys

12.1 GENERAL

Construction surveys provide the horizontal and vertical layout for every key component of a construction project. This provision of "line and grade" can only be accomplished by experienced surveyors familiar with both the related project design and the appropriate construction techniques. A knowledge of related design is essential to interpret effectively the design drawings for layout purposes, and a knowlege of construction techniques is required to ensure that the layout is optimal for both line and grade transfer and construction scheduling.

12.2 ACCURACY AND MISTAKES

The elimination of mistakes and the achievement of required accuracy have been stressed in this text. In no area of surveying are these qualities more important than in construction surveying. All field measurements and calculations are suspect until they have been verified by independent means or by repeated checks. Mistakes have been known to escape detection in as many as three independent, conscientious checks by experienced personnel.

Unlike other forms of surveying, construction surveying is often associated with speed of operation. Once a contract has been awarded, the contractor may wish to commence construction immediately as he will probably have definite future commitments for his men and equipment. He will not accept delay. A hurried surveyor is more likely to make mistakes in measurements and calculations, and thus even more vigilance than normal is required. Construction surveying is not an occupation for the faint of heart; the responsibilities are great, and the working conditions often less than ideal. However, the sense of achievement possible can be very rewarding.

12.3 CONSTRUCTION CONTROL

Depending on the size and complexity of the project, the survey crew should arrive on site one day or several days prior to the commencement of construction. The first on-site job for the construction surveyor is to relocate the horizontal and vertical control used in the preliminary survey (see Chapter 7). Usually, a number of months, and sometimes even years, have passed between the preliminary survey, the project design based on the preliminary survey, and the budget decision to award a contract for construction.

It may be necessary to reestablish the horizontal and vertical control in the area of proposed construction. If this is the case, extreme caution is advised, as the design plans are based on the original survey fabric, and any deviation from the original control could well lead to serious problems in construction. If the original control (or most of it) still exists in the field, it is customary to check and verify all linear and angular dimensions that could directly affect the project.

The same rigorous approach is required for vertical control. If local bench marks have been destroyed (as is often the case), the bench marks must be reestablished accurately. Key existing elevations shown on design drawings (e.g., connecting invert elevations for gravity-flow sewers, or connecting beam seat elevations on concrete structures) must be resurveyed to ensure that (1) the original elevation shown on the plan was correct, and (2) the new and original vertical control are, in fact, both referenced to the same vertical datum. In all these areas, absolutely nothing is taken for granted.

Extension of Control

Once the original control has been reestablished or verified, the control must be extended over the construction site to suit the purposes of each specific project. This operation and the giving of "line and grade" are discussed in detail for most types of projects in subsequent sections.

12.4 MEASUREMENT FOR INTERIM AND FINAL PAYMENTS

On most construction projects, partial payments are made to the contractor at regular time intervals, as well as a final payment upon completion and acceptance of the project. The payments are based on data supplied by the project inspector and the construction surveyor. The project inspector records items, such as daily progress, manpower and equipment in use, and materials used, whereas the construction surveyor records items that require a surveying function (e.g., excavation quantities, concrete placed in structures, placement of sod).

The discussion in Chapter 1 covered field book layout and stressed the importance of diaries and thorough note taking. These functions are important in all survey work but especially so in construction surveying. Since a great deal of money can depend on the integrity of daily notes and records, it is essential that the construction surveyor's notes and diaries be complete and accurate with respect to dates, times, locations, quantities, method of measurement, personnel, design changes, and so on.

12.5 FINAL MEASUREMENTS FOR AS-BUILT DRAWINGS

Upon completion of a project, it is essential that a final plan be drawn showing the actual details of construction. The final plan, known as the *as-built drawing*, is usually quite similar to the design plan, with the exception that revisions are made to reflect changes in design that invariably occur during the construction process. Design changes result from problems that become apparent only after construction is underway. It is difficult, especially on complex projects, to plan for every eventuality that may be encountered; however, if the preliminary surveyor and the designer have both done their jobs well, the design plan and the as-built plan are usually quite similar.

12.6 MUNICIPAL ROADS CONSTRUCTION

12.6.1 Classification of Roads

The plan shown on Figure 12.1 depicts a typical municipal road pattern. The local roads shown have the primary purpose of providing access to individual residential lots. The collector roads (both major and minor) provide the dual service of lot access and traffic movement. The collector roads connect the local roads to arterial roads; the main purpose of the arterial roads is to provide a relatively high level of traffic movement service.

Municipal works engineers base their road design on the level of service to be provided. The proposed cross sections and geometric alignments vary in complexity and cost from the fundamental local roads to the more complex arterials. The highest level of service is given by the freeways, which provide high-velocity, high-volume routes with limited access (interchanges only), ensuring continous traffic flow when design conditions prevail.

12.6.2 Road Allowances

The road allowance varies in width from 40 ft (12 m) for small locals to 120 ft (35 m) for major arterials. In parts of North America, including most of Canada, the local road allowances originally were 66 ft wide (one Gunter's chain), and when widening was required due to increased traffic volumes, it was common to take 10-ft widenings on each side, initially resulting in an 86-ft road allowance for major collectors and minor arterials. Further widenings left major arterials at 100- and 120-ft widths. In Canada, new subdivision registrations now require metric dimensions, the local road allowance now being 20 m wide, and the other road allowance classes being changed accordingly.

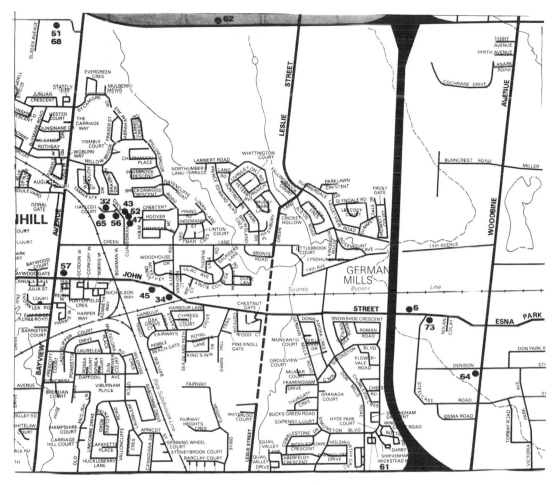

FIGURE 12.1 Municipal road pattern. (Courtesy Fine Line Graphic and Cartographic Services)

12.6.3 Road Cross Sections

A full service municipal road allowance will usually have asphalt pavement, curbs, storm and sanitary sewers, water distribution pipes, hydrants, catch basins, and sidewalks. Additional utilities such as natural gas pipelines, electrical supply cables, and cable TV are also often located on the road allowance. The essential differences between local cross sections and arterial cross sections are the widths of pavement and the quality and depths of pavement materials. The construction layout of sewers and pipelines is covered in subsequent sections. See Figure 12.2 for a typical municipal road cross section.

The cross fall (height of crown) used on the pavement varies from one municipality to another, but is usually close to a 2% slope. The curb face is often 6 in. (150 mm) high except at driveways and crosswalks, where the height is restricted to about 2 in. (50 mm) for vehicle and pedestrian access. The slope on the boulevard from the curb to the streetline usually rises at a 2% minimum slope, thus ensuring that roadway storm drainage does not run onto private property.

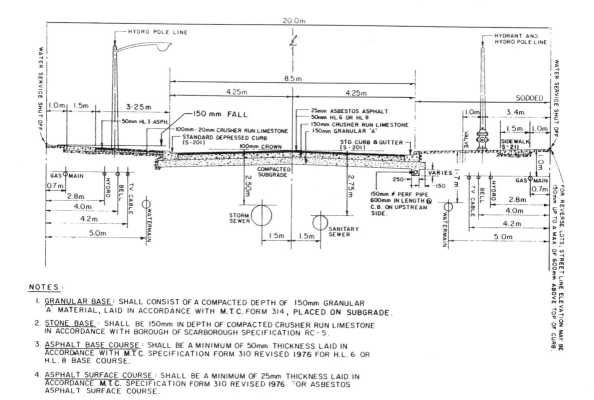

NOTES:

1. GRANULAR BASE: SHALL CONSIST OF A COMPACTED DEPTH OF 150mm GRANULAR 'A' MATERIAL, LAID IN ACCORDANCE WITH M.T.C. FORM 314, PLACED ON SUBGRADE.

2. STONE BASE: SHALL BE 150mm IN DEPTH OF COMPACTED CRUSHER RUN LIMESTONE IN ACCORDANCE WITH BOROUGH OF SCARBOROUGH SPECIFICATION RC-5.

3. ASPHALT BASE COURSE: SHALL BE A MINIMUM OF 50mm THICKNESS LAID IN ACCORDANCE WITH M.T.C. SPECIFICATION FORM 310 REVISED 1976 FOR H.L. 6 OR H.L. 8 BASE COURSE.

4. ASPHALT SURFACE COURSE: SHALL BE A MINIMUM OF 25mm THICKNESS LAID IN ACCORDANCE M.T.C. SPECIFICATION FORM 310 REVISED 1976 FOR ASBESTOS ASPHALT SURFACE COURSE.

NOTE: THIS DRAWING TO BE READ IN CONJUNCTION WITH THE SPECIFICATIONS FOR PAVED ROADS IN SUBDIVISIONS IN SEWER AREAS, BOROUGH OF SCARBOROUGH.

FIGURE 12.2 Typical cross section of a local residential road.

12.6.4 Plan and Profile

A typical plan and profile are shown in (Figure 12.3). The plan and profile, which usually also show the cross section details and construction notes, are the "blueprint" from which the construction is accomplished. The plan and profile, together with the contract specifications, spell out in detail precisely where and how the road (in this example) is to be built.

The plan portion of the plan and profile gives the horizontal location of the facility, including curve radii, whereas the profile portion shows the key elevations and slopes along the road center line, including vertical curve information. Both the plan and profile relate all data to the project stationing established as horizontal control.

12.6.5 Establishing Center Line (℄)

Using the example of a ditched residential road being upgraded to a paved and curbed road, the first job for the construction surveyor is to reestablish the center line (℄) of the roadway. Usually this entails finding several property markers delineating street line (℄). Fence and hedge lines can be used initially to guide the surveyor to the approximate location of the property markers. When the surveyor finds one property marker, he can measure out frontage distances shown on the property plan (plat) to locate a sufficient number of additional markers. Usually, the construction surveyor has with him the notes from the preliminary survey showing the location of property markers used in the original survey. If possible, the construction surveyor will utilize the same evidence used in the preliminary survey, taking the time of course, to verify the resultant alignment. If the evidence used in the preliminary survey has been destroyed, as is often the case when a year or more elapses between the two surveys, the construction surveyor will take great care to see to it that his results are not appreciably different than those of the original survey, unless, of course, a blunder occurred on the original survey.

The property markers can be square or round iron bars (including rebars), or round iron or aluminum pipes, magnetically capped. The markers can vary from $1\frac{1}{2}$ to 4 ft in length. It is not unusual for the surveyor to have to use a shovel, as the tops of the markers are often buried. The surveyor can use a magnetic metal detector to aid in locating buried markers.

Sometimes even an exhaustive search of an area will not turn up a sufficient number of markers to establish ℄. The survey-

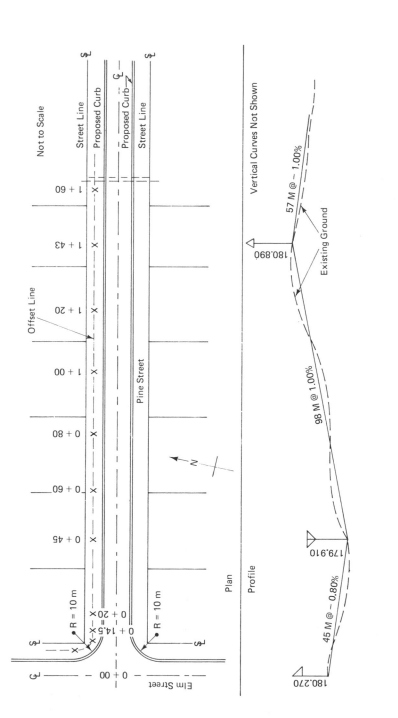

Plan

Profile

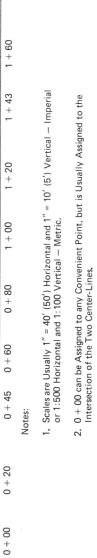

FIGURE 12.3 Plan and profile.

Notes:

1. Scales are Usually 1" = 40' (50') Horizontal and 1" = 10' (5') Vertical — Imperial or 1:500 Horizontal and 1:100 Vertical — Metric.

2. 0 + 00 can be Assigned to any Convenient Point, but is Usually Assigned to the Intersection of the Two Center-Lines.

510

or must then extend his search to adjacent blocks or backyards in order to reestablish the missing markers. Alternatively, the surveyor can approach the home owners in the affected area and inquire as to the existence of a "mortgage survey plan" (see Figure 12.4) for the specific property. Such a plan is required in most areas before a financial institution will provide mortgage financing. The mortgage survey plan shows dimensions from the building foundation to the street line and to both sidelines. Information thus gained can be used to narrow the search for a missing marker or can be used directly to establish points on the street line.

Once a number of points have been established on both sides of the roadway, the ₵ can be marked from each of these points by measuring (at right angles) half the width of the road allowance. The surveyor then sets up his transit on a ₵ mark near one of the project extremities and sights in on the ₵ marker nearest the other project extremity. He can then see if all his markers line up in a straight line (assuming tangent alignment) (see Figure 12.5). If the markers do not all line up, the surveyor will check the affected measurements; if discrepancies still occur (as often is the case), the surveyor will make the "best fit" of the available evidence. Depending on the length of roadway involved and the quantity of ₵ markers established, the number of markers lining up perfectly will vary. Three well-spaced, perfectly aligned markers is the absolute minimum number required for the establishment of ₵. The reason that all markers do not line up is that over the years most lots are resurveyed; some lots may be resurveyed a number of times. The landsurveyor's prime area of concern is that area of the plan immediately adjacent to his client's property, and he must ensure that the property stake-out is consistent both for evidence and plan intentions. Over a number of years, cumulative errors and mistakes can significantly affect the overall alignment of the ₵'s.

If, as in this example, the ₵ is being marked on an existing road; the surveyor will use nails with washers and red plastic flagging to establish his marks. The nails can be driven into gravel, ashphalt, and in some cases concrete surfaces. The washers will keep the nails from sinking below the road surface, and the red flagging will help in relocation.

If the project had involved a new curbed road in a new subdivision, the establishment of the ₵ would have been much simplified. The recently set property markers would for the most part be intact, and discrepancies between markers would be minimal (as all markers would have been set in the same comprehensive sur-

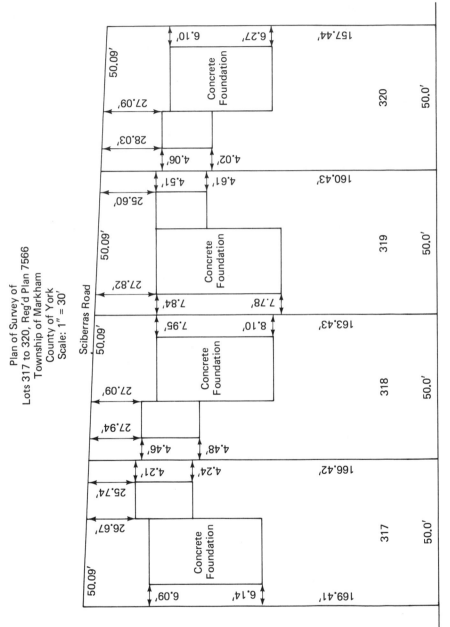

FIGURE 12.4 House survey (plat).

512

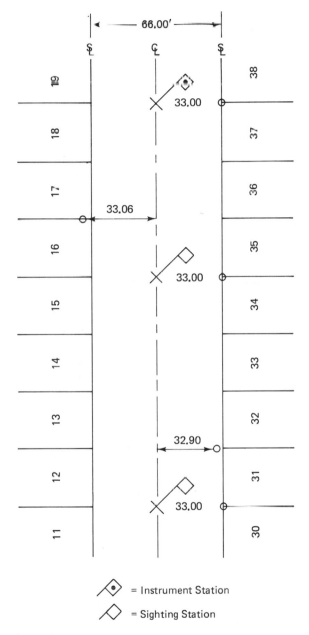

FIGURE 12.5 Property markers used to establish ℄.

vey operation). The ₵ in this case would be marked by wood stakes $2'' \times 2''$ or $2'' \times 1''$ and 18 in. long.

12.6.6 Establishing Offset Lines

In this present example, the legal fabric of the road allowance as given by the property markers constitutes the horizontal control. Construction control would consist of offset lines referenced to the *proposed curbs* with respect to line and grade. In the case of ditched roads and most highways (see Section 12.7) the offset lines are referenced to the *proposed center-line* with respect to line and grade.

The offset lines are placed as close to the proposed location of the curbs as possible. It is essential that the offset stakes not interfere with machinery and formwork; and it is also essential that the offset stakes be far enough removed so that they are not destroyed during cut or fill operations. Ideally, the offset stakes, once established, will remain in place for the duration of construction. This ideal can often be realized in municipal road construction, but it is seldom realized in highway construction due to the significant size of cuts and fills. If cuts and fills are not too large, offset lines for curbs can be 3 to 5 ft (1 to 2 m) from the proposed face of curb. An offset line this close allows for very efficient transfer of line and grade.

In the case of a ditched gravel road being upgraded to a curbed paved road, the offset line will have to be placed far enough away on the boulevard to avoid the ditch filling operation and any additional cut and fill that may be required. In the worst case, it may be necessary to place the offset line on the street line, an 18- to 25-ft offset (6 to 8 m).

12.6.7 Determining Cuts and Fills

The offset stakes (with nails or tacks for precise alignment) are usually placed at 50-ft (20-m) stations and at any critical alignment change points. The elevations of the tops of the stakes are determined by rod and level, based on the vertical control established for the project. It is then necessary to determine the proposed elevation for the top of curb at each offset station.

Referring to Figure 12.3, the reader can see that elevations and slopes have been designed for a portion of the project; the plan and profile have been simplified for illustrative purposes, and offsets are shown for one curb line only.

Given the ℄ elevation data, the construction surveyor must calculate the proposed curb elevations. He can proceed by calculating the relevant elevations on ℄ and then adjusting for crown and curb height differential, or he can apply the differential first and work out his curb elevations directly.

To determine the difference in elevation between ℄ and top of curb, the surveyor must analyze the appropriate cross section. In Figure 12.6, it can be seen that the cross fall is $4.5 \times 0.02 = 0.090$ m (90 mm). The face on the curb is 150 mm; therefore, the top of curb is 60 mm above the ℄ elevation.

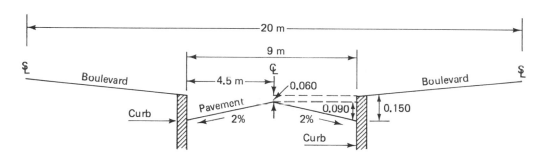

FIGURE 12.6   Cross section showing the relationship between ℄ and top of curb elevations.

A list of key stations (see Figure 12.3) is prepared and ℄ elevations at each station are calculated. The ℄ elevations are then adjusted to produce curb elevations. Since superelevation is seldom used in municipal design, it is safe to say that the curbs on both sides of the road are normally parallel in line and grade. A notable exception to this can occur when intersections of collectors and arterials are widened to allow for turn lanes.

Station	℄ Elevation		Curb Elevation
0 + 00	180.270		
	−0.116		
BC 0 + 14.5	180.154	+0.060	180.214
	−0.044		
0 + 20	180.110	+0.060	180.170
	−0.160		
0 + 40	179.950	+0.060	180.010
	−0.040		
0 + 45	179.910	+0.060	179.970
	−0.150		
0 + 60	180.060	+0.060	180.120
	+0.200		
0 + 80	180.260	+0.060	180.320
	+0.200		
1 + 00	180.460	+0.060	180.520
	+0.200		
1 + 20	180.660	+0.060	180.720
	+0.200		
1 + 40	180.860	+0.060	180.920
	+0.030		
1 + 43	180.890	+0.060	180.950
etc.			

The construction surveyor can then prepare a grade sheet (see Table 12.1), copies of which are given to the contractor and project inspector. The top of stake elevations, determined by level and rod, are assumed in this example.

TABLE 12.1   *Grade Sheet*

Station	Curb Elevation	Stake Elevation	Cut	Fill
0 + 14.5	180.214	180.325	0.111	
0 + 20	180.170	180.315	0.145	
0 + 40	180.010	180.225	0.215	
0 + 45	179.970	180.110	0.140	
0 + 60	180.120	180.185	0.065	
0 + 80	180.320	180.320	On Grade	
1 + 00	180.520	180.475		0.045
1 + 20	180.720	180.710		0.010
1 + 40	180.920	180.865		0.055
1 + 43	180.950	180.900		0.050
etc.				

The grade sheet, signed by the construction surveyor, also includes the street name, date, limits of the contract, and most importantly the offset distance to the face of the curb.

It should be noted that the construction grades (cut and fill) refer only to the vertical distance to be measured down or up from the grade stake to locate the proposed elevation. Construction grades do not define with certainty whether the contractor is in a cut or fill situation at any given point. For example (see Figure 12.7), at 0 + 20, a construction grade of cut 0.245 is given, whereas the contractor is actually in a fill situation (i.e., the proposed top of curb is *above* the existing ground at that station). This lack of correlation between construction grades and the construction process can become more pronounced as the offset distance lengthens. For example, if the grade stake at station 1 + 40 had been located at the street line, the construction grade would have been cut, whereas the construction process is almost entirely in a fill operation.

When the layout is performed using foot units, the basic station interval is 50 ft. The dimensions are recorded and calculated to the closest one hundredth (0.01) of a foot. Although all survey measurements are in feet and decimals of a foot, for the contractors purposes the final cuts and fills are often expressed in feet and inches. The decimal–inch relationships are soon committed to memory by surveyors working in the construction field (see Table 12.2). Cuts and fills are usually expressed to the closest $1/8$ in. for concrete, steel, and pipelines, and to the closest $1/4$ in. for highway, granular surfaces, and ditchlines.

The grade sheet in Table 12.3 illustrates the foot–inch relationships. The first column and the last two columns are all that is required by the contractor, in addition to the offset distance, to construct the facility properly.

TABLE 12.2 *Decimal Foot/Inch Conversion:*

$1' = 12''$, $1'' = 1/12' = 0.083$ ft

$1'' = 0.08 (3)'$	$7'' = 0.58'$	$1/8'' = 0.01'$
$2'' = 0.17'$	$8'' = 0.67'$	$1/4'' = 0.02'$
$3'' = 0.25'$	$9'' = 0.75'$	$1/2'' = 0.04'$
$4'' = 0.33'$	$10'' = 0.83'$	$3/4'' = 0.06'$
$5'' = 0.42'$	$11'' = 0.92'$	
$6'' = 0.50'$	$12'' = 1.00'$	

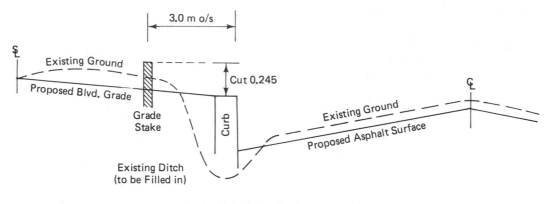

Station 0 + 20 Cut Grade

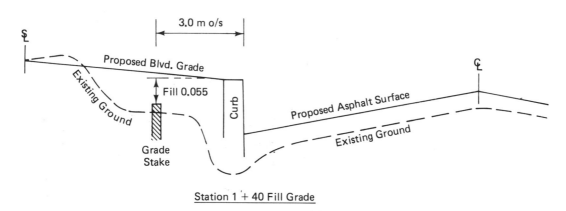

Station 1 + 40 Fill Grade

FIGURE 12.7   Cross sections showing cut and fill grades.

TABLE 12.3

Station	Curb Elev.	Stake Elev.	Cut	Fill	Cut	Fill
0 + 30	470.20	471.30	1.10		1'1¼"	
0 + 50	470.40	470.95	0.55		0'6⅝"	
1 + 00	470.90	470.90	On Grade		On Grade	
1 + 50	471.40	471.23		0.17		0'2"
2 + 00	471.90	471.46		1.44		1'5¼"
2 + 50	472.40	472.06		0.34		0'4⅛"

In some cases the cuts and fills (grades) are written directly on the appropriate grade stakes. This information, written with lumber crayon (keel) or felt markers, is always written on the side of the stake facing the construction. The station is also written on each stake and is placed on that side of each stake facing the lower chainage.

Grade

Unfortunately, the word *grade* has several different meanings. In construction work alone, it is often used in three distinctly different ways:

1. To refer to a proposed elevation.
2. To refer to the slope of profile line (i.e., gradient).
3. To refer to cuts and fills, vertical distances below or above grade stakes.

The surveyor should be aware of these different meanings and always note the context in which the word *grade* is being used.

12.6.8 Treatment of Intersection Curb Construction

Intersection curb radii usually range from 35 ft (10 m) for two local roads intersecting to 60 ft (18 m) for two arterial roads intersecting. The angle of intersection is ideally 90°; however, the range from 70° to 110° is allowed for sight design purposes.

The curb elevation at the BC (180.214) is determined from the plan and profile of Pine Street (see Figure 12.3 and Table 12.1). The curb elevation at the EC is determined from the plan and profile of Elm Street (assume EC elevation = 180.100). The length of curb can be calculated:

$$L = \frac{\pi R \Delta}{180}, \ldots \text{ equation 10.5}$$
$$= 15.708 \text{ m}$$

The slope from BC to EC can be determined:

$$180.214 - 180.100 = 0.114 \text{ m}$$

The fall is 0.114 over an arc distance of 15.708 m, which is −0.73%.

These calculations indicate that a satisfactory slope (0.5% is the usual minimum) joins the two points. The intersection curve is located by four offset stakes, BC, EC, and two intermediate points. In this case, 15.708 ÷ 3 = 5.236 m, the distance measured from the BC to locate the first intermediate point, the distance measured from the first intermediate point to the second intermediate point, and the distance used as a check from the second intermediate point to the EC.

In actual practice, the chord distance is used rather than the arc distance. Since $\triangle/2 = 45°$, and we are using a factor of $1/3$, the corresponding deflection angle for one-third of the arc would be 15°.

$$C = 2\,R\,\text{Sin (deflection angle)}$$
$$= 2 \times 10 \times \text{Sin } 15° = 5.176 \text{ m}$$

These intermediate points can be deflected in from the BC or EC, or they can be located by the use of two tapes, one tape at the radius point (holding 10 m in this case) and the other tape at the BC or intermediate point (holding 5.176), while the stakeman holds the zero point of both tapes. The latter technique is the most often used on these small-radius problems. The only occasions when these curves are deflected in by transit occurs when the radius point (curve center) is inaccessible (fuel pump islands, front porches, etc.).

The proposed curb elevations on the arc are as follows:

$$
\begin{array}{rl}
\text{BC} \quad 0 + 14.5 = & 180.214 \\
& -0.038 \\
\text{No. 1} = & 180.176 \\
& -0.038 \\
\text{No. 2} = & 180.138 \\
& -0.038 \\
\text{EC} = & 180.100
\end{array}
$$

$$
\begin{array}{rl}
\text{Arc interval} = & 5.236 \text{ m} \\
\text{Difference in elevation} = & 5.236 \times 0.0073 = 0.038
\end{array}
$$

Grade information for the curve can be included on the grade sheet.

The offset curve can be established in the same manner, after making allowances for the shortened radius (see Figure 12.8).

For an offset (o/s) of 3 m, the radius will become 7 m. The chords required can be calculated in one of two ways:

1. $C = 2R \text{ Sin (deflection)}$
    $= 2 \times 7 \times \text{Sin } 15°$
    $= 3.623 \text{ m}$

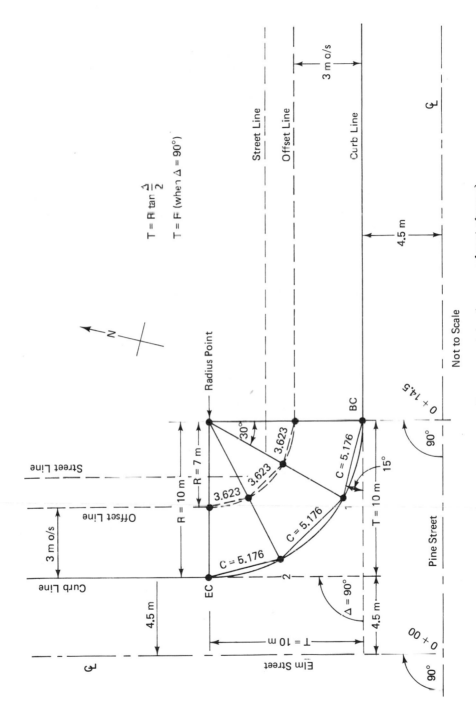

FIGURE 12.8 Intersection geometrics (one quadrant shown).

521

2.

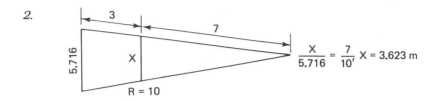

$$\frac{X}{5.716} = \frac{7}{10'} \quad X = 3.623 \text{ m}$$

12.6.9 Sidewalk Construction

The sidewalk is constructed adjacent to the curb or at some set distance from the street line. If the sidewalk is adjacent to the curb, no additional layout is required as the curb itself gives line and grade for construction. In some cases the concrete for this curb and sidewalk is placed in one operation.

When the sidewalk is to be located at some set distance from street line ($\pounds$), a separate layout is required. Sidewalks located near the $\pounds$ give the advantages of increased pedestrian safety and boulevard space for the stockpiling of a winter's accumulation of plow snow in northern regions. See Figure 12.9, which shows the typical location of sidewalk on road allowance.

Sidewalk construction usually takes place after the curbs have been built and the boulevard has been brought to sod grade. The offset distance for the grade stakes can be quite short (1 to 3 ft). If the sidewalk is located within 1 to 3 ft of the $\pounds$, the $\pounds$ is an ideal location for the offset line. In many cases only line is required for construction as the grade is already established by boulevard grading and the permanent elevations at $\pounds$ (existing elevations on private property are seldom adjusted in municipal work). The cross slope (toward the curb) is usually given as being ¼ in./ft (2%).

When the sidewalk is being located as near as 1 ft to the $\pounds$,

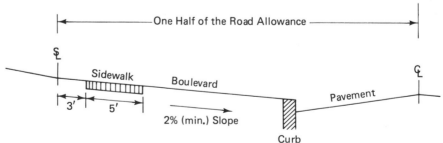

Note: The Sidewalk is Always Constructed so that it Slopes Toward the Road — Usually
       @ 1/4" per foot (2%)

FIGURE 12.9   Typical location of sidewalk on the road allowance.

greater care is required by the surveyor to ensure that the side-walk does not encroach on private property throughout its length. (Due to numerous private property surveys performed by different surveyors over the years, the actual location of ₵ may no longer conform to plan location.)

12.7 HIGHWAY CONSTRUCTION

Highways, like municipal roads, are classified as locals, collectors, arterials, and freeways. The bulk of the highways, in mileage, are arterials that join towns and cities together in a state or provincial network.

Unlike municipal roads, highways do not usually have curbs and storm sewers, relying instead on ditches for removal of storm drainage. Whereas in municipal work the construction layout and offsets are referenced to the *curb lines,* in highway work, the layout and offsets are all referenced to the *center line of construction.*

The construction surveyor must first locate the right of way (ROW) legal markers and set up the construction ₵ in a similar manner to that described in the previous section. Whereas in municipal practice the ₵ of construction is normally the ₵ of the road allowance, in highway work the ₵ of construction can often be shifted from ROW ₵ due to sidehill or other constraints. The construction surveyor is cautioned in this work when establishing the ROW lines; it is often the case that the ₵ of construction has been referenced by offset monuments that are located close to the ROW limit. In the past, serious problems have occurred when the construction surveyor has mistakenly used the construction monuments to establish the ROW line. Mistakes can be eliminated if the surveyor, armed with a complete set of construction and legal plans, takes the time to verify all evidence by checking plan measurements against field measurements.

The construction surveyor next reestablishes the chainage used for the project. Chainage established in the preliminary survey can be reestablished from reference monuments or cross-road intersections. Chainages are reestablished from at least three independent ties to ensure that verification is possible.

Highways are laid out at 100 ft (30 or 40 m) stations, and additional stations are required at all changes in horizontal alignment (e.g., BC, EC, TS, ST) and at all changes in vertical direction (e.g., BVC, EVC, low points). The horizontal and vertical curve sections of highways are often staked out at 50 ft (15 to 20 m) intervals to ensure that the finished product closely conforms to the design.

When using foot units, the full stations are at 100-ft intervals (e.g., 0 + 00, 1 + 00). In metric units, municipalities use 100-m full station intervals (0 + 00, 1 + 00), whereas most highway agencies use 1000-m (kilometer) intervals for full "stations" (e.g., 0 + 000, 0 + 100, . . . , 1 + 000).

The ℄ of construction is staked out using a steel tape, plumb bobs, and so on, using specifications designed for 1/3000 accuracy (minimum). The accuracy of ℄ layout can be verified at reference monuments, road intersections, and the like.

Alternately, highways can be laid out from random coordinated stations using polar layouts, rather than rectangular layouts. Most interchanges are now laid out using polar methods, whereas most of the highways between interchanges are laid out using rectangular layout offsets. The methods of polar layout are covered in Chapter 11.

The profile grade, shown on the contract drawing, can refer to the top of granular elevation or it can refer to the top of asphalt elevation; the surveyor must ensure that he is using the proper reference before calculating subgrade elevations for the required cuts and fills.

12.7.1 Clearing, Grubbing, and the Stripping of Topsoil

Clearing and grubbing are the terms used to describe the cutting down of trees and the removal of all stumps. The full highway width is staked out, approximating the limits of cut and fill, so that the clearing and grubbing can be accomplished.

The first construction operation after clearing and grubbing is the stripping of topsoil. The topsoil is usually stockpiled for later use. In cut sections, the topsoil is stripped for full width, which extends to the points at which the far-side ditch slopes intersect the original ground surface (OG). See Figure 12.10. In fill sections the topsoil is usually stripped for the width of the highway embankment (see Figure 12.11). Most highways agencies do not strip the topsoil where heights of fill exceed 4 ft (1.2 m), believing that this water-bearing material cannot damage the road base below that depth.

The bottom of fills (toe of slope) and the top of cuts (top of slope) are marked by slope stakes. These stakes, which are driven in angled away from ℄, delineate not only the limits of stripping, but also indicate the limits for cut and fill, such operations taking

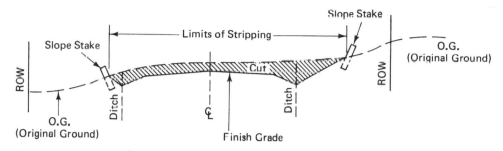

FIGURE 12.10 Highway cut section.

place immediately after the stripping operation. Lumber crayon (keel) or felt markers are used to show station and s/s (slope stake).

12.7.2 Placement of Slope Stakes

Figure 12.12 shows typical cut and fill sections in both foot and metric dimensions. The side slopes shown are 3 : 1, although most agencies use a steeper slope (2 : 1) for cuts and fills over 4 ft (1.2 m). To locate slope stakes, the difference in elevation between the profile grade at $\mathcal{C}$ and the invert of ditch (cut section) or the toe of embankment (fill section) must first be determined. In Figure 12.12a, the difference in elevation consists of

$$
\begin{aligned}
\text{Depth of granular} &= 1.50 \text{ ft} \\
\text{Subgrade crossfall at 3\% over 24.5 ft} &= 0.74 \text{ ft} \\
\text{Minimum depth of ditch} &= \underline{1.50 \text{ ft}} \\
\text{Total difference in elevation} &= 3.74 \text{ ft}
\end{aligned}
$$

The $\mathcal{C}$ of this minimum depth ditch would be 29.0 ft from the $\mathcal{C}$ of construction.

In cases where the ditch is deeper than minimum values, the additional difference in elevation and the additional distance from

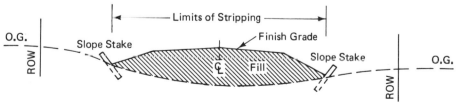

FIGURE 12.11 Highway fill section.

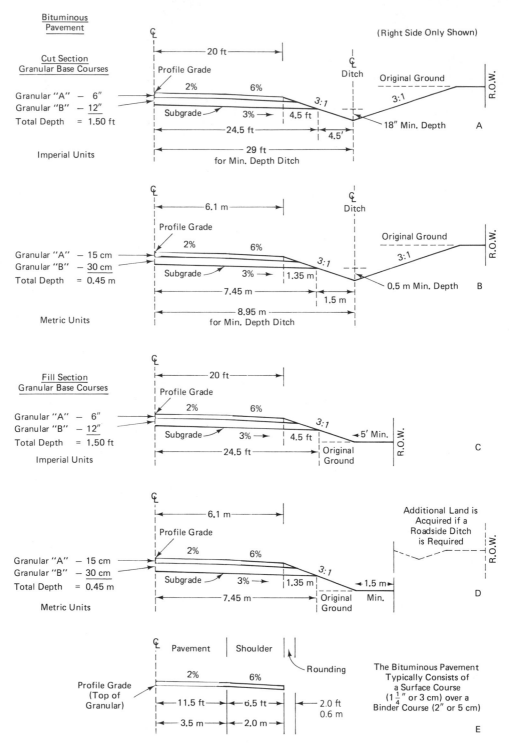

FIGURE 12.12 Typical two-lane highway cross section.

℄ of construction can be easily calculated using the same slope values.

In Figure 12.12b, the difference in elevation between ℄ and the invert of the ditch consists of

$$
\begin{array}{ll}
\text{Depth of granular} = 0.45 \text{ m} \\
\text{Fall at 3\% over 7.45 m} = 0.22 \text{ m} \\
\underline{\text{Minimum depth of ditch} = 0.50 \text{ m}} \\
\text{Total difference in elevation} \quad = 1.17 \text{ m}
\end{array}
$$

The ℄ of this minimum depth ditch would be 8.95 m from the ℄ of construction.

In these two examples, only the distance from highway ℄ to ditch ℄ has been determined. See Example 12.1 for further treatment.

In Figure 12.12c, the difference in elevation consists of

$$
\begin{array}{ll}
\text{Depth of granular} = 1.50 \text{ ft} \\
\underline{\text{Fall at 3\% over 24.5 ft} - 0.74 \text{ ft}} \\
\text{Total difference in elevation} \quad = 2.24 \text{ ft}
\end{array}
$$

The difference from ℄ of construction to this point where the subgrade intersects the side slope is 24.5 ft.

In Figure 12.12d, the difference in elevation consists of

$$
\begin{array}{ll}
\text{Depth of granular} = 0.45 \text{ m} \\
\underline{\text{Fall at 3\% over 7.45 m} - 0.22 \text{ m}} \\
\text{Total difference in elevation} \quad = 0.67 \text{ m}
\end{array}
$$

The distance from ℄ of construction to this point where the subgrade intersects the side slope is 7.45 m.

In the last two examples, the computed distance locates the slope stake.

Example 12.1 *Location of a Slope Stake in a Cut Section*

Refer to Figure 12.13. Given that the profile grade (top of granular) is 480.00 and that HI = 486.28.

$$
\begin{array}{l}
\text{Ditch invert} = 480.00 - 3.74 = 476.26 \\
\text{Grade rod} = 486.28 - 476.26 = 10.02 \\
\text{Depth of cut} = \text{grade rod} - \text{ground rod}
\end{array}
$$

The following equation must be satisfied by trial-and-error ground rod readings:

$$
x = (\text{depth of cut} \times 3) + 29.0
$$

The rodman, holding a cloth tape as well as the rod, estimates the desired location and gives a rod reading. For this example, assume the rod reading was 6.0 ft at a distance of 35 ft

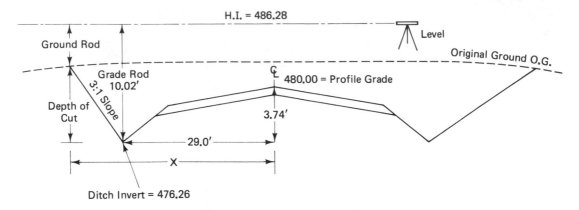

FIGURE 12.13 Location of a slope stake in cut section.

from ₵.

$$\text{Depth of cut} = 10.02 - 6.0 = 4.02$$
$$x = (4.02 \times 3) + 29.0 = 41.06 \text{ ft}$$

Since the rodman was only 35 ft from the ₵, he must move farther out.

At the next point, 43 ft from ₵, a reading of 6.26 was obtained.

$$\text{Depth of cut} = 10.02 - 6.26 = 3.76$$
$$x = (3.76 \times 3) + 29.0 = 40.3$$

Since the rod man was at 43 ft, he was too far out.

The rod man moves closer in, and gives a rod reading of 6.10 at 41 ft from ₵. Now

$$\text{Depth of cut} = 10.02 - 6.10 = 3.92$$
$$x = (3.92 \times 3) + 29 = 40.80$$

This location is close enough for placing the slope stake; the error of 0.2 ft is not significant in this type of work. Usually, two or three trials are required to locate the slope stake properly.

Example 12.2 *Location of a Slope Stake in a Fill Section*

Refer to Figures 12.14 and 12.15. If there is a possibility that the slope stakes may be destroyed by the cut and fill operations, the slope stakes can be offset further from the ₵. As was noted earlier, this method is essentially one of trial and error. Once the work has commenced, the surveyors can proceed more quickly. However, even with experience, it is still a time-consuming endeavor.

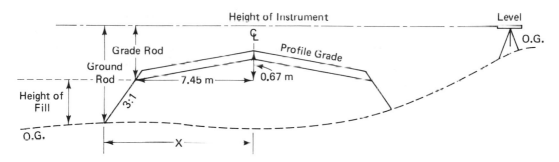

FIGURE 12.14 Location of slope stakes in a fill section. Case I: instrument HI above subgrade.

Alternately, the slope stake distance from center line can be scaled from cross sections or topographic plans. In most cases, cross sections (see Chapter 9) are drawn at even stations (100 ft or 30 to 40 m). The cross sections are necessary to calculate the volume estimates used in contract tendering. The location of the slope stakes can be scaled from the cross-section plan. In addition, highway contract plans are now usually developed photogrammetrically from aerial photos; these plans show contours which are precise enough for most slope stake purposes. One can usually scale off the required distances from ₵ by either using the cross-section plan or the contour plan to the closest 1.0 ft or 0.3 m. The cost savings effected by having this information determined in the office should usually outweigh any resultant loss of accuracy. It is now possible to increase the precision through advances in computers and photogrammetric equipment. Occasional field checks can be used to check on these "scale" methods.

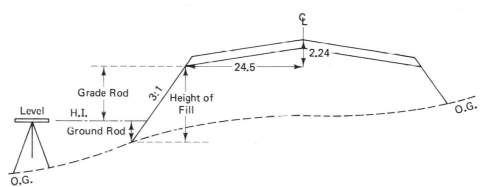

FIGURE 12.15 Location of slope stakes in a fill section. Case II: instrument HI below subgrade.

If scale methods are employed, trigonometric leveling or EDMs and vertical angles can be used to establish the horizontal distance from ℄ to the slope stake. These methods will be more accurate than using a cloth tape on deep cuts or high fills when "breaking tape" may be required several times.

12.7.3 Layout for Line and Grade

In municipal work, it is often possible to put the grade stakes on offset, issue a grade sheet, and then go on to other work. The surveyor may be called back to replace the odd stake knocked over by construction equipment, but usually the layout is thought to be a one-time occurrence.

In highway work, the surveyor must accept the fact that the grade stakes will be laid out several times. The chief difference between the two types of work is the large values for cut and fill. For the grade stakes to be in a "safe" location, they must be located beyond the slope stakes. Although this location is used for the initial layout, as the work progresses this distance back to ℄ becomes too cumbersome to allow for accurate transfer of alignment and grade.

As a result, as the work progresses the offset lines are moved ever closer to the ℄ of construction, until the final location for the offsets is 2 to 3 ft (1 m) from each edge of the proposed pavement. The number of times that the layout must be repeated is a direct function of the height of fill or depth of cut involved.

In highway work, the center line is laid out at the appropriate stations. The center-line points are then individually offset at convenient distances on both sides of ℄. For the initial layout the ℄ stakes, offset stakes, and slope stakes are all put in at the same time. The cuts and fills are written on the grade stakes, referenced either to the top of the stake or to a mark on the side of the stake that will give even foot (even decimeter) values. The cuts and fills are written on that side of the stake facing ℄, whereas, as previously noted, the stations are written on that side of the stake facing back to the 0 + 00 location.

As the work progresses and the cuts and fills become more pronounced, care should be taken in breaking tape when laying out grade stakes so that the horizontal distance is maintained. The center-line stakes are offset by turning 90°, either with a right-angle prism or, more usually, by the swung arm method. Cloth tapes are used to lay out the slope stakes and offset stakes.

Once a ℄ station has been offset on one side, care is taken when offsetting to the other side to ensure that the two offsets and the ℄ stake are all in a straight line.

As in cross sectioning for preliminary surveys, the optimal size of the survey crew is four: the party chief to take notes, the instrument-man to run the instrument, one rodman (chainman) to hold one end (zero) of the tape on ℄, and the other rodman (chainman) to establish the stakes using the other end of the tape and to give rod readings for slope and grade stakes.

When the cut and/or fill operations have brought the work to the proposed subgrade (bottom of granular elevations), the subgrade must be verified by cross sections before the contractor is permitted to place the granular material. Usually, a tolerance of 0.10 ft (30 mm) is allowed. Once the top of granular profile has been reached, layout for pavement (sometimes a separate contract) can commence.

The final layout for pavement is usually on a very close offset (3 ft or 1 m). If the pavement is to be concrete, more precise alignment is provided by nails driven into the tops of the stakes.

When the highway construction has been completed, a final survey is performed. The final survey includes cross sections and locations that are used for final payments to the contractor and for a completion of the as-built drawings. The final cross sections are taken at the same stations used in the preliminary survey.

The description here has referred to two-lane highways. The procedure for layout of a four-lane divided highway is very similar. The same control is used for both sections; grade stakes can be offset to the center of the median and used for both sections. When the lane separation becomes large and the vertical alignment is different for each direction, the project can be approached as being two independent highways.

The layout for elevated highways, often found in downtown urban areas, follows the procedures used for structures layout (see Section 12.10).

12.7.4 Grade Transfer

The grade stakes can be set so that the tops of the stakes are at "grade." Stakes set to grade are colored red or blue on the top to differentiate them from all other stakes. This procedure is time consuming and often impractical except for final pavement layout.

Generally, the larger the offset distance, the more difficult it is to drive the tops of the stakes to grade.

As noted earlier, the cut and fill can refer to the top of the grade stake or to a mark on the side of the grade stake referring to an even number of feet (decimeters) of cut or fill. The mark on the side of the stake is located by sliding the rod up and down the side of the stake until a value is read on the rod that will give the cut or fill to an even foot (decimeter). This procedure of marking the side of the stake is best performed by two men, the rodman and one other to steady the bottom of the rod and then to make the mark on the stake. The cut or fill can be written on the stake or entered on a grade sheet, one copy of which is given to the contractor.

To transfer the grade (cut or fill) from the grade stake to the area of construction, a means of transferring the stake elevation in a horizontal manner is required. When the grade stake is close (within 6 ft or 2 m), the grade transfer can be accomplished using a carpenters level set on a piece of sturdy lumber (see Figure 12.16). When the grade stake is far from the area of construction, a string line level can be used to transfer the grade (cut or fill) (see Figure 12.17). In this case, a fill of 1 ft 0 in. is marked on the grade stake (as well as the offset distance). A guard stake has been placed adjacent to the grade stake, and the grade mark is transferred to the guard stake. A 1-ft distance is measured up the guard stake, and the grade elevation is marked.

A string line is then attached to the guard stake at the grade elevation mark; then a line level is hung from the string (near the halfway mark), and the string is pulled taut so as to eliminate most of the sag (it is not possible to eliminate all the sag). The string line is adjusted up and down until the bubble in the line level is centered; with the bubble centered, it can be quickly seen at ℄ how much more fill may be required to bring the highway, at that point, to grade.

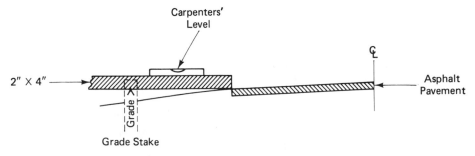

FIGURE 12.16   Grade transfer using a carpenters' level.

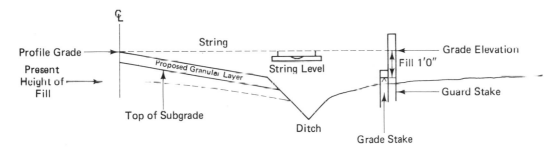

FIGURE 12.17 Grade transfer using a string level.

It can be seen in Figure 12.17 that more fill is required to bring the total fill to the top of subgrade elevation. The surveyor can convey this information to the grade foreman so that the fill can be properly increased. As the height of fill approaches the proper elevation (top of subgrade), the grade checks become more frequent.

In the preceding example, the grade fill was 1 ft 0 in.; had the grade been cut 1 ft, the procedure with respect to the guard stake would have been the same; that is, measure up the guard stake 1 ft so that the mark now on the guard stake would be 2 ft *above* the ℄ grade. The surveyor or grade man would simply measure down 2 ft (using a tape measure from the level string at ℄). If the measurement down to the "present" height of fill exceeded 2 ft, it would indicate that more fill was required; if the measurement down to the "present" height of fill were less than 2 ft, it would indicate that too much fill had been placed and that an appropriate depth of fill must be removed.

Another method of grade transfer used when the offset is large and the cuts or fills significant is the use of boning rods (batter boards). See Figure 12.18. In the preceding example, a fill grade is transferred from the grade stake to the guard stake; the fill grade is measured up the guard stake and the grade elevation is marked (any even foot/decimeter cut or fill mark can be used so long as the relationship to profile grade is clearly marked).

A crosspiece is nailed on the guard stake at the grade mark and parallel to ℄. A similar guard stake and crosspiece are established on the opposite side of ℄. The surveyor or grademan can then sight over the two crosspieces to establish profile grade datum at that point. Another worker can move across the section with a rod, and the progress of the fill (cut) operation can be visually checked.

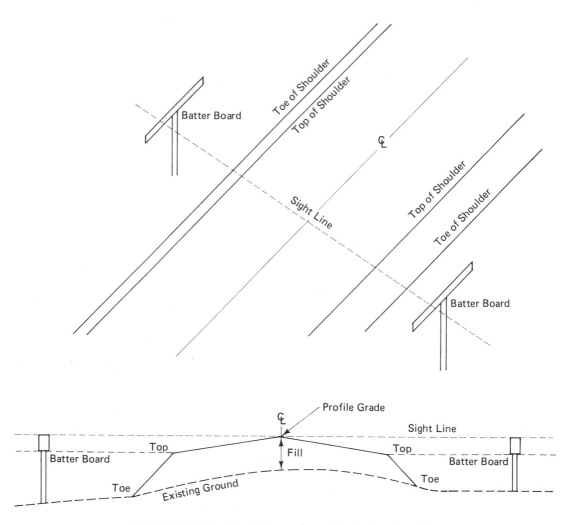

FIGURE 12.18   Grade transfer using batter boards.

In some cases, two crosspieces are used on each guard stake, one indicating ₵ profile grade and the lower one indicating the shoulder location.

12.7.5 Ditch Construction

The ditch profile often parallels the ₵ profile, especially in cut sections. When the ditch profile does parallel the ₵ profile, no additional grades are required to assist the contractor in construction.

However, it is quite possible to have the ₵ profile at one slope (even 0%) and the ditch profile at another slope (0.3% is often taken as a minimum slope to give adequate drainage). If the ditch grades are independent of ₵ profile, the contractor must be given these cuts or fill grades, either from the existing grade stakes or from grade stakes specifically referencing the ditch line.

In the extreme case (e.g., a spiralled highway going over the brow of the hill), the contractor may require five separate grades at one station (i.e., ₵, two edges of pavement, and two different ditch grades); it is even possible in this extreme case to have the two ditches flowing in opposite directions for a short distance.

12.8 SEWER AND TUNNEL CONSTRUCTION

Sewers are usually described as being in one of two categories. Sanitary sewers collect domestic and industrial liquid waste and convey these wastes (sewage) to a treatment plant. Storm sewers are designed to collect runoff from rainfall and to transport this water (sewage) to the nearest natural receiving body (e.g., creek, river, lake). The rainwater enters the storm sewer system through ditch inlets or through catch basins located at the curb line on paved roads.

The design and construction of sanitary and storm sewers are similar in the respect that the flow of sewage is usually governed by gravity. Since the sewer grade lines (flow lines) depend on gravity, it is essential that the construction grades be precisely given.

Figure 12.19 shows a typical cross section for a municipal roadway. The two sewers are located 5 ft (1.5 m) either side of ₵. The sanitary sewer is usually deeper than the storm sewer, as it must be deep enough to allow for all house connections. The sanitary house connection is usually at a 2% (minimum) slope. If sanitary sewers are being added to an existing residential road, the preliminary survey must include the basement floor elevations. The floor elevations are determined by taking a rod reading on the window sill and then, after getting permission to enter the house, measuring from the window sill down to the basement floor. As a result of deep basements and long setbacks from ₵, sanitary sewers often have to be at least 9 ft (2.75 m) below ₵ grade.

The depth of storm sewers below ₵ grade depends, in the southern United States on the traffic loading, and in the northern United States and most of Canada on the depth of frost penetration. The minimum depth of storm sewers ranges from 3 ft (1 m)

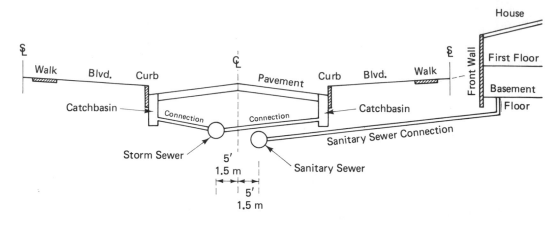

FIGURE 12.19 Municipal road allowance showing typical service locations (also see Figure 12.2).

in some areas of the south to 8 ft (2.5 m) in the north. The design of the inlets and catch basins depends on the depth of sewer and the quality of the effluent.

The minimum slope for storm sewers is usually 0.50%, whereas the minimum slope for sanitary sewers is often set at 0.67%. In either case, the designers try to achieve self-cleaning velocity 2.5 to 3 ft/s (0.8 to 0.9 m/s) to avoid excessive sewer maintenance costs.

Manholes are located at each change in direction, slope, or pipe size. In addition, manholes are located at 300 to 450 ft (100 to 140 m) intervals maximum.

Catch basins are located at 300-ft (100-m) maximum intervals; they are also located at the high side of intersections and at all low points. The 300 ft (100 m) maximum interval is reduced as the slope on the road increases.

For construction purposes, sewer layout is considered only from one manhole to the next. The stationing (0 + 00) commences at the first (existing) manhole (or outlet) and proceeds only to the next manhole. If a second leg is also to be constructed, that station of 0 + 00 is assigned to the downstream manhole and proceeds upstream only to the next manhole. Each manhole is described by a unique manhole number to avoid confusion with the stations for extensive sewer projects. Figure 12.20 shows a section of sewer pipe. The *invert* is the inside bottom of the pipe. The invert grade is the controlling grade for construction and design. The sewer pipes may consist of vitrified clay, transite, metal, some of the newer "plastics," or, as usually is the case, concrete.

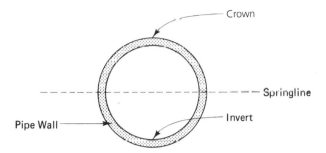

FIGURE 12.20 Sewer pipe section.

The pipe wall thickness depends on the diameter of the pipe. For storm sewers, 12 in. (300 mm) is usually taken as minimum diameter. The *spring line* of the pipe is at the halfway mark, and connections are made above this reference line. The *crown* is the outside top of the pipe. Although this term is relatively unimportant (sewer cover is measured to the crown) for sewer construction, it is important for pipeline (pressurized pipes) construction as it gives the controlling grade for construction.

12.8.1 Layout for Line and Grade

As in all other construction work, offset stakes are used to provide line and grade for the construction of sewers. Before deciding on the offset location, it is wise to discuss the matter with the contractor. The contractor will be excavating the trench and casting the material to one side or loading it into trucks for removal from the site. Additionally, the sewer pipe will be delivered to the site and positioned conveniently alongside its future location. The position of the offset stakes should not interfere with either of these operations.

The surveyor will position the offset line as close to the pipe center line as possible, but seldom is it possible to locate the offset line closer than 15 ft (5 m) away. 0 + 00 is assigned to the downstream manhole or outlet, the chainage proceeding upstream to the next manhole. The center line of construction is laid out with stakes marking the location of the two terminal points of the sewer leg. The surveyor will use survey techniques giving 1/3000 accuracy as a mimimum for most sewer projects. Large-diameter (6 ft or 2 m) sewers require increased precision and accuracy.

The two terminal points on ℄ are occupied by transit or theodolite, and right angles are turned to locate precisely the terminal points at the assigned offset distance. 0 + 00 on offset is occupied by a transit or theodolite, and a sight is taken on the other terminal offset point. Stakes are then located at 50 ft (20 m) intervals, checking in at the terminal point to verify accuracy.

The tops of the offset stakes are surveyed and their elevations determined, the surveyor taking care to see that his leveling is accurate; the invert elevation of MH 3, (Figure 12.21) shown on the contract plan and profile, is verified at the same time. The surveyor next calculates the sewer invert elevations for the 50 ft (20 m) stations. He then prepares a grade sheet showing the stations, stake elevations, invert grades, and cuts. The following examples will illustrate the techniques used.

Assume that an existing sewer (Figure 12.21) is to be extended from existing MH 3 to proposed MH 4. The horizontal alignment will be a straight-line production of the sewer leg from MH 2 to MH 3. The vertical alignment is taken from the contract plan and profile (Figure 12.21).

The straight line is produced by setting up the transit or theodolite at MH 3, sighting MH 2, and double centering to the location of MH 4. The layout then proceeds as described previously.

The stake elevations, as determined by differential leveling, are shown in Table 12.4.

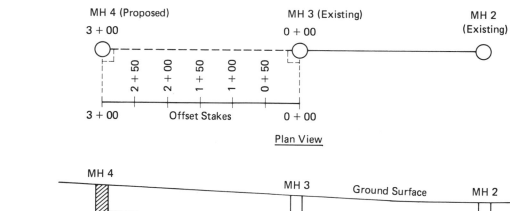

FIGURE 12.21   Plan and profile of a proposed sewer (English units).

TABLE 12.4 *Sewer Grade Sheet, English Units*[a]

Station		Invert Elev.	Stake Elev.	Cut	Cut
MH 3	0 + 00	420.86	429.27	8.41	8'4⅞"
	0 + 50	421.46	429.90	8.44	8'5¼"
	1 + 00	422.06	430.41	8.35	8'4¼"
	1 + 50	422.66	430.98	8.32	8'3⅞"
	2 + 00	423.26	431.72	8.46	8'5½"
	2 + 50	423.86	431.82	7.96	7'11½"
MH 4	3 + 00	424.46	432.56	8.10	8'1¼"

[a]Refer to Section 12.6.7 for foot-inch conversion.

At station 1 + 50, the cut is 8'3⅞" (see Figure 12.22). To set the cross-trench batter board at the next even foot, measure up 0'8⅛" to top of batter board. The offset distance of 15 ft can be measured and marked at the top of the batter board over the pipe ℄, and a distance of 9 ft measured down to establish the invert elevation. This even foot measurement from the top of the batter board to the invert is known as the grade rod distance. A value can be picked for the grade rod so that it is the same value at each station. In this example, 9 ft appears to be suitable for each station as it is larger than the largest cut. The arithmetic shown for station 1 + 50 is performed at each station so that the batter boards can be set at 50-ft intervals. The grade rod (a 2 × 2 in. length of lumber held by a worker in the trench) has a foot piece attached to the bottom at a right angle to the rod so that the foot piece can be inserted into the pipe and allow measurement precisely from the invert.

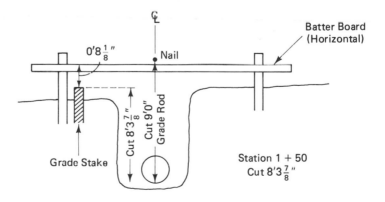

FIGURE 12.22 Use of cross-trench batter boards.

This method of line and grade transfer has been shown first because of its simplicity; it has not, however, been widely used in the field in recent years. With the introduction of larger and faster trenching equipment, which can dig deeper and wider trenches, this method would only slow the work down as it involves batter boards spanning the trench at 50-ft intervals. Most grade transfers are now accomplished by free-standing offset batter boards or laser alignment devices. Using the data from the previous example, it can be illustrated how the technique of free-standing batter boards is utilized (see Figure 12.23).

These batter boards (3 ft or 1 m wide) are erected at each grade stake. As in the previous example, the batter boards are set at a height that will result in a grade rod that is an even number of feet (decimeters) long. However, with this technique, the grade

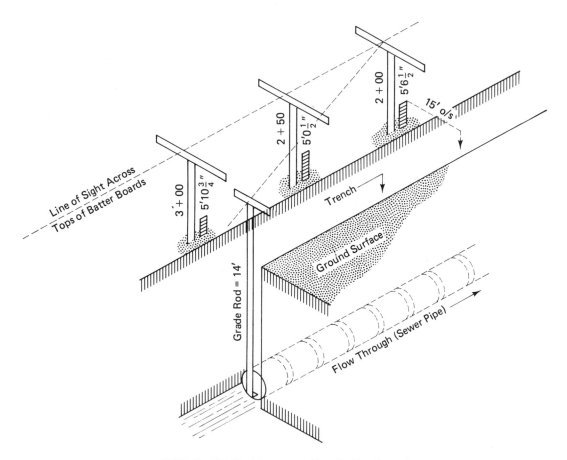

FIGURE 12.23 Free-standing batter boards.

rod distance will be longer as the top of the batter board should be at a comfortable eye height for the inspector. The works inspector usually checks the work while standing at the lower chainage stakes and sighting forward to the higher chainage stakes. The line of sight over the batter boards is a straight line parallel to the invert profile, and in this example (see Figure 12.21) the line of sight over the batter boards is rising at 1.20%.

As the inspector sights over the batter boards, he can include in his field of view the top of the grade rod, which is being held on the most recently installed pipe length. The top of the grade rod has a horizontal board attached to it in a similar fashion to the batter boards. The inspector can visually determine whether the line over the batter boards and the line over the grade rod are at the same elevation. If an adjustment up or down is required, the workman in the trench makes the necessary adjustment and has the work rechecked. Grades can be checked to the closest $1/4$ in (6mm) in this manner. The preceding example is now worked out using a grade rod of 14 ft (see Table 12.5).

The grade rod of 14 ft requires an eye height of 5'7$1/8$" at 0 + 00; if this is considered too high, a grade rod of 13 ft could have been used, which would have resulted in an eye height of 4'7$1/8$" at the first batter board. The grade rod height is chosen to suit the needs of the inspector.

In some cases an additional station is put in before 0 + 00 (i.e., 0 − 50). The grade stake and batter board refer to the theoretical pipe ₵ and invert profile produced back through the first manhole. This batter board will of course line up with all the others and will be useful in checking the grade of the first few pipe lengths placed.

It should also be noted that many agencies use 25 ft (10 m) stations, rather than the 50 ft (20 m) stations used in this example. The smaller intervals allow for much better grade control.

TABLE 12.5

Station		Invert Elev.	Stake Elev.	Cut	Stake to Batter Board	Stake to Batter Board
MH 3	0 + 00	420.86	429.27	8.41	5.59	5'7$1/8$"
	0 + 50	421.46	429.90	8.44	5.56	5'6$3/4$"
	1 + 00	422.06	430.41	8.35	5.65	5'7$3/4$"
	1 + 50	422.66	430.98	8.32	5.68	5'8$1/8$"
	2 + 00	423.26	431.72	8.46	5.54	5'6$1/2$"
	2 + 50	423.86	431.82	7.96	6.04	5'0$1/2$"
MH 4	3 + 00	424.46	432.56	8.10	5.90	5'10$3/4$"

One distinct advantage to the use of batter boards in construction work is that an immediate check is available on **all** the survey work involved in the layout. The line of sight over the tops of the batter boards (which is actually a vertical offset line) must be a straight line. If, upon completion of the batter boards, all the boards do not line up precisely, it is obvious that a mistake has been made. The surveyor will check his work by first verifying all grade computations and, second, by releveling the tops of the grade stakes. Once the boards are in alignment, the surveyor can move on to other projects.

See Figure 12.24 and Table 12.6 for an additional sewer grade sheet example—stated in metric units.

12.8.2 Laser Alignment

Laser devices are now in use in most forms of construction work. Lasers are normally used in a fixed direction and slope mode or in a revolving, horizontal pattern. One such device (Figure 12.25) can be mounted in a sewer manhole, aligned for direction and slope, and used with a target for the laying of sewer pipe. Since the laser beam can be deflected by dust or high humidity, care must be taken to overcome these factors.

Lasers used in sewer trenches usually are accompanied by blowers that remove the humid air. To overcome the humidity problem, the laser can be used above grade with a signal-sensing target rod. Working above ground not only eliminates the humidity factor; it allows for more accurate and quicker horizontal alignment. These devices allow for setting slope within the limits of

TABLE 12.6   *Grade Sheet: Metric Units*

Station	Invert Elev.	Stake Elev.[a]	Cut	Stake to Batter Board, GR = 5.0 m
MH 12A 0 + 00	173.650	177.265	3.615	1.385
0 + 20	173.890	177.865	3.975	1.025
0 + 40	174.130	177.200	3.070	1.930
0 + 60	174.370	178.200	3.830	1.170
0 + 80	174.610	178.005	3.395	1.605
MH 13A 1 + 00	174.850	178.500	3.650	1.350

[a]Stake elevations and computations are normally carried out to the closest 5 mm.

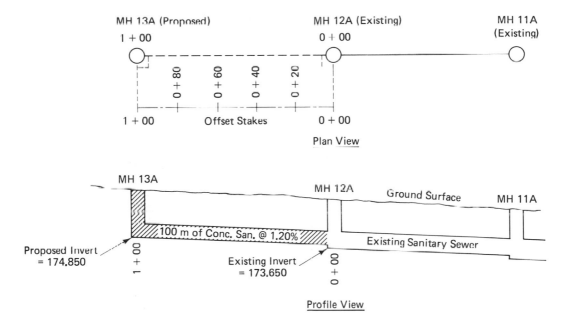

FIGURE 12.24 Plan and profile of a proposed sewer (SI units).

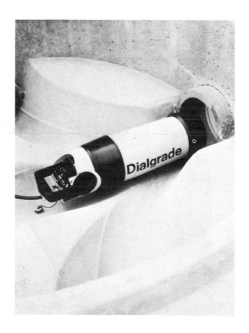

FIGURE 12.25 Pipeline laser mounted in storm-pipe manhole. (Courtesy Spectra-Physics, Inc.)

$-10°$ to $30°$. Some devices have automatic shutoff capabilities when the device is disturbed from its desired setting. Figures 12.26 and 12.27 illustrate additional laser applications.

(a)

(b)

FIGURE 12.26 Rotating lasers: (a) site grade control; (b) laser sensor mounted on backhoe gives a proper trenching depth for field tile installation. (Courtesy Laser Alignment, Inc.)

FIGURE 12.27 Laser transit supplies reference laser beam and a reference laser plane (horizontal or vertical). Used here to control a dredging operation. (Courtesy Spectra-Physics, Inc.)

12.8.3 Catch Basin Construction Layout

Catch basins are constructed along with the storm sewer or at a later date, just prior to curb construction. Usually, the catch basin is located by two grade stakes, one on each side of the CB. The two stakes are on curb line and are usually 5 ft (2 m) from the center of the catch basin. The cut or fill grade is referenced to the CB grate elevation at the curb face. The ₡ pavement elevation is calculated, and from it the crown height is subtracted to arrive at the top of grate elevation (see Figures 12.28 and 12.29).

At low points, particularly at vertical curve low points, it is usual practice for the surveyor to lower arbitrarily the catch-basin grate elevation to ensure that ponding does not occur on either

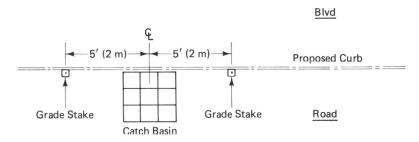

FIGURE 12.28 Catch basin layout (plan view).

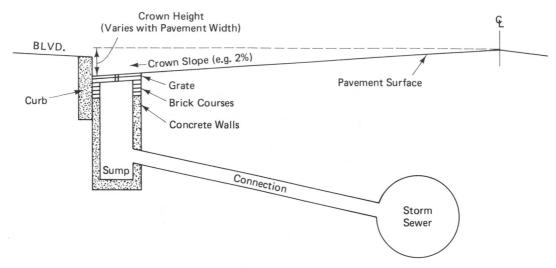

FIGURE 12.29   Typical catch basin (with sump).

side of the completed catch basin. It was noted in Chapter 10 that
the longitudinal slope at vertical curve low points is virtually flat
for a significant distance. The catch basin grate elevation can be
arbitrarily lowered as much as 1 in. (25 mm) to ensure that the
gutter drainage goes directly into the catch basin without
ponding. The catch basin, which can be of concrete poured in
place, but is more often prefabricated and delivered to the job site,
is set below finished grade until the curbs are constructed. At the
time of curb construction, the finished grade for the grate is
achieved by one or more courses of brick laid on top of the concrete walls.

12.8.4 Pipeline Construction

Pipelines are designed to carry water, oil, natural gas, and so on,
while under pressure. Because pressure systems do not require
close attention to grade lines, the layout for pipelines can proceed
at a much lower order of precision than is required for gravity
pipes. Pipelines are usually designed so that the cover over the
crown is adequate for the loading conditions expected and also
adequate to prevent damage due to frost penetration, erosion,
and the like.

The pipeline location can be determined from the contract
drawings; the grade stakes are offset an optimal distance from the
pipe ℄ and are placed at 50 to 100 ft (15 to 30 m) intervals.

When existing ground elevations are not being altered, the standard cuts required can be simply measured down from the ground surface (required cuts in this case would equal the specified cover over the crown plus the pipe diameter plus the bedding if applicable). See Figure 12.30.

In the case of proposed general cuts and fills, grades must be given to establish suitable crown elevation so that **final** cover is as specified. Additional considerations and higher precisions are required at major crossings (e.g., rivers, highways, utilities).

Final surveys show the actual location of the pipe and appurtenances (valves and the like). As-built drawings, produced from final surveys, are especially important in urban areas where it seems there is no end to underground construction.

12.8.5 Tunnel Construction

Tunnels are used in road, sewer, and pipelines construction when the cost of working at or near the ground surface becomes prohibitive. For example, sewers are tunneled when they must be at a depth that would make open cut too expensive (or operationally unfeasible), or sewers may be tunneled to avoid disruption of services on the surface such as would occur if an open cut was put through a busy expressway. Roads and railroads are tunneled through large hills and mountains in order to maintain optimal grade lines.

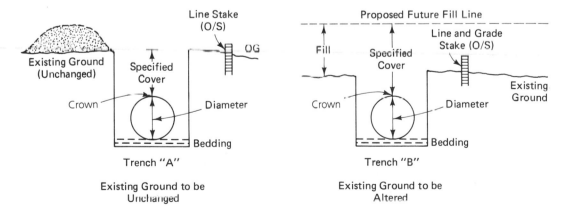

FIGURE 12.30 Pipeline construction: (a) existing ground to be unchanged; (b) existing ground to be altered.

Control surveys for tunnel layouts are performed on the surface, joining the terminal points of the tunnel. These control surveys use triangulation or precise traverse survey methods and allow for the computation of coordinates for all key points (see Figure 12.31).

In the case of highway (railway) tunnels, the ℄ can be run directly into the tunnel and is usually located on the roof either at ℄ or at a convenient offset (see Figure 12.32). If the tunnel is long, intermediate shafts could be sunk to provide access for materials, ventilation, and alignment verification. Conventional engineering theodolites are illustrated in Figure 12.32; for cramped quarters, a suspension theodolite (Figure 12.33) can be used. Levels can also be run directly into the tunnel and temporary bench marks are established in the floor or roof of the tunnel.

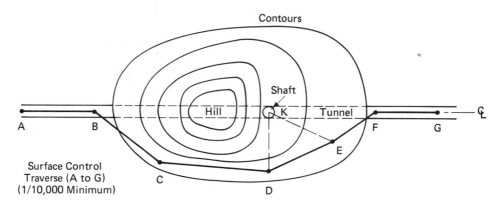

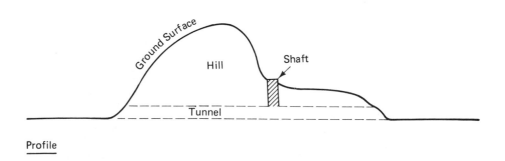

FIGURE 12.31 Plan and profile of tunnel location.

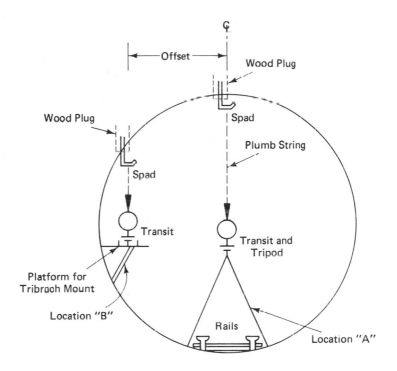

Tunnel ₵ is Usually Located in the Roof (Location "A") and then can be
Offset (Location "B") to Provide Space for Excavation and Materials Movement.
Line and Grade can be Provided by a Single Laser Beam which has been
Oriented for Both Alignment and Slope.

FIGURE 12.32 Establishing "line" in a tunnel.

In the case of long tunnels, work can proceed from both
ends, meeting somewhere near the middle. Constant vigilance
with respect to errors and mistakes is of prime importance.

In the case of a deep sewer tunnel, mining surveying techniques must be employed to establish line and grade (Figure 12.34). The surface ₵ projection *AB* is carefully established on
beams overhanging the shaft opening. Plumb lines (piano wire)
are hung down the shaft, and the tunnel ₵ is developed by
overaligning the transit or theodolite in the tunnel. A great deal of
care is required in overaligning as this very short backsight will
produce relatively long distances, thus magnifying any sighting
errors.

The plumb lines usually employ heavy plumb bobs (capable
of taking additional weights if required). Sometimes the plumb
bobs are submerged in heavy oil in order to dampen the swing os-

FIGURE 12.33 Breithaupt mining suspension theodolite; can be used with a tripod or can be suspended from a steel punch in the ceilings of cramped galleries or drifts. (Courtesy Keuffel & Esser Co.)

cillations. If the plumb line swing oscillations cannot be eliminated, the oscillations must be measured and then averaged.

Some tunnels are pressurized in order to control ground water seepage; the air locks associated with pressure systems will cut down considerably on the clear dimensions in the shaft, making the plumbed line transfer even more difficult.

Transferring line from ground surface to underground locations by use of plumb lines is an effective-although out-dated—technique. Modern survey practice favours the use of precise optical plummets (see Figure 12.45) to accomplish the line transfer. These plummets are designed for use in zenith or nadir directions, or—as illustrated in figure 12.45—in both zenith and nadir directions. The accuracy of this technique can be as high as one or two millimetres in 100 metres. In addition, telescope—mounted penta-prisms will also give precise location of a zenith or nadir line—

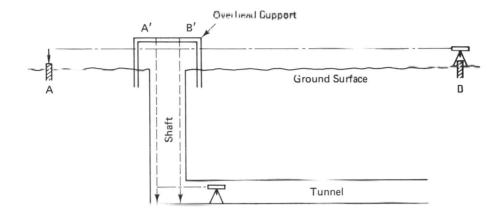

Tunnel ₵ AB is Carefully Marked on the Overhead Support @ A' and B'. These Two Marks are Set as Far Apart as Possible for Plumbing into the Shaft.

The Transit in the Tunnel Over-aligns the Two Plumb Lines (Trial and Error Technique); ₵ is then Produced Forward using Double-Centering Techniques.

Precision can be Improved by Repeating this Process when the Tunnel Excavation has Progressed to the Point where a much Longer Back-Sight is Possible.

FIGURE 12.34 Transfer of surface alignment to the tunnel.

two marks are set with the alidade revolved 180° making the plumb point the midpoint of the two set marks.

In addition to tunnel shaft applications, these instruments are very useful in controlling high-rise construction through elevator shafts or through—floor ports.

Precise optical plummets, to-gether with laser light directed through the eye-piece, have been used with success in positioning target helicopters over control points for angle and distance (EDM) measurements.

12.9 BRIDGE CONSTRUCTION

Accuracy requirements for structure construction are generally of the highest order for survey layouts. Included in this topic are bridges, elevated expressways, and so on. Accuracy requirements for structure layouts range from 1/3000 for residential housing to 1/5000 for bridges and 1/10000 for long-span bridges. The accuracy requirements depend on the complexity of the construction, the type of construction materials specified, and the ultimate design use of the facility. The accuracy required for any specific project could be specified in the contract documents, or it could be left to the common sense and experience of the surveyor.

12.9.1 Contract Drawings

Contract drawings for bridge construction typically include general arrangement, foundation layout, abutment and pier details, beam details, deck details, and reinforcing steel schedules and could also include details on railings, wing walls, and the like. Whereas the contractor must utilize all drawings in the construction process, the construction surveyor is only concerned with those drawings that will facilitate the construction layout. Usually, the entire layout can be accomplished using only the general arrangement drawing (Figures 12.35 and 12.36) and the foundation layout drawing (Figure 12.37). These plans are analyzed to determine which of the many dimensions shown are to be utilized for construction layout.

The key dimensions are placed on foundation layout sketches (Figures 12.38 and 12.39), which will be the basis for the layout. The sketches show the location of the piers and the abutment bearing ℄, the location of each footing, and in this case the skew angle. Although most bridges have symetrical foundation dimensions, the bridge in this example is asymmetrical due to the spiraled pavement requirements. This lack of symmetry does not pose any problems for the actual layout, but it does require additional work in establishing dimension check lines in the field.

12.9.2 Layout Computations

Chainage at ℄ of bridge	588 + 37.07	Chainage at ℄ of bridge	588 + 37.07
	+ 24.50		− 24.50
℄ North pier	588 + 61.57	℄ South pier	588 + 12.57
	+ 35.50		− 35.50
℄ North abut. brg.	588 + 97.07	℄ South abut. brg.	587 + 77.07

$(588 + 97.07) - (587 + 77.07) = 120.00$ ft Check

Assume, for this example, that the offset lines are 5 ft south from the south footing faces of the north abutment and both piers, and 5 ft north from the north face of the south abutment. For this example, the offset stakes (1-in. reinforcing steel, 2 to 4 ft long) will be placed opposite the ends of the footings and a further

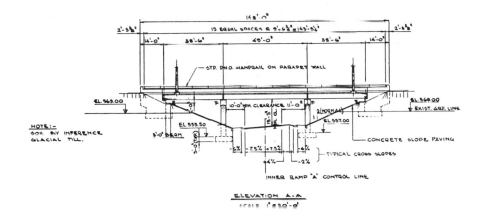

ELEVATION A-A
SCALE 1"=20'-0'

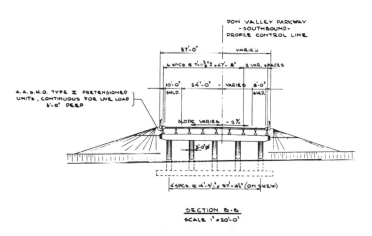

SECTION B-B
SCALE 1"=20'-0'

FIGURE 12.35 General arrangement plan for Bridge No. 5, Don Valley Parkway.

25 ft away on each offset line. The stakes can be placed by right-angle ties from ℄ or by locating the offset line parallel to the footings.

In the latter case, it will be necessary to compute the ℄ stations for each offset line. Referring to Figure 12.38, it can be seen that the distance from abutment ℄ of bearings to face of footing is 4′8″ (4.67′). Along the N–S bridge ℄, this dimension becomes 4.84′ (4.67 × secant 15°31′13″). Similarly, the dimension from pier ℄ to face of pier footing is 5.00′; along the N–S bridge ℄ this dimension becomes 5.19′ (5.00 × secant 15°31′13″). Along the N–S bridge ℄, the 5′ offset distance also becomes 5.19′.

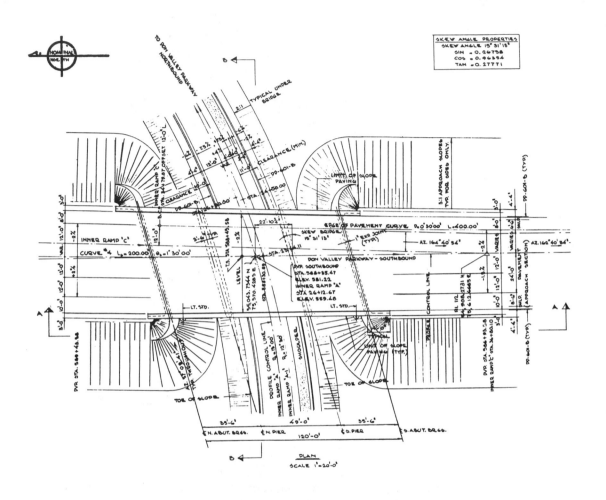

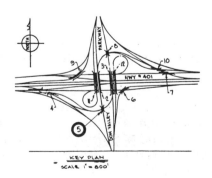

FIGURE 12.36 Key plan and general layout for Bridge No. 5, Don Valley Parkway.

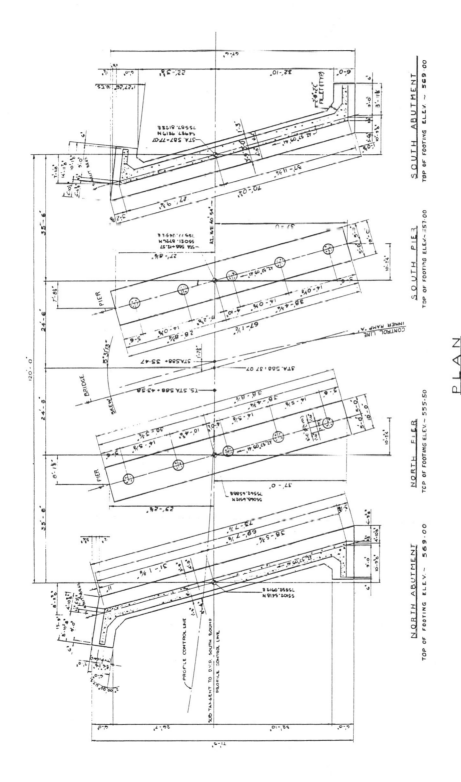

FIGURE 12.37 Foundation layout plan for Bridge No. 5, Don Valley Parkway.

555

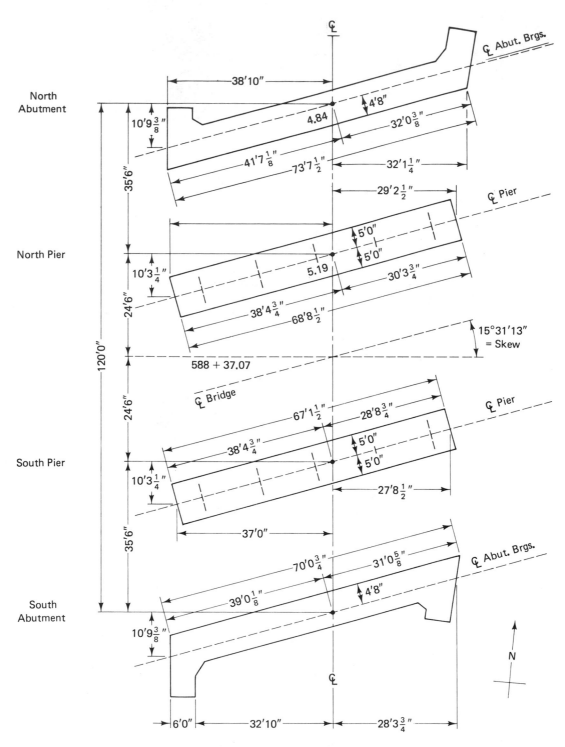

FIGURE 12.38 Foundation layout sketch.

556

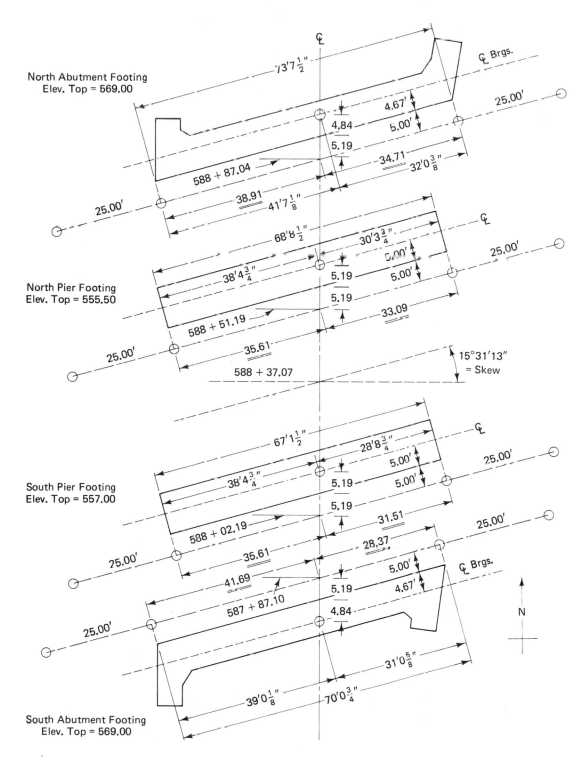

FIGURE 12.39 Foundation layout and offsets sketch.

Accordingly, the stations for the offset lines at the N–S bridge ℄ become (refer to Figure 12.39)

$$
\begin{aligned}
\text{North abutment offset} &= 588 + 97.07 - 10.03 = 588 + 87.04 \\
\text{North pier offset} &= 588 + 61.57 - 10.38 = 588 + 51.19 \\
\text{South pier offset} &= 588 + 12.57 - 10.38 = 588 + 02.19 \\
\text{South abutment offset} &= 587 + 77.07 + 10.03 = 587 + 87.10
\end{aligned}
$$

12.9.3 Offset Distance Computations

The stations at which the offset lines intersect the bridge ℄ are each occupied with a transit (theodolite). The skew angle ($15°31'13"$) is turned and doubled (minimum) and the appropriate offset distances are measured out each side of the bridge ℄.

For the north abutment offset line, the offset distance left is

$$41.59' \ (41'7\tfrac{1}{8}") - 10.03 \times \text{Sin } 15°31'13" = 38.91'$$

The offset distance right is $32.03 + 2.68 = 34.71'$.

$$38.91 + 34.71 = 73.62' = 73'7\tfrac{1}{2}" \text{ Check (see Figure 12.38)}$$

For the north pier offset line, the offset distance left is

$$38.39 \ (38'4\tfrac{3}{4}") - 10.38 \text{ Sin } 15°31'13" = 35.61'$$

The offset distance right is $30.31 + 2.78 = 33.09'$.

$$35.61 + 33.09 = 68.70 = 68' \ 8\tfrac{1}{2}" \text{ Check (see Figure 12.38)}$$

For the south pier offset line, the offset distance left is

$$38.39 - 10.38 \text{ Sin } 15°31'13" = 35.61'$$

The offset distance right is $28.73 \ (28' \ 8\tfrac{3}{4}") + 2.78 = 31.51'$.

$$35.61 + 31.51 = 67.12' = 67'1\tfrac{1}{2}" \text{ Check (see Figure 12.38)}$$

For the south abutment offset line, the offset distance left is

$$39.01 \ (39'0\tfrac{1}{8}") + 10.03 \times \text{Sin } 15°31'13" = 41.69'$$

The offset distance right is $31.05 - 2.68 = 28.37'$.

$$41.69 + 28.37 = 70.06' = 70'0\tfrac{3}{4}" \text{ Check (see Figure 12.38)}$$

All these distances are shown on Figure 12.39 double underlined. Once these offsets and the stakes placed 25 ft farther on each offset line have been accurately located, the next step is to verify the offsets by some independent means.

12.9.4 Dimension Verification

For this type of analysis, the technique described in Section 6.13 (omitted measurements) is especially useful. Bearings are assumed that will provide the simplest solution. See Figure 12.40.

Computation for Check Diagonal 1–4

Line	Bearing	Distance	Lat.	Dep.
1–2	N 74°28′47″ E (90°-skew angle chosen for conven ience)	66.69	+17.84	+64.26
2–3	Due north	99.94	+99.94	+ 0.0
3–4	N 74°28′47″ E	59.81	+16.00	+57.63
1–4			+133.78	+121.89

Distance 1–4 = $\sqrt{133.78^2 + 121.89^2}$ = 180.98 ft
Allowable error = ±0.03 ft

Computation for Check Diagonal 5–6

Line	Bearing	Distance	Lat.	Dep.
5–2	S 74°28′47″ W	53.37	−14.28	−51.42
2–3	Due north	99.94	+99.94	− 0.0
3–6	S 74°28′47″ W	63.81	−17.07	−61.48
5–6			+68.59	−112.90

Distance 5–6 = $\sqrt{68.59^2 + 112.90^2}$ = 132.10 ft
Allowable error = ±0.02 ft

These values are shown in Figure 12.40. The same techniques can be used to calculate any other series of diagonals if sight lines are not available for the diagonals shown.

In addition to diagonal check measurements, or even in place of them, check measurements can consist of right-angle ties from ℄, as shown in Figure 12.38. Additional calculations are required for the chainage and distance to the outside stakes.

As in all construction layout work, great care must be exercised while placing the grade stakes. The grade stakes (iron bars) should be driven flush with the surface (or deeper) to prevent de-

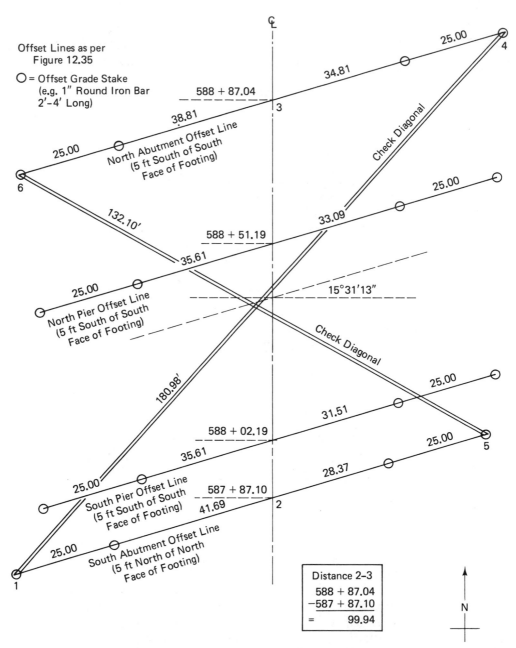

FIGURE 12.40 Layout verification sketch.

560

flection if heavy equipment were to cross over them. The grade
stakes are protected by placing substantial guard stakes adjacent
to them. The surveyor must bear in mind that, for many reasons,
it is more difficult to place a marker in a specific location than it
is to locate a set marker by field measurements. For example, it is
difficult to drive a 3- or 4-ft steel bar into the ground while keeping it in its proper location with respect to line and distance. It is
often the case that, as the top of the bar nears the ground surface,
it becomes evident that the top is either off line or off distance.
The solution to this problem is either to remove the bar and replace it or to wedge it into its proper location by driving a piece of
rock into the ground adjacent to the bar. The latter solution will
often prove unsuccessful, as the deflected bar can in time push
the rock aside and return to near its original position. These problems can be avoided by taking great care during the original placement; continuous line and distance checks are advised.

In this example, an offset line was established 5 ft from the
face of the footings. As foundation excavations are usually "neat,"
the value of 5 ft is quite realistic; however, the surveyor, in consultation with the contractor, must choose an offset that is optimal.
As in all construction work, the shorter the offset distance, the
easier it is to transfer line and grade to the structure accurately.

12.9.5 Grades

The 5-ft offset line in this example can be used not only to place
the footings, but also as the work proceeds above ground with the
abutment walls and pier columns; the formwork can be easily
checked as the offset lines will be clear of all construction activity.

The proposed top of footing elevations are shown on the general arrangements plan (Figure 12.35) and in Figure 12.39. The
elevation of the grade stakes opposite each end of the footings is
determined by differential leveling, and the resultant "cuts" are
given to the contractor either in the form of a grade sheet or batter
boards.

A word of caution: it is often the case that a bench mark is
located quite close to the work, and probably one instrument setup is sufficient to take the backsight, grade stake sights, and the
foresight back to the bench mark. When this is the case, two potential problems should be considered. One problem is that after
the grade stake sights have been determined the foresight back to
the bench mark will be taken—the surveyor, knowing the value
of the backsight taken previously, will be influenced in reading
the foresight as he will expect to get the same value. In this case,

the surveyor could well exercise less care in reading the foresight, using it only as a quick check reading; blunders have been known to occur in this situation. The second potential problem involves the use of automatic (self-leveling) levels as sooner or later all automatic levels will become inoperative due to failure of the compensating device. This device is used to keep the line of sight horizontal, and often relies on wires or bearings to achieve its purpose. If a wire breaks, the compensating device will produce a line of sight that is seriously in error. If the surveyor is unaware that the level is not functioning properly, the leveling work will be unacceptable, in spite of the fact that the final foresight will agree with the backsight.

The only safe way to deal with these potential problems is to always use two bench marks when setting out grades. Start at one BM and check into the second BM. Even if additional instrument setups are required, the extra work is a small price to pay for this accuracy check. Each structure location should have three bench marks established prior to construction so that if one is destroyed the minimum of two will remain for the completion of the project.

12.9.6 Cross Sections for Footing Excavations

Original cross sections will have been taken prior to construction. As the excavation work proceeds, cross sections are taken, keeping the structural excavations (higher cost) separate from other cut and fill operations for the bridge site. When all work is completed, final cross sections are taken to be used for payments and final records.

The structural excavation quantities are determined by taking preliminary and final cross sections on each footing. The offset line for each footing is used as a base line for individual footing cross sections.

12.9.7 Culvert Construction

The plan location and invert grade of culverts are shown on the construction plan and profile. The intersection of the culvert ₵ and the highway ₵ will be shown on the plan and will be identified by its highway chainage. In addition, when the proposed culvert is not perpendicular to the highway ₵, the skew number will be shown (see Figure 12.41).

The construction plan will show the culvert location (₵ chainage), skew number, and length of culvert; the construction

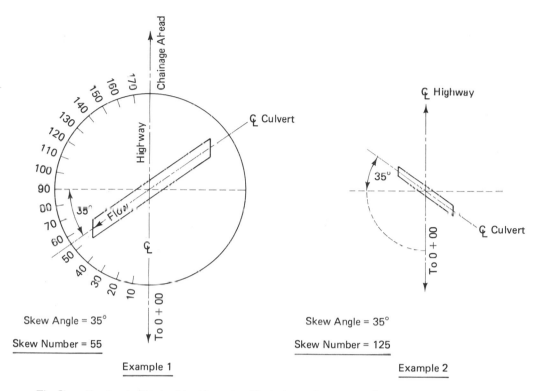

The <u>Skew Number</u> is Obtained by Measuring Clockwise to the Nearest 5°, the Angle Between the Back Tangent ₵ of the Highway and the ₵ of the Culvert.

FIGURE 12.41 Culvert skew numbers, showing the relationship between the skew angle and the skew number.

profile will show the inverts for each end of the culvert. One grade stake will be placed on the ₵ of the culvert, offset a safe distance from each end (see Figure 12.42).

The grade stake will reference the culvert ₵ and will give the cut or fill to the top of footing for open footing culverts, to top of slab for concrete box culverts, or to the invert of pipe for pipe culverts. If the culvert is long, intermediate grade stakes may be required. The stakes may be offset 6 ft (2 m) or longer distances if site conditions warrant.

In addition to the grade stake at either end of the culvert, it is customary when laying out concrete culverts to place two offset line stakes to define the end of the culvert. These end lines are normally parallel to the ₵ of construction or perpendicular to the culvert ₵. See Figures 12.43 for types of culverts.

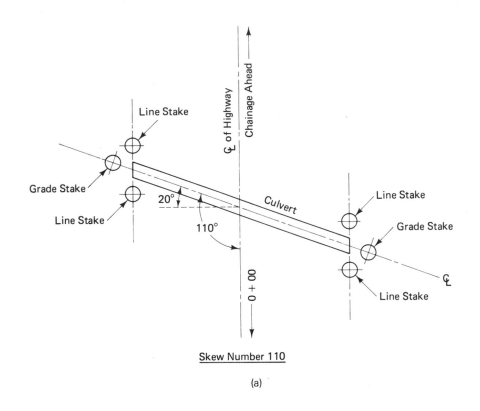

Line Stake

Grade Stake

Line Stake

Ҁ of Highway

Chainage Ahead

20°

110°

Culvert

Line Stake

Grade Stake

Ҁ

Line Stake

0 + 00

Skew Number 110

(a)

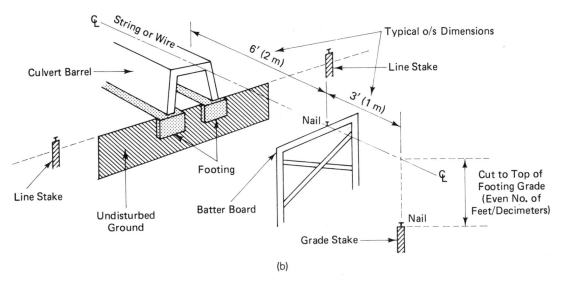

Ҁ

String or Wire

Culvert Barrel

6' (2 m)

Typical o/s Dimensions

Line Stake

3' (1 m)

Nail

Line Stake

Footing

Ҁ

Undisturbed
Ground

Batter Board

Nail

Cut to Top of
Footing Grade
(Even No. of
Feet/Decimeters)

Grade Stake

(b)

FIGURE 12.42 Line and grade for culvert construction: (a) plan view; (b) perspective view.

564

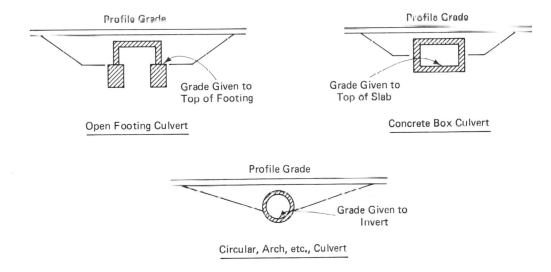

FIGURE 12.43 Types of culverts.

12.10 BUILDING CONSTRUCTION

All buildings must be located with reference to the property limits. Accordingly, the initial stage of the building construction survey involves the careful retracement and verification of the property lines. Once the property lines are established, the building is located according to plan, with all corners marked in the field. At the same time, original cross sections will be taken, perhaps using one of the longer wall lines as a base line.

12.10.1 Batter Boards

The corners already established will be offset an optimum distance and batter boards will be erected (see Figure 12.44). The cross pieces will either be set to first floor elevation (finished) or to a set number of feet (decimeters) above or below the first floor elevation. The contractor is always informed as to the precise reference datum.

The batter boards for each wall line will be joined by string or wire running from nails or sawcuts at the top of each batter board. It is the string (wire) that is at (or referenced to) the finished first floor.

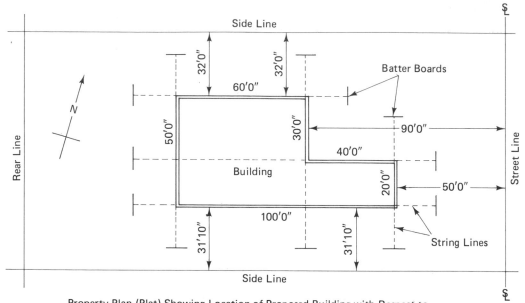

Property Plan (Plat) Showing Location of Proposed Building with Respect to the Property Lines.

This Plan also Shows the Location of the Batter Boards and the String Lines for Each Building Wall Footing.

FIGURE 12.44   Building layout.

After the excavation for footings and basement has been completed, final cross sections can be taken to determine excavation quantities and costs.

In addition to layout for walls, it is usual to stake out all columns or other critical features, giving the location and proposed grade.

Once the foundations are complete it is only necessary to check the contractor's work in a few key dimension areas on each floor. Optical plummets, lasers, and theodolites can all be utilized for high-rise construction (see Figures 12.45 and 12.46).

12.11 OTHER CONSTRUCTION SURVEYS

The techniques for survey layout of heavy construction (i.e., dams, port facilities, and other large-scale projects) are similar to those already described. Key lines of the project will be located according

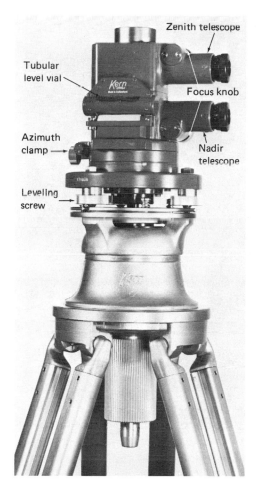

Zenith telescope

Tubular
level vial

Focus knob

Azimuth
clamp

Nadir
telescope

Leveling
screw

FIGURE 12.45 Kern OL precise optical plummet. SE in 100 m for a single measurement (zenith or nadir) = ±1 mm (using a coincidence level). Used in high-rise construction, towers, shafts, and the like. (Courtesy Kern Instruments, Inc.)

to plan and referenced for the life of the project. All key points will be located precisely, and in many cases these points will be coordinated and tied to a state or province coordinate grid.

The two key features of construction control on large-scale projects are (1) high precision and (2) durability of the control monuments and the overall control net.

Prior to construction, horizontal and vertical controls are established over the project site. Most of the discussion in this chapter has dealt with rectangular layout techniques, but as noted in Chapter 11, some projects lend themselves in whole or part to polar layout techniques utilizing coordinated control and coordinated construction points. Heavy construction is one project area usually well suited for polar layouts, and the reader is referred to Chapter 11 for specific illustrations.

FIGURE 12.46 Rotating laser positioned to check plumb orientation of construction wall. (Courtesy Laser Alignment, Inc.)

12.11.1 Construction Survey Specifications

By far the bulk of all construction work is laid out using $1/3000$ (e.g., sewers, highways) or $1/5000$ (e.g., curbed roads, bridges) specifications. The foregoing is true for rectangular layouts, but as noted in Chapter 11, the control for layout by polar techniques will be accomplished using survey techniques allowing for $1/10000$ to $1/20000$ accuracy ratios. Section 11.11 gives specifications now being used internationally in construction surveying.

PROBLEMS

12-1 A new road is to be constructed beginning at an existing road— ℄ elevation = 472.70 ft.—for a distance of 600 ft. The ℄ gradient is to rise at 1.18%. The elevations of the offset grade stakes are as follows: 0 + 00 = 472.60; 1 + 00 = 472.36; 2 + 00 = 473.92; 3 + 00 = 475.58; 4 + 00 = 478.33; 5 + 00 = 479.77; 6 + 00 = 480.82. Prepare a grade sheet (see Table 12–3) showing the cuts and fills in feet and inches.

12-2 A new road is to be constructed to connect two existing roads. The ℄ intersection with the east road (0 + 00) is at an elevation of 210.666 m. and the ℄ intersection at the west road (1 + 32.562) is at an elevation of 209.446 m. The elevations of the offset grade stakes are as follows: 0 + 00 = 210.831; 0 + 20 = 210.600; 0 + 40 = 211.307; 0 + 60 = 210.114; 0 + 80 = 209.772; 1 + 00 = 209.621; 1 + 20 = 209.308; 1 + 32.562 = 209.400. Prepare a grade sheet (see Table 12–1) showing cuts and fills in meters.

12-3 In Figure 12.12A, if the ditch invert is to be 3′6″ deep (below subgrade), how far from ℄ would be the ℄ of the ditch?

12-4 In Figure 12.12B, a) if the ditch invert is to be 2.0 m deep (below subgrade), how far from ℄ would be the ℄ of the ditch? b) if the original ground is level with ℄ right across the width of the highway, how far from ℄ would be the slope stake marking the far side of ditch at original ground?

12-5 A storm sewer is to be constructed from existing M.H. #8 (invert elevation = 360.21) @ +1.32% for a distance of 240 ft to proposed M.H. #9. The elevations of the offset grade stakes are as follows: 0 + 00 = 368.75; 0 + 50 = 368.81; 1 + 00 = 369.00; 1 + 50 = 369.77; 2 + 00 = 370.22; 2 + 40 = 371.91. Prepare a grade sheet (see Table 12.4) showing stake-to-batterboard distances in feet and inches; use a 14-ft grade rod.

12-6 A sanitary sewer is to be constructed from existing manhole #4 (invert elevation = 150.810) @ + 0.68% for a distance of 115 meters to proposed M.H. #5. The elevations of the offset grade stakes are as follows: 0 + 00 = 152.933; 0 + 20 = 152.991; 0 + 40 = 153.626; 0 + 60 = 153.725;

0 + 80 = 153.888; 1 + 00 = 153.710; 1 + 15 = 153.600. Prepare a grade sheet (see Table 12–6) showing stake-to-batterboard distances in meters. Use a 4-m grade rod.

13 *Land Surveying*

13.1 GENERAL

Land surveying involves the establishment of boundaries for public and/or private properties. It includes both the measurement and laying out of the boundaries.

Land surveys are made for one or more of the following reasons:

1. To subdivide the public lands (United States) or crown lands (Canada) into townships, sections or concessions and quarter-sections or lots, thus creating the basic fabric to which all land ownership and subsequent surveys will be directly related.

2. To attain the necessary information for writing a legal description and for determining the area of a particular tract of land, the boundaries of the property being defined by occupation lines, such as fences; natural linear features, such as river banks and rock bluffs; and/or singular features, such as individual trees and rock or boulder formations.

3. To reestablish the boundaries of a parcel of land that has been previously surveyed and legally described.

4. To subdivide a parcel of land into two or more smaller units in agreement with a plan that dictates the size, shape, and dimensions of the smaller units.

5. To establish the position of particular features such as buildings on the parcel with respect to the boundaries.

Land surveys are required whenever real property or real estate is transferred from one owner to another. The location of the boundaries must be established to ensure that the parcel of land being transferred is properly located, acceptably close to the size and dimensions indicated by the owner, and free of encroachment by adjacent buildings, roadways, and the like. The boundary must be related directly to a preestablished point in the township survey fabric, or *tie line*, to determine properly the location of the land tract. It is often necessary to determine the location of buildings on the property relative to the boundaries for the purpose of arranging mortgages and ensuring that building bylaws concerning location restrictions have been satisfied.

This section describes the techniques employed and the modes of presentation used for various types of land surveys. Familiarization with the methods employed is essential for anyone involved in any aspect of surveying. For example, construction surveying (see Chapter 12) and layout for new road construction within an allowance for road on the original township survey necessitates the reestablishment of the legal boundaries of the road allowance. Bends and/or jogs in the road allowance alignment, attributable to the methods employed in the original township survey, occur frequently. An understanding of the survey system used not only explains why these irregularities exist, but also warns the surveyor in advance where these may occur.

A recently surveyed subdivision provides ample evidence of property boundary locations. However, as soon as any construction commences, the lot markers are removed, bent, misplaced, or covered with earth fill. Therefore, reestablishment of the street and lot pattern becomes an essential yet demanding task. It must be carried out before further work may proceed. The use of surveys positioning buildings on the lots, known as *title* or *mortgage* surveys, greatly assists in reestablishing the property boundaries.

13.1.1 Surveyor's Duties

The land surveyor must be knowledgeable in both the technical and legal aspects of property boundaries. Considerable experience is also required, particularly with regard to deciding on the "best evidence" of a boundary location. It is not uncommon for the surveyor to be exposed to conflicting physical as well as legal evidence of a boundary line. Consequently, a form of apprenticeship is often required to be served, in addition to academic qualifications, before the surveyor is licensed to practice by the state, province, or territory.

Regulations are usually required, by law, to assist in standardizing procedures and requirements. Most states and provinces have organizations or associations, which are corporate bodies operating under conditions set out in legislative acts. The objectives of the Association of Ontario Land Surveyors are, for example, as specified in the Surveyors Act:

1. To regulate the practice of professional land surveying and to govern the profession in accordance with this act, the regulations and the bylaws;
2. To establish and maintain standards of knowledge and skill among its members; and
3. To establish and maintain standards of professional ethics among its members, in order that the public interest may be served and protected.

13.1.2 Historical Summary

The first surveys of townships were made in Ontario, Canada, in 1783 and north of the Ohio River in the United States in 1785. Federal legislation in the United States dictated the establishment of section corners and the intervals at which markers were to be placed. Originally, only the township lines were run, and section corners were established along these lines at intervals of 1 mile. In 1805, a congressional act directed that each section corner should be marked and that the public lands should be divided into quarter-sections on the township maps, or *plats*, as they were termed legislatively. In 1832, the Congress ruled that the public lands should be subdivided into quarter quarter-sections. In 1909, the secretary of the interior was directed to resurvey the boundaries of previously surveyed public lands at his discretion. This applied only to lands that were still public and thus had no effect upon the boundaries of privately owned lands.

Prior to 1910, when the contract system of surveying the public lands was abolished, most of the land surveying had been done by private surveyors under government contract. These early surveys were made with crude instruments such as a compass and a chain, often under adverse field conditions. Consequently, some were incomplete and others showed field notes for lines that were never run in the field. Therefore, the corners and lines are often found in other than their theoretical locations. Regardless, the original corners as established during the original survey, however inaccurate or incomplete, stand as the true corners. The surveyor must therefore use these corners for all subsequent survey work.

In 1946, the Bureau of Land Management in the Department of the Interior was made responsible for public land surveys, which had been administered by the General Land Office in the Treasury Department since 1812.

In 1789 the surveyor general of Canada was directed to prepare plans of each district using townships 9 miles wide by 12 miles deep. Prior to 1835, a variety of nonsectional or "special" township survey systems were authorized. Each presented particular techniques for the establishment of individual lines and generated difficulties for resurveys and access along public road allowances. These difficulties are discussed in this section. This confusion was partially ended in 1859, when sectional townships 6 miles square divided into 36 sections each 1 mile square without road allowances were authorized through federal legislation.

In 1906, the 1800-acre sectional system, having townships 9 miles square with 12 concessions and lots of 150 acres each, was authorized by the federal government. The methods of surveys in sectional townships and aliquot part divisions were changed. The techniques for establishing the boundaries of aliquot parts were altered by an amendment to the Surveys Act in 1944. The long and somewhat turbulent history of land surveying in both the United States and Canada is evident from the preceding historical summary.

The complexities of survey systems based on French common law, used in states such as Louisiana and provinces like Quebec, are beyond the scope of this section. These systems are based on the importance of frontage ownership along water bodies, such as rivers, for purposes of water access and transportation. Consequently, the irregularities resulting from following natural boundaries present a totally different resurvey concept to the systems based on English common law.

Since various regions of the United States and Canada have been surveyed under different sets of instructions from 1783 to the present time, important changes in detail have occurred. Therefore, the local surveyor or engineer should become familiar with the exact methods in use at the time of the original survey, before commencing the retracement of land boundaries.

13.2 PUBLIC LAND SURVEYS

13.2.1 General

The most common methods for public land surveys in the United States and Canada provided for townships that are 6 miles square, each containing 36 sections. Each section was thus 1 mile square and was numbered from 1 to 36, as illustrated in Figure 13.1. Section 1 is in the northeast corner of the township. The sections are then numbered consecutively from east to west and from west to east alternately, ending with section 36 in the southeast corner, as illustrated in Figure 13.1.

Initially, only the exterior boundaries of the township were surveyed and mile corners were established on the township lines. However, the *plats* or original township maps showed subdivisions into sections 1 mile square, numbered as just described. Under these conditions, subsequent surveys had to be carried out to establish the corners and boundaries of sections in the interior of the townships. Changes in the procedures for township surveys led successively to each section corner being marked, then to each quarter section corner being established. As the methods of resurvey differ depending on the corners that were originally established, it is essential that the surveyor or engineer understand these techniques.

Other systems were used for original township surveys, particularly in Ontario, Canada. These systems are discussed herein

6	5	4	3	2	1
7	8	9	10	11	12
18	17	16	15	14	13
19	20	21	22	23	24
30	29	28	27	26	25
31	32	33	34	35	36

FIGURE 13.1 Numbering of sections.

as nonsectional or special. One example of these systems, termed *single front*, is illustrated in Figure 13.2. The usual practice in the original survey involved surveying the township boundaries and the *fronts* of the concessions. The concessions are usually oriented east–west, although the bearing of the front of a concession line is often oriented up to 20° off due east or due west.

Each concession represents a band 100 chains (6600 ft, 2000 m) wide, stretching between the township boundaries on both sides, as illustrated in Figure 13.2. The lots are numbered starting at the township boundary where the survey commenced (the southerly boundary in Figure 13.2). Road allowances 1 chain

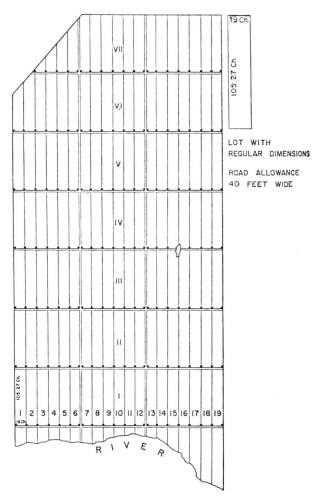

FIGURE 13.2 Original single-front system.

(66 ft, 20 m) wide are established along the fronts of the concessions and at regular intervals between the lot (every five lots in Figures 13.2). The principles involved in resurveying these non-sectional townships are also applicable to rural land surveys, discussed in Section 13.5.

13.2.2 Standard Lines

Initial Points. The point at which a survey commences in any area is known as the *initial point.* A meridian, called the *principal meridian,* and a parallel of latitude, called the *base line,* are run through the initial points as illustrated in Figure 13.3. After the initial point has been established, the latitude and longitude of the point are determined using field astronomical observations.

Many of the original township surveys were carried out simultaneously. Therefore, a large number of initial points have been established. The principal meridian through an initial point is given a name, and the original surveys governed by each initial point are recorded. For example, the Mount Diablo principal meridian, governing surveys in the states of California and Nevada, has a longitude of 121°54′47″ and a latitude of 37°52′54″. The original surveys referred to a particular initial point are shown on a map entitled "United States, Showing Principal Meridians, Base Lines, and Areas Governed Thereby," published by the Bureau of Land Management (Government Printing Office, Washington, D.C.).

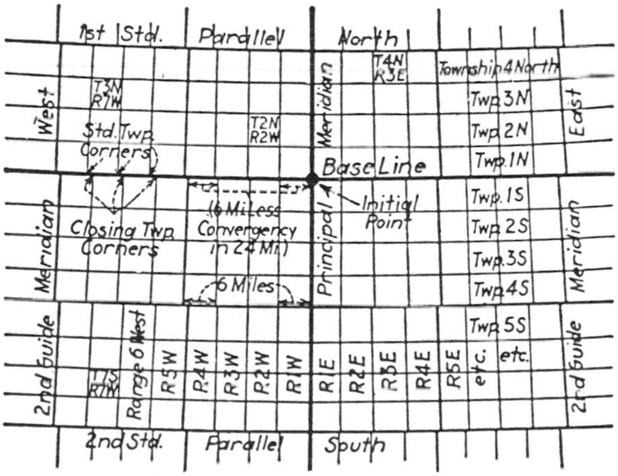

FIGURE 13.3 Initial point, parallels, and meridians.

Meridians The principal meridian is a true north–south line that is extended in either direction from the initial point to the limits of the area covered by the township surveys. This line is monumented at 40-chain ($^1/_2$ mile, 800 m) intervals in both the United States and Canada. Additional monumentation was provided in the United States at intersections with streams 3 chains (198 ft, 60 m) or more wide, navigable water bodies, and lakes having an area of 25 acres (10 hectares) or more.

In the United States, two independent linear measurements of the meridian are made. When the difference in these measurements is greater than 20 links (13.2 ft, 4.0 m) per 80 chains (1 mile, 1600 m), the line must be remeasured until the difference between two of the measurements is less than 20 links. The corners are placed at the mean distances. If the alignment is found to be over 3′ off the north–south course, the alignment must be corrected.

In Canada, no accuracy tolerances were given in the early instructions and no penalties were levied if the work were incorrect. There was no suggestion that the work should be redone. A common source of error was miscounting the number of chains (66 ft, 20 m) across the front of a section or lot. Consequently, it is not uncommon to find the older township surveys differing from the intended dimensions by approximate multiples of 66 ft. As the only equipment available to carry out the early surveys was a compass and chain, differences much greater than those stated previously will be found in doing resurveys.

Guide meridians are located at intervals of 24 miles (38.5 km) east and west of the principal meridian. These lines are true meridians and extend north of the base line to their intersection with the standard parallel, as illustrated in Figure 13.4. These guide meridians are established in the same manner as the principal meridians. Due to the convergence of meridians, the distance between these lines will be 24 miles (38.5 km) only at the starting points along the base line. As illustrated in Figure 13.4, the distance between meridians at all other points is less than 24 miles (38.5 km). As a new monument is established where the guide meridian meets the next standard parallel to the north, two sets of monuments are located on the standard parallels. Those established when the parallel was first located are called *standard corners* and apply to the lands north of the parallel. Those established by the intersection with the parallel of the guide meridians from the south are called *closing corners*. The distance between the standard and closing corners should be recorded in the field notes.

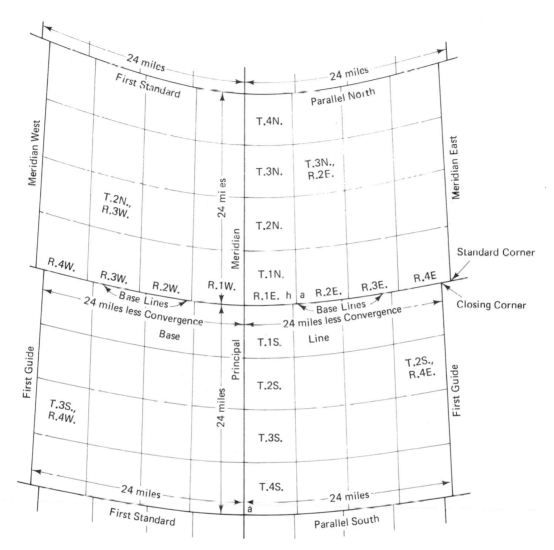

FIGURE 13.4 Standard parallels and guide meridians.

Similar accuracy requirements apply to the guide meridians as to the principal meridians. Monuments are placed at half-mile (40 chain, 800 m) intervals. All measurement discrepancies are placed in the last half-mile. Consequently, the distance from the first monument south of the standard parallel to the closing monument on the parallel may differ from 40 chains (½ mile, 800 m).

Convergence of Meridians. As all meridians form a great circle
through both the north and south poles, these lines converge toward a pole as they proceed northerly or southerly from the equator. Consequently, a line at right angles (90°) with a meridian will
be an east–west line for an infinitely small distance from the meridian. If the line at 90° to the meridian is extended, it will form a
great circle around the earth that will not be an east–west line except along the equator.

The true east–west line is called a parallel of latitude, which
is represented by a small circle, illustrated by line *AB* in Figure
13.5. The true parallel, due to convergence, will gradually curve to
the north (when located north of the equator) of the great circle at
right angles to the meridian.

In Figure 13.5, *DAG* and *EBG* represent two meridians. *P* is

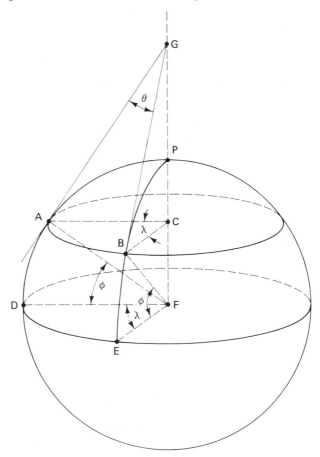

FIGURE 13.5 Convergence of meridians.

the earth's north pole; F is the center of the earth; DF is an arc of the equator; AB is the arc of a parallel of latitude at latitude Φ; λ is the difference in longitude between the meridians. The angular and linear convergency of the two meridians is to be determined.

The difference in longitude between the meridians is

$$\lambda \text{ (radians)} = \frac{AB}{BC} \therefore AB = BC \cdot \lambda \text{ (radians)} \qquad (13.1)$$

The latitude of the circle, of which AB is an arc, is

$$\phi - \angle BFE = \angle BGC \qquad (13.2)$$

Therefore,

$$\text{Sin } \phi = \frac{BC}{BG} \text{ and } BG = \frac{BC}{\text{Sin } \phi}$$

The angle of convergency, for all practical purposes, is

$$\theta \text{ (radians)} = \frac{AB}{BG}$$

Substituting for AB and BG from Eqs. (13.1) and (13.2) gives

$$\theta = \frac{BC \cdot \lambda}{BC/\text{Sin } \phi} = \lambda \text{ Sin } \phi \qquad (13.3)$$

where θ and ϕ are in radians.

If AB, the distance measured along a parallel between two meridians is d, and the radius of the earth at the parallel is R, then from Eq. (13.1),

$$\lambda = \frac{AB}{BC} = \frac{d}{R \cos \phi}$$

Substitution of this equation into Eq. (13.3) gives

$$\theta \text{ (radians)} = \frac{d \text{ Sin } \phi}{R \text{ Sin } \phi} = \frac{d \text{ Tan } \phi}{R} \qquad (13.4)$$

If d is in miles and $R = 20{,}890{,}000$ ft (the mean radius of the earth), then from Eq. (13.4)

$$\theta \text{ (seconds)} = 52.13 \, d \text{ Tan } \phi \qquad (13.5)$$

If d is in kilometers, then

$$\theta \text{ (seconds)} = 32.392 \, d \text{ Tan} \qquad (13.6)$$

If the distance between parallels along the meridians, represented by AD and BE in Figure 13.5, is y, and if the difference in arc length along the two parallels, that is, $DE - AB$ in Figure 13.5, is p, then for all practical purposes

$$\theta = \frac{p}{y}$$

where θ is the mean angle of convergency of the two meridians. If the mean latitude is ϕ, substituting in Eq. (13.4) gives

$$p = \frac{dy \, \text{Tan} \, \phi}{R} \tag{13.7}$$

Therefore, the reduction in arc distance along the northerly parallel due to convergence can be calculated using Eq. (13.7).

If d and y are in miles and R is the mean radius of the earth, then

$$
\begin{aligned}
p \quad &\text{(feet)} &&= 1.33 \, dy \, \text{Tan} \, \phi & (13.8) \\
p \quad &\text{(chains)} &&= 0.0202 \, dy \, \text{Tan} \, \phi & (13.9) \\
p \quad &\text{(meters)} &&= 0.4055 \, dy \, \text{Tan} \, \phi & (13.10)
\end{aligned}
$$

If d and y are in kilometers, then

$$p \quad \text{(meters)} = 0.1565 \, dy \, \text{Tan} \, \phi \tag{13.11}$$

Example 13.1

Find the angular convergency between two guide meridians 24 miles (38.63 km) apart at latitude 47°30′:

1. Using Eq. (13.5), $\theta'' = 52.13 \times 24 \times \text{Tan} \, 47°30′ = 1365'' = 0°22′45''$.
2. Using Eq. (13.6), $\theta'' = 32.38 \times 38.63 \times \text{Tan} \, 47°30′ = 1365 = 0°22′45''$.

Example 13.2

Find the convergency in feet, chains, and meters of two guide meridians 24 miles apart and 24 miles long if the *mean* latitude is 47°30′.

1. Using Eq. (13.8) $p = 1.33 \times 24 \times 24 \times \text{Tan} \, 47°30′ = 836$ ft
2. Using Eq. (13.9) $p = 0.0202 \times 24 \times 24 \times \text{Tan} \, 47°30′ = 12.70$ chains
3. Using Eq. (13.10) $p = 0.4055 \times 24 \times 24 \times \text{Tan} \, 47°30′ = 255$ m

4. Using Eq. (13.11) $p = 0.1505 \times 38.63 \times 38.63 \times$ Tan 47°30' = 255 m

As discussed in the previous section, two sets of monuments are established on the standard parallels due to convergence of the meridians. Using Example 13.2, the distance between these monuments would be 12.70 chains.

Parallel of Latitude. Due to this convergence, the base line, being a true parallel, must be run as a curve having chords 40 chains ($\frac{1}{2}$ mile, 800 m) long. There are three methods of establishing a parallel of latitude: (1) the solar method, (2) the tangent method, and (3) the secant method, which is the most commonly used.

Solar Method

The sun is used to determine the true meridian every 40 chains using the methods set out in Chapter 11. The true parallel is then established by turning 90° from the meridian. A slight error will be incurred using this method because observations are taken at 40-chain intervals rather than at small increments along the parallel. However, the line defined using this method will be well within acceptable accuracies for original surveys.

Tangent Method

The direction of the tangent is determined by turning a horizontal angle of 90° to the east or west with the meridian. Points on the parallel are established at 40-chain ($\frac{1}{2}$-mile, 800-m) intervals by offsets to the north from the tangent. The establishment of a base line for a latitude of 47°30' is illustrated in Figure 13.6. The illustration is exaggerated to illustrate properly the offset lengths. In fact, the small magnitude of the offset distances compared with the distances along the tangent leads to the conclusion that distances measured along the tangent are essentially equal to those measured along the parallel within the accuracy requirements for base-line surveys.

The values of the offsets are dependent on the latitude and the distance from the starting meridian. These values are proportional to the square of the distances from the starting meridian, as illustrated in Table 13.1.

The values of the tangent offsets significantly exceed the width of normal cut lines, as illustrated in Figure 13.6. Consequently, as the true parallel must be blazed and the landmarks noted, it may be necessary to clear two lines: the tangent and the true parallel. Therefore, this method is more costly and time consuming than the secant method.

TABLE 13.1 *Offsets from Tangent to Parallel (Feet) for Latitudes from 30° to 50° Inclusive*

Latitude	1 Mile	2 Miles	3 Miles	4 Miles	5 Miles	6 Miles
30°	0.38	1.54	3.46	6.15	9.61	13.83
31°	0.40	1.60	3.60	6.40	10.00	14.40
32°	0.42	1.66	3.74	6.65	10.40	14.97
33°	0.43	1.73	3.89	6.91	10.80	15.56
34°	0.45	1.80	4.04	7.18	11.22	16.16
35°	0.47	1.86	4.19	7.45	11.65	16.77
36°	0.48	1.93	4.35	7.73	12.09	17.40
37°	0.50	2.01	4.51	8.02	12.53	18.05
38°	0.52	2.08	4.68	8.32	12.99	18.71
39°	0.54	2.15	4.85	8.62	13.47	19.39
40°	0.56	2.23	5.02	8.93	13.95	20.09
41°	0.58	2.31	5.20	9.25	14.46	20.82
42°	0.60	2.40	5.39	9.58	14.97	21.56
43°	0.62	2.48	5.58	9.92	15.50	22.33
44°	0.64	2.57	5.78	10.28	16.06	23.12
45°	0.66	2.66	5.98	10.64	16.62	23.94
46°	0.69	2.75	6.20	11.02	17.21	24.79
47°	0.71	2.85	6.42	11.41	17.83	25.67
48°	0.74	2.95	6.65	11.81	18.46	26.58
49°	0.76	3.06	6.88	12.24	19.12	27.53
50°	0.79	3.17	7.13	12.68	19.81	28.52

Secant Method

The secant used for laying out a parallel of latitude passes through the 1- and 5-mile points on the parallel, as illustrated in Figure 13.7. The offset is measured southerly from the initial point to the secant and the angle turned to determine the direction of the secant easterly. The secant is the line actually run, the points on the parallel being located at 40-chain ($^{1}/_{2}$-mile, 800-m) intervals by offsets from the secant. As illustrated in Figure 13.7, all offsets to the true parallel are southerly and those at the $^{1}/_{2}$-, $5^{1}/_{2}$-, and 6-mile points are to the north.

The secant method is advantageous because the offsets are much smaller than those required for the tangent method, as can be seen by comparing Figures 13.6 and 13.7. Therefore, a cleared line of reasonable width will contain both the secant and the parallel. The measurements made to landmarks along both lines will be essentially the same, thus requiring no modifications in the field notes.

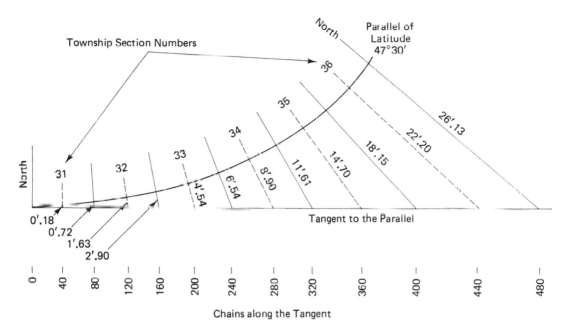

FIGURE 13.6 Baseline using tangent offsets for latitude of 47°30' (offsets from the tangent to the parallel of latitude are shown in feet).

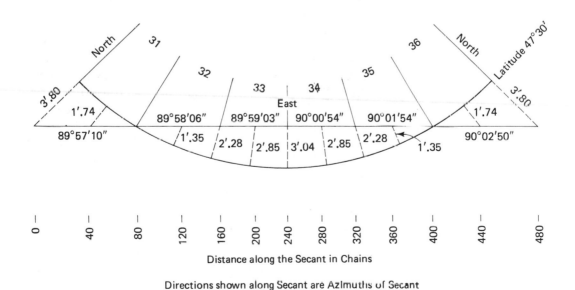

Directions shown along Secant are Azimuths of Secant determined from Table 13.2

FIGURE 13.7 Base line using secant offsets for latitude of 47°30' (offsets from the secant to the parallel of latitude are shown in feet).

Both the tangent and secant are straight lines on the plan view. However, due to the convergency of meridians, the azimuths of these lines vary along the line. The changes in azimuths along these lines are determined using Eq. (13.5), where d is the distance along the parallel from the starting meridian.

The tangent commences at an azimuth of 90° from the meridian and bends gradually southerly, as shown on Figure 13.6. The azimuth of the tangent 6 miles from the starting meridian is therefore 90°05′41″, the additional 5′41″ being the convergence for a d of 6 miles. Using the tangent method, the 90° azimuth would be reestablished every 6 miles and the process of laying out the parallel repeated as illustrated in Figure 13.6.

The azimuths of the secant are shown in Table 13.2. Assuming that the secant is laid out toward the east as illustrated in Figure 13.7, the direction of the secant from the starting meridian to the end of the third mile is north of true east. From the 3-mile to the 6-mile points, the azimuth is south of true east. At the 3-mile point, the azimuth of the secant is 90°. At the starting meridian, the secant has an azimuth of 89°57′10″ at a latitude of 47°30″, as illustrated in Figure 13.7 and interpolated from Table 13.2. The difference between the starting azimuth and that at the 3-mile point is 02′50″, the angular convergency for meridians 3 miles apart. By the same reasoning, the secant azimuth at the 6-mile point is 90°02′50″. The transit is set up at the 0-mile offset point, and the azimuth of the secant line is laid off using the azimuths given in Table 13.2. For our example at a latitude of 47°30′, the azimuth is 89°57′10″. The secant is then directed in a straight line for 6 miles and the points established at ½-mile (40-chain, 800-m) intervals using the offsets as described previously. At the end of 6 miles, the succeeding secant line may be established by either (1) establishing the true meridian and laying off the same azimuth angle as before or (2) turning off a deflection angle, from the preceding secant to the succeeding secant, equal to the convergency of meridians 6 miles apart. These deflection angles are given in the last column of Table 13.2.

TABLE 13.2 *Azimuths of the Secant*

Lat.	0 Mile	1 Mile	2 Miles	3 Miles	Deflection Angle 6 Miles
°	° ′	° ′	° ′		′ ″
25	89 58.8	89 59.2	89 59.6	90°	2 25
26	58.7	59.2	59.6	E or W.	2 32
27	58.7	59.1	59.6		2 39
28	58.6	59.1	59.5		2 46
29	58.6	59.0	59.5		2 53

TABLE 13.2 *(continued)*

	6 miles	5 miles	4 miles	3 miles	
30	58.5	59.0	59.5		3 00
31	58.4	59.0	59.5		3 07
32	58.4	58.9	59.5		3 15
33	58.3	58.9	59.4		3 23
34	58.2	58.8	59.4		3 30
35	58.2	58.8	59.4		3 38
36	58.1	58.7	59.4		3 46
37	58.0	58.7	59.3		3 55
38	58.0	58.6	59.3		4 04
39	57.9	58.6	59.3		4 13
40	57.8	58.5	59.3		4 22
41	57.7	58.5	59.2		4 31
42	57.7	58.4	59.2		4 41
43	57.6	58.4	59.2		4 51
44	57.5	58.3	59.2		5 01
45	57.4	58.3	59.1		5 12
46	57.3	58.2	59.1		5 23
47	57.2	58.1	59.1		5 34
48	57.1	58.1	59.0		5 46
49	57.0	58.0	59.0		5 59
50	56.9	57.9	59.0		6 12
51	56.8	57.9	58.9		6 25
52	56.7	57.8	58.9		6 39
53	56.6	57.7	58.8		6 54
54	56.4	57.6	58.8		7 09
55	56.3	57.5	58.8		7 25
56	56.2	57.4	58.7		7 42
57	56.0	57.3	58.7		8 00
58	55.8	57.2	58.6		8 19
59	55.7	57.1	58.6		8 38
60	55.5	57.0	58.5		8 59
61	55.3	56.9	58.4		9 22
62	55.1	56.7	58.4		9 46
63	54.9	56.6	58.3		10 11
64	54.7	56.5	58.2		10 38
65	54.4	56.3	58.1		11 08
66	54.2	56.1	58.1		11 39
67	53.9	55.9	58.0		12 13
68	53.6	55.7	57.9		12 51
69	53.2	55.5	57.8		13 31
70	89°52'.9	89°55'.3	89°57'.6		14' 55"
	6 miles	5 miles	4 miles	3 miles	

Source: U.S. Bureau of Land Management

13.2.3 Township Boundaries

Most township boundaries in the United States and Canada were established using the sectional systems, wherein the boundary lines were oriented north–south and east–west. As this was the most common method, it is emphasized in this section. Nonsectional or special systems were used in certain areas due primarily to dominant geographical features combined with questionable regulations and ill-conceived legislation.

Sectional Systems. The normal method of establishing township boundaries is best understood by reference to Figure 13.8. The procedure is set out as follows:

1. Commencing at point 1 located on the first standard parallel south (a base line) 6 miles to the east of the principal meridian (or guide meridian), a line is run due north for 6 miles to point 2. Monuments are established at intervals of 40 chains ($^1/_2$ mile, 800 m).

2. From point 2, a line is run westerly to intersect the principal meridian (or guide meridian) at or near point 3. As this is a trial line, only temporary monuments are set at 40-chain ($^1/_2$-mile, 800-m) intervals. Since point 3 has previously been set during the running of the principal (or guide) meridian, the amount by which the trial line fails to intersect point 3 is measured.

3. The line from point 3 to point 2 is run along the correct course, and the temporary monuments are replaced by permanent monuments in the correct pattern. The true line is properly blazed, and the topographic and vegetative features are recorded along the true line. Due primarily to the convergence of meridians, the length of the line between points 2 and 3 will be less than 6 miles. This difference resulting from convergence and measurement errors is placed in the most westerly 40 chains ($^1/_2$ mile, 800 m). All other distances are therefore 40 chains ($^1/_2$ mile, 800 m).

The northern boundary should close on point 3 within 3 chains for line minus the convergence (see Section 13.2.2). If the error exceeds these limits, the lines are retraced until the error is located and corrected.

4. The eastern boundary of the next township to the north is run northerly from point 2 to point 4.

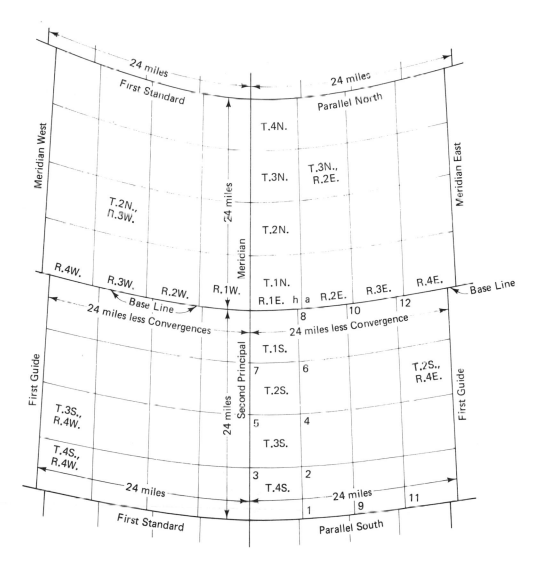

FIGURE 13.8 Township boundaries.

5. The procedure in step 2 is repeated to establish the north boundary of the next township between points 4 and 5. Any discrepancies are again left in the most westerly 40 chains (1/2 mile, 800 m).

6. The preceding procedures are continued, establishing easterly and northerly township boundaries in that order, that is, points 6 and 7.

7. From point 6, the meridian is run to its intersection with the base line (or standard parallel) at point 8. A closing corner is established at point 8. The distance from the closing corner to the nearest standard corner is measured and recorded. Since all errors in measurement up the meridian from point 1 to point 8 are placed in the line from point 6 to point 8, the last 40 chains ($\frac{1}{2}$ mile, 800 m) on the meridian may be less or more than 40 chains.

8. The two other meridians, points 9 to 10 and 11 to 12 in Figure 13.8, in the 24-mile-square block are run in a manner similar to that described for points 1 to 8. Therefore, any east–west discrepancies are placed in the westerly 40 chains ($\frac{1}{2}$ mile, 800 m) of each township as before, and the northerly 40 chains of each meridian will absorb the measurement errors.

An understanding of this procedure is important to the surveyor because of the realization that all 36 sections in a township are not of equal size. Therefore, the legal requirements relating to the rectangular surveys of the public lands should be examined. Due to the procedures employed in carrying out the township boundary surveys, combined with the effect of convergence of meridians, certain sections within each township will vary in areal extent. This is illustrated in Figure 13.9. Of the 36 sections in each standard township, 25 are considered to contain 640 acres (approximately 250 hectares), as these sections are not affected by the convergence of meridians or the placing of measurement errors. The sections containing less than 640 acres (approximately 250 hectares) are located along the westerly township boundary, primarily due to convergence of meridians, and along the northerly boundary, primarily due to the measurement errors being concentrated in the northerly 40 chains ($\frac{1}{2}$ mile, 800 m) along the meridian. Therefore, sections 1 to 5, 7, 18, 19, 30, and 31 will contain lesser acreages. The differences between the area of these sections from the standard 640 acres (approximately 250 hectares) will be minimal for section 31 and will increase northerly along the westerly township boundary to section 7 because of the increasing effect of convergence toward the north. Sections 1 to 5 inclusive will contain 640 acres (approximately 250 hectares) *more or less* depending upon the accuracy of the field measurements. Section 6, as illustrated in Figure 13.9, is most effected by convergence and is also subject to measurement errors. Therefore, in a standard township, section 6 should have the smallest area.

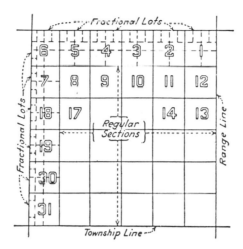

FIGURE 13.9 Relative sizes of sections in sectional townships.

These variations in section size and the reasons causing them are important to understand before we discuss the subdivision of townships in Section 13.2.4.

Nonsectional and Special Systems. The situation in the Province of Ontario best illustrates the myriad of special survey systems employed. The two qualities required of the land before it was considered for being opened into a township were accessibility and suitability for agriculture. Accessibility dictated the early opening of riverfront and lakefront townships. Therefore, most of the early townships were situated along the shores of the St. Lawrence, Ottawa, and Niagara rivers and Lakes Ontario and Erie. The urgency to provide the early settlers with river or lake frontage resulted in the township boundaries paralleling these major waterways. The result was that the township boundaries were oriented in almost any other direction than north–south or east–west. The best example is illustrated by those township boundaries established parallel with the general direction of the north shore of Lake Ontario. This bearing of N 74° E governs a large number of the township boundaries in Southern Ontario, as the original townships established along the shoreline were extended northerly and thus have the same orientation.

When it was decided to survey a township in Southern Ontario, instructions for the survey were prepared by the surveyor general. The details included the locations and dimensions of the township. Dimensions varied considerably from 9 miles square to 10 miles square to 9-mile by 12-mile townships. These sizes, as well as the special survey system to be used, varied with the opinions of the surveyor general, legislative changes, and dissatisfaction of the settlers with previous systems. The point where the surveyor would commence his township boundary survey was described either by reference to a natural feature, such as a creek mouth or point of land, or to a previously established survey monument.

When the base line was established across the front to the township, the usual practice was for the surveyor to run the perimeter of the township and finally some or all of the interior concession lines (see Section 13.2.5).

Although all surveyors were urged to carry out their work accurately, no tolerances were given on the early instructions and no penalty was levied if the work were done incorrectly. If lines were found to be run incorrectly, the line was not rerun and corrected. The survey instructions simply encouraged the surveyor to adjust his instruments so that his future work would be better.

As a result of these procedures (or lack thereof), townships surveyed parallel with or back from a natural water boundary would intersect with another group established parallel with a different water boundary. The result in such cases was the creation of "filler" townships. These filler townships have highly irregular boundaries, one example being Toronto Gore Township, which has roughly the configuration of a triangle with all sides vastly unequal in length. The map of Southern Ontario (Figure 13.10) provides a vivid illustration of the number of survey systems used and the vast array of bearings of township boundaries.

Monumentation requirements along the township boundaries was dependent upon the survey system used (see Figure 13.10). In several instances, once the base line had been established, the township boundaries off each end of the line were run short distances, possibly only 1 mile, to satisfy only the immediate needs of the settlers. The running of the remainder of the boundaries was then left to future surveys. Therefore, by the time these future surveys were completed, monumentation instructions were subject to change, which created further irregularities in an already confusing and highly variable township survey.

The subdivision of certain survey systems is discussed in Section 13.2.5. A complete discussion of all nonsectional and special survey systems is well beyond the scope of this text. No doubt a voluminous seperate text could be written on this subject alone.

Numbering and Naming of Townships. In a sectional system, the townships of a survey district are numbered into ranges and tiers in relation to the principal meridian and the base line used for the

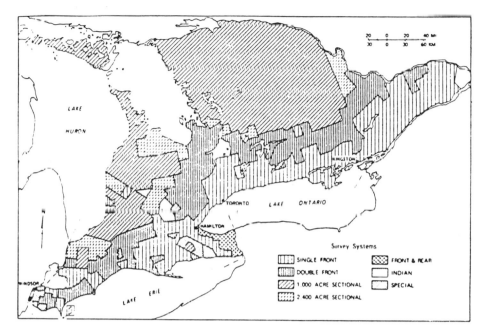

FIGURE 13.10 Township survey systems in Southern Ontario. (From Sebert, 1980)

district. As illustrated in Figure 13.8, the fourth township south of the base line is in tier four south (T.4.S.). The fourth township west of the principal meridian is located in range 4 west. Using this method of numbering, any township is located if its range, tier, and principal meridian are stated. The example shown in Figure 13.8 and discussed previously illustrates this system: tier 4 south, range 4 west of the second principle meridian. The abbreviation for this township would be T.4.S., R.4.W., 2nd P.M.

The variety of shapes and sizes in a nonsectional or special survey system, described in the previous section, rendered an organized numbering system impossible. Consequently, the townships are randomly named after prominent politicians of the day, local lakes and rivers, historical cities, towns in the British Isles, and so on. Counties consist of an arbitrary group of townships, usually between 6 and 12 in number. Therefore, the township is located as, for example, Innisfil Township, Simcoe County. As a result, to locate a specific township, not only the township and county names must be known, but also the nearest city or town. The county is then located on a provincial map and the township within that county on either a county map or topographic map (usually at a scale of 1:50 000). This system is cumbersome and very awkward, but there is no alternative.

13.2.4 Subdivision of Townships: Sectional Systems

In subdividing the township into sections, the primary objective involves laying out as many sections as possible that will be 1 mile on each side and thus contain 640 acres (approximately 250 hectares). As discussed previously, the monument locations during the township boundary surveys reflect the effects of convergence and accumulated measurement errors. Convergence has the effect of reducing the areas of sections adjacent to the westerly township boundary as the north–south lines are run parallel to the easterly township boundary. Accumulated measurement errors are placed on the sections adjacent to the northerly township boundary, as all measurements to the north commence at the southerly township boundary. Therefore, for a regular township, where the boundaries have been established within the accuracy limits previously described in this section, the subdivision proceeds in order with the south and east boundaries being the governing lines. If the township boundaries are irregular, the variations in the subdivision procedures are too numerous to be discussed in this section.

The subdivision survey is begun on the southern township boundary at the section corner between sections 35 and 36, as illustrated in Figure 13.11. This figure sets out the order in which the lines are run to subdivide the township. This order is indicated by the numbers on the lines within the township. The subdivision process takes place using the following procedure, referring to Figure 13.11.

1. Line 1 between sections 35 and 36 is run northerly, parallel to the eastern township boundary. The quarter-section corner between sections 35 and 36 is set at 40 chains ($1/2$ mile, 800 m).

2. Line 2 between sections 25 and 36 is run easterly parallel to the southern boundary of section 36, and a temporary quarter-section corner is set at 40 chains ($1/2$ mile, 800 m). If this line intersects the eastern township boundary at the corner of sections 25 and 36 within the allowable closure limits of 50 links ($1/2$ chain, 10 m), it is blazed and established as a true line. The temporary quarter-section corner is made permanent by adjusting it to the midpoint of line 2 using the field measurements.

3. The process in steps 1 and 2 is repeated until the northern boundary of section 12 is established.

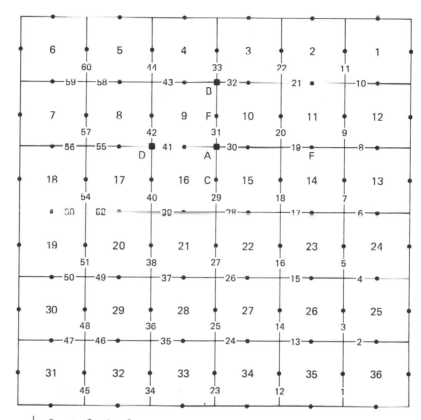

●   Quarter Section Corner

◆■ Section Corner Established for Resurvey Example

FIGURE 13.11   Subdivision of township sectional system.

4. The northern boundary of the township may or may not be a
   base line or standard parallel. The procedures under each
   condition are as follows:

   a. If the north township boundary is a base line or a stand-
       ard parallel, line 11 between sections 1 and 2 is run as
       a true line parallel to the east township boundary. A per-
       manent quarter-section corner is set at 40 chains
       (½ mile, 800 m). A closing corner is set at the point of
       intersection of line 11 and the base line or standard par-
       allel. The distance from this closing corner to the nearest
       standard corner is then measured and recorded.

   b. If the north township boundary is not a base line or stand-
       ard parallel, line 11 between sections 1 and 2 is run
       north as a random line parallel to the eastern township

boundary. A temporary quarter-section corner is set at 40 chains (½ mile, 800 m) from the start of the line. The distance to its point of intersection with the north township boundary is measured. If the random line intersects the northerly township boundary east or west of the section corner for sections 1 and 2, a common occurrence, the error is measured and the bearing of the true course of line 11 is calculated. The true line for line 11 is established, and the temporary quarter-section corner is made permanent by adjusting its location to the true line, while maintaining the full 40-chain (½-mile, 800-m) distance from the south end of line 11. Any difference in linear measurement along line 11 is therefore placed in the portion of the line between the permanent quarter-section corner and the northern township boundary.

5. The successive series of north–south oriented sections (termed *ranges*) are surveyed using the same procedure until section 32 (Figure 13.11) is reached.

6. Lines 45 and 46, section 32, are established using the same procedure set out previously for lines 1 and 2.

7. Line 47 is run as a random line parallel to the southerly township boundary from the westerly end of line 46 to the western township boundary. The difference in the north–south direction of the intersection of line 47 with the true corner of sections 30 and 31 is measured. The course of the true line 47 is calculated and the true line is blazed and established. The permanent quarter-section corner is placed on the true line 40 chains (½ mile, 800 m) from the easterly end of line 47 (corner of sections 29 through 32). Therefore, the differences due to convergence and/or errors in linear measurements are thrown into the most westerly portion of line 47.

8. The survey of the remainder of the two most westerly ranges is continued using the same procedure and in the order of lines shown in Figure 13.11.

It can be seen from the preceding procedures that the surveyor must be aware of the section number in which the subsequent work is to be carried out, due to the variations of subdivision methods with the township.

Subdivision of Sections: Regular. The official government township original survey maps (or plats) show the interior boundaries of quarter-sections as dashed straight lines connecting quarter-section corners, as illustrated in Figure 13.12. The sections adjacent to the north and west boundaries of a regular township, with the exception of section 6, are further subdivided into parts containing two regular half-quarter sections and four *lots*.

The southerly limits of the half-quarter sections are determined by joining points *A*, *B*, and *C* in the case of sections 7 and 8 and points *D*, *E*, and *F* in the case of sections 5 and 6, all shown in Figure 13.12. The southerly boundaries of the fractional lots are determined by laying off 20 chains (¼ mile, 400 m) to the north of line *DEF*. The boundaries between lots 2 and 3, section 5, are determined by the midpoint *K* of the quarter-section, established by equal proportioning between *EF* and *MP* (Figure 13.12).

In section 6, the quarter-section interior corner *Q* is located along a north–south line positioned by joining two points *X* and *Y* located 40 chains (½ mile, 800 m) westerly of the east limit of the

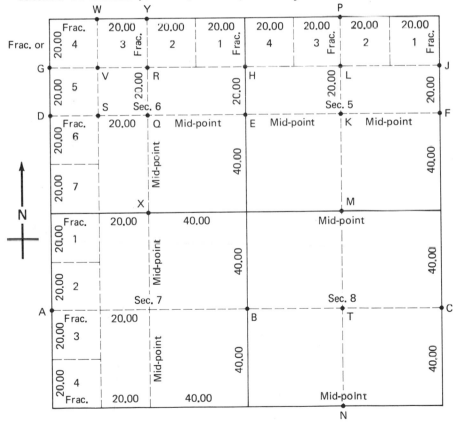

FIGURE 13.12 Subdivision of sections in sectional system.

section, as shown in Figure 13.12. The two interior quarter quarter-section corners, R and S, on the boundaries of the fractional northwest quarter are located. R is 20 chains ($\frac{1}{4}$ mile, 400 m) north of Q, and S is 20 chains ($\frac{1}{4}$ mile, 400 m) west of Q. As points Y and D have already been established during the township boundary survey, points R and S are located on lines QY and QD, respectively. Therefore, the fractional portions of QY and QD lie in the lines RY and VG (the latter primarily due to convergence).

As illustrated in Figure 13.12, the fractional lots bordering on the northern township boundary are numbered successively westerly beginning with 1 at the east corner of each section. The fractional lots bordering on the western township boundary, except section 6, are numbered in succession southerly beginning with 1 at the north section corner. As section 6 is common to both north and west boundaries, its fractional lots are numbered in succession beginning with 1 in the northeast corner and ending with 7 in the southwest corner, as illustrated in Figure 13.12.

The variations in original survey procedures, particularly along the north and west township boundaries, are significant to the modern surveyor because the existing property boundaries are positioned by reference to these township subdivisions. Reestablishment of towhship subdivision boundaries is often necessary to establish the nearest appropriate line in the survey fabric, as discussed further in this section. The numbering of fractional lots, particularly in township section 6, must also be understood by the surveyor.

The Canadian government adopted a system of township subdivision similar to the American model in 1859. This system, known as the "640 Acre Sectional Township, Pattern 1," was used for the lands north of Lakes Huron and Superior in Ontario and throughout the western provinces. A 1-chain (66 ft, 20 m) road allowance was provided on all sides of each section. The north and east sides of the road allowances were monumented. The "640 Acre Sectional Township, Pattern 2" was identical with the American model, the sections being numbered in the same fashion. The "640 Acre Sectional System, Pattern 3," shown in Figure 13.13, differs from patterns 1 and 2 in that half-sections, not quarter-sections, were used. Rather than dividing the square mile section into 160-acre quarter-sections, it was divided into two 320-acre lots by a line laid out northerly from a monument placed in the south section boundary. As illustrated in Figure 13.13, the lines from the south boundary posts were located 40 chains ($\frac{1}{2}$ mile, 800 m) from the southeast section corner, thus placing

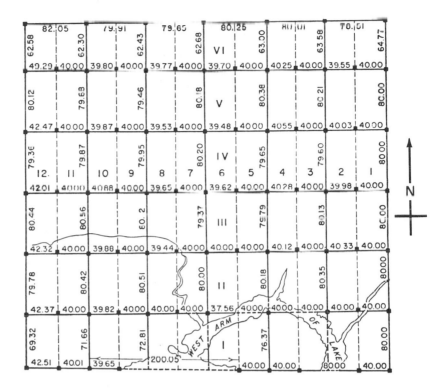

FIGURE 13.13 The 640-acre sectional system (Canadian, pattern 3). (Sebert, 1980)

all measurement errors in the westerly portion of each section. The numbering system was totally changed from section numbers to *concessions* (I, II, III, etc., Figure 13.13), each concession representing six sections in an east–west strip. Lots were usually numbered commencing with the 320-acre lot adjacent to the easterly boundary and were numbered consecutively across the township to the west township boundary. Consequently, lands within these pattern 3 townships are described by lot and concession, for example, lot 7, concession IV in Figure 13.13, rather than by fractional lot numbers and section numbers in the regular sectional system (American and patterns 1 and 2, Canadian). This system was used primarily in Ontario for township surveys from 1874 to 1906.

While the American and Canadian systems are similar in many ways, four important differences are notable. First, in most cases, particularly for patterns 2 and 3, each north–south subdivision line was run as an independent meridian. Therefore, the crrors due to convergence were distributed across the township rather than accumulated adjacent to the westerly township boundary as described previously for the American system. Second, the measurement errors were not accumulated and placed in the sections adjacent to the northerly township boundary, as illustrated by the variations from the intended lot depths of 80 chains (½ mile, 800 m) in Figure 13.13. These variations are attributable to not only a different procedure of check measurements and closing lines, but also changes in Canadian survey regulations under different governmental agencies responsible for township surveys. Third, the base lines were not always run along a parallel of latitude as discussed in Section 13.2.2, but rather directly easterly or westerly from the initial points. Fourth, the lot depths in concession VI and occasionally in concession V were significantly influenced by township boundaries previously run to suit the nonsectional or special systems discussed in Section 13.2.5. These variations are well illustrated in Figure 13.13, the net result being a great variation in individual lot areas.

While the end result of the American and Canadian sectional system subdivisions was similar, the variations in procedural details is of considerable significance to the surveyor. This is particularly true regarding reestablishment of boundaries and retracement surveys as discussed further in this section.

The plats or original township maps are available from the appropriate state and provincial government agencies responsible for retaining survey maps and records. The measurements shown for quarter-sections and fractional lots are in the form of boundary dimensions in chains, boundary bearings, and areas in acres. Some or all of these measurements will be shown depending upon the survey regulations in effect when the township subdivision was carried out. It is important to realize that these surveys were done using primitive equipment, the compass and chain. Therefore, the measurements shown on the original plat or map should not be taken literally, and considerable allowance for error must be made. This is particularly true when the boundaries are remeasured using modern survey equipment. Consequently, comparisons of new measurements with those shown on the original survey plat should be used only as a rough check.

Acreages shown on plats or original township maps are usually calculated using the relationships that 1 acre = 10 square chains = $10 \times 66 \times 66$ = 43,560 ft^2. For example, a regular sized quarter-section would be 40 chains by 40 chains, and thus the area would be $(40 \times 40)/10$ or 160 acres (64 hectares).

Subdivision of Sections: Irregular and Fractional. Sections made fractional through being adjacent to the north and west boundaries were discussed previously in this section. Sections are also made fractional due to meandering water bodies, areas within their limits such as Indian reserves, and mining claims. These sections are subdivided by proportioning and the other principles described previously for regular sections wherever possible. However, certain procedures are necessary to describe these lands separately from the water bodies, mining claims, and so on.

Figure 13.14 illustrates a section made fractional due to meandering bodies of water. The fractional lots are numbered beginning with 1 in the northeast corner. If the water body severs the lot, as is the case in Figure 13.14, the southerly portion is number 2. This process continues across the northerly tier, westerly, up to number 6 in the example shown. In the next tier to the south, the numbering is carried out from west to east (numbers 8 and 9). The number for the next tier south is successive for the fractional lots from east to west (numbers 10 to 13). Lot 14 is numbered because the water body forms part of the boundary of this lot.

Meandering is the term applied to establishing the high-water mark of all important lakes and rivers. The high-water mark is defined as the line between aquatic and terrestrial vegetation. Therefore, this line determined the areas of land that could be used and sold for agricultural purposes.

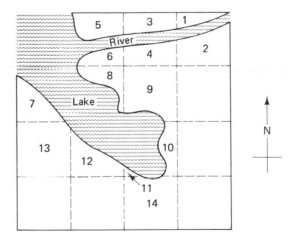

FIGURE 13.14 Fractional lots caused by meandering water bodies

The U.S. Supreme Court has set out the principles governing the purpose of meandering shorelines and the ownership regulations. The decision reads as follows:

> Meander lines are run in surveying fractional portions of the public lands bordering on navigable rivers, not as boundaries of the tract, but for the purpose of defining the sinuosities of the banks of the stream, and as a means of ascertaining the quantity of land in the fraction subject to sale, which is to be paid for by the purchaser. In preparing the official plat from the field notes, the meander line is represented as the border line of the stream, and shows to a demonstration that the water-course, and not the meander line as actually run on the land, is the boundary.

During the township subdivision process, the regular section lines are run to an intersection with the high-water marks of navigable water bodies. The points established at these intersections are called *meander corners*. The traverse that is run between meander corners is termed a *meander line*. The process of establishing the traverse lines is called *meandering*. The traverse commences at a meander corner and follows the bank or shoreline as closely as convenience permits. The traverse consists of a series of straight meander lines, which ends at the next meander corner. The length and bearing of each meander line are measured with precision. For the purposes of computing areas and plotting on the plats, the meander lines are laid out to the nearest quarter of a degree, and the stations are plotted to the nearest one chain or 10 links wherever possible. The latitudes and departures of the boundaries of the fractional lot boundaries are calculated. If the error of closure exceeds the allowable (maximum error for either latitude or departure of 1:640), the boundaries are remeasured until the allowable error is satisfied. Rivers are meandered on both banks. For example, the river through lots 1 to 6 in Figure 13.14 would be traversed on both the north and south sides.

All lakes having a water area of 25 acres (10 hectares) or greater must be meandered. Figure 13.15 illustrates situations where no regular meander corners would exist. Where the lake is contained entirely within a section, such as Lake A, section 16, a *special meander corner* is located, as illustrated in Figure 13.15, near the shore of the lake along a line connecting opposite quarter-section corners. The distance from the quarter-section corner is measured. Subsequently, the lake shoreline is measured by running a traverse around the shoreline, these meandering lines beginning and ending at the special meander corner. Where a lake is contained within a quarter-section, such as Lake B, section 20,

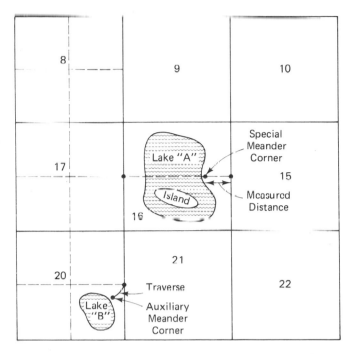

FIGURE 13.15 Locations of special and auxiliary meander corners.

an *auxiliary meander corner* is placed conveniently on the shoreline and connected by traverse to a regular corner, as illustrated in Figure 13.15.

Islands within bodies of water over 25 acres (10 hectares) are located in relation to regular section corners, meandered, and shown on the plat or original township map. Islands that have formed since the state involved joined the union are not measured and shown on the plats.

Reestablishing Boundaries. The aim of both the American and Canadian governments was to monument the corners using the procedures discussed in Sections 13.2.3 and 13.2.4. However, for a multitude of reasons, some the more common being listed later, many of these corner marks have been obliterated. Others were never set, although the plats or original township maps indicate otherwise. Consequently, the modern surveyor is required to relocate a missing corner. The basic requirements for this process are a thorough understanding of the methods used in establishing the original boundaries of the particular township in question, combined with good judgment regarding the best evidence of the corner location.

A *lost corner* is a survey point whose position cannot be determined because no reasonable evidence of its location exists. The only means of reestablishment involves clearly defined surveying procedures using the closest related existing corners.

The main reasons for lost corners are as follows:

1. Obliteration of the marked lines and wooden corner posts by forest fires, often covering large areas.
2. Improper marking of the lines and/or monuments during the original township or subdivision surveys (see Section 13.2.6.).
3. Inaccurate placing of lines and monuments.
4. Lack of concern for preservation of the original corners and lines, particularly by the early settlers.

The surveyor must first determine if the survey corner is really lost. If the original monuments were wood, the decaying of the stake will often leave a brownish rust color in the soil. This is investigated by carefully slicing the earth horizontally using a thin shovel blade. Blazes on trees used as witness points for survey corners, also called *bearing trees,* as well as blazes marking the survey lines (see Section 13.2.6), will become totally covered by subsequent tree growth and enlargement of tree diameter. For example, the author was reestablishing a corner located along a township boundary, along which much of the evidence had been obliterated by a forest fire. Using the original survey field notes and the terrain features described thereon (see Section 13.2.7), a trial line was run for a distance of $3^3/4$ miles (approximately 6 km) where a bearing tree (witness point) had been blazed and inscribed BT during the original survey. As this tree was located outside the forest fire affected area, the most likely tree based on the original notes was selected for investigation. No evidence was visible on the outside of this tree, now 4 ft (1.3 m) in diameter. After the tree was felled, the most likely location of the inscribed BT was determined through determining the date of the original survey from the field notes. As the survey was carried out in mid-winter, the assumption was made that the BT would be inscribed approximately 7 ft (2.2 m) off the ground rather than the normal 4 ft (1.3 m) due to snow cover. It was necessary to remove thin slabs of wood starting from the bark of the appropriate face recorded on the original survey notes. This laborious process was rewarded by finding the initials BT approximately 1 ft (0.3 m) inside the tree. Based on direction and distance from the true corner to this witness post, the original corner was reestablished. Perserverance had paid off and was appropriately celebrated.

Fence lines placed by the early settlers were commonly erected using a combination of post and wire and wire stapled to trees on or close to the intended line. Some line fences were installed using only the wire stapled on trees depending upon the density of growth. With the passing years, the original fences disintegrated by rusting and were not renewed for a variety of reasons. Therefore, under the worst conditions, the only remaining evidence is the wire stapled to the tree preserved due to subsequent growth. The presence of the wire causes a slight bulge in the outside bark of the tree, which diminishes in size in relation to the elapsed time since the fence was stapled. The surveyor should look for these bulges in the trees along the suspected location of the line up to approximately 4 ft (1.3 m) above the ground surface. In addition, dip needles or the more accurate battery-operated bar detectors should be used to verify the presence of fence wire within the trees. If the wire is detected, it is necessary to notch the tree where indicated and uncover the wire. The number of growth rings outside the wire should be carefully counted as this will determine the year of the fence installations. The use of these investigative methods has recently been instrumental in the reestablishment of an original survey line by the authors. The net result was the proving of ownership of 6 acres (approximately 2.5 hectares) of valuable land ($30,000 Canadian, $25,000 U.S.).

The preceding examples illustrate the intensity of investigation required to determine if the survey corner is truly lost. If all avenues of best evidence are not explored in their entirety, the surveyor will likely be put in the embarrassing and expensive position of defending his lack of thoroughness in a court of law. Experience in reestablishing original survey points and boundaries is essential. A young and/or inexperienced surveyor would be well advised, under these circumstances, to carry out such surveys under the direction of an experienced surveyor. This is often arranged by hiring the experienced surveyor in a consultative capacity for the particular survey, searching for evidence under his direction, accepting his analysis of the results, and requiring his explanation leading to these decisions. To understand the reestablishment process, the surveyor must be familiar with the original survey procedures described in this chapter, as the reestablishment of original survey points is based on the intent of the original survey.

For sectional townships, the field process required for the

restoration of lost corners is given next. The most important consideration involves re-creating the techniques employed in the original survey. The standard lines discussed in Section 13.2.2 must be given the greatest weight. For example, the following priorities exist:

1. Standard parallels over township boundaries.
2. Township boundaries over township subdivision lines.
3. Township subdivision lines over quarter-section corners.
4. Quarter-section corners over fractional lot boundaries.

Detailed instructions for the relocation of lost corners are given in the manual of the Bureau of Land Management for the United States and the Dominion Land Surveyors (now known as the Canadian Land Surveyors) manual, as well as in manuals produced by provincial land surveying associations in Canada. Some examples will illustrate the principles of the process involved. Reference should be made to Figures 13.11 and 13.12.

1. Section corner common to four sections in township interior (point A, Figure 13.11). Assume that all section lines were run in the original township subdivision and that points B, C, D, and E are the closest established corners. Lines BC and DE are run straight through, and point A is located at the intersection of these two lines.

2. Quarter-section corner in township interiors (point F, Figure 13.11). Line AB is measured, and F is located on line AB midway between points A and B.

3. Quarter-section corner at center of section (Point T, section 8, Figure 13.12). The lines between the opposite quarter-section corners, both east–west (BC) and north–south (MN), are run, and point T is placed at their intersection. Thus, point T is located at the midpoint of both lines BC and MN.

4. Fractional lot corner in northwest section (point R, section 6, Figure 13.12). Having established the quarter-section corner at the center of the section, point Q, by measuring 40 chains ($^1/2$ mile, 800 m) northerly from point X along the line XY, point R is established by measuring 20 chains ($^1/4$ mile, 400 m) along the line XY from Q to R. In this way, the line between fractional lots 2 and 3 (RY) may be 20 chains ($^1/4$ mile, 400 m) but will likely be greater or less depending upon the accuracy and care used in running the original township subdivision. Similarly, to establish point V, Figure 13.12, it would be lo-

cated 20 chains north of point S on the line SW if these points have been previously established. If no evidence of points S and W can be established but points R and G are known, point V is set 20 chains ($\frac{1}{4}$ mile, 400 m) west of point R on the line RG.

These examples illustrate the variations in techniques required for reestablishment in a sectional system originally surveyed using the American model. The surveyor should note that reestablishment techniques vary depending upon the section(s) being surveyed, particularly in sections adjacent to north and/or west boundaries.

The three Canadian systems described previously, known as the 640 Acre Sectional Townships, use patterns 1, 2, and 3. Patterns 1 and 2 are similar to the American system, and therefore the techniques for reestablishment described here are similar. Where road allowances have been established in the original survey, such as pattern 1, the width of these allowances (usually 1 chain) cannot be changed. All other measurements are adjusted using the methods discussed previously. As previously discussed, the north–south lines were sometimes run as individual meridians, particularly for patterns 2 and 3. In these cases, all boundaries are reestablished by proportion between two known corners, regardless of the location of the section within the township. These techniques are similar to example 1 (point A, Figure 13.11) in the preceding list.

The road allowances established along each section boundary, such as pattern 1 described previously, meant that the road locations on publicly owned allowances, 1 chain in width, were permanently set. As a result, if these allowances were along lines where road building would be very difficult and/or expensive, two alternatives to the settlers were usually available. First, the road would not be constructed. Consequently, the land adjacent to these road allowances was not opened up for settlement and the settlers in the area were forced to use other constructed or "opened" roads, thus increasing their access distances to market. Second, the roads would be constructed on the privately owned lands on either side of the road allowance, in order to avoid the most difficult terrain conditions for construction. Consequently, the traveled road was constructed mainly on private lands, usually following a sinuous path that crossed and recrossed the actual road allowance. Usually these roads were initially built through the cooperation of the land owners on each side of the allowance.

As time passed, these roads would become public roads through usage for access by settlers in the area. The result was the creation of two road allowances between two original survey points, the original and the actual, sometimes called the *forced road allowance.*

As a result, the modern surveyor must determine the following if surveying lands adjacent to a forced road:

1. Whether the road is *legally* a public road or private road, usually determined through consultation with the municipality and/or land ownership records.
2. The location of both road allowances, the original through boundary reestablishment techniques, and the forced through traverse lines along the road center line connecting two or more established township subdivision corners.

These procedures result in the creation of small, irregularly shaped parcels of land between the two road allowances owned by the appropriate landowner on the side of the original allowance. The results of the survey of such lands, which are greatly increased in complexity and cost because of these circumstances, are shown on the appropriate plans by the surveyor. Subsequently, any questions involving the ownership of the irregular parcels created are usually left to legal councilors. In addition, the local municipality often becomes involved through arranging to "close" the original allowance in exchange for the forced road allowance on which the road is actually located.

Sectional townships surveyed without road allowances usually provide for a 5% allocation for road allowances in both the United States and Canada. The roads are then located across logical terrain originally as settlement proceeds, thus largely avoiding the complications caused by the establishment of road allowances in the original survey.

The necessity of determining the type of survey system originally employed before the surveyor carries out any reestablishment work must again be emphasized. Otherwise, the wrong techniques may be used, resulting in incorrect reestablishment.

13.2.5 Nonsectional or Special Systems

The survey systems shown on the legend in Figure 13.10 illustrate the variety used on the original surveys for Southern Ontario alone. A detailed description of the township subdivision systems and methods for reestablishing the original boundaries is beyond the scope of this text. The systems described in this section have been selected based on the following criteria:

1. Significant differences in procedures for reestablishment of boundaries in comparison with the techniques used in the sectional systems discussed in Section 13.2.4.
2. Problems primarily associated with the alignment, location, and access connections between and among original road allowances and the resulting practical difficulties encountered by both the modern surveyor and engineer.

Three township subdivision systems are discussed in the following subsections. In each case the following are described: techniques used in the original survey; irregular and fractional lots; reestablishment of boundaries; and legal and engineering problems caused by road allowance locations.

The same general rules of evidence discussed previously regarding the relocation of missing corners and lines still apply.

Front and Rear System. This system is illustrated in Figure 13.16. The usual practice was to survey the side lines of each lot. For example, in Figure 13.16, line *AB* was run parallel with the eastern township boundary and monumented at 50-chain (5/8-mile, 1000-m) intervals with a road allowance 1 chain in width every 50 chains. The line *BE* to the lake was run and marked, and the distance *BE* to the high-water mark was recorded. Note that no monumentation was placed at the lake shoreline. The next line run was *BC* for a distance of 20 chains (1/4 mile, 400 m) parallel with the southern township boundary along the northerly road allowance limit between concessions *A* and *B*. Line *CD* was then run parallel with the easterly township boundary, the monumentation being placed only along the westerly side of the 1-chain-wide road allowance, making due allowance to monument both sides of the 1-chain-wide road allowances between concessions. Consequently, the cumulative errors in line *CD* are concentrated in the westerly limit of lot 3, concession VIII. Point *F* was set on the township boundary, distant 20 chains (1/4 mile, 400 m) from point *D*. Line *FG* was then run in the same fashion as line *AB*. This process was continued with alternative lines being run south to north, then north to south, until the subdivision had been completed to the western township boundary. As shown in Figure 13.16, road allowances were placed every two lots across the township.

As a result of the preceding subdivision procedures, the following circumstances normally occur:

1. If the western township boundary were established prior to subdivision, the areas of lot 23, Figure 13.16, vary considerably. All measurement error, particularly east–west, is also thrown into the lots in the various concessions adjacent to the westerly boundary, in this case. Consequently, the modern surveyor should be aware of this significant variation in lot size for the last numbered lot in the township, regardless of the dimensions shown on the original plat or township survey map.

2. The road allowances between concessions will vary in direction at each lot corner because the lot lines are run alternatively south to north and north to south. For example, it is highly unlikely that point C would lie on the line between B and G, Figure 13.16. As a result, the modern surveyor would have to establish the limits of the road allowance between concessions at each lot line. This involves considerable time and effort as the original monuments are only 20 chains ($1/4$ mile, 400 m) apart.

3. The lot frontages along the concession roads in the central concessions will vary from the intended width of 20 chains ($1/4$ mile, 400 m). This is due to directional error in running the north–south lot lines. For example, the frontages of lot 4, concession 4, may be either greater or smaller than 20 chains ($1/4$ mile, 400 m) as originally intended. Consequently, the surveyor should not be surprised to find this variation and should rely on the best evidence on the ground rather than on the original plats or township survey map dimensions.

4. The depths of lots in the concession adjacent to a preestablished township boundary will be variable. This is due to the fact that the accumulated measurement error in the north–south (lot lines) occurs on the concession adjacent to the township only on alternative side lines. For example, in Figure 13.16, lot line 1–2, concession VIII, on line AB, would be closer to the intended distance than lot line 2–3, concession VIII, on line CD. The net result is a variation in depths of these lots. The original plat or original township survey map may or may not indicate these variations, depending on the diligence of the original surveyor in accurately recording distance measurements.

Legal and engineering problems associated with road allowances are the same for this survey system as described in Section

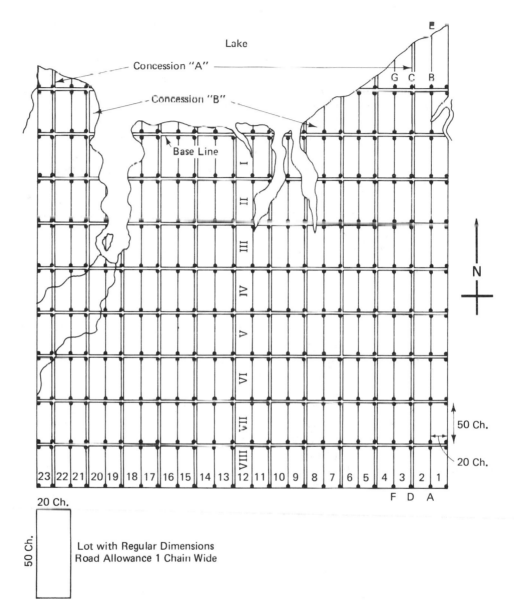

FIGURE 13.16 Front and rear system. (A.O.L.S., 1973)

13.2.4. Bends of small magnitude occur along the concession road allowances and occasionally along the side line road allowances. Assuming that the terrain along the original road allowances is suitable for economical construction, access is available to all lots as shown on the original township survey plan. Jogs in side line road allowances, as described in this section, do not exist. Also, road allowances dead-ending at half-lot lines, as discussed in this section do not occur.

Reestablishment of original subdivision boundaries in the front and rear system is carried out through proportional measurements between established monuments. Road allowances are not proportioned and are fixed at a width of 1 chain (66 ft, 20 m).

In summary, the front and rear system resulted in a sufficient number of monuments being set that all reestablishment is carried out by proportional measurements between or among found corners. The large number of original monuments increases the cost of reestablishment of road allowances along the concession lines, as well as creating small changes in direction at each lot corner. The side-line road allowances may vary slightly in direction, but do not have jogs or potential dead ends.

Single-Front System. The principles of this system are illustrated in Figure 13.17. After the township boundary was established, the subdivision was carried out by running each concession line. The monuments were placed at the front corners of each lot on one side of the road allowance only, commonly at 20-chain (¼-mile, 400-m) intervals. For example, the line *AB* in Figure 13.17 was run and monumented first, representing the "front" of concession I. Note that a 1-chain-wide road allowance was left every five lots. Line *BC* was then run at the same direction as line *AD* on the southern or, in this case, the first township boundary established. The direction of line *AD* was termed the *governing bearing,* an important influence on the township subdivision. Line *CD* was then run and monumented, thus throwing all measurement errors into the frontage of lot 1, concession II. The depth of concession II (line *DE*) was then measured off along the township boundary a distance of 100 chains (1¼ miles, 2000 m), plus a 1-chain (66-ft, 20-m) road allowance, as shown in Figure 13.17. It should be noted that, while it was intended that *DE* would be on the same bearing as *AD,* this was seldom the case. Line *EF* was then run in the same fashion as *AB,* the accumulated measurement errors being thrown into lot 37, concession III. This process of running and monumenting the fronts of the remaining concessions in the township continued until the township subdivision was completed.

The side lines of the lots were determined by running the lines from the monument at the front of the concession on the same bearing as the governing bearing for that concession. For example, the northerly limit of lot 5, concession 1, would be run

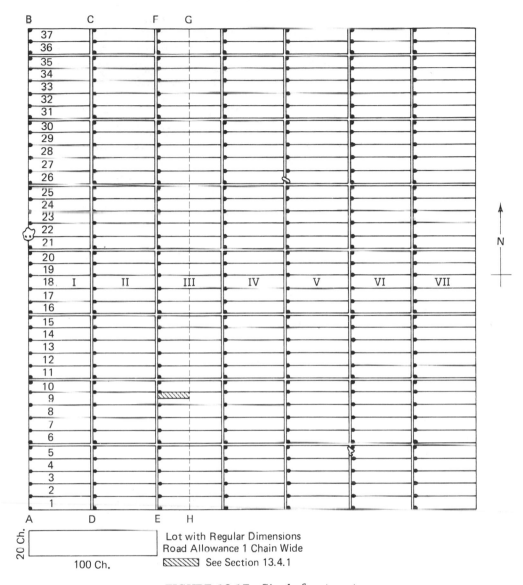

FIGURE 13.17 Single-front system.

on the same bearing as line *AD*, a distance of 100 chains (1¼ miles, 2000 m), to establish the northeast corner of lot 5, concession 1. As the northerly limit of this lot is along the southerly limit of the side-line road allowance between lots 5 and 6, concession 1, this procedure also located the road allowance from the front to the rear of concession I. The northerly limit of lot 5, concession II, is established from the monument at the northwest corner of lot 5, concession II, and by running the side line parallel to the governing bearing line *DE* for the lot lines in concession II.

As result of the preceding subdivision procedures described, the following circumstances normally occur:

1. If the northerly township boundary were established prior to township subdivision, the areas of lots 37, Figure 13.17, will vary considerably for two reasons. First, these lots consist of the remaining lands in the township after the first 36 lots have been monumented, in this example. Second, the accumulated errors are placed in lot 37 in every second concession due to the concessions lines being run north to south and then south to north alternatively.

2. The road allowances along the fronts of each concession will vary in direction at each front lot corner due to directional error during the original survey. As these are placed every 20 chains ($1/4$ mile, 400 m), considerable reestablishment work is required by the modern surveyor to locate the road allowance accurately. The location of the front of the concession carries more weight than the rear of the concession due to the original subdivision survey techniques employed.

3. Both the frontage and area of lots 1 will vary as the accumulated error is thrown into these lots in alternative concessions. Therefore, the modern surveyor should not be surprised that his field measurements vary up to approximately 3 chains (198 feet, 60 m) from those shown on the original plat or township survey plan.

4. Significant jogs on the sideroad allowances will occur. For example, the method of positioning the road allowance between lots 5 and 6, concession I, and the road allowance between lots 5 and 6, concession II, described previously will cause this problem for three reasons. First, the accumulated error is placed alternatively in lots 1 and lots 37. Second, the governing bearings for each concession are usually different. Third, the counting of the number of chains surveyed was carried out in a very primitive fashion, such as dropping stones into a container after each 1-chain measurement, thus creating some lots with either 19 or 21 chain frontages versus the intended 20 chains. The problems caused by these jogs are significant to the modern surveyor for several reasons. First, the existence of these jogs is totally unexpected unless the system of township subdivision is understood.

Second, confusion results when the original plat or township survey map does not indicate these jogs, thus leading the modern surveyor to misbelieve that a gross error exists in the retracement survey. Third, if the primary access route to property on which the surveyor is working, such as a proposed plan of subdivision, is a sideroad, large traffic loads must be dealt with. Consequently, careful consideration must be given to arranging for acquiring two triangles of land from adjacent land owners at the jog to provide a simple reverse curve transition along the sideroad versus two sharp right-angle turns. In other words, privately owned land must be acquired to "straighten" the road to permit higher traffic volumes and speeds. This property acquisition involves significant capital cost, sometimes borne by the municipality and often by the proposed land development company responsible for the increased traffic flow. Naturally, the situation is most critical when the only access to the proposed land development is along the sideroad of a single-front township.

The modern surveyor or engineer working for either the municipality or the developer must be aware of the jogs usually caused by the single-front system. Aside from the potential costs for land acquisition, considerably more land surveying is required to reestablish the lot boundaries along the front of the concession road on each side of the jog. In addition, powers of expropriation for acquiring the triangles of land may be required.

Reestablishment of monumentation along the fronts of the concessions is carried out by proportioning. Road allowances are a fixed width of 1 chain (66 ft, 20 m). It is important to note that the monumentation or other evidence along the road allowance at the front of the concession has priority over the evidence on the rear of the adjacent concession, which is directly across the concession road.

Where a side line of a lot is being reestablished and no evidence of the line's position exists, the following procedure is used. The monument at the front of the concession on the line is first reestablished. For example, if the lot side line in question is the lot 27–28 line in concession II, Figure 13.17, the northwest corner of lot 27, concession II, is either found or reestablished. The direction of line *DE* must then be determined to establish the governing bearing. Subsequently, the 27–28 lot line in concession II is run easterly from the northwest corner of lot 27, concession II, on the governing bearing determined from the direction of line *DE*.

The additional work required for side line reestablishment for this system is therefore considerable.

The original lot sizes were 20 chains by 100 chains, and thus consisted of 200 acres (2000 square chains at 10 square chains = 1 acre) or 80 hectares. These lots were commonly sold or "patented" in half-lots of 20 chains by 50 chains, as illustrated by the dashed line *GH,* concession III, in Figure 13.17. The *half-lot line* is established by determining the midpoint of the side line between, for example, the front monuments of concessions III and IV.

In summary, the single-front system resulted in the minimum number of monuments required to achieve the desired township subdivision. Reestablishment of concession lines are carried out by proportioning, while lot lines are established by governing bearings from the lot corners set on the concession fronts. The net result is a high cost for reestablishment for both concession roads and sideroad allowances. The sideroads jog at each concession crossing, every 100 chains (1 1/4 miles, 2000 m). This causes additional survey difficulties for reestablishment and incurs additional survey and land acquisition costs for smoothing out the jogs for high speeds.

Double-Front System. The double-front township subdivision system is carried out using similar techniques to those just described for the single-front system. However, two important differences have a lasting effect on reestablishment of boundaries and on modern surveying. As shown on Figure 13.18, the fronts of lots on both sides of the concessions were monumented, compared with only one side in the single-front system. These lot corners were placed 1 chain (66 ft, 20 m) apart at right angles to the center line of the concession, which was the actual run line. Road allowances were left along the side lines, usually every five lots as shown in Figure 13.18.

The lot frontages were 30 chains (3/8 mile, 600 m) and the lot depths were 66.67 chains (1340 m). Each lot contained approximately 200 acres (80 hectares). The parcels were sold or "patented" by the government in half-lots of approximately 100 acres (40 hectares). Therefore, the rear boundary of half-lots, as shown in Figure 13.19, had to be established shortly after the original township subdivision had been completed. As the concessions were surveyed alternatively north to south and south to north, as discussed in detail in the preceding section, the accumulated measuring errors were placed in the lots at opposite ends of adjacent concessions. Therefore, as the lot lines were all run in the same di-

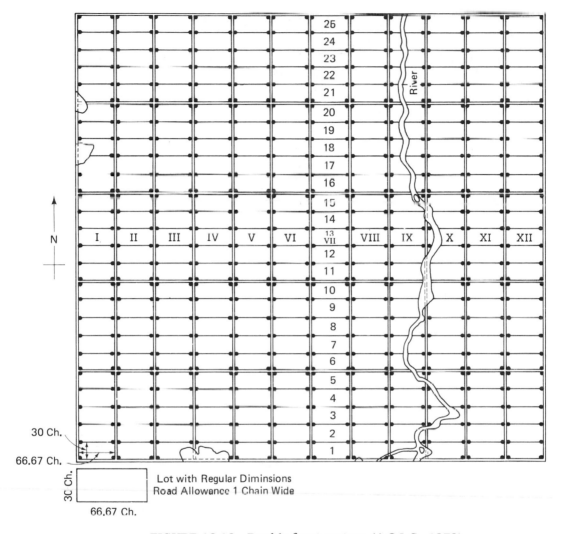

FIGURE 13.18 Double-front system. (A.O.L.S., 1973)

rection as the governing bearing for each concession and were terminated at the half-lot line, the jog in the side-line road allowances
occurred in the middle of the concession. As the road allowance is
1 chain (66 ft, 20 m) wide, this represents the maximum shift in
the road allowance boundaries to preserve any width of road allowance between the two halves of the concession. An examination
of Figure 13.19 clearly illustrates the narrowing of the side-line
road allowances between lots 5 and 6 and lots 11 and 12.

This situation can and does result in the creation of two

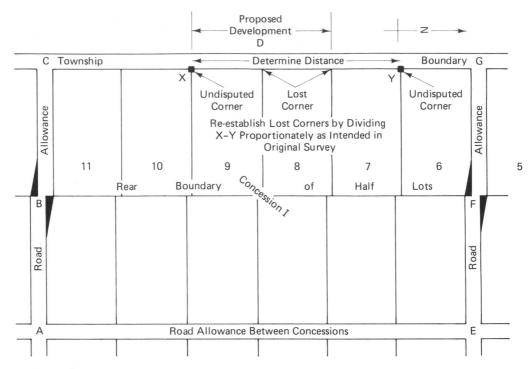

FIGURE 13.19 Double-front system; effect of half-lot line on road allowances.

dead-end road allowances, each terminating at the half-lot line. In the authors' experience, this causes a number of serious problems. The most critical are discussed, with examples where appropriate.

1. Assume that there is no existing traveled road surface within the side-line road allowance, and that the land on both sides of the road allowance is privately owned. If the nearest existing road to the proposed development shown on Figure 13.19 is from the south to point *A,* the most direct and, more importantly, the most economical route is *A–B–C–D.* Most municipalities require a continuous road allowance width of between 50 ft (15 m) and 66 ft (20 m) to provide sufficient space for the traveled road, hydro lines, ditches, sewers, watermains, and so on. As this is not available at point *B,* negotiations must be carried out with the adjacent owners (lots 11 and 12, concession I) to acquire one of the triangles of land shown on Figure 13.19. If this cannot be accomplished at point *B*, the only option is to negotiate for a triangle of land at point *F*. Even if the latter is successful, the only route to the proposed development is *A–E–F–G–D*. This represents an increase in required road construction of approximately 180 chains (2¼ miles, 3600 m), plus the cost of land acquisition at point *F*. A similar situation occurred on a project in which the authors were surveying and engineering consultants. The key factor was the surveyor's prior knowledge that the subdivision was the double-front system. Therefore, the potential access problems to the proposed development had to be examined before any detailed surveying and engineering work was carried out on the property. Consequently, the developer was informed by the consultants that an additional $400,000 U.S. ($500,000 Canadian) would have to be spent for road construction. As a result, the developer placed the development on hold until road access was constructed that did not involve sideroad access, such as new construction along the township boundary. Therefore, the original township survey system was directly responsible for excessive surveying costs for investigative work, costly and time consuming yet unsuccessful negotiations to acquire the required width of road allowance, and postponement of a viable development concept. Had the modern surveyor initially involved in this project not realized the implications of the double-front system, the surveying and engineering required for the development proposal could easily have been carried out at high cost but to no avail.

2. Assume that a traveled road surface exists within the traveled road allowance and that the land on both sides of the road allowance is privately owned. If access is required to the proposed development shown on Figure 13.19, the existing road from *A–B–C* provides a false sense of security to the unwary surveyor. Further investigation should be carried out to determine whether the road is private (agreement between the owners of lots 11 and 12, concession I) or public. If the road is private, the same negotiations required in situation 1 are required and the same problems may prevail. If the road has been used by the public with no interruption for sufficient time under provincial or state law to be deemed as public road, this should be established. Normally, legal advice is required, as well as the cooperation of the municipal government, to document the status of the traveled road. In addition, the rear boundary on both sides of the jog (point *B*, Figure 13.19) and the side lines *AB* and *BC* must be reestab-

lished to map the "new" road allowance. Under the best of conditions, such as a proven public road, excessive surveying and legal fees are incurred.

3. The same situation regarding variable size lots as described for single-front systems in Section 13.2.5.

4. The same variation in direction along the concession roads as described for single-front systems in Section 13.2.5. The only difference is the distance between the lot corners: 20 chains ($^1/_4$ mile, 400 m) in the single-front system and 30 chains ($^3/_8$ mile, 600 m) in the double-front system. Therefore, only two-thirds the reestablishment of monuments along the concession roads is required for the same length of road. In addition, twice the number of monuments were originally set, so the chances of finding direct field evidence is increased. As evidence on each side of the concession road has equal weight, it is common to set the road center line midway between two opposite monuments. If the distance between the monuments is different than 1 chain (66 ft, 20 m), the road allowance boundaries are set at $^1/_2$ chain (33 ft, 10 m) at right angles to the center line in each direction.

Reestablishment of monumentation along the fronts is established using proportionings, as illustrated for points X and Y in Figure 13.19. The reestablishment of lot lines is carried out using the same governing bearing system described for single-front systems with one important difference. The side lines are run back to the half-lot line on the governing bearing for the concession, as discussed previously in this section and as illustrated on Figure 13.19. The presence of the half-lot line complicates the survey work considerably over the single-front system as lines from two monuments must both be run toward the half-lot line from each side of the concession.

In summary, the double-front system has created and will continue to create serious problems related to access along sideline road allowances. The necessity of running side lines on governing bearings, particularly to half-lot lines, is one of the most ill-conceived ideas encountered by the authors in the history of township subdivision surveying. For this reason, the modern surveyor must, for the sake of his sanity, pocketbook, and professional reputation, determine immediately if the area in which work is to be carried out is any type of double-front system. The client should then be warned immediately of the potential problems, increased cost of the survey reestablishment, and probable requirements of legal assistance.

13.2.6 Corner Monumentation and Line Marking

Manuals are available from the Bureau of Land Management, Washington, D.C., U.S.A.; Surveys and Mapping Branch, Department of Energy, Mines and Resources, Ottawa, Canada; and provincial and state associations such as the Association of Ontario Land Surveyors, Toronto, Ontario. In addition to specifying the methods of township subdivision and reestablishment techniques for each survey system, details on corner monumentation and line marking are given.

The information given in this section has been selected to assist the modern surveyor in locating the corner monuments and the survey lines for purposes of reestablishment.

Corners. It is unfortunate that most of the public lands were surveyed before the present monumentation regulations went into effect. Consequently, most of the monuments used consisted normally of wooden posts. If nothing else were available, mounds of earth, earth-covered charred stakes, or charcoal, boulders, and the like were used. These monuments were not of a permanent nature, and considerable skill is required to relocate them. For example, as many of the wooden posts have either rotted or burnt in forest fires, careful slicing of the soil in the area with a sharp shovel will sometimes reveal a rust-colored stain from the underground portion of the post.

The existing regulations in the United States require the use of iron-pipe monuments about 3 ft (1 m) long and 2 in. (5 cm) in diameter. The pipe is split at one end for about 4 in. (10 cm) and the two halves are spread to form a solid footing. A brass cap is riveted to the opposite end and the pipe is then filled with concrete. The caps are plainly marked to indicate their location within the township survey fabric as discussed later in this section. If authorized, durable native stone may be used instead of the iron posts. The stone must be at least 20 in. (50 cm) long and at least 6 in. (15 cm) in both minor dimensions. The stone is marked by the appropriate letters and numbers chiseled on the faces near the top, as about three-quarters of its length is in the ground. Where bedrock is close to the surface, the corner point is marked by a cross ($\times$) cut in the rock. If stones are readily available, the corner monument, either an iron post or stone, is supported by a mound of stones known as a *stone cairn*.

The Canadian regulations for township subdivision surveys, termed *Crown Land* surveys, permit several types of monumentation depending on ground conditions. A *Crown Lands iron post*, shown as C.L.I.P. on survey plans, and its installation are the same as the American iron pipe described previously, except that it is 1 in. (2.5 cm) in diameter instead of 2 in. (5 cm).

Additional monuments, described in the regulations as *Crown Land monuments,* include a standard iron bar, 4 ft (1.2 m) long and 1 in. (2.5 cm) square, designated on the survey plan as S.I.B.; a concrete monument, 4 ft. (1.2 m) long, 5 in. (12.5 cm) square at the top and 8 in. (20 cm) square at the bottom (C.M.); a rock bar, 1 in. (2.5 cm) square, 6 in. (15 cm) long, cemented or leaded into bedrock so that the base of the bar is at least 5 in. (12.5 cm) into solid bedrock (C.L.R.B.); a Crown Lands rock post, a bronze indentification cap mounted on a stem 3 in. (7.5 cm) long, which is leaded or cemented into solid bedrock so that the cap is flush with the ground level (C.L.R.P.); an iron bar, 2 ft (0.6 m) long and 5/8 in. (1.5 cm) square driven into the ground until the top of the bar is flush with the ground level (I.B.); and a short standard iron bar, 2 ft (0.6 m) long, 1 in. (2.5 cm) square, driven flush with the ground level (S.S.I.B.).

Stone mounds at least 3 ft (1 m) in diameter at the base and at least 1 1/2 ft (0.4 m) high (S.M.) and a wooden guide post at least 4 ft (1.2 m) long and 4 in. (10 m) square, beveled at the top on all sides to form a central point and driven in the ground so that the point is at least 3 ft (1 m) above ground level, may be used under certain conditions and if permission is obtained. This bewildering array of monumentation choices means that some are easy to mark, such as those with a bronze cap on top and wooden posts, while the others are difficult to mark, if not impossible.

Corner markings for sectional survey systems are complex, and the reader is referred to the "Manual of Surveying Instructions" of the Bureau of Land Management in Washington, D.C., or the Canadian equivalent from the Department of Energy, Mines, and Resourses, Ottawa, Ontario. As the variety of markings is beyond the scope of this text, only general characteristics will be discussed and two typical examples given.

All monuments are marked using a system that will provide accurate information regarding the type of monument and its location. Metal caps placed on top of iron posts are marked with capital letters and numbers. The township, section, range numbers, and the year of the survey are all inscribed. Stone monuments were marked with notches and grooves on the edges, which indicate the distances in miles from the township boundaries. For example (see Figure 13.11), the common corner for sections 8, 9, 16, and 17 would have four notches on the east side and four notches on the south side. This system minimized the carving of letters and numbers in stone, a task not welcomed by the original surveyors in a then primitive country. If the corner point fell exactly at the location of a healthy living tree, too large to be removed, the tree was appropriately marked as the corner.

The markings on the caps of iron post monuments are illustrated in Figure 13.20. The sketch in part c illustrates the position of the points in parts a and b. Point a is a township corner marked by the symbol SC (standard corner). As the markings are set to be viewed from the south, the township to the north is shown: T20N. The ranges on each side of the township line are shown, in this case range 6 east (R6E) and range 7 east (R7E). The section numbers to the northwest (S36) and to the northeast (S31) are shown below the respective ranges, thus fixing the point. The date of survey (1908) is shown as the bottom number, which assists the modern surveyor in obtaining the original field notes from the appropriate state government authority. Point b is a section corner. The township number (T20N) is shown on the upper left of the symbol, and the section to the northwest is indi-

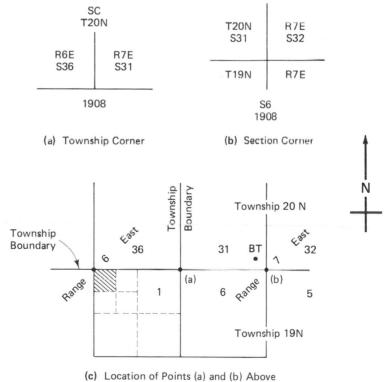

(a) Township Corner

(b) Section Corner

(c) Location of Points (a) and (b) Above

▨▨▨ See Section 13.4.1

FIGURE 13.20 Example markings on iron monuments.

cated below the township number (S31). The range number, in this case range 7 east (R7E) is shown on the right side, with the northeast section corner (S32) indicated below. The township (T19N) and the range (R7E) to the south of the corner marker are shown as indicated in Figure 13.20b.

The nonsectional or special systems, used mainly in Canada, have simpler corner marking systems, providing that the surveyor knows the township and county name. One example is Bastard Township, County of Leeds, Province of Ontario, originally surveyed using the single-front system as recorded in "Manual for the Association of Ontario Land Surveyors, 1973." The corner markings in Bastard Township are based on the lot, concession, and road allowances, shown in Figure 13.21c. As illustrated in Figure 13.21a, a monument at the intersection of concession and sideline road allowances was marked either on the top of a cap or on the sides of a stone monument. The road allowances are shown as R; the concession is in roman numerals and the lot number is shown in arabic numbers (XI and 16 in Figure 13.21a). A lot cor-

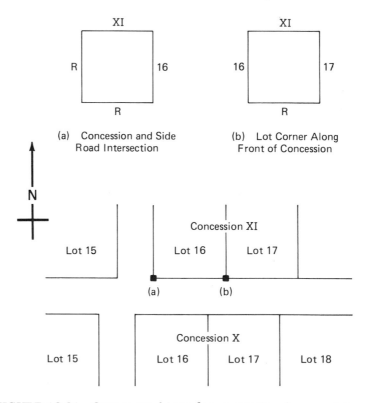

FIGURE 13.21 Corner markings for nonsectional or special systems.

ncr along a concession road is marked as illustrated in Figure 13.21b. The lot numbers are shown on each side of the monument, and the concession number and road allowance are the same as Figure 13.21a.

In addition to setting the monuments on the actual section or lot corners in the original survey, additional monuments known as *witness points* were within 5 chains (330 ft, 100 m) of the monuments. These consist of trees, known as *bearing trees*, designated as BT, and *bearing objects*, designated as BO, used where trees were not existing.

Trees used as witnesses are blazed (cut to the smooth wood surface), and the smooth wood surface on the blaze is inscribed with letters and figures to aid in the identification of the location. A tree used as a witness for a section corner, such as point b in Figure 13.20, would be inscribed T20N R7E S31 BT, which indicates that the bearing tree (BT) is located as shown on Figure 13.20c. The true bearing and horizontal distance are determined, including a description of the tree, and recorded in the field notes. Normally, at least one and usually two bearing trees are established for each corner set in the original survey. Due to changing regulations and/or lack of diligence by the original surveyor, the only markings on a bearing tree are often BT. However, this does not create a practical problem, as the surveyor doing the retracement usually has a good idea where the original monument was located. Therefore, he can usually deduce the location of the found bearing tree referred to in the original field notes. This becomes a problem only when two bearing trees close together have been marked to witness the same monument. Two important facts regarding bearing trees should be appreciated by the modern surveyor:

1. The tree continues to grow around the original blaze, thus covering the blaze and the inscriptions thereon. Consequently, it is necessary to cut carefully into the tree until the inscription BT is found.

2. The blaze and inscription may be up to 5 ft (1.6 m) higher on the tree if the original monument were set and witnessed during winter. Therefore, it is necessary to determine the month of survey in snowfall areas to determine that portion of the tree vertically from the ground where the BT was originally carved.

Bearings objects, within 5 chains (330 ft, 100 m) of the monument, are used where no substantial trees exist. These consist of one or more of the following: (1) significant cliffs or large boul-ders, (2) stone mounds, and (3) pits dug into the ground. Where the bearing object consists of rock, the point used for the meas-urements to the monument is marked using a chisel by a cross (×), the letters BO, and the section number, the latter only for sectional survey systems. Stone mounds, described previously in this subsection, are used where loose stone is readily available. Where no rocks, stones, or trees are available, the corners may be witnessed by digging pits 18 in. (45 cm) square and 12 in. (30 cm) deep. One side (not corner) of the pit should face the monument. Depending upon the country (United States or Canada) and the regulations in force at the time, up to four pits are sometimes required (placed on all four sides of the monument), and the dis-tance of the pits from the monument varies, although it is nor-mally 3 ft (1 m). As the pit will usually fill in gradually with a different soil, and often revegetates with a different species, the lo-cation of the pit(s) may be identified many years or decades later.

Lines. The marking of the survey lines on the ground was done to preserve the lines between monuments. In addition to setting the monuments and witness points as described in the previous sec-tion, natural terrain features along the line were recorded in the field notes, as discussed in Section 13.2.7, and blazes or "hack marks" were made on living timber at regular intervals along the line.

The regulations pertaining to public land surveys in the United States and Crown Land surveys in Canada have always re-quired blazing and hack marks along lines through timber. Trees directly on the line, called *line trees*, are marked with two horizon-tal V-shaped notches, called *hack marks*, on each side of the tree facing along the line in both directions. Trees along the line with-in one-half chain (50 links, 33 ft, 10 m) are blazed at breast height, the flat side of the blaze being placed parallel to and facing the line. The frequency of blazes along the lines varies widely depending on the density of the trees and the thoroughness of the original surveyors.

The modern surveyor should be aware of the following prac-tical aspects of line reestablishment:

1. Forest fires and settlers have destroyed much of the timber. Therefore, the surveyor should check the age of a mature tree in the area by counting the growth rings. If the tree exam-ined could not have existed at the time of the last known original or retracement survey, there is no point in wasting time looking for evidence of hack marks or blazes on the trees near the suspected line location.

2. If the original blazed trees still exist, a keen eye is required to observe evidence of old blazes. After determining the expected height of the blazes above the ground surface through knowing the month of the year of the last survey and allowing for snow depths, if applicable, the surveyor should carefully look for any vertical bark scar, unnatural flat spots, or bulges on the outside of the tree. Any unnatural marking at the appropriate height should be investigated.

13.2.7 Field Notes and Survey Records

Field notes along all run survey lines are often of great value in reestablishing lines where no physical ground evidence exists. In many cases, no other evidence exists. The present-day surveyor should bear in mind the conditions under which the original field notes were made. The chain may have been missing one or even two links. The number of chains to a particular feature may have been miscounted, thus creating potential discrepancies of approximately 66 ft (20 m) or 132 ft (40 m). The use of aerial photgraphs and comparison with the field notes, prior to attending in the field, are usually of great assistance in identifying the discrepancies mentioned (see Chapter 15). This is particularly true for streams, rivers, swamps, and geological features such as ridges or cliffs. Office and subsequent field adjustments of the line on the aerial photographs, bearing the possible discrepancies in mind, will save the modern surveyor considerable time in line reestablishment.

The information to be included in field notes, according to the manuals, is similar in both the United States and Canada. A summary list is given next:

1. Course and length of each line run, including offsets, the reason, and the method used.
2. The type, diameter, and bearing and distance from the monument of all bearing trees and bearing objects (see Section 13.2.6), including the material used for the monuments and depth set into the ground.
3. Line trees: their species name, diameter, distance along the line, and markings.
4. Line intersections with either natural or person-made features, such as the line distance to Indian reservations, min-

ing claims, railroads, ditches, canals, electric transmission lines; changes of soil types, slopes, vegetation, and geological features such as ridges, fractures, outcrops, and cliffs. All pertinent information such as the bearings of intersecting boundaries including the margin of heavy timber lines are recorded.

5. Line intersections with water bodies such as unmeandered rivers, creeks, and intermittent water courses such as ravines, gullies, and the like. The distance along the line is measured to the center of the smaller water courses and to each bank for larger rivers.

6. Lakes and ponds on line, describing the type and slopes of banks, water quality (clear or stagnant), and the approximate water depth.

7. Towns and villages; post offices; Indian occupancy; houses or cabins; fields; mineral claims; mill sites; and all other official monuments other than survey monuments.

8. Stone quarries and rock ledges, showing the type of stone.

9. All ore bodies, including coal seams, mineral deposits, mining surface improvements, and salt licks.

10. Natural and archeological features, such as fossils, petrifactions, cliff dwellings, mounds, and fortifications.

Two types of field notes were kept. One type listed the chainages to pertinent points down the left side of the page with each pertinent feature listed shown on the right side opposite the appropriate chainage. The other type was termed *split line*, where the center line was widened to accommodate the chainage figures, and the features to the left or right of the line as run were shown on the side of the page where they were located. A typical example of the latter is shown in Figure 13.22.

Copies of the field notes and the township plat or original survey maps are kept on record at the Bureau of Land Management, Washington, D.C.; the Department of Energy, Mines, and Resources at Ottawa, Ontario; and at the majority of the state and provincial government offices. The departments storing these records in the state or provincial governments have a variety of names (Auditor of State, Public Survey Office, Register of State Lands, Natural Resources, Lands, and Forests, etc.) Consequently, the present-day surveyor is advised to determine the appropriate government agency, in the state or province where the records are kept.

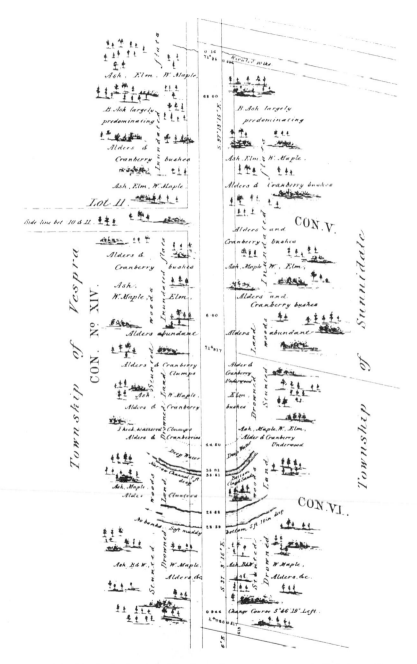

FIGURE 13.22 Split-line field notes for township subdivision. (Courtesy Ontario Ministry of Natural Resources)

Resurveys and Real Property Boundaries. The storage of township monument and line resurveys varies considerably depending on the regulations of the federal, state, and provincial jurisdictions in which the property is located. The normal situation is as follows. The original field notes and plans must, by law, be made available to other surveyors or members of the public for a reasonable fee. The maximum fee is usually stipulated by the federal, state, or provincial land surveyors' association. This fee is reasonable, for example, $25 to $50, depending on the amount of information required. The fee is determined by the time required to locate the proper file and plans in the surveyors' office, combined with the costs of reproduction for the notes and plans.

The same system is generally used for field notes and plans relating to real property boundary surveys. Plans of subdivision are kept on record in the county registry office. These are numbered consecutively as received by the registrar. The degree to which these plans are examined for reliability and accuracy varies greatly between jurisdictions. Some county registry offices will accept plans of subdivision without any investigation, while others will go to great lengths to verify or check key measurements through field investigation.

In certain jurisdictions, copies of field notes and plans are required to be registered with the appropriate county office. In many cases, the field notes and plan are shown on the same reproducible transparency. Therefore, the registry office has a complete record of the survey details, which is available to any interested parties at the minimal cost of reproduction for copies. This is, in the authors' opinion, the ideal situation. It avoids contacting several surveyors who may have worked on the lands under consideration and obtaining copies of all their individual records, a time-consuming and expensive task. The registry office should act as a central depository of field notes and plans. While this may be the ideal situation, it has been achieved in a relatively small number of county, state, and provincial jurisdicitions for reasons that are difficult to understand. However, the trends are in this direction and it is hoped this practice will accelerate.

Records of transfers of land from one owner to another are usually kept in the registry office, registry of deeds office, land titles office, or the like. It is therefore possible to trace the history of ownership, mortgages, rights-of-way, easements, and so on, for a specific land area. Usually, each section or lot within each concession has an individual page in books, known as the registry books, deed books, or abstracts. The deeds, mortgages, and the like are registered in order and given a registry number. Consequently, a *search* of the history of title is commonly conducted by a surveyor and/or a lawyer to identify numbers of the documents, known as deeds and/or instruments, relating to the property being examined. As these deeds and/or instruments are registered by number, copies and a description of the property boundaries or a plan of the same can be obtained.

It is essential that the modern surveyor understand that this information is available, the level of detail to which it has been documented, the completeness of the recordings, and the location of other pertinent data, such as the offices of individual surveyors who have worked in the area. As the previously described factors vary widely from county to county, the modern surveyor is advised to visit the county registry and/or land titles office to determine the information available. This should be a first step in conducting any land survey, as the subsequent work requirements and therefore costs are directly related to the information available at the county registry office.

Organizations, such as title companies, and the land titles office in Ontario, will guarantee the title against all prior claims. This guarantee varies from defects in property description, right-of-way claims, and the like, to the market value of the property should a valid ownership claim be presented. In certain cases, previously unregistered deeds have a prior claim to the property. When this happens, the property is evaluated at market price and the "owner" is recompensed accordingly, even though he may have to yield the property to the recipient who holds unregistered, but legal, deeds.

If the modern surveyor is dealing with property whose value is guaranteed, previous field notes and plans will have been vigorously examined and may generally assumed to be correct. If no guarantees are involved and property plans and descriptions are merely recorded and not checked, the records may be totally or partially in error, thus complicating subsequent surveys and resulting in significant cost increases.

Legal survey plans for state and provincial highways are kept on file with the appropriate highway departments or ministries. These plans will be registered on the title of the property if land has been acquired for a widening of the highway road allowance. The original surveys of the highway road allowances in the vicinity of the property being surveyed have usually involved reestablishment of portions of the original township subdivision. This is also true for utility companies, such as hydro power and gas pipelines regarding easement or right-of-ways surveys. The modern surveyor should check all these sources before proceeding with the required survey work.

13.3 PROPERTY CONVEYANCE

When any piece of land changes ownership, some type of survey is usually required. The purchaser normally wants to ensure that he or she is in fact acquiring the land(s) described in the offer to purchase. It is also important to establish that any structures included in the offer are located entirely on the land. Knowledge of any right-of-ways, easements, highway widenings, and the like are also essential. Such title *encumbrances* are important as the owner's rights on these lands are usually severely curtailed. For example, erection of any structure such as even a small building is usually not allowed even though the land is still owned by the owner of the lands adjacent to the encumbrance.

The principles of establishing boundaries are well established. These are based on established definitions of legal terms, deed descriptions, riparian rights (for lands bordering on water bodies), and adverse possession. The authors have selected these topics for discussion in this section because they are common to most property boundary surveys. Many other legal principles exist, but do not usually affect the actual establishment of property boundaries. In addition, these rarer principles are usually dealt with by the legal profession during the land transfer process. Therefore, the authority and liability of the surveyor are discussed as these concerns arise.

The survey techniques for carrying out rural land surveys, urban land surveys, and city surveys are discussed in Sections 13.4, 13.5, and 13.6, respectively.

13.3.1 Definitions of Legal Terms

Some of the most common legal terms relating to conveyance of land are defined next. The definitions are presented as a practical interpretation of the formal definitions presented in legal dictionaries. The real meanings of the terms as they affect the modern surveyor are more important than those couched in formal legal language.

1. *Adverse possession*: When land is used by a person other than the owner for an extended period of time (usually 20 years), the land may be claimed from the owner by the user under certain circumstances. This term is discussed in more detail in Section 13.3.4.

2. *Alluvium*: When land along the bank of a river or shore of a lake or sea is increased by any form of wave or current action or by dropping of the water level, the additional land is called alluvium. This usually is a gradual process and the rate of addition cannot be accurately or easily determined in relation to small increments of time. As the owner of the lands along the river bank or shoreline may gain or lose land by these processes, the additional lands become the property of the owner of the shoreline. The gain or loss of land through alluvial processes is often a contentious issue. It may have been caused by the installation of shore- or bank-related structures upstream or downstream from the property. As the surveyor's evidence is often critical to determining the cause and therefore recoverable damages, this aspect is discussed in further detail in Section 13.3.3.

3. *Avulsion*: This involves the sudden removal of land of one owner to the land of another caused by water forces. When this happens the transferred property belongs to the original owner. The main difference between avulsion and alluvium is the rate at which the transfer of soil and land occurs. Avulsion can be attributed to a sudden movement caused by an identifiable event such as a violent storm or flood.

4. *Deed description*: The deed is the legal document transferring land(s) from one owner to another. The most important documentation in the deed from the surveyor's aspect is the description of the property being transferred. The description is intended to describe the details of the property boundaries, including bearings, distances, and appropriate corner markers. As the interpretation of the description is of great importance, this topic is covered in detail in Section 13.3.2.

5. *Fee simple*: The word "fee" indicates that the land and structures thereon belong to its owner and may be transferred to those heirs that the owner chooses or, if necessary, that the law appoints. The word "simple" means that the owner may transfer the property to whomever he chooses. Therefore, to hold the title of land in "fee simple" is the most straightforward and simplest form. It places no restrictions on the owner regarding the sale or disposition of his land.

6. *High-water mark*: The high-water mark occurs where the vegetation changes from aquatic species to terrestrial species. If no vegetation is present, such as a bare rock shoreline, the high-water mark is usually clearly indicated by a distinctive change in the color or tone of the rock along a level line representing the appropriate water level. This line may appear either above or below the existing water level at the time of the survey, depending on local water-level conditions. Regarding sea shorelines, the elevation to which the highest tides reach is termed the high-water mark. The information relating to high tides is usually available from tidal charts and local measurements at tidal gauging stations. Significant tidal fluctuations occur locally depending primarily on shoreline configuration. The reader is referred to Chapter 14 for a more detailed description of tidal measurements.

7. *Mortgage*: The purchaser usually obtains a mortgage, involving financing by a bank or loan corporation for the remainder of the purchase price of the property and structures thereon, if applicable. The mortgage is usually registered as a document on the title of the land at the county registry office. When it has been paid in full, a document called a *discharge of mortgage* is registered on title as well. Most mortgage lending institutions require survey documentation to ensure that the buildings are not only entirely on the property but are also within the acceptable clearances from the property lines as established by the local municipality. These surveys are discussed in Section 13.5.

8. *Patent*: When property is conveyed by federal, state, or provincial governments to institutions, companies, or individuals, it is called a "patent." This is usually the first entry made in the registry office books kept for sections and lots or concessions.

13.3.2 Deed Descriptions

Deed descriptions include the directions and distances of all lines along the property boundaries of the parcel of land. The types of corner monumentation and the area of the parcel may or may not be included.

The description is usually in written form, rather than a survey plan. The property is described as starting or commencing at a point, and the description continues either clockwise or counter clockwise around the property, returning to the starting point.

The starting point or *point of commencement* must be directly related by both distance and direction to a point established in the original township subdivision survey or a lot corner in a registered plan of subdivision. In the latter case, the plan of subdivision is already located with respect to the township subdivision fabric.

It is important to realize that deed descriptions have been written by people from every walk of life, particularly during the early development of the United States and Canada. Consequently, the deed description may contain errors and mistakes in measurement, direction, and calculations of areas. The units used in descriptions vary considerably. Chains (66 ft, 20 m), links (0.01 chains), and rods (16.5 ft, 5 m) were commonly used in older deeds in both the United States and Canada. The measurements were subsequently given in feet in both countries and recently in meters in Canada.

As a result, inconsistencies with the deed description and the boundary evidence on the ground are difficult and sometimes impossible to resolve. When these conflicts occur, the law states that the surveyor should establish the boundaries to match the intentions of the parties involved in the property conveyances.

Figure 13.23 shows a typical property that requires a detailed deed description. It also provides examples of mistakes and omissions common to many deed descriptions.

The deed description of the property might read as follows:

All and singular that certain parcel, tract of lands and premises situate, lying and being in lot 17, concession 6, township of Bastard, County of Leeds, being more particularly described as follows:

Commencing at a point in the southerly limit of lot 17, concession 6, distant 200.00 feet measured easterly from the southwest corner of the said lot 17; thence N 9°00′ W, 400.00 feet, more or less to a point in a post and wire fence, along a course parallel with the westerly limit of lot 17; thence N 74°00′ E along a post and wire fence a distance of 150.00 feet; thence S 9°00′ E, 400.00 feet, more or less, to a point in the southerly limit of the said lot 17; thence S 74°00′ W, 150.00 feet more or less to the point of commencement.

The following comments on this example description will explain some of the wording, establish the priorities for the surveyor, and point out common errors in description writing. Reference is made to the corners and boundaries lettered from *A* to *E* inclusive in Figure 13.23.

1. Point *A* is a *fixed point* as it was originally established during the township subdivision. As mentioned previously, the

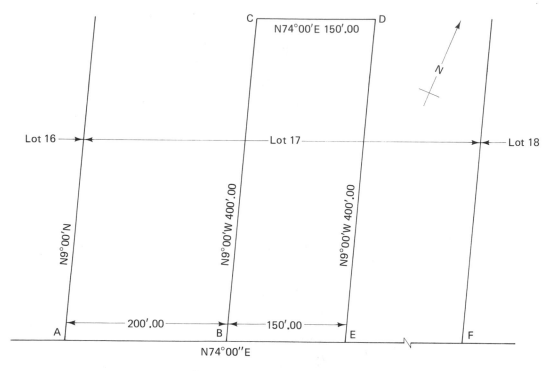

FIGURE 13.23 Typical land parcel requiring detailed deed description.

deed description must refer to an established corner in the township fabric.

2. Point *B* is the point of commencement, which is always the first point mentioned in the deed description that is an actual corner of the property being described.

3. The distance *BC* is intended to be 400.00 ft. The "more or less" is shown after the measurement in the description because the intent is that point *C* should lie at the base of the fence. Therefore, the fence, not the 400.00 ft, governs.

4. The line *CD* runs along the post and wire fence, as stated in the description, and is therefore not necessarily a straight line from *C* to *D*. Consequently, as the fence bends, the line does accordingly.

5. The distance along the easterly boundary of the parcel, line *DE*, is governed by the distance from the fence to the south limit of the township lot, for the same reasons in comment 3.

The bearing is indicated as S 9°00' E without any other qualifications. This indicates that the line *DE* may not be parallel with the line *BC* and hence the westerly lot line. Therefore, point *E* could actually be east or west of the point shown in Figure 13.23.

6. The southerly boundary *BE* was likely intended to be 150.00 ft, but this distance may be greater or less depending on the actual location of point *E*. The "more or less" shown after this measurement in the description does not have as much significance as the "more or less" phrase previously mentioned. The reason for this is complex and beyond the scope of this text. Suffice it to say that "more or less" is commonly attached to the last measurement mentioned in the deed description.

One of the most common errors in deed description writing is failing to change the bearing directions from, for example, northwest to southeast for line *DE*. This means that, taken literally, no parcel of land exists. This is considered a mistake and is usually not critical as long as the surveyor and legal council involved in the property transaction agree that the intent of the survey is not violated. If necessary, a *correction deed* incorporating the correct description might be registered.

References in the description to monuments, at property corners, such as "an iron post planted," "a wooden stake," "a stone monument," add little value from the surveyor's standpoint. These monuments are not necessarily permanent. It is also assumed that all field evidence regarding the location of any corner will be thoroughly investigated by the surveyor. Permanent markings such as cut crosses (×) inscribed in bedrock or solid concrete structures may be useful additions to the written description as they represent more unusual forms of monumentation, which the surveyor may find more difficult to locate in the field unless forewarned by the description.

Land bordering on a highway or street may be assumed to convey ownership to the center of the road allowance in certain jurisdictions. As it is usually intended that the state, province, or municipality will own the road allowance, the description must clearly state the ownership to the street boundary.

In Figure 13.23, the distance *AB* is known as the *tie distance* as it represents the length from the point of commencement to point *A*, which is fixed within the township survey fabric. It is not uncommon for the tie distance to be stated incorrectly in the description. This problem becomes compounded as subsequent descriptions, such as the parcel immediately east of line *DE,* would use a tie distance of 200.00 ft, plus 150.00 ft, or 350.00 ft. This perpetuation of the error in tie distance has been known to continue over a large number of parcels as the lands are sold over the years. One particular occurrence encountered by the authors resulted in the described property lines being misplaced 66 ft (20 m) along a consecutive line of 18 parcels, having frontages ranging between 100 ft (30 m) and 150 ft (45 m). This resulted in the described property lines running through 12 of the 18 residences. As no monumentation was present along the parcel frontages, no fence lines or occupational boundaries existed, and the township section corner had been established beyond doubt, this resulted in a serious situation. The main difficulty was mutual trespassing by all concerned. Also, none of the parcels could be transferred without the registration of correction deeds, which required the consent of all 18 owners. The six property owners whose residences were luckily, for them, not interfered with by the property boundaries felt that they should receive financial remuneration from the other 12 for signing the correction deeds. This totally unpleasant situation was entirely due to two factors: the incorrect tie distance to the first parcel, and the lack of thoroughness of subsequent surveys of adjacent parcels in actually confirming the tie distance.

13.3.3 Riparian Rights

Riparian rights refer to those rights of a property owner of land that borders on a water body. The rights include the use of the shore and ownership of land under the water surface and therefore use of the water. The main difficulties in surveys involving properties along water bodies are as follows:

1. The boundaries are both irregular and subject to change with time due to alluvial processes (see Section 13.3.1).
2. The ownership may extend only to the high-water mark (see Section 13.3.1) or to the center of a stream or river. This depends on the laws of the state or province in which the land is located. These vary widely and are very inconsistent.
3. Certain survey systems incorporated a publicly owned strip of land, usually 1 chain (66 ft, 20 m) wide, parallel with the shoreline or high-water mark. As the owner normally wants to locate buildings as close to the water as possible, the net result is that all or a portion of a private residence is located on public lands. As these public strips of land are termed

"shoreline road allowances," the implications range from the public right to camp, barbecue, drink, and the like, in front of the helpless owner's residence or cottage, to a land developer constructing a road in front of or through existing buildings to provide access to proposed development lands.

The difficulties briefly outlined establish the complexities of establishing the property lines and rights of riparian landowners. However, certain principles usually apply. These are illustrated in the following situations:

1. *Nonnavigable and navigable waters*: As discussed in Section 13.2.4, meander lines were run at convenient locations adjacent to the shorelines of water bodies as defined by the original township survey regulations. The meander line is *not* a land boundary, as it was used only as a traverse line to record the variations in the shoreline. The ownership is usually either to the high-water mark for navigable waters or to the center of the stream or river for nonnavigable waters. Therefore, the first step in establishing the shoreline boundary involves the determination of whether the waters are *legally* considered to be navigable or otherwise. This determination if far from simple. Normally, acts or statutes relating to navigable waters that are subject to recent updating must be consulted. These laws usually list by name the water bodies that are legally navigable. Observations of large vessels using the waters for transportation routes are meaningless in many cases. Nothing may be assumed in this regard. For example, the Mississippi River is considered by some bordering states to be navigable and by others to be nonnavigable. The surveyor is therefore obliged to determine the definition of the waters adjacent to the property being surveyed through reference to the most recent acts and statutes only. Common sense and logic are immaterial.

2. *Public reserves along shorelines*: The situation regarding lakes requires reference to the original survey regulations for specific townships. For example, a township boundary may pass through the middle of a lake, as illustrated in Figure 13.24. This information is available through careful examination of the plats or original township maps. For the property owners along the west side of the lake, ownership extends only to the dashed line 1 chain (66 ft, 20 m) from the shoreline, usually the high-water mark of the lake, as

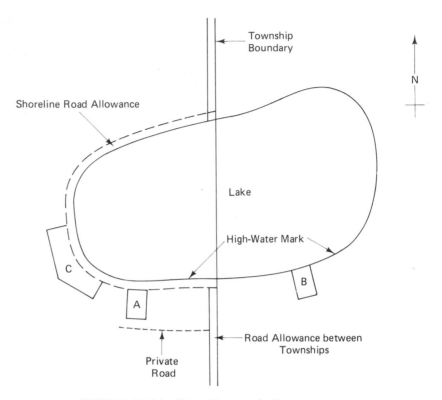

FIGURE 13.24   Shoreline road allowances.

most lakes are presumed to be navigable, particularly if their surface area is over 25 acres (10 hectares). The ownership boundaries of parcels *A* and *B* are shown on Figure 13.24. The difference is the road allowance along the shoreline, which has a direct effect upon parcel *A*. The owner of parcel *A* is often totally unaware that the property does not extend to the shoreline (usually high-water mark) of the lake, usually due to lack of communication between the purchaser and his legal council. Aside from the problems of the owner of parcel *A* building on the shoreline road allowance and/or the public utilizing his shoreline at will, the owner of parcel *C* may decide to subdivide his shoreline. As a private road is the only land access to parcel *A*, the owner of parcel *C* has every right to build an access road along the shoreline road allowance in front of parcel *A*. The owner of parcel *A* will not take kindly to this proposal nor will his neighbours to the east. The only reasonable solution to this problem is for the owner of parcel *A* and his neighbors to agree to allowing the owner of parcel *C* to use the private road for his development access. This means that negotiations must be carried out with the owner of the lands behind parcel *A* to allow this private road to become a public road allowance. The unpleasantries caused by this situation are evident. However, the owner of parcel *B*, on the same lake but in the adjacent township, shares none of these concerns. The surveyor's responsibility in this situation involves previous investigation to establish the existence of the shoreline road allowance; surveying the property correctly; marking the road allowance "public lands" on the plan of survey; and ensuring to the best of his ability that the purchaser, in this case of parcel *B*, is aware of the possible consequences.

3. *Property lines in areas of alluvium*: Where the shoreline location has been changed by alluvial processes, the existing property boundaries intersect the new shoreline. This is illustrated by Figure 13.25. The following survey procedure is carried out:

   a. The distances along the frontage where the shoreline has been altered are determined. Distance *AF* along the original shoreline will usually have to be determined from either existing monumentation evidence or from the deed

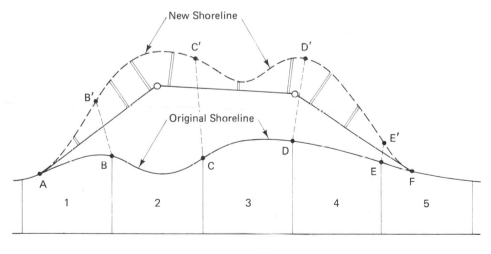

FIGURE 13.25   Riparian boundaries for areas added by alluvium.

descriptions. The distance AF along the new shoreline can be measured using a traverse and offset measurements to bend points in the shoreline, as illustrated in Figure 13.25. Stadia techniques, discussed in Chapter 7, may also be used. The new shoreline is plotted from this information, and the distance AF is scaled as accurately as possible. Where the frontage of alluvium is large and/or difficult to survey on the ground, existing or newly acquired aerial photographs will provide an accurate and economical positioning of the new shoreline (see Chapter 15).

b. The ratio between the distances AF (new shoreline) to *AF* (original shoreline) is determined. In this example, the ratio is 1.25.

c. Locate point *B'* so that *AB'* : *AB* is 1.25. This same ratio is used to determine the locations of points *C'*, *D'*, and *E'*. Therefore, the new shoreline is proportioned among the five parcels affected by the alluvium.

d. Note that the sidelines of the affected lots now bend at the original shoreline where they did not previously. Thus the original points *A* to *F* must be located on the ground and the angles measured to determine the directions of the lines crossing the alluvium, such as *BB'* and *CC'*.

13.3.4 Adverse Possession

The use of land by other than the registered owner for a sufficiently long (usually 20 years) uninterrupted time may lead to a land transfer to the user from the registered owner. The legalities of adverse possession vary considerably among states and provinces. The general principles are discussed herein to warn the surveyor of his responsibilities and the investigative techniques to be employed.

Some common legal aspects are given next. These do not hold in all jurisdictions and it is the surveyor's responsibility to investigate their validity for the property under investigation.

1. Public property, such as unpatented lands, streets and highways, cannot be acquired by adverse possession.

2. Usually the land use, whether it involves agriculture, erection of buildings, or fencing, is known to the owner of the land. In many cases, the landowner is not aware that adverse possession exists, and if these other land uses on his land do not interfere with the owner, he commonly chooses to ignore them.

3. These land uses under adverse possession must be against the interests of the owner under most jurisdictions. This has to be interpreted carefully. At the beginning of the adverse land uses, the owner could not have been hostile or he would have put a stop to the adverse uses of his land. Therefore, the owner becomes hostile when he decides to stop the adverse land uses and finds out that, by law, he cannot.

4. The adverse possession may consist of public use of a private right-of-way on private lands. In other words, private land can become public if extended usage is permitted by the owner.

5. If the land owner informs those carrying out the adverse land uses on his land that he is the owner of these lands before the adverse land use has continued for the number of years required by law, the time period for adverse possession starts over again. For example, if the legal period for continuous use is 20 years, the adverse land uses started in 1975, and the owner informs those conducting the adverse land uses that he is the owner of the lands in 1983, no action can be undertaken for adverse possession until 2003, assuming that the owner does not mention his ownership to the user again. The best form of notification is by registered mail as this means that a record of having received the letter exists. Verbal notification should be carried out in front of witnesses and the date noted.

Three examples drawn from the author's experience are discussed next to illustrate these principles.

1. A fence was built approximately 20 ft (6 m) over on the adjacent owner's property parallel with the true boundary line, and this strip of land was cultivated for over 20 years. As no mention of ownership was made, nor was any formal survey conducted, the strip of land became the property of the adverse land user, regardless of the true property boundary.

2. The owner of a large shoreline lot on a lake, having approximately 500 ft (150 m) of frontage, allowed a friend (at that time) to build a small residence at the other end of the lot and use the property. The friend learned of the appropriate statute and the owner was not aware of it until it was too late. The ex-friend now legally owns the property consisting of 150 ft (45 m) of valuable shoreline frontage.

3. The owner lived on a residential street in a large city close to a public school. As the houses were close together and all

were fenced, there was no shortcut for the children attending the school. This owner, out of kindness, opened a 6 ft (2 m) wide gap in his back fence, thus allowing the students to cross his back yard and proceed down his driveway. After this had continued for approximately 22 years, the owner decided that he no longer wanted the herds of students crossing his property, so he rebuilt the fence, thus cutting off access. One student's parent was a lawyer who instigated legal action. The result today is a public right-of-way, or easement, across the rear yard and down the driveway, all of which is registered on the title of the land. As a result, the market value of the land and residence has dropped very substantially, and the public continues to use it for access.

In the authors' opinion, laws relating to adverse possession do not receive enough publicity. The possible ramifications to the rightful landowner are truly horrendous and financially destructive. The surveyor's responsibility involves not only knowing the details of applicable laws, but also inquiring of the landowner about any adverse land uses that are being practiced, how long they have continued, and what notice of ownership, if any, has been given and the dates involved. The surveyor is responsible to the extent that he must show certain examples of adverse land uses on his plans. The simplest common situation is the relationship of an existing fence line to the true property line.

13.4 RURAL LAND SURVEYS

Rural land is considered to be land outside the boundaries of cities, towns, villages, and the like. It usually consists of relatively large areas, in excess of 2 and commonly 10 acres (4 hectares). These larger land tracts usually differ from urban lands and city lands in the following ways:

1. Many of the original land grants in the United States and Canada were not regular in shape, as the boundaries often were natural features such as ridges, streams, or river banks. Person-made features and names of adjacent owners were often used as property boundaries. The property descriptions using these features and names did not contain any definitive bearings and distances. Many of the features have since disappeared and, of course, the adjacent owners have changed many times since.

2. Rural lands are not as valuable as urban or city on a per unit basis. Therefore, the property descriptions and surveys have not been done with the same care, because to do so would incur high survey costs disproportionate to the land values.

3. Control survey networks, such as those discussed in Chapter 11, have not been extended into many rural areas at this time, because the priorities for control systems are for urban and city surveys. Therefore, the boundary markers will not be incorporated into the statewide or provincewide horizontal and vertical coordinate system. As a result, each rural land survey basically stands on its own. Errors accumulate and no overall control system is available to correct or proportion the errors between coordinate control points.

13.4.1 Descriptions

Descriptions of rural lands take several forms, mainly depending on when the description was written and the value of the land. These different types are briefly discussed next.

1. *Descriptive only*: These relate all property locations to natural and person-made features, as discussed in Section 13.4.

2. *Metes and bounds*: A typical metes and bounds description is given in Section 13.3.2 and illustrated in Figure 13.23, wherein only the bearings and lengths of the sides are given. This example is given to the nearest 0.01 ft and the degrees to the nearest minute, thus indicating that the survey was done with a transit and tape. However, many property descriptions were based on compass and chain surveys and are described using chains and links for measurements, and directions are given to the nearest degree. The meanings of the wording used in deed descriptions and some of the common errors are discussed in Section 13.3.2.

3. *Township subdivisions*: Part of lots, called *aliquot parts*, are easily described for both sectional and nonsectional or special systems (see Section 13.2.4). Examples of each are given next:

   a. *Sectional system*: The legal description of a 40-acre (16-hectare) tract for the hatched area in Figure 13.20c and the example principal meridian used in Section 13.2.2 is as follows:

   The northwest quarter of the northwest quarter of section six (6), Township nineteen (19) North, Range six (6) East, of the

Initial Point of the Mount Diablo Meridian, containing forty (40) acres, more or less, as set out in the United States Public Lands Survey.

b. *Nonsectional or special system*: The legal description of a 50-acre (20-hectare) tract for the hatched area in Figure 13.17 is as follows:

The north half of the west half of lot 9, concession 3, Bastard Township, County of Leeds, Province of Ontario.

4. *Coordinates*: For economic reasons previously mentioned in this subsection, coordinates have not achieved common usage for rural land surveys. One exception to this is the use of coordinate systems by state or provincial highway departments. Therefore, lands bordering a newly constructed highway may be sufficiently close to coordinated monuments to incorporate this information into the deed description. The description is similar in wording to the example given in Section 13.3.2, with the addition of north and east coordinates stated in brackets after the point location of each corner. For example, point *B* in Figure 13.23 would show "N 638,014.08, E 160,269.69" in brackets after the first mention of point *B*. The coordinates quoted are in feet, and the metric equivalent would be used in countries where the metric system actually exists or when the conversion process has actually been completed and sanctioned by law. The other corners on the property boundary would have the appropriate coordinates indicated in brackets after the first mention of the corner in the description. The use of coordinates related to a state or province-wide system is, in the authors' opinion, the ultimate solution to presenting an accurate description of a parcel of land. The description would state or province, as follows: "The lands enclosed by the following coordinates," followed by a simple tabulation of north and east coordinates. This would result in truly fixing the property corners and would avoid references to oblique statements such as the post and wire fence discussed in Section 13.3.2. The technology to achieve this simple solution exists. The savings in lawsuits between or among adjacent landowners would be tremendous. The only real requirements are an extension of the state or provincial coordinates to establish enough coordinated monuments, combined with the laws to ensure that all descriptions are directly related to the state or provincial coordinate systems.

13.4.2 Boundary Surveys

Two general types of boundary surveys are carried out in rural lands. One is the original survey whereby new property lines are created. The other is a resurvey, which results in relocating property lines that have been previously surveyed. The principles for both types are similar in that both have to be located with respect to the original township survey fabric. However, the procedures are somewhat different. Due to the similarities of approach, the reader is urged to read both subsections.

Original. An original survey is necessary when a tract of land, which has not previously been surveyed, is being transferred from one owner to another. The land is usually defined in an informal description in the offer to purchase. The form of description may be set out using the general format of any one of the four description types discussed in Section 13.4.1.

The following procedure is necessary to carry out an original survey properly:

1. A copy of the offer to purchase or any other document relating to the property boundaries should be obtained. For example, if the westerly portion of the property illustrated in Figure 13.23 is to be surveyed, it is important to determine if the parcel is "the westerly half of the parcel" or "the westerly 75 ft (22 m) or the parcel." The difference will be apparent from examining the wording of the description and subsequent explanation given in Section 13.3.1. Half the parcel will be 75 ft (22 m) more or less, and the westerly 75 ft (22 m) will be exactly 75 ft. At this time, it is also wise for the surveyor to establish the intent of the purchaser (who is usually the client) regarding what he or she thinks the property should consist of.

2. If the description in the offer mentions any registered deed numbers, such as the "westerly half of the lands described in instrument number 45792," a copy of this document must be obtained from the county registry office as it bears direct reference to the survey results. It may also be necessary to obtain deed descriptions of adjacent properties on each side of the parcel being surveyed. This will indicate the field measurements that should be held and those that should be proportioned.

3. The reference monument or corner in the township subdivision fabric must be located or reestablished, for example,

point *A* in Figure 13.23. Often, this point is not easily located, and the techniques discussed in Section 13.2.4 must often be used. These methods are time consuming and
therefore costly. Discussions with local residents regarding
the corner location are useful only if all else fails. The survey-
or should be aware that the locals seldom have any real con-
ception of distance and/or direction in terms of measurement, have little or no knowledge of surveying, and may have
a vested interest in the location of property boundaries. A
properly signed, witnessed, and legally worded affidavit (statement) of the local person's length of residence in the area and
any other pertinent facts relating to why he or she should
know the location of the corner should definitely be obtained
by the surveyor. This applies only where this local knowledge
is to be used in any way to assist in defining the location of
the corner.

4. The property line adjacent to the street or highway, commonly called the *street line*, must be determined, for example, the
line on which points *B* and *E*, Figure 13.23, are located. This
requires knowledge of the township subdivision system originally used. For the sectional system, where 5% of the area
was allotted for roads, the best evidence may be in adjacent
boundary surveys or deed descriptions, as the roads may follow a circuitous route. If the road allowance happens to follow along a section boundary in sectional systems, or along a
concession or sideroad in nonsectional or special systems.
The surveyor must be aware that the corner to be reestablished (if possible) is the adjacent section, quarter-section, or
lot corner on the opposite side of the parcel, for example,
point *F* in Figure 13.23. In certain cases, the only evidence
existing is the location of the traveled road. In this case, the
center of the road is determined by measurement, and the
two street lines are located 33 ft (10 m) on the either side.

5. The property corners on the street line are established and
monumented. As fences, trees, and other obstructions are
usually located on this line, it is common to use an offset line
parallel with the street line for ease of measuring distances
and angles. This is illustrated in Figure 13.26. Points *A'* and
B' are set, using temporary markers such as nails with flagging, on the offset line, which usually is located along the
shoulder or ditch of the road. The property corners on the
street line *A* and *B* are set by turning the appropriate angle

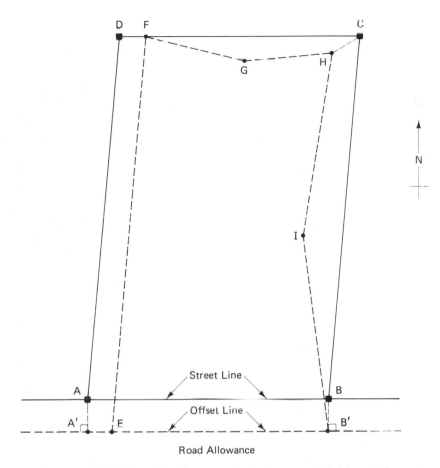

FIGURE 13.26   Rural land survey techniques and field note format.

(90° is safest) and measuring the offset distance (*A'A* and *B'B*.)

6. The other corners of the property are established by a variety of field techniques, most of which are illustrated in Figure 13.26. Based on the boundary requirements and existing evidence as discussed later, the surveyor normally uses a combination of offset lines incorporated within a closed traverse (*E–F–G–H–I–B'–E*). This assumes that obstacles to running the property boundaries directly exist. Some examples relating to Figure 13.26 are as follows:

    a. In the distance *AD* is fixed, then the offset line *EF* can be run to point *F*, and *D* is subsequently set by turning the appropriate angle *EFD*.

b. If point *C* is located at an identifiable point in the field, such as a fence intersection, the monument is set at point *C* and tied to the traverse by measuring the angle off the reverse lines to point *C* and the distance *HC.* This distance should be measured at least twice and verified, as subsequent calculations of the position of point *C* will not provide a check on this measurement.

7. The distances and angles measured are carefully recorded in the field notes. The traverse is then closed for latitudes and departures using the techniques described in Chapter 6. The governing bearing to which all other traverse bearings are related must be established. Most present-day survey regulations require that this be related to true north, referenced as passing through a well-defined point nearby, such as a main section or lot corner. The true north or "astronomic" bearing of *AB* or *A'B'* may have been established by a survey of adjacent lands involving this same line. If this information is not available, it may be necessary to determine the astronomic bearing through observations on Polaris, as described in Chapter 11. If state or provincial coordinate control points are within a reasonable distance, an open traverse measuring angles only could be used to transfer, through azimuth or bearing calculations, the bearing along the line between the control points to one of the property boundaries. The latter is unlikely in rural surveys. If the survey area is located a long distance from the surveyor's office, serious consideration should be given to calculating the traverse closure before leaving the area. Easily portable, battery-operated calculators are available, and the surveyor is well advised to be equipped with one for surveys in remote areas. There is no cost comparison between this procedure, where potential error lines can be remeasured on the spot, and the time and money wasted in having to return to the site.

8. The rectangular coordinates of each property boundary corner are calculated using the methods given in Chapter 6. Subsequently, the bearing and distance of each property boundary are calculated, for example, *AB, BC, CD,* and *DA* in Figure 13.26.

9. A plan of the property boundaries is drawn, showing the bearings and distances of each line, the monumentation is set, and any relevant factors such as existing fences are noted. Depending on the degree of supervision within the local jurisdiction, the surveyor is advised to contact the registrar of

deeds regarding such details as drawing size and whether an Imperial scale or metric (in Canada) is required.

10. The survey plan is submitted to the client and, if required, to the local county registry office. If the latter is necessary, the inspector of surveys at the registry office will examine the plan for errors and omissions, and the exchange of revised plans will continue until the survey plan is approved. These details often delay the acceptance of the survey, thus possibly postponing the closing date of the property transaction. The surveyor is well advised to warn his client beforehand of this possibility.

11. The area is calculated, if required, using the methods given in Chapter 6.

12. A deed description, if required, is prepared. Metes and bounds and/or coordinates are the preferred systems depending on the jurisdiction. It is becoming common practice to attach the survey plan to the deed and clearly identify the property and its boundaries. The description in the deed can then merely indicate "the property shown as Part 1 on the attached reference plan No. 59R 4047," for example. In the province of Ontario, the 59 is the registry office number, R indicates reference, and 4047 is the survey plan number. This system avoids most of the disadvantages of deed descriptions and the common errors and misunderstandings discussed in Section 13.3.2.

Resurvey. A resurvey means that previously surveyed boundaries have to be reestablished in their original locations. Resurveys are required for a variety of reasons. Transfer of land from one owner to another is a common reason, as the new owner usually requires documentation of his holdings. Also, the owner may wish to subdivide his land for financial gain or family concerns, such as dividing the land among his children. The location of a new highway, railway, hydro power line, or gas distribution line also requires resurveys.

The surveyor should be aware that many rural properties have never required survey work since the original township subdivision. For example, if the property has remained in the same family for several generations, the need for a resurvey has never arisen.

Resurveys are considerably more challenging to the surveyor than original surveys. In fact, resurveys in rural lands require more experience, investigative capabilities, imagination, and perseverance than any other type of survey. Many of the factors

involved are discussed previously. The main reasons for the difficulties in resurveys are as follows:

1. Most of the evidence has been destroyed by forest fires, disin-
   tegration of wooden posts, removal of monuments, and so on.
2. The original surveys contained many errors, particularly if
   the survey were carried out using a compass and link chain.
   The magnetic declination (difference between magnetic north
   and true north) at the time of the survey was usually different than at present. Although link chains originally contained 100 links, either wear at the link connections and/or
   the loss of one or more links off one end during the original
   survey resulted in chains of other than 66 ft (20 m) long. Occasionally, the counting of the number of chains actually
   measured along a line in the original survey was in error,
   thus creating an immediate error of plus or minus 66 ft
   (20 m).
3. Information transferred from one record to another, such as
   from the field notes to the plat or original survey plan, was
   often in error.
4. Evidence from persons presumably familiar with the location
   of corners or lines often conflicts. Personal selfish interests
   and conflicts with adjoining property owners are but two
   common causes.

The surveyor must also be a detective and judge in carrying
out resurveys. Conflicting evidence both in the field and on the
original records is the norm. The surveyor is responsible for sorting through all available evidence; accepting all, portions, or none
of each; and being able to defend his decisions to all involved
parties, including the presentation of expert evidence in a court of
law. The surveyor is required to report his decisions to his client.
Therefore, he is well advised to keep a detailed log of his decisions
and the reasons for each, as it should be assumed from the beginning that the evidence may be examined by a formal body, such
as a court, at any future time.

The following steps are usually required in carrying out a
resurvey. It is impossible to cover all the possibilities within the
scope of this text. However, the principles are essentially the
same.

1. The surveyor obtains the descriptions of the property, as well
   as descriptions of all adjacent lands. He reviews the descriptions critically for gross errors (see Section 13.3.2) and cor-

rects them, where possible, to match the apparent intent of the description.

2. The descriptions are plotted to scale on a plan. At this point the property boundaries in conflict with adjacent lands as described become apparent. For example, the same property lines may overlap or leave gaps due to discrepancies in the descriptions. This identifies the described boundaries requiring further intensive investigation both through reviewing the deed descriptions and fieldwork. Those lines along which the descriptions of adjacent lands agree are identified as the most reliable starting points for the field survey work. Some boundary lines may be found to disagree by multiples of chains (66 ft, 20 m), in which case the prior dated deed usually takes precedence. One adjacent property may be described as proportionately larger or smaller in all distance measurements, thus indicating that the chain used for one of the adjacent properties was the wrong length.

3. If the bearings shown on the records are magnetic, the magnetic declination at the time of the survey is determined through past records, and the bearings are corrected to true north. This provides a crude approximation, particularly in areas of magnetic anomalies, such as rocks containing iron compounds. If the date of the survey is unknown and at least one boundary can be identified, astronomical observations are made to determine the true bearing of this line. All other lines are then referred to this bearing.

4. A number of situations may confront the surveyor when the detailed field investigations have been undertaken. The three usual circumstances and the procedures to be followed are discussed next.

   a. *At least one boundary established*: The established boundary line and the remaining boundary lines are measured and compared with the measurements stated in the deed descriptions. The proportionate lengths of the other sides of the parcel are calculated based on this comparative value. The property boundaries are then rerun as described and/or mapped on the original township plat. The approximate locations of each corner are thus established. The surveyor carefully looks for field evidence of, first, the corner and related witness points and, second, the lines originally run between monuments, using the methods described in Section 13.2.4. If positive evidence is found regarding the location of a corner, the old monument is left if in good condition. If it is not, it is replaced by a new monument. If no evidence of either the parcel corners or lines exists beyond the one established boundary, the corner is temporarily established based on the deed description. The area around the temporary corner is thoroughly examined for any evidence of the monuments and/or lines. If no evidence is found, the temporary corner becomes the permanent corner, providing that the distance measurements have been proportioned correctly prior to setting the temporary corner, as this is the best evidence available.

b. *If only one corner is established*: The surveyor has no choice but to establish the true bearings of the lines in the deed description from the original survey using the magnetic declination at the time of the original survey if the date can be established. The survey is rerun on the basis of the single corner and the surveyor's best estimate of the bearings from or to that point. Proportional correction is impossible to determine; therefore, the other parcel boundaries must be established based only on deed descriptions.

c. *If no corners or lines are evident*: The situation is extremely difficult. Descriptions of adjacent properties must be given greater reliance under these conditions. The plotting of the description of the parcel and adjacent tracts becomes more significant. As previously mentioned, the evidence of local residents is highly questionable for the reasons stated. It is important that the surveyor realize that agreement on the corners and boundaries by all affected owners is critical under these conditions. After having carried out the detailed investigation described previously and having reached no conclusion, the surveyor is well advised to call a meeting of all involved parties. Notice should be given by registered mail and preferably through personal delivery, if possible. The meeting should be held in the field at each potentially contentious corner, and a resolution of the location resolved, if possible. It is important to prepare carefully the required documents in advance, stating that those involved agree to the applicable corner location. This procedure involves all concerned parties through formal notice. If a consensus is achieved, the surveyor has solved his or her problem, and the required documentation is summarily signed and witnessed. One or more property owners may not attend, but given reasonable notice, the majority rules. The authors have found this method to be extremely effective, and the results are more than gratifying. As all parties are or should be involved, the resulting agreements carry considerable weight. Properly witnessed signatures agreeing to the common boundaries established provide valuable legal evidence. The surveyor must realize that, in this type of situation, the key word is agreement, not theory. In the authors' experience, the owners involved should be convinced that compromising on boundary locations will avoid both hard feelings and expensive lawsuits in the future.

5. Once the corners and boundaries have been established by field evidence and/or agreement of owners, the surveyor should continue with procedure 6, as described in Section 13.4.2.

6. The remaining procedure consists of implementing steps 7 to 12 inclusive, as described in Section 13.4.2.

The key factor in rural surveying is the acquisition of sufficient evidence, field or otherwise, to designate the boundaries as agreed upon by the owners involved.

13.4.3 Subdivisions

Rural land is subdivided for reasons other than proportional or aliquot parts of quarter-sections or lots. The rules governing the establishment of boundaries for aliquot parts are discussed in Sections 13.2.4 and 13.2.5. When rural land is irregularly subdivided, the purpose is commonly the creation of large lots, usually between 0.5 acre (0.2 hectare) and 2 acres (0.8 hectare). The houses subsequently built on these lots use wells for water supply and have sewage disposal systems for each house located on the lot. In areas where this type of subdivision is being developed, the rural road system is established, and most original township subdivision boundaries have been fenced.

Certain principles are involved in surveying the irregular boundaries of subdivisions in rural lands. The surveyor should be aware of the more important, which are discussed next. Fences are defined as those that have existed long enough to meet the requirements of adverse possession as defined in Section 13.3.4.

1. Fences along either side of road allowance boundaries are not used as property lines if these road allowances were established as part of the township subdivision. Therefore, lines between quarter-section corners, between lot corners along concession roads, and side-line allowances between concessions are governed by the lines between adjacent corners regardless of the fence location. Consequently, the monumentation along a fence adjacent to a road allowance is often located some distance from the fence. This fact is also useful in relocating previous monumentation along road allowance fence lines.

2. Fences along lines other than road allowances which form boundaries of these subdivisions usually predominate over the direct corner to corner boundaries. This fact is important as subdivision lot corners intersecting these lines should be set in the fence line, regardless of the number of bends in the fence. The surveyor should be aware that the fence line represents an informal, yet agreed upon, boundary between the property owners.

A typical situation regarding this principle arose within the authors' experience. Approximately 12 iron bars, each 1 in. (2.5 cm) square and 4 ft (1.2 m) long, had been placed along the straight line between property corners, due to the surveyor not being aware of this principle. The result was that all 12 bars had to be removed using power equipment and reset within the fence line at distances of up to 5 ft (1.5 m) from their original locations. Also, each subdivision lot boundary had a different bearing and distance, which had to be determined. Removal and resetting of monuments and the additional survey measurements are inconvenient and costly.

13.5 URBAN LAND SURVEYS

Urban land surveys differ from rural land surveys as follows:

1. The lands are located within or adjacent to the city or town boundary. A more practical definition of location would be the proximity of the lands to existing or potential water supply and sewage collection systems. The costs of these sewers dictate that the lot sizes must be much smaller than those in rural land surveys to permit the distribution of costs over a

larger number of lots for a comparable land area. The lot sizes are usually between 0.10 acre (0.04 hectare) and 0.20 acre (0.08 hectare).

2. The land value is greater than rural lands because of the smaller dimensions per lot. Therefore, the surveys must be carried out with greater precision. In a rural survey, fractions of inches or centimeters are not critical due to the large dimensions involved. In urban surveys, the precision of measurements is more critical. For example, the positioning of a house on a lot may not satisfy the requirements of a mortgage company or municipality. As the lot sizes are small, a slight variation in the subdivision lot boundary has more effect.

3. Resurveys are simpler and easier than in rural lands because monumentation is more recent and more permanent.

13.5.1 Descriptions

The boundaries of a tract of land can be described within or adjacent to city or town boundaries in relation to registered plans, blocks, or lots. The original township subdivision boundaries have usually been well established, as well as subsequent aliquot parts and/or irregular subdivisions discussed in Section 13.4. Consequently, deed descriptions of urban lands usually relate to street locations and lots within a registered plan of subdivision. Examples of such situations and the deed descriptions are stated next and illustrated in Figure 13.27.

1. *Boundaries coincide with lot(s) on plan of subdivision*: If the subdivision is registered and given a plan number, the description may read "Lot 6, Registered Plan No. 696, Bastard Township, County of Leeds, Province of Ontario," or "Lot 16, Registered Plan No. 969, City of Toronto, Formerly County of York," or "Lot 8 in Block B, as said lots and blocks are delineated upon a map entitled Townsend Shores, filed in the office of the County Recorder for the County of Sarasota, State of Florida."

2. *Boundaries do not coincide with lot(s) on plan of subdivision*: The properties are described by metes and bounds, with the point of commencement referred to a lot or block corner shown on the plan (parcel A, Figure 13.27), by aliquot parts of lots such as half-lots (parcel B, Figure 13.27, east half of lot 4 and west half of lot 5), or by fixed portions of

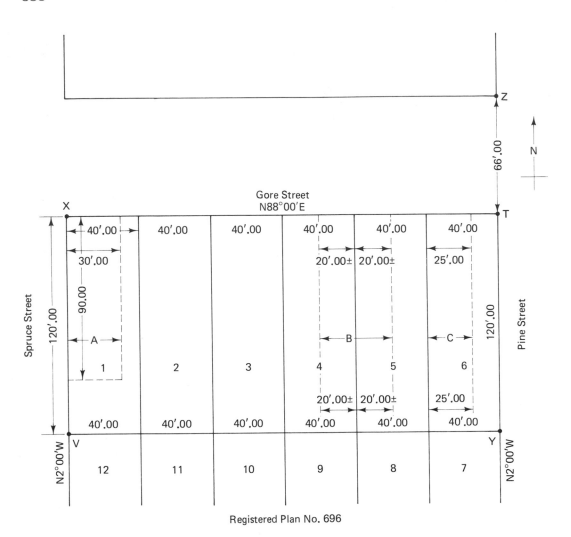

Registered Plan No. 696

FIGURE 13.27 Urban land descriptions.

lots, such as the westerly 25 ft (8 m) of a certain lot (parcel C, Figure 13.27).

3. *Boundaries related to coordinated monuments*: Control surveys (see Chapter 11) have been completed over larger American and Canadian cities. The monuments resulting from such a survey are coordinated precisely, with respect to either a state or provincial coordinate system or to an arbitrarily selected initial point within the city. These initial points are given large enough coordinates for both north and east directions that negative coordinates are not encountered. The boundaries may be defined by the following:

a. Positioning the point of commencement with respect to a coordinated reference point by bearing and distance (or calculated coordinates) and describing the property by metes and bounds (see Section 13.3.2).

b. The same procedure as (a) with a corner on the boundary, other than the point of commencement, described in relation to another coordinated reference point in the city control network. This provides a useful check on the property location.

c. Positioning all property boundary corners by listing the coordinates of each in order (clockwise or counterclockwise) around the property. The surveyor should be aware that transposition of numbers in listing these coordinates is not uncommon. He is well advised to plot the coordinates as given in the description to scale and use this as a check for the intended property dimensions.

d. Deeds or certificates of title avoid written description through the attachment of survey plans that show the bearings and distances of the property boundaries. Where coordinate systems, either state, provincial, or city based have been established, these may also be shown on the plan for each property corner.

13.5.2 Boundary Surveys

Original. Original boundary surveys fall into two categories. The first is the establishment of the boundary of the area proposed for subdivision, such as lots 1 to 12 inclusive in Figure 13.27. In this case, the original survey would have defined the boundaries of Gore, Spruce, and Pine streets. This would have been done by determining the overall boundaries of the lands to be subdivided and relating the street boundaries to these limits. The original boundary is established using the best-evidence rules discussed in Section 13.4.2. The surveyor should note that much more evidence usually exists for urban surveys than for rural surveys.

The second is the establishment of the lot lines such as those shown in Figure 13.27. These are usually laid out on paper by a town or urban planner and the approximate dimensions are shown on a proposed or "draft" plan of subdivision. The planner normally fits as many lots as possible into the available area at this stage. As this planning work is done based on the overall subdivision boundaries and minimum lot frontages and areas as regulated by the municipality, the surveyor is often faced with the problem of fitting the number of lots shown on the draft plan into the final surveyed plan. This fact should be made clear to the planner and client before the survey work begins. In addition, the usual required minor modifications in lot sizes and pattern should be discussed with the client and planner as the project proceeds. This avoids the possibility of moving lot boundary monuments after these have been set.

Resurveys. Before techniques of resurveying urban boundaries are discussed, the surveyor should be aware of the lack of permanence of the original street line and lots boundaries created. The urban subdivision is monumented under close to ideal conditions. The land is vacant at the time and no construction has taken place. However, once construction of roads and houses, as well as the installation of water mains and sewers, begins, the survival rate of lot corner monumentation decreases substantially. The great percentage of the survey bars are either removed, severely bent, or covered by up to 3 ft (1 m) of fill. Consequently, the surveyor will usually be confronted with an incredible lack of survey evidence. This is very disconcerting when the survey plan of subdivision indicates, neatly and clearly, the position of original monuments set at each lot corner. Such is not the case and considerable time and effort must be spent in reestablishing the lot boundaries. Without this understanding, the surveyor will expect to find monuments that either do not exist or have been misplaced.

Resurveying a subdivision is a considerable challenge, requiring perseverance and understanding. In the authors' experience, it is common to locate only between 5% to 15% of the lot corners in their original position. For example, in Figure 13.27, the surveyor would be relatively fortunate to find undisturbed monuments at points *X, Y,* and *Z.* If the object of the resurvey involves the boundary establishment of parcel *B,* the entire boundaries of lots 4 and 5 must be determined. In this example, a procedure similar to the following would have to be undertaken:

1. The line *YZ* would be measured. This distance should be 186.00 ft (120 ft, plus 66 ft) according to plan but measures 186.04 ft in the field.
2. Point *T* is set 66.00 ft south of point *Z* on the line *YZ.* Road allowances always retain their original width when resurveyed.
3. The distance *XT* is measured at 240.12 ft, although it is 240.00 ft according to the plan. This means that the frontage

of the lots on Gore Street must be adjusted by proportioning to 240.12 ÷ 6, or 40.02 ft.

4. The street line locations of lots 4 and 5 are set using the frontage of 40.02 ft for each.

5. As parcel B consists of the east half of lot 4 and the west half of lot 5 by description, the midpoints of lots 4 and 5 are set along the street line (line XT) a distance of 20.01 ft both west and east of the 4–5 lot boundary.

6. As no further evidence of the easterly boundary of Spruce Street other than point Z exists, the plan angle of 90° is turned off at X from T to establish this street line.

7. Line YV is run parallel to Gore Street to intersect the easterly boundary of Spruce Street at point V. Therefore, the actual lengths of both YT and VX are 120.04 ft, rather than the plan distance of 120.00 ft.

8. The rear corners of parcel B are set proportionately in the same way as the front corners. Thus, the boundaries of parcel B have been established using resurvey techniques based on the best evidence available. The dimensions of this parcel are 40.02 ft by 120.04 ft by proportioning.

The example described is quite straightforward. If horizontal curves, including reverse curves (see Chapter 10), exist between two found points on a street line, the establishment of the street line involves considerable calculations, more fieldwork, and some trial and error.

Resurveys must often be carried out for lots or parts thereof on urban subdivisions originally monumented by perishable markers such as wooden stakes. Under these conditions, resurveys become more difficult due to lack of monumentation but are easier because occupational boundaries such as fence lines have been firmly established.

Resurveys in urban lands are seldom as easy as they may appear. The existence of a neatly presented registered plan of subdivision does not, in any way, guarantee the present existence of the monuments shown. Title or mortgage surveys, discussed next, often assist greatly in establishing street and property lines.

13.5.3 Title or Mortgage Surveys

These surveys involve the detailed positioning of the existing buildings on a parcel of land in relation to the boundaries. The surveys are performed for two primary reasons.

1. The finance company or bank granting a mortgage on a property and usually the buildings thereon requires proof that the buildings are within the property boundaries.

2. Most municipalities have bylaws stating minimum distances to the street line and the side lines of the lots. The title or mortgage survey shows whether these bylaws have been satisfied or not. Surveys for this purpose are usually carried out as soon as the house's basement has been constructed up to the first floor level. Hence, if the building location does not satisfy the bylaws and negotiations for exemptions from these regulations with the municipality are unsuccessful, the building can be moved before construction is completed.

Figure 13.28 illustrates a typical title or mortgage survey. The steps in conducting such a survey are the following:

1. The lot boundaries as discussed in Section 13.5.2 are re-established. Note that the lot patterns in Figures 13.27 and 13.28 are identical.

2. The distances from the outside of the basement walls to the street line and both lot lines are measured. This is usually done by reading the measurement on the tape through the theodolite telescope. The distances required are those at right angles to the lot line, in other words, the shortest distance between the lot line and the building corner.

3. The measurements to the street line, called *setbacks*, and those to the lot lines, called *sideyards*, are also measured for the buildings on both sides of the lot or parcel being surveyed.

4. Occupation lines such as fences are positioned and shown on the survey plan. If the fence is not on the property line such as the southerly fence in Figure 13.28, its position is noted.

5. All lot distance measurements are shown. It is common to show both the "plan" and "measured" distances for each boundary as illustrated in Figure 13.28.

It is necessary to measure the outside dimensions of the buildings in order to plot them to scale. The corner markers may or may not be replaced. Often they are not, as this increases the cost of the survey.

The surveyor should note that title or mortgage survey measurements are very useful for reestablishing property boundaries, as the distances from the buildings are measured to the true property lines. Often, the residents have a copy of the survey among

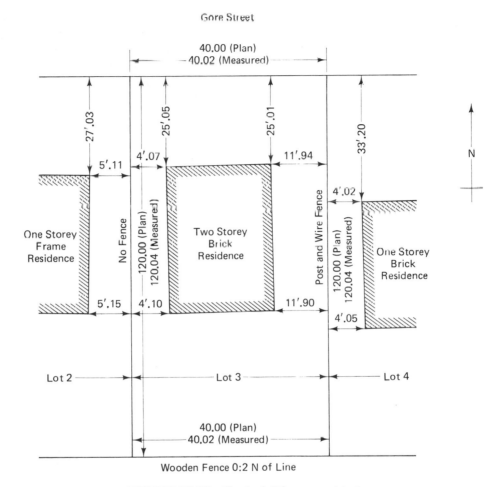

Gore Street

40.00 (Plan)
40.02 (Measured)

27'.03

25'.05

25'.01

33'.20

N

5'.11 4'.07

11'.94

4'.02

One Storey
Frame
Residence

No Fence

120.00 (Plan)
120.04 (Measured)

Two Storey
Brick
Residence

Post and Wire Fence

120.00 (Plan)
120.04 (Measured)

One Storey
Brick
Residence

5'.15 4'.10

11'.90

4'.05

Lot 2

Lot 3

Lot 4

40.00 (Plan)
40.02 (Measured)

Wooden Fence 0:2 N of Line

FIGURE 13.28 Typical title or mortgage survey.

their valuable papers in the house. It is well worth the surveyor's time to ask local residents on nearby properties for a loan of these surveys. The surveyor's name and date of the title or mortgage survey should be recorded and shown in the resurvey field notes.

13.6 CITY SURVEYS

City surveys differ from the urban surveys described in Section 13.5 in subtle ways. The "boundary" between city and urban surveys is constantly changing as the highly developed inner core expands outward to encompass urban areas. The direction in specific areas of this expansion cannot be accurately predicted, as it depends upon factors such as planning decisions and major public transit route locations (subways, for example). Also, as this boundary progresses, bylaws relating to development density, usually defined by the total floor area permitted in relation to the land area, are changed, thus permitting a greater density. These bylaw changes result in higher land values and thus require greater precision in land surveys. In addition, the city should anticipate these changes and set up a citywide coordinate system. This should be based on a state or province wide system where possible, as this avoids costly transference of coordinates from a temporary system to a permanent one.

The large spectrum of city surveys can be divided into two major categories:

1. Property surveys, which define boundaries with greater precision and use different techniques from either rural or urban surveys.
2. Municipal surveys, which define the location of services both above and below ground.

These categories are discussed in the following two subsections.

13.6.1 Property Boundaries

The definition of property boundaries in intensely developed areas differs from urban lands in that more or less permanent objects such as building corners, sidewalks, manhole covers, and street-line intersections are used as points of reference for corner monuments. Also, the center lines of building walls common to two adjacent buildings often form a property boundary. The best illustration of a typical city survey based on the authors' experience is shown in Figure 13.29. The method of survey for the boundaries of property A is as follows:

1. The legal street boundaries of both College Street and Spadina Avenue are determined by a thorough search of all evidence only within the block, contained by four streets in total, in which property A is located. This involves examination of descriptions and survey records in the registry office, as well as field points. The surveyor should note that cut crosses, chiseled into concrete sidewalks, are common survey markings in city surveys, as the entire property, particularly

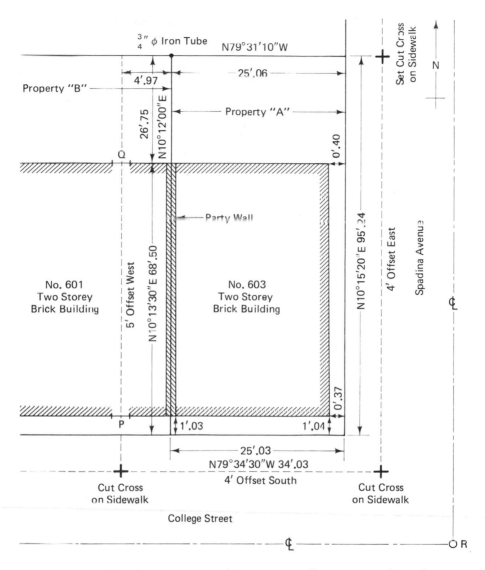

FIGURE 13.29 City survey involving party wall as property boundary.

along the street lines, is entirely covered by buildings and concrete.

2. The westerly boundary of property A is defined in the description as "the center line of the common wall shared by the premises at street numbers 601 and 603 College Street." The next step then is to establish the length and bearing of

this wall and thus calculate the dimensions of its center line. In this case, it was necessary to run an offset line 5 ft west of the center line of the party wall. The offset for this line is selected to permit a line of sight through the building, in this case a door at point P and a window that can be opened at point Q. This line is carried through the house, usually with some difficulty, to the rear property line. The distances perpendicular from this line to the party wall are measured to determine any significant bends in the wall, although these seldom occur.

3. The offset line intersects the rear property line 4.97 ft from an iron tube previously established from a nearby survey. The westerly limit of property A therefore has a slight bend at the northern end of the party wall, as indicated by the difference in bearings on Figure 13.29.

4. The rear boundary is accurately measured, and a cut cross is chiseled into the sidewalk on the offset line, parallel with the Spadina Avenue street boundary, as indicated.

5. The measurements from the building on property A are taken to the offset lines, which are located parallel with each street line. For example, the two measurements at the southeast building corner would be 5.04 ft and 4.37 ft. The distances of the offset lines from the street lines, both 4 ft in this case, are subtracted to give the measurements shown of 1.04 and 0.37 ft.

6. The building is measured accurately for plotting on the plan to scale. The thickness of the party wall is often difficult to establish as no doors or windows through this wall usually exist. It is sometimes necessary to run another offset line through the building on property A, similar to that through the one on the adjacent building. Further measurements perpendicular to the offset line are taken, and the thickness of the wall is established by calculation.

Property surveys such as that illustrated in Figure 13.29 require considerable research, patience, and familiarity with all survey techniques. The surveyor or a person in his employ has the right, by law, to enter and pass over the land of any person or enter the building at any time suitable to the occupant. The latter requirement means that the surveyor must be very diplomatic when he has to run an offset line through a building. If the occupant is not home, is hostile, or is involved in personal relations at the

time, a return visit may be required at considerable extra cost. If
the timing of running the offset line through the building cannot
be agreed upon within a reasonable length of time, the occupants
can be ordered to cooperate through law enforcement agencies.

In Figure 13.29, point R is located at the intersection of the
center lines of the road allowances for both streets. If a city coordinate system has been established, either citywide or state or province wide, this would be the most logical location for the control
point. This could consist of a cut cross, center point of manhole
cover, or the like. In this example, point R would control if it were
coordinated, and all property corners would be coordinated in relation to point R.

13.6.2 Municipal Surveys

These surveys involve the location of all municipal services, utility
lines, and related terrain features such as topography. This information is required by government agencies, planners, engineers,
architects, and land developers.

A simple example of the complications caused by lack of
such an overall municipal survey system is the construction of the
first subway in Toronto, Ontario. This occurred in the 1950s before the city had established either a citywide control system or a
central depository for survey maps showing the location of
services, particularly buried lines. The subway was located using
the best available planning and engineering considerations. As the
topography was relatively level, the basic design involved excavation to an average depth of 25 ft (8 m), construction of the facility,
installation of roof support beams, minimal backfill, and repairing
of the overlying road surface. The contracts were awarded essentially on a lump-sum bid, whereby the contractor stated a maximum price for the work, thus eliminating most of the opportunities for legitimate claims or "extras" over this limit. The results
of this contractual arrangement plus the lack of service location
records were disastrous. As the project proceeded, numerous buried servicing facilities were constantly encountered, such as water
mains, sanitary and storm sewers, natural gas pipes, and hydro
power lines. The newly discovered services had to be replaced
and/or rerouted. Claims by the contractor for extra payments, all
justifiable, were extremely high. The original project budget suf-
fered severe overruns, which had to be explained to the populace,
as the contract funds were public money from taxes. The opening
of the subway was significantly delayed. These problems could

have been avoided if an accurate municipal survey had been available. All locations on such a map should be shown in relation to a citywide coordinate control system, preferably based on the state or provincial coordinate system, as described in Chapter 11.

Surface Features. Maps should be produced to show all relevant surface features such as public property lines (road allowance street lines), traveled roads, laneways, sidewalks, railroad lines and boundaries of their rights-of-way, hydro distribution lines and poles, fire hydrants, and manhole covers, as a bare minimum. As most of these features are concentrated within public or utility owned rights-of-way, the resulting maps become extremely crowded, even when drawn at scales of 1 in. = 50 ft (1:600). Three solutions to this problem should be considered:

1. Each category can be shown on the maps as a different color. The advantage of this system is that it is relatively easy to implement and, although cluttered, is still readable. The disadvantage is the high cost of reproducing multi-colored copies.

2. Each category can be mapped on a transparent overlay that fits exactly over the original base map. The base map normally shows street lines (road allowance boundaries) and rights-of-way boundaries for railroads and hydro transmission. Using this system, whiteprints can be produced that show only those surface features that are of concern to the user. The difficulty is that, in the case of complex projects, all surface features are significant and the same cluttered mappings result. However, the advantages of low-cost reproduction and flexibility do exist.

3. Digital mapping techniques, such as the system recently developed by Nortway-Gestalt Survey Corporation in Toronto, provide considerable flexibility. The coordinates of all surface (and underground) features are determined from existing drawings and maps. These are stored in a computer, which mathematically adjusts the scales of previous maps to a common base. As features are added, the coordinate information from recent surveys is subsequently stored. The system is highly flexible, as the data can be selectively recalled at any time, and the required scale of the maps produced can be easily changed to suit the user requirements. As computer-assisted drafting (CAD) is an integral part of the system, the user can examine the information on a video display unit pri-

or to actual map production. Therefore, if the map will be too cluttered, the scale can be adjusted appropriately prior to the expense of map production. These systems are initially expensive to install, but the subsequent advantages to the municipality are evident.

Additional surface features that may be included in the city mapping are laneways on public lands; locations of public buildings; boundaries of registered subdivisions; wooded areas, including individual large trees; topographic contour lines and spot elevations; drainage systems, including streams, rivers, and ponds; and, possibly, intermittent drainage channels.

The necessity of mapping this additional information is evident in most cases. However, mapping of individual large trees may be considered to be too detailed to the uninitiated. The usefulness of this information was pointed out to the authors during the preparation of a tender cost for construction of a sanitary sewer. The specifications for this particular project stated that the sewer had to be installed by tunneling under the roots of trees over 2 ft (0.6 m) in diameter, thus preserving the trees. As this project consisted of 10 miles (16 km) of sewer line throughout a new city subdivision, the existing mapping was invaluable in determining the cost estimate.

Underground Features. Features such as sanitary sewers, storm sewers, water distribution lines, gas mains, hydroelectric conduits, tunnels, underground walkways, and subways should be mapped. The horizontal location of each should be shown, as well as elevations at critical points such as changes in horizontal and vertical alignment. For example, manholes for either sanitary or storm sewers would be logical points.

The primary sources of information for underground features are the public utility companies who were responsible for the original installation of the buried services. The surveyor should be aware that two sets of drawings usually exist. One set is the original construction design drawings, which indicate the *intended* locations of these underground features. However, as construction proceeds, the location of these features is often changed due to unexpected problems, such as existing services (location previously unknown) or adverse soil conditions. A second set of drawings is usually prepared after construction has been completed, which shows the *actual* location of these features. These drawings are known as *as-built* or *as-constructed*. It is critical that underground features be located on municipal maps using these as-built drawings. If a surveyor involved in compiling municipal maps cannot obtain the as-built drawings, he or she should be prepared to do the necessary survey work to determine the actual locations of the buried features.

The maps of surface features are normally used as the base map for underground features. Permanent lines, such as street lines (legal limits of road allowances) and property boundaries, provide a solid reference base.

13.7 CADASTRAL SURVEYING

Cadastral surveying is a general term applied to a number of different types of surveys. It is mentioned here only to make the reader aware of the expression and the broad aspects of its use. A rigid definition of a cadastral survey involves only the information required to define the legal boundaries of a parcel of land, whether rural, urban, or city. Therefore, the monumentation, bearings, distances, and areas would be shown. This definition has now been expanded through common usage to include cultural features, such as building location, drainage features, and topographic information, such as either spot elevations or contours.

PROBLEMS

13-1 Calculate the angular convergency for 2 meridians, 6 miles (9.66 km) apart at latitudes 36°30′ and 46°30′.

13-2 Find the convergency in feet, chains, and meters of 2 meridians 24 miles apart if the latitude for the southern limit is 40°20′ and the latitude for the north limit is 40°41′.

13-3 If the tangent offset is 3.28 ft (1.00 m) at 3 miles (4.83 km) from the meridian, calculate the offsets for 5 miles (8.05 km), 7 miles (11.27 km) and 10 miles (16.10 km) from the same meridian.

13-4 For a 6-mile-square sectional township, calculate the area of section 6, if the mean latitude through the middle of the township is 46°30′. Take only the effects of convergence into account in this calculation.

13-5 In Figure 13–11, if the southwest corner of section 35 and the northwest corner of section 11 arc the only established points along this north-south line and points D and E have been established, describe the step by step procedure to re-establish the northwest corner of section 14.

13-6 In Figure 13–16, state which side of lot 9, concession VIII is likely to be closest to the intended 50 ch, and give your main reason.

13-7 In Figure 13–17, the northwest corners of lots 17 and 22, concession V have been determined. There is no other evidence available between these 2 corners. The total measured distance between the 2 corners is 0811.20 feet. Calculate the frontage to be allowed for lots 18 to 22 inclusive in feet, meters and chains.

13-8 In Figure 13–19, if no evidence of the township boundary existed other than points X and Y as shown, state how you would establish the southeast corner of the western half of lot 11, concession I.

13-9 In Figure 13–20, show how the iron monument would be marked for

    a The northwest corner of township T.20.N., R.7.E.
    b The southeast corner of section 32 of township T.20.N., R.6.E.

13-10 In Figure 13–21, show how the corner would be marked for the northwest corner of lot 18, concession X.

13-11 Plot the following description (the property is located in a standard U.S. sectional system) to a scale of either 1: 5000 or 1 in. = 400 ft.

    "Commencing at the south west corner of Section 35 in Township T.10.N., R.3.W.,

    Thence N 0°05′W along the westerly boundary of Section 35, 2053.00 feet to a point therein;

    Thence N 89°45′E, 1050.00 feet;

    Thence southerly, parallel with the westerly limit of Section 35, 670.32 feet;

    Thence N 89°45′E, 950.00 feet;

    Thence southerly, parallel with the westerly limit of Section 35, 1381.68 feet, more or less, to the point of intersection with the southerly boundary of Section 35;

    Thence westerly along the southerly boundary of Section

35, 2000.00 feet, more or less, to the point of commencement.

13-12 The locations of the original and new shoreline are shown on the following sketch.

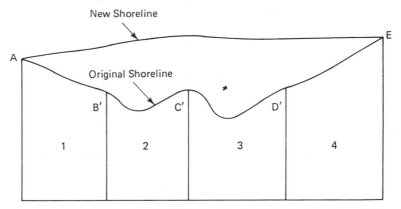

a Describe how you would establish points *B*, *C*, and *D* along the new shoreline.

b Sketch the locations of these points along the new shoreline on a clear plactic overlay.

13-13 Prepare a legal description for parcel B in Figure 13–27.

13-14 If the measured bearing of Pine Street in Figure 13–27 was N 1°50′W instead of N 2°00′W and the distance XT remains the same as shown in the figure, calculate the rear dimensions of lots 1, 2, 3, 4, 5 and 6.

13-15 In Figure 13–29, if the iron tube along the northern boundary of property A was located 504 ft east of the 5-ft offset line through property B (instead of the 4.97 ft shown), calculate:

a The length and bearing of the rear boundary of property A.

b The length and bearing of the portion of the westerly boundary of property A from the north end of the party wall to the rear boundary.

14 *Hydrographic Surveys*

14.1 GENERAL

Offshore engineering and the shipping industry have progressed rapidly since World War II. Drilling rigs, subjected to violent storms, located up to 125 mi (200 km) offshore, search for resources, particularly oil and gas. Offshore islands are constructed of dredged material, sometimes under severe weather conditions, to support marine structures. Harbor depths up to 80 ft (25 m) are required to accommodate larger ships and tankers. Containerization has become an efficient and preferred method of cargo handling. The demand for recreational facilities ranges from large pleasure cruise ships to small sailboats.

Hydrographic surveys are made to acquire and present data on oceans, lakes, bays, or harbors. In addition to harbor construction and offshore drilling, these surveys are carried out for one or more of the following industrial activities:

1. Water depths and location of rocks, sandbars, and wrecks for navigation channel and salvage operations.
2. Dredging for harbor deepening, maintenance, mineral recovery, and navigation channel access.

673

3. Evaluation of areas of sedimentation and erosion for coastline protection and offshore structures.

4. Measurement of areas subject to siltation or scouring to determine effects on water quality and existing structures, such as bridge abutments and storm sewage outfalls.

5. Provision of recreational facilities such as beaches and marinas.

6. Site locations for submarine cables and underwater pipelines.

7. Determination of pollution sources.

8. Evaluation of the effects of corrosion, particularly in salt water.

This section describes the procedures required to obtain the necessary data for hydrographic surveys.

14.2 OBJECTIVES OF HYDROGRAPHIC MAPPING

The primary requirement involves showing the topographic configuration of the underwater features, both natural and manmade. The resulting product is therefore similar to a topographic map of land areas. However, the methods required to obtain the information for hydrographic surveys are vastly different.

In the horizontal plane, the position of a survey vessel must be fixed to the required accuracy. The location of the vessel is complicated by weather conditions, particularly winds, waves and fog. The depth of the seabed below the survey vessel, known as a *sounding*, is subject to variations caused by wave and tidal action. After the original sounding has been corrected, the resulting depth is termed a *reduced sounding*.

The hydrographic survey is more costly than its land survey equivalent because of the preceding factors. The surveyor must also be aware that the accuracies of hydrographic surveys are normally not comparable to land surveys. The comparison is valid only under ideal conditions, which are rare. While the land surveyor can see the features, the hydrographic surveyor cannot, except in shallow, clear waters. This may result in the omission of important features, such as rocks or wrecks, that vitally affect the proposed undertaking.

The objective of the hydrographic survey must be considered in light of these elements. For example, compromises in the accuracy of individual soundings may be justified to locate important underwater features through an increased number of soundings. Therefore, the proper planning of a hydrographic survey, as discussed in the following section, is critical for the final product to satisfy the project requirements economically.

14.3 PLANNING

Careful planning and preparation are essential for any undertaking. The considerations required for hydrographic surveying further emphasize the importance of this phase of the operation. Flexibility in the plan is desirable to allow for delays due to weather, equipment breakdowns, sickness, and the like, to permit immediate remedial actions. Each aspect is discussed in the following sections.

14.3.1 Specifications

If the client is familiar with the survey requirements, the specifications will be supplied. These should be examined in detail regarding the use or uses to which the hydrographic information will be put by the client or other companies or agencies. Therefore, the specifications may have to be modified accordingly regarding the level of detail of the survey. In addition, the hydrographic survey maps should be qualified as to their accuracy and/or applications in the map legend. Many serious problems have resulted from lack of concern over specifying these limitations. For example, preliminary data used for marine facility site selections may be used for detailed volumes and costs for underwater excavations and become parts of the detailed contract plans and specifications, a situation to definitely be avoided.

If the client is unacquainted with hydrographic surveys, some education is required, particularly relating to expected accuracies and costs. In addition to the concerns expressed in the preceding paragraph, each item discussed in this planning section must be explained in detail. The limitations of the hydrographic survey must be clarified. In particular, the reasons for the higher costs of hydrographic surveys must be carefully explained. Contingency cost allowances must be clarified, primarily due to the dependence of the survey on weather conditions.

The specifications should clearly describe the type of vessel, sounding equipment, position-fixing methods and horizontal control, size of sounding grid and total number of soundings, accuracy of soundings, corrections required for reduced soundings, personnel requirements, costs, timing for completion, and final presentation of data.

14.3.2 Review of Existing Information

All available and applicable information should be examined and copies should be obtained whenever possible. Hydrographic charts of the general area provide indications of the sounding depths and depth variations that are likely to be encountered. This information is useful for sounding equipment selection and design of the sounding pattern.

Topographic maps indicate the configuration of the shoreline and natural or man-made features that may be used as control stations for horizontal position fixing. Aerial photographs viewed stereoscopically (three dimensions) will provide this information in greater detail, as well as offshore bar and shore formations. Color aerial photographs are particularly useful because of their high penetration capabilities through water.

Navigational directions and boating restrictions, including channel locations, normal and storm wave heights, prevailing wind directions, and areas of restricted vessel use, assist in sounding vessel selection and in identifying sources to obtain permits in restricted areas.

Tide tables are essential for the design of recording gauges. If existing gauges are operational, their locations, frequency of water level recordings, base datum used, and availability of data to the surveyor should be known in advance.

Horizontal control data from previous land and/or hydrographic surveys will reduce the effort and cost of the survey measurements among shoreline control stations. Previous knowledge of existing coordinate control networks will allow the survey data to be tied to the survey system if required.

If possible, a field reconnaissance, prior to planning the survey details, is useful for a number of reasons. Availability of survey vessels, docking facilities, local personnel, equipment, and supplies; access arrangements to horizontal control stations or private lands; location of boat and equipment repair centers; and arrangements for the cooperation of governmental agencies are all important aspects of the work.

14.3.3 Equipment

The sounding vessel, position-fixing equipment, and depth-measuring devices are the most vital components of a hydrological survey. Equipment breakdowns are a common cause of preventing or delaying the work. Not only are repair facilities for depth-measuring devices, theodolites, and the like uncommon, but the resulting downtime delays the survey and increases costs.

All equipment should be thoroughly checked for performance prior to commencing the survey. A backup array of essential equipment should be arranged. Anticipation of potential equipment failures and corrective alternatives is one of the simplest yet most important planning functions.

An example would be a nearshore hydrographic survey using a 6-m (20-ft) inboard or outboard boat as the survey vessel, simultaneous double sextant angles for position fixing, and a continuous recording echo sounder for depth determinations. A logical backup system in this case would be an outboard motor mounted on a transom bracket; an extra sextant aboard the boat; and an additional inexpensive, simplified echo sounder, not equipped with continuous recording capabilities. The additional cost of this backup system is nominal when compared with the wages and expenses of four persons even for only one nonproductive day.

14.3.4 Horizontal Control

The horizontal control consists of a network of shoreline stations located to satisfy the selected method of position fixing, as discussed in Section 14.6. Horizontal control procedures and accuracy requirements are set out in Section 14.8.

The control stations must be intervisible between the shoreline and the survey vessel. This fact is not always appreciated by the onshore observer, particularly in areas where the background is densely wooded and/or the sun will be behind the station during position fixing. As the station under these circumstances may be impossible or difficult to see from the survey vessel, it is useless. Possible solutions to this problem may involve moving the station away from the shoreline to a local typographic high; selecting a man-made feature away from the shoreline, yet along the intended lines of sight; or placing the station at the shoreline above the expected high-high-water mark. In the latter case, the background as seen from the survey vessel should be carefully examined for contrast with the intended station target design.

Targets to be set over shoreline stations are subject to disturbance by wind and vandalism. The difficulty caused by wind can be partially solved by supporting the target securely by rocks and/or mortar at the base and three other points connected to the target just above the halfway height. Vandalism can be controlled, but not prevented, by clearly indicating the name of the target owner and/or removing the targets at the end of the day's soundings.

14.3.5 Operational Plan and Logistics

The operation plan includes the details of position fixing; instrumentation and techniques to be used for soundings and tidal gauge location; and personnel, equipment, and logistical requirements.

The position-fixing techniques to be used depend upon the range from shoreline, the accuracies required, the type and possible locations of shoreline targets, the size of survey vessel available, the experience of the personnel, and the equipment available. These techniques are discussed in detail in Section 14.6. The hydrographic surveyor should consider combinations of techniques to satisfy the specifications. The formulation of alternative means of position fixing during the planning process will achieve the necessary flexibility.

Tidal gauge locations and backup equipment systems have previously been discussed in Section 14.3.3.

14.3.6 Survey Team Requirements

Personnel requirements should be evaluated carefully from the standpoint of operational flexibility. Individual members of the survey team should be able to carry out more than one of the key functions. Serious consideration should be given to hiring a local resident who is familiar with the offshore physical and navigational characteristics.

Logistics depend primarily upon the location of the base in relation to the survey area. The wages and expenses of land travel from the base to the site versus vessel transportation only should be carefully analyzed. Experience has shown that, if this cost comparison is competitive, land travel to the site is preferable. The uncertainties of travel on water should never be underestimated.

14.3.7 Sounding Plan

The locations of the individual soundings to meet the specifications are largely dependent on the *variation* of known depths within the project area. This may be determined from the review of existing information, discussed in Section 14.3.2, particularly hydrographic charts and aerial photographs. The design specifications for the sounding plan are detailed in Section 14.7.

Variations in depths are primarily dependent on the bedrock and surficial geology. For example, horizontally bedded sedimentary rocks result in relatively uniformly increasing depths offshore. Conversely, igneous intrusives, steeply dipping sedimentaries, or metamorphic rocks will result in large depth variations within short horizontal distances. The number, size, and particularly the angle with the shoreline of offshore bars are a secondary cause of depth variations.

The spacing between the sounding lines, their angle with the shoreline, and the frequency of soundings along each line should be planned according to the conditions discussed previously.

14.3.8 Processing and Presentation of Data

The hydrographic surveyor is well advised to plot all data before leaving the project site. Preferably, this should be carried out immediately after each sounding or during the following evening or day. The immediacy with which data plotting can be carried out depends upon the method(s) of position fixing used. Provision should therefore be made for plotting aboard the survey vessel and/or at a temporary office facility near the site. The specific field and office procedures required are detailed in Section 14.9.

The planner of a hydrographic survey should recognize the importance of reviewing the data before leaving the site. This provides the opportunity of detecting and checking anomalies while the equipment and personnel are still available on site.

14.4 SURVEY VESSELS

The vessel may be owned by the surveyor, thus committing its use if its requirements satisfy the specifications and logistics of the survey. In many cases, it is necessary to charter a vessel in the locality of the survey area.

General considerations to be considered should include the following:

1. Overall purpose of the survey, particularly the need for geophysical survey equipment or additional survey requirements.
2. Weather conditions, such as wave heights.
3. Size of the survey team and whether they are to live on the craft.

Specific conditions that will always apply include the following:

1. Sufficient space for position fixing and plotting. The plotting board should be under cover and relatively free from engine vibration.
2. An all-around view for visual position-fixing techniques.
3. Sufficient electrical power at the required voltages for all equipment needs.
4. Compatibility of fuel capacity and storage for supplies within the range and operational requirements.
5. Stability and maneuverability at slow speeds (up to 6 km/h or 4 knots).
6. Cruising speeds of at least 15 km/h (10 knots) to minimize time loss from the base to the survey area and to provide sufficient speed to return safely to port in the event of sudden storms.

14.5 VERTICAL CONTROL: DEPTH AND TIDAL MEASUREMENTS

The depth of the point below the water surface or *sounding* must be related to the desired datum or reference level. This will normally involve corrections for seasonal water levels under nontidal conditions and for tidal variations where the latter is a factor. The relationship between *soundings* and corrected or *reduced soundings* is shown in Figure 14.1.

14.5.1 Depth Measurements

A weighted line, graduated in meters, feet, or both, is used only for programs involving a small number of soundings. However, it is a

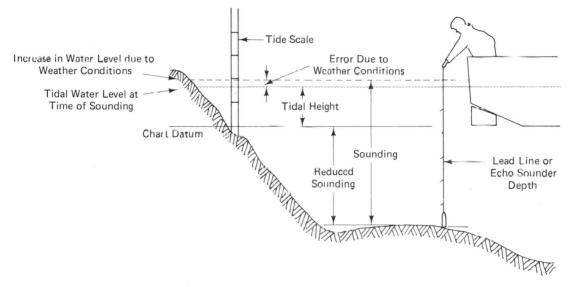

FIGURE 14.1 Reduced soundings and errors due to wave action and water level fluctuations.

valuable addition to the surveyor's equipment for calibrating echo sounders and also as a backup system.

The echo sounder provides depth measurements by timing the interval between transmission and reception of an acoustic pulse, which travels to the bottom and back at a rate of approximately 1500 m/s (5000 ft/s). Separate transmissions are made at rates of up to six per second. The beam width that emanates from the vessel is typically about 30°, as illustrated in Figure 14.2a. The portion of the seabed within the beam width is termed the *isonified* area.

In depth measurement, the most significant point is directly below the transducer, which is the vibratory diaphragm that controls the frequency of transmission. However, the echo sounder records the earliest return from its transmission (that which has traveled the shortest distance). Within a beam of 30°, this return may be from a target not directly beneath the transducer, but from seabed anomalies within the isonified area (see Figure 14.2a). This will lead to anomalies in the soundings, which may be differentiated by an experienced observer. Highly reflective targets such as bare rock and shipwrecks, located near the edges of the beam, show as narrow, clearly defined bands on the readout compared with thick, poorly defined bands over soft sediments, weedbeds, and the like, as illustrated in Figure 14.3. Constant monitoring of the transmission returns and notes of anomalies should be incorporated into the hydrographic survey.

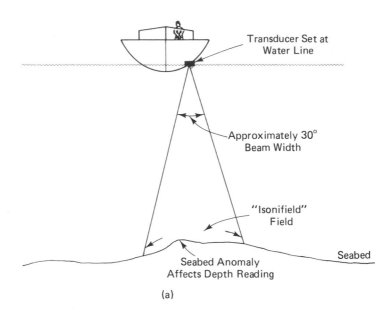

(a)

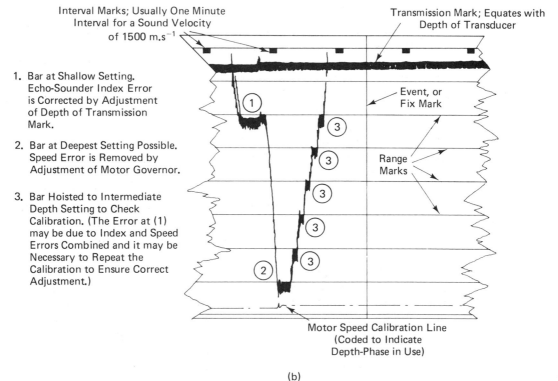

Interval Marks; Usually One Minute
Interval for a Sound Velocity
of 1500 m.s⁻¹

Transmission Mark; Equates with
Depth of Transducer

1. Bar at Shallow Setting.
Echo-Sounder Index Error
is Corrected by Adjustment
of Depth of Transmission
Mark.

2. Bar at Deepest Setting Possible.
Speed Error is Removed by
Adjustment of Motor Governor.

3. Bar Hoisted to Intermediate
Depth Setting to Check
Calibration. (The Error at (1)
may be due to Index and Speed
Errors Combined and it may be
Necessary to Repeat the
Calibration to Ensure Correct
Adjustment.)

(b)

FIGURE 14.2 Isonified area and bar check for echo-sounder depths: (a)
transducer and isonified area; (b) typical bar check.

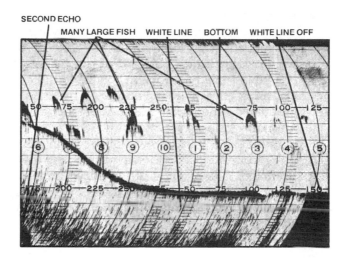

FIGURE 14.3 Variations in echo-sounder readout characteristics. Note that poorly defined bottom readout in topographic depression indicates soft bottom sediments.

Sound velocity in water is a function of temperature, salinity, and density. These factors vary daily, seasonally, and as a result of periodic occurrences such as heavy rainfalls and tidal streams. Any attempt to correct the soundings with respect to these variables is both unsatisfactory and cost ineffective. As a result, it must be recognized that accuracies in acoustic measurements in seawater will not be better than 1 part in 200.

Calibration of echo sounders is carried out either by comparison with direct measurements using weighted lines or by a bar check in depths less than 30 m, as illustrated in Figure 14.2b. The latter involves setting a bar or disc horizontally beneath the transducer at various depths. The echo-sounder recorder is adjusted to match the directly measured depths. If the sounder is not adjustable, the differences are recorded for regular depth intervals and the resulting corrections applied to each sounding. It is advisable to calibrate the echo sounder at the beginning and end of each day's use, particularly at 10%, 50%, and 80% of the maximum depths being measured.

Sounding rods may be used in depths less than 4 m (13 ft) under low current velocities. A conventional survey level rod will suffice, providing that a metal shoe is placed over the lower end to prevent sinking into soft underwater sediments.

Weighted (hand) sounding lines are seldom used for depths over 30 m (100 ft). The lines may be small linked steel chain, wire, cotton, hemp cord, or nylon rope. A weight, usually lead, is attached to one end. Markers are placed at intervals along the line for depth reading. Lines constructed of link chains are subject to wear through abrasion. Wire lines will stretch moderately when suspended, depending upon the size of the bottom weight. The weights vary from 2.3 kg (5 lb) to 32 kg (75 lb), although 4.5 kg (10 lb) is usually sufficient for moderate depths and low velocities. Cotton or hemp lines must be stretched before being used and graduated when wet. These must be soaked in water for at least an hour before use to allow the rope to assume its working length. Nylon lines stretch appreciably and unpredictably and are not recommended for other than very approximate depth measurements. The hydrographic surveyor is well advised to calibrate his sounding lines against a steel survey tape regularly under the conditions most similar to actual usage, such as hemp lines after soaking.

14.5.2 Tidal Gauging

Observations of tidal variations from the datum are required throughout the sounding operation for the reduction of soundings. Any datum may be selected, as discussed previously. However, it is common practice to use the level of the lowest predicted tide, known as the lowest astronomical tide (LAT). This datum can be obtained from available information and existing bench marks in developed countries.

A tide pole or recording tide gauge is required to obtain the necessary information during the sounding period. A simple graduated pole, erected with its zero mark below the lowest expected water level and of sufficient length to cover the tidal range, may be used. However, an observer is required to record the water levels and corresponding times, which may be uneconomical. The observer must also be trained to allow properly for local disturbances, such as wake from passing boats, in order that only the tidal variations are measured.

The recording tide gauge involves higher capital costs, which may be justified on the grounds of not requiring an observer. A typical tidal gauge is illustrated in Figure 14.4.

The difficulties of tidal observation primarily relate to the location of the observation gauges. The following factors are involved:

1. Proximity to the survey area.
2. Configuration of the coastline. Indentations or embayments result in tidal variations between the back of the bay and the

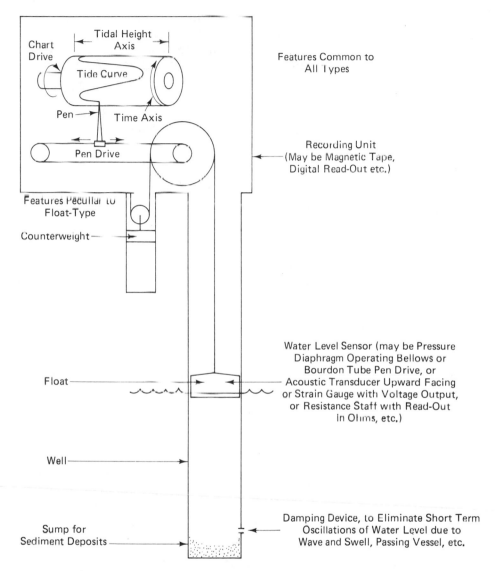

FIGURE 14.4 Typical tidal gauge.

straighter portion of the shoreline. Therefore, a survey encompassing such an embayment should provide tidal recording stations at both locations.

3. Slope of seabed. Gently sloping seabeds immediately offshore will affect the real time of tidal variations. This is not significant in most cases and can only be corrected by placing the measuring pole in deeper waters outside this zone. This is usually not practical.

4. Impounding of water due to offshore bars and/or islands. The measuring rod or tidal gauge should be placed outside the area thus affected. Tidal variation measurements taken where impoundment occurs will be worthless for reducing soundings.

Location of the tide measuring device is extremely important in acquiring meaningful data for reduction of soundings. Accurate time records of these measurements are essential for correlation with the actual time of each sounding.

14.6 POSITION-FIXING TECHNIQUES

The location of the survey vessel in the horizontal plane when a particular sounding has been measured is a fundamental requirement for the hydrographic survey. Until the method(s) to be used have been selected, the time schedules, horizontal control requirements, and the like cannot be defined. Directional control of the vessel along the sounding lines is an important factor to ensure that the survey area is covered sufficiently to meet the specifications.

Optical methods are described in detail, as the great majority of surveys undertaken by the reader will fall in this category. Electronic and miscellaneous techniques are listed and their range, accuracy, equipment requirements, and future potential are briefly discussed.

Experience has confirmed that real times should be recorded for all soundings regardless of the techniques employed. This practice will not only ensure appropriate tidal corrections, but will resolve most problems where members of the survey team are stationed at different locations and are not in direct communication.

14.6.1 Double Sextant Angles Observed from Vessel

Simultaneous horizontal sextant angles are observed between three shore stations as illustrated in Figure 14.5. This is the most common and versatile method overall. The accuracies attained at distances from shore of between 200 m and 7 km are normally ±3 to 5 m depending on the distances from the shore stations.

Not only are the required two sextants inexpensive, but this method has the added advantage that all members of the survey team are together in the survey vessel. The importance of the con-

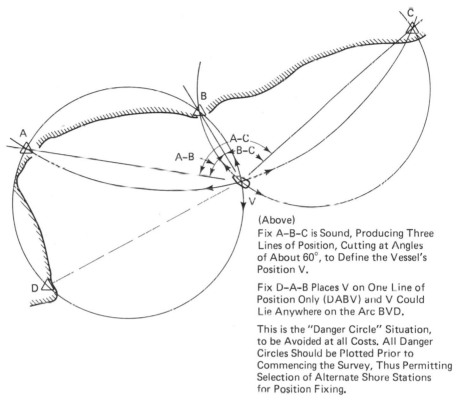

(Above)

Fix A–B–C is Sound, Producing Three Lines of Position, Cutting at Angles of About 60°, to Define the Vessel's Position V.

Fix D–A–B Places V on One Line of Position Only (DABV) and V Could Lie Anywhere on the Arc BVD.

This is the "Danger Circle" Situation, to be Avoided at all Costs. All Danger Circles Should be Plotted Prior to Commencing the Survey, Thus Permitting Selection of Alternate Shore Stations for Position Fixing.

FIGURE 14.5 Double sextant method.

venience of direct communication and immediate plotting of all fixes should be recognized by the hydrographic surveyor.

The vessel should be equipped with a plotting board and appropriate equipment for fixing the position of the sounding as soon as the two angles have been read by the observers. A 360° protractor with three legs pivoting about the center point, as illustrated in Figure 14.6, is easily constructed. By setting the angles between the three legs to match the observed sextant angles and then placing all three legs on the plotting sheet over the shore stations observed, the position of the vessel is at the center of the protractor. As it is common to allow an interval of one minute between position fixes, the operator of this plotting instrument (sometimes termed a *station pointer*) should be able to read the depth and plot the fix and depth within this period.

The vessel is kept on the course predetermined by the sounding plan (see Section 14.7) through a combination of compass bearings and minor course corrections after each fix is plotted.

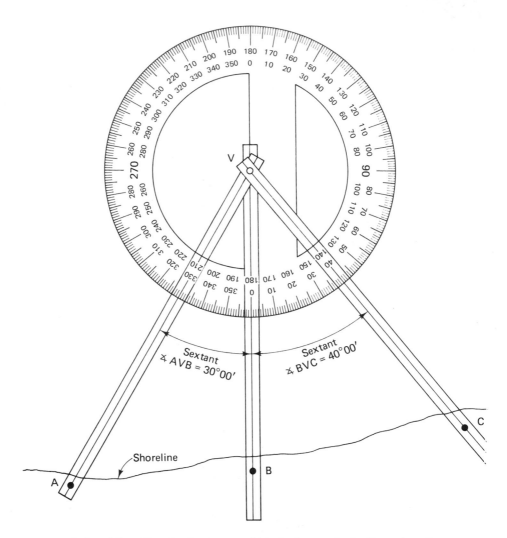

A, B and C are Shoreline Stations on which the Sextant Angles Shown have Been
Observed. The Location of Each Station is Accurately Plotted on Milar and the
Fix is Carried Out on the Vessel At "V".

Three Clear Plastic Strips with Center Marking Lines as Shown Above are Pivoted
About the Center Point of the Protractor.

FIGURE 14.6 Station pointer for double sextant angle fixing.

Continuous plotting of fixes aboard the vessel thus ensures the
intended coverage of the sounding area with the least number of
soundings, resulting in maximum efficiency. The survey team requirements are normally met by four persons: the driver of the
vessel, two sextant angle observers, and a plotter and depth recorder.

The correct speed of the vessel to satisfy the sounding plan is normally precalculated (see Section 14.7). This speed can be adjusted along the sounding lines to compensate for wind and wave directions. As the soundings are taken on the basis of one each minute, the driver should be responsible for the timing of fixes.

The sextant angle observers should stand close together and be positioned over the depth-measuring instrument to minimize positioning errors. Immediately after reading the sextant angle for one fix, each observer should realign the sextant with the shore stations and keep the images constantly superimposed. This will ensure correct readings at any time that a fix is requested by the driver.

Each sextant angle should be between $20°$ and $110°$. To ensure further the accuracy of the fix, the sum of the two angles should exceed $50°$. The danger circle occurs when the vessel and all three shore stations lie on the circle's perimeter. This situation must be avoided, as the boat's position may lie at any point on the circle and is therefore indeterminate. Figure 14.5 illustrates this situation using shore stations D, A, and B. When the vessel is on or near this circle, an alternative shore station such as C should be used to solve the problem. All danger circles should be preplotted before commencing the survey.

14.6.2 Single Vertical Sextant Angle Subtense

A graduated subtense board similar to that illustrated in Figure 14.7 is observed with a sextant held vertically. This technique is used for large-scale surveys within 600 ft (200 m) of the shore stations. The best approach involves graduating the subtense board in equal intervals, such as even meters or yards. Therefore, the rod shown in Figure 14.7 would be either 5 m or 5 yd long. The graduations should be clearly numbered and differentiated by contrasting colors, such as red and white. A fixed vertical angle is then calculated for one interval as shown on Figure 14.7. For example, if the area to be surveyed extends to 150 m from the shore station and there are six metric graduations, a position fix should be obtained every $150 \div 5$, or every 30 m. The tangent of the fixed angle to be set on the sextant would be $1 \div 30$. Similarly, if the board is graduated in yards and the area to be surveyed extends 150 yards, the tangent of the angle would be $1 \div 30$, or 0.033. Therefore, the angle used would be $1°55'$.

The survey vessel moves toward the shore station along the sounding line at speeds of 2 to 4 mph (3 to 6 Km/h). The sextant angler keeps the board constantly in view and requests a sound-

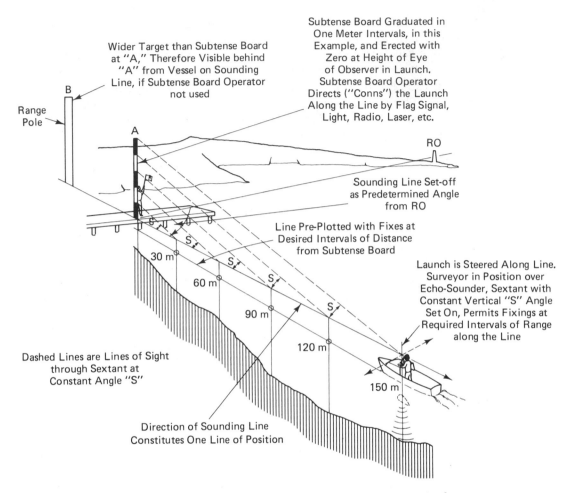

FIGURE 14.7 Single vertical sextant method.

ing when the images of the lower extent of adjacent graduations on the board are superimposed when viewed through the sextant.

The survey vessel is best kept on the sounding line by placing a range pole behind the subtense board as shown on Figure 14.7, thus permitting direct observation from the boat. If such a target arrangement is not feasible due to existing land uses or steep topography along the shoreline, a subtense board operator may be used, as illustrated in Figure 14.7. The vessel is kept on the sounding line by using a theodolite and accompanying flag signals, a directional light beam held by the operator, or radio communication or laser beams. The difficulties with radio communications are discussed in Section 14.6.3. Laser beams are expensive for a relatively straightforward survey method.

14.6.3 Intersection from Theodolite Stations Ashore

The position of the vessel is determined by two simultaneous horizontal angles measured by theodolites set up on shore stations. The theodolites are zeroed on any visible shore station that is part of the horizontal control network.

As the boat proceeds along the sounding lines, the theodolite observers track the vessel's path, sighting a target on the boat mounted over the echo sounder. At the instant that a fix is required, the boat driver raises a prearranged signal; the sounding is recorded on the vessel; and each shore observer records the measured angle. Each fix is consecutively numbered in the field notes by both theodolite operators and the echo-sounder reader in the vessel. Plotting of the fixes is undertaken after the data from the vessel and shore station have been correlated.

This method has many pitfalls and is not recommended. However, availability of equipment and personnel capabilities may leave no other choice. Therefore, the operational problems and partial solutions are discussed next.

Separation of the Personnel between the Vessel and Shore. Direct communication by CB radio, for example, is becoming increasingly more difficult in populous areas due to overuse of the channels. The visual signaling must be clearly defined beforehand and be unique. Raising a life preserver in the air when a fix is required is definitely not recommended as this may happen inadvertently. Accurate synchronization of watches among the crew personnel prior to the start of the survey and recording of the time of each fix to the nearest 10 seconds is absolutely essential in attempting to coordinate vessel and shore data.

Theodolite Tracking of the Vessel. The boat target must be continuously tracked to ensure the accuracy of the horizontal angles. Therefore, the observer's eye must never leave the theodolite eyepiece during the sounding runs. Consequently, the slightest diversions of attention will result in a missed signal. The dedication of the theodolite operator is further tested by the limited range of the tangent (slow motion) adjustment on the theodolite. It requires some prior practice in tracking a moving object to ensure that latitude of movement in the tangent screw exists at the moment of fix. A prearranged signal from the theodolite operator to the boat, in case of instrument problems such as an inadvertently moved tripod leg or turning the lower motion tangent screw, significantly reduces downtime.

Plotting of Data. The fix numbers and, if necessary, the recorded times should be compiled by the survey team after completion of a maximum of four range lines or 50 fixes. At this time, the farthest points offshore on each sounding line should be plotted to ensure that the spacing between the lines satisfies the specifications. The plotting of fixes using this technique is time consuming at best. Positioning of a properly oriented 360° protractor over each theodolite station and a leg fixed at the center of each accelerates the plotting of the angle intersection points. The plotting apparatus is therefore of similar construction to the station pointer illustrated in Figure 14.6. Two protractors, each having two legs, are required, rather than one protractor with three legs, as illustrated in the figure.

14.6.4 Known Range Line and Single Angle

The range line markers control the course of the vessel as shown in Figure 14.8. The fix is obtained by taking a horizontal angle from shore with a theodolite or from the vessel using a sextant. The latter is preferable for the reasons set out in Section 14.6.3.

A three-person team is required: a boat driver using range poles for line, a sextant angle observer, and a echo-sounder reader. Two lines are normally fixed and the targets then are moved similar to the manner discussed in Section 14.6.2.

This method is usually restricted to within 3 km (approximately 2 miles) of the shoreline depending on the size of the targets and the size of the horizontal angles. The accuracy of the fix drops below normally acceptable standards when the angle from the vessel is below 10°. This should be considered in the location of shore stations, discussed in Section 14.8.

14.6.5 Constant Vessel Velocity

The vessel travels at a constant velocity between two shore stations. After an initial trial run to determine the optimum speed, soundings are taken at regular time intervals varying from 15 to 60 s depending upon the level of detail required for the survey. The boat is kept on course by a team member situated at the station being approached by the vessel. The line of sight between the two stations on the line is visual, and signal flags are used for direction. The distances offshore from the stations at each end of the lines where the boat starts and stops should be measured. A driver and echo-sounder reader are also required. If the sounder has an automatic readout, only two persons are required for the

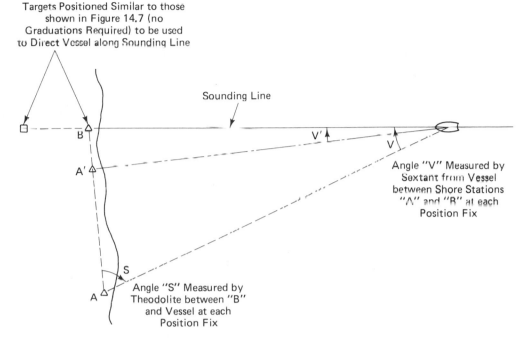

Targets Positioned Similar to those shown in Figure 14.7 (no Graduations Required) to be used to Direct Vessel along Sounding Line

Sounding Line

Angle "V" Measured by Sextant from Vessel between Shore Stations "A" and "B" at each Position Fix

Angle "S" Measured by Theodolite between "B" and Vessel at each Position Fix

Note: If Shore Station Located at A' Instead of A, Angle V' is less than 10° thus Reducing Accuracy of Fix Below Acceptable Survey Standards.

Only One of the Two Angles Shown is Required for the Position Fix, that is Either "S" or "V."

FIGURE 14.8 Range line and single angle.

survey team. This method is particularly applicable to lake surveys and center-line profiles of rivers and streams, as illustrated in Figure 14.9.

It is usually important to locate the deepest areas in lake surveys. As these may be missed in the original sounding line layout, cruising the lake at higher vessel velocities or sweeping (see Section 14.6.12) while constantly monitoring the soundings will locate deep areas as well as shoals.

14.6.6 Intersecting Range Lines

This method is used when it is necessary to repeat the soundings at the same points. The determinations of dredging quantities and of changes in the bottom due to scour of silt or sand are two common applications.

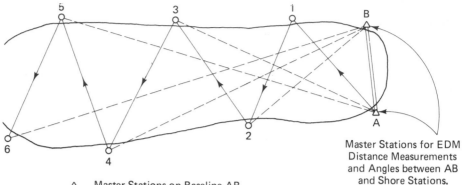

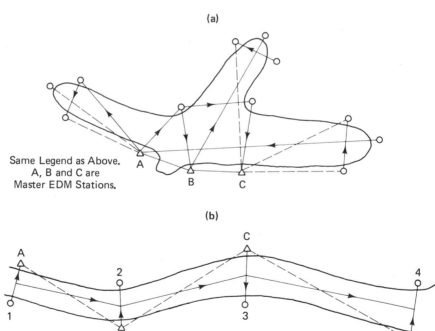

△   Master Stations on Baseline AB

○   Shore Stations

→   Sounding Lines

---   Angle and Distance Tie Lines to Stations "A" or "B"
to Fix Location of Shore Stations

Master Stations for EDM
Distance Measurements
and Angles between AB
and Shore Stations.

(a)

Same Legend as Above.
A, B and C are
Master EDM Stations.

(b)

Same Legend as in (a) Above
A, B, C and D are
Master EDM Stations.

(c)

FIGURE 14.9 Layout for constant vessel velocity soundings: (a) inland lake with regular shoreline; (b) inland lake with irregular shoreline; (c) river.

Fixed range lines are established on shore and located to intersect as closely as possible to right angles, as shown on Figure 14.10. The shore stations are permanently marked, usually by iron bars driven into the ground, mortared stone cairns, or painted crosses on bedrock. Targets are erected over the shore stations during the survey and are stored for reuse.

The boat proceeds to the intersection and takes the sounding as required. A common difficulty is keeping the vessel stationary long enough to obtain an accurate sounding at the intersection point, particularly under strong wind conditions. It is therefore advisable to use a boat capable of good maneuverability at low

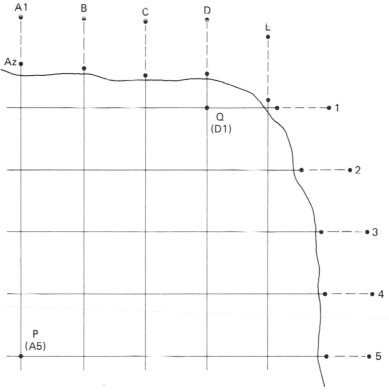

---- Measured Distances Between Range Poles

—— Sounding Lines

• Range Poles set over Permanent Shore Stations

Position of "Q" Located with Greater Accuracy than "P"
due to Proximity of Shoreline Stations.
Position Fixes Designated by Letter + Number System,
for Example "P" is "A5" and "Q" is "D1."

FIGURE 14.10 Intersecting range lines.

speeds. The point of intersection should be approached into the direction of the wind.

Heavy weather conditions must be anticipated using this method, as the depth measurements are taken at preselected intervals: monthly, weekly, and so on. Since the surveyor cannot wait for more suitable weather conditions and the soundings must be compared with previous readings, the following procedures will be of assistance:

1. Swells caused by wave action mean that the sounding referred to the mean water level or mean low water level could be in error by as much as half the wave height. This error will be minimized by recording at least three soundings at both the wave crest and the wave trough and averaging the figures. The vessel may have to reapproach the intersection point each time to achieve this due to rapid drifting.

2. Strong winds blowing in one direction over periods of over 24 hours will cause the mean water level to raise or lower during and after this period. The resulting error in all soundings taken under these conditions must be corrected. Permanent marking of the mean datum elevation should be established, preferably on a shoreline having a slope toward the water of over 25%. This will minimize the error caused by breaking waves offshore. The change in water level is measured by a graduated rod, taking the average of the trough and crest readings and comparing this to the mean datum elevation. All soundings are corrected accordingly.

3. Tidal variations, discussed in Section 14.5.2, must be accounted for in addition to the preceding considerations.

The precision of this method depends primarily on the success of locating the intersection points. This is, in turn, dependent on the distance between the two range poles at each shore station as shown on Figure 14.10. The desired accuracy for surveys of this type is 1:1000. To achieve this, experience and practice have shown that the offshore distance to the point of intersection (*A2* to *P*) should be not greater than 10 times the distance between the two range poles (*A1* to *A2*). This factor is often limited by existing land uses on shore and/or shoreline topography.

14.6.7 Transit and Stadia

The basic principles of location by horizontal angle and stadia rod readings are detailed in Chapter 7. This method is only applicable to smooth waters using a heavy, flat-bottomed boat. The transit or theodolite is set up over a shore station, oriented on another shore station, and the rod is held on the bottom of the boat as close to the depth-measuring instrument as possible. If the water depths are less than the length of the rod, a weight can be attached to the bottom and the rod used for soundings as well. It is important to set up the transit on a station near the water to avoid reading vertical angles.

This method is limited to soundings within 660 ft (200 m) of the shore. If the waters are very calm, this technique is highly effi cient both in the field and for plotting with a 360° protractor or a drafting machine.

14.6.8 Compass and Range Finder

The vessel compass is used for maintaining the course along the sounding line. A range finder is used to determine the distance offshore when the sounding is taken. Range finders are crude instruments that are available in a variety of designs. Essentially, the common principle of each involves a hand-held scope containing a number of graduated cross hairs, similar to stadia hairs in a transit or theodolite. The viewer estimates the distance from shore along the sounding line by reading the interval on the scope between the ends of a measured target on the shoreline or existing features such as buildings (height, length, distances between, etc.).

Both the compass and range finder are crude instruments for position fixing. Therefore, the accuracy of this method is between 1:20 and 1:50. However, the survey can be carried out with a two-person team, thus reducing costs. This technique is most applicable to preliminary surveys in areas where available information, discussed in Section 14.3.2, is either minimal or nonexistent.

14.6.9 Electromagnetic

Electromagnetic position-fixing (EPF) systems determine the vessel's location by the intersection of a minimum of two range distances measured to shore stations (distances 1 and 3, Figure 14.11). The systems are classified into short, medium, and long range based on the different characteristics of radio-wave propogation being used. Each is described briefly, but greater emphasis is placed on the short-range systems as these are the tools more commonly available to the hydrographic surveyor and also provide

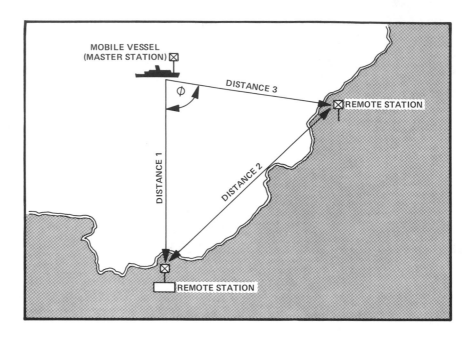

FIGURE 14.11 Tracking of a remote mobile unit. The base station calculates the position of the vessel using the distance shown. Master measures to two remote stations (distances 1 and 3), distance between remotes measured using standard survey techniques (distance 2).

greater accuracies. The wavelengths used penetrate through rain, thus making these systems relatively weather independent.

Before discussing any EPF system, the various motions of the vessel should be considered. Roll (sideway movement), pitch (stern to bow movement), heave (vertical displacement), and combinations thereof are present except for the rare condition of "dead calm" seas. Consequently, the antenna on the vessel masthead is subject to constant movement, thus affecting the accuracy of the position determination. Therefore, to achieve the greatest accuracies, the shore station antenna should be elevated rather than the vessel antenna, while still minimizing the height of the shore stations for reasons explained later in this section.

Short range (up to 70 km offshore) and medium range (up to 700 km offshore) are described together due to the similarities in equipment, operation, and environmental effects. Long-range systems are briefly discussed subsequently, as their present application is primarily navigational rather than for hydrographic surveying.

Short- and Medium-Range Systems

System Types. Two general system types are presently in use: continuous wave and pulse. The operation and advantages of each system are discussed next.

The continuous wave system employs both a carrier frequency and a measuring (modulation) frequency, while the pulse uses only a carrier frequency. All frequencies are within the microwave portion of the electromagnetic spectrum, utilizing wavelengths from 10 cm to 100 m. These long wavelengths are subject to minimal interference by adverse weather conditions such as heavy rain and/or fog. This represents a significant advantage over the position-fixing techniques previously discussed in this section, all of which rely upon good visibility.

The carrier frequency essentially establishes contact between or among the two or more shore stations and the master station in the vessel. Its characteristics are similar to high-frequency radio waves. The carrier frequency carries the measuring (modulation) frequency, which, although having a much longer wavelength, is implanted in the carrier wave. In essence, the carrier wave provides the transportation route through the atmosphere for the measuring wave.

It is important to note that the higher the carrier frequency, the more straight the line through the atmosphere between the two stations becomes. For example, a carrier frequency of 9 GHz (such as that used by the trisponder VHF system) will curve less to fit the earth's surface than a carrier frequency of 400 MHz (such as that used in the trisponder system). The "straightness" of the transmission line has a direct effect upon the antenna heights required for both shore and vessel stations in order to achieve the necessary range offshore. The continuous-wave systems typically send the carrier frequency and a measuring frequency from the master station to the remote. The remote returns a slightly different carrier frequency back to the master. As the master station is attuned to the remote station frequency and the return frequency is slightly different, the measurement of the distance between the two stations is achieved by recording the phase difference between the two transmissions, each in the form of a sine wave. This process results in a continuous loop of microwave energy between the two stations, a situation similar to an FM radio station. When the station is turned off and the tuner (your radio) is turned on, static results indicating that no carrier frequency is being transmitted. When the station is on and the tuner is properly locked onto the station, there is no static, thus indicating that a carrier frequency is being transmitted. When the announcer talks, the voice modulates the carrier frequency. This example is a one way system while continuous-wave is two way. A typical station for continuous-wave is illustrated in Figure 14.12.

The pulsed system operates on a different principle than the continuous wave. Pulses of microwave energy are sent by the master station in Figure 14.11 at regular intervals, usually in small fractions of a second. These pulses are similar in concept to the leading edge of an ocean wave front. The receiving station (a remote station in Figure 14.11) magnifies the power of the pulse and retransmits it to the master station. The distance measurement is obtained by recording the time lapse for the pulse to travel the distance from one station to the other.

This requires a very precise clock in the master station (see Figure 14.13). The time lapses are measured in nanoseconds, a minute period of time. For example, the ratio between 1 nanosecond and 1 second is approximately the same as the ratio between 1 second and 33 years. As the single wave front is distorted by atmospheric conditions such as water vapor and inversions, averages of return pulses are used to determine a reliable distance measurement. This factor has a direct effect upon the accuracy.

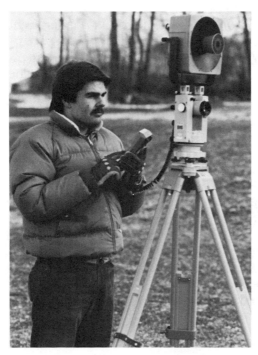

FIGURE 14.12 Setup of a microwave range/angle shore station. In this case, a microfix unit is mounted on the theodolite. On the vessel, an identical microfix is typically mast mounted.

FIGURE 14.13 Pulse-type transponder, the key component at the master station. (Courtesy Davis Canada Engineering Products, Toronto)

The precision of the clock in the master station is a critical factor. For example, the use of an atomic clock would increase the accuracy substantially. However, the cost of these clocks is prohibitive at present. A comparison of the two systems is summarized next:

1. *Cost*: Pulsed system presently cheaper by approximately 30%. This difference is expected to be negligible by 1986.
2. *Accuracy*: Continuous wave typically more accurate (0.2 m $\pm$ 3 to 5 ppm) versus (1 to 2 m $\pm$5 ppm) for pulsed.
3. *Flexibility*: Pulsed system is presently more flexible because it is not locked into one remote station for a long period and can therefore be measuring from several remote stations on shoreline simultaneously. This is achieved through coding the remote stations for the microcomputer accompanying the master station, thus permitting differentiation of the returning impulses. The continuous wave is measuring one master to 3 remotes simultaneously and is thus locked in for this time period.
4. *Location of remote*: Very little difference regarding antenna heights and beam widths.
5. *Finite position fixing for dredged volumes or drill rigs*: Continuous wave is preferred due to greater accuracy.

A third type of EPF system having a range up to 700 km is based on a hyperbolic mode versus a range/range mode. The system used is typically the continuous wave system. One type of configuration involves shore stations transmitting control pulses between or among each other. This results in timing coordination between prepositioned shore stations. A passive receiver on the vessel is typically capable of identifying four or more separate shore station pulses. These signals are separately coded for each shore station, and the resulting distances are calculated using a microcomputer. The particular advantage of this configuration is that there is no limit to the number of vessels able to use the system simultaneously. A second configuration can be operated through the addition of a transmitter to the passive receiver aboard the vessel. The operation then has the option of either a hyperbolic or range-range system.

Practical Considerations. The majority of areas where hydrographic surveying is to be undertaken involve aboveground natural or cultural features along the shoreline. The microwave signals are transmitted in large angular cones. Published cone sizes or beam widths by manufacturers are not to be taken as the limit of the signal. The chances for reflections from these shoreline features are significant, thus having the potential of causing errors. Consequently, the proper selection of shore remote stations and certain operational procedures become critical. Signal fading due to atmospheric conditions; cancellations, commonly known as range holes; and multipath effects are discussed next. In addition, the applications and selections of range–range versus range–angle techniques are discussed.

Figure 14.14 presents a situation not uncommon in areas requiring hydrographic surveys. Reflections of the carrier beams by the water, ground surface near the ray paths, and/or buildings, as illustrated in Figure 14.14, cause position-fixing errors. These effects are commonly known among hydrographers as *multi-path*. Due to the beam width, the outer portions of the beams are reflected by the earth surface and nearby objects.

The reflected parts of the beam take longer paths than the direct beam, thus causing a phase difference or difference in pulse time between the master and remote stations that will be different from that of the direct wave beam. In more cluttered areas, such as the buildings shown in Figure 14.14, reflections can cause serious problems. Preliminary runs should be carried out to determine the severity of the reflection problems. The simplest solutions to alleviate this problem involve lowering the antenna of the remote shore stations and/or modifying their locations to minimize background reflection.

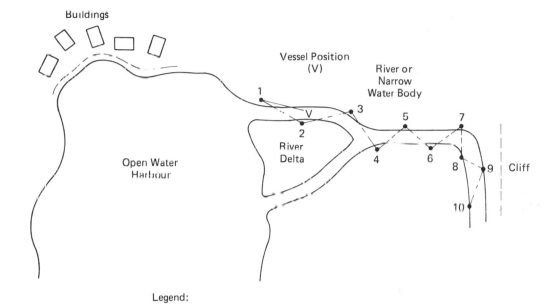

FIGURE 14.14 Reflection due to natural and cultural features and example of range-angle technique applications.

Surveying over smooth water causes adverse effects as the reflection coefficient of the water surface may be similar to that of the direct beam. Experience has shown that slight adjustment of the antenna of the remote stations skyward may minimize this problem provided that this procedure does not increase reflections behind the remote due to surrounding objects.

Random reflections caused by ships, buildings, topographic features, and surface water conditions can be significant factors in hydrographic surveying, particularly in harbors. Gross effects can occur and care should be taken to examine the geometry of the remotes and master stations. It is also advisable to have an independent checking system available, such as a third range.

As mentioned previously, some components of a received signal will have been reflected and will therefore have traveled a longer path than that of the direct beam. However, if the difference in path lengths equals an even number of half-wavelengths of the carrier wave, the energy content of that beam component cancels that of the direct beam. The net effect is the lack of any distance reading, commonly known as *range holes, cancellation zones,* or *signal fade.* The position of these zones or holes can be calculated using formulae provided in the manufacturer's operating manual. The two most critical variables are antenna height and carrier frequency. The carrier frequency is usually fixed within a small range. Consequently, it may be necessary to adjust antenna heights correctly initially to prevent cancellation zones falling within the intended survey area. Antenna heights may also have to be adjusted during the survey as the operating area changes location in order to avoid these same effects.

Based on the angular limitations at the vessel illustrated in Figure 14.11, weak triangular configurations sometimes called *triangles of uncertainty* will usually be present within the survey area offshore. On many coastlines, it is not possible to position remote stations to maintain proper triangle configuration. Therefore, if accuracy is important, consideration should be given to placing one or more of the remote stations further inland or aboard a solidly moored vessel or other offshore platform. For example, the angle $\emptyset$ in Figure 14.11 should be between 30° and 150°.

Under certain conditions, it is sometimes required that a series of remote stations be placed progressively along the coast or river shoreline, mainly to increase survey efficiency, as illustrated by stations 1 to 10 inclusive, Figure 14.14. Thus the vessel may proceed undelayed instead of waiting while one set of remote stations is removed and repositioned.

Weather conditions affect instrument accuracy, primarily the refractive index of air to cause errors as much as 40 p.p.m. This index is dependent upon the prevailing atmospheric pressure, temperature, and humidity along the beam path. It is normal practice to measure periodically these variables at each end of the line, average the results, and make the necessary measurement corrections using charts or nomographs supplied by the manufacturer. Certain weather conditions are not favorable for accurate measuring. One example is a warm offshore wind, which will disproportionately distort the effects of normal air temperatures over the water.

The position of the vessel is usually determined by the two or more distances or multi–range method. For example, this technique would be used in the open-water harbor shown in Figure 14.14. Hydrographic surveying along narrow bodies of water such as rivers requires a different approach, known as the range–angle method. The remote station locations shown in Figurc 14.14 are located close to the banks to minimize vegetation cutting for clear lines of site. The stations are located using angle and distance, employing the type of instrument setup illustrated over a theodolite (see Figure 14.12).

As the vessel proceeds along the river it is located by determining the distance $1 - V$ and the angle $V12$ as illustrated in Figure 14.14. This method requires a master station aboard the vessel, constant tracking of the vessel by the theodolite, and a prearranged signal system or time interval for fixing. The remote stations are usually prelocated, and several units are set up in groups and then moved farther along as the survey proceeds.

Long-Range Systems and Satellites

The permanent long-range EPF systems are more useful for navigation than hydrographic surveying. Worldwide coverage is available using eight stations, transmitting at wavelengths approximately 100 times those of the short-range systems. The system is known as Omega, and a vessel can expect to be positioned within 2 km (approximately 1.3 miles). Therefore, the applications are primarily navigational and for very preliminary location surveys for marine features such as areas having sufficient water depths for large oil tankers.

The satellite system used by the U.S. Navy, known as TRANSIT, shows the greatest promise for use by the hydrographic surveyor. Accurately coordinated stations located worldwide track a large number of navigational satellites. The Doppler frequency shifts of the transmissions from the satellites are recorded on a central computer. A vessel equipped with a receiver and computer receives the navigation message from the central computer and measures the Doppler frequency shift as an appropriate satellite rises, crosses the point of closest approach, and sets. These measurements are compared with the true position of the satellite, and the ship's location is obtained through these comparisons. The accuracy of the system is suitable only for navigation, unless passes of a large number of satellites over a period of at least a week from one vessel position are made. Offshore position fixes within 50 m are possible under these circumstances. To obtain greater accuracy (±3 m) requires very expensive equipment, which is sensitive to vibrations and difficult to maintain.

The use of satellites for position fixing is expensive and does not presently provide sufficient accuracy for most hydrographic survey requirements. However, the rapid advances being made daily in the fields of electronics and computers, as well as significant cost reductions for equipment, should be of interest to the surveyor. There is little doubt that satellite systems will replace many of the existing position-fixing systems in the foreseeable future.

14.6.10 Miscellaneous Methods

Radar has been used for many years for position fixing. This system is based on a polar coordinate system, thus fixing the point by azimuth and distance from a shore station. It is not suitable for other than very preliminary surveys due to the low accuracy, particularly along the azimuth direction. However, the accuracy can be increased to satisfy survey standards up to 40 km (24 miles) offshore with the addition of specialized ranging attachments and directional beacons.

The laser has been used for both horizontal position fixing and shallow depth measurements. Its greater potential is for automatic directional control of the vessel along a sounding line. However, the movement of the vessel and attached reflectors causes problems with the beam return of the highly focused light beam. Nevertheless, the potential for laser use on hydrological surveying is very high.

14.6.11 Combined Techniques

The surveyor is well advised to be familiar with the capabilities of all techniques described herein. As some are more effective from the standpoints of cost, range, and accuracy than others, combining two techniques on the same survey should be seriously considered. For example, double sextant angles may be used as the primary system and single vertical sextant angles used for nearshore location. This combination is particularly useful if greater accuracy and more detail are required near the shoreline, not an uncommon requirement.

14.6.12 Sweeping

Sweeping is the term applied to taking additional soundings to locate underwater features in areas not covered by the original sounding plan. Figure 14.15 illustrates the problem. The isonified area of the seabed below the echo sounder is proportional to the water depth. As discussed in Section 14.5.1, the cone beneath the transducer of the echo sounder is approximately 30°. Consequent-

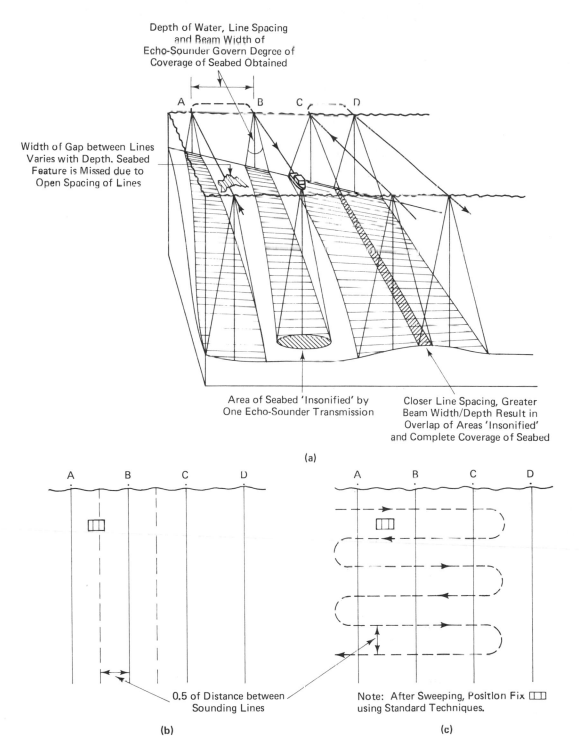

Depth of Water, Line Spacing and Beam Width of Echo-Sounder Govern Degree of Coverage of Seabed Obtained

Width of Gap between Lines Varies with Depth. Seabed Feature is Missed due to Open Spacing of Lines

Area of Seabed 'Insonified' by One Echo-Sounder Transmission

Closer Line Spacing, Greater Beam Width/Depth Result in Overlap of Areas 'Insonified' and Complete Coverage of Seabed

(a)

0.5 of Distance between Sounding Lines

(b)

Note: After Sweeping, Position Fix ⊞ using Standard Techniques.

(c)

FIGURE 14.15   Sweeping to ensure complete seabed coverage.

ly, the shallower depths along sounding lines *A* and *B* on Figure 14.15 leave a gap between the two lines that has not been sounded. As illustrated, an important seabed feature can be missed during the survey. A gap exists between lines *B* and *C*, while the area between lines *C* and *D* is adequately covered.

The difficulty is created because the surveyor does not know the water depths until he has completed the survey. Therefore, the sweeping operation is the last requirement of the project. It is necessary to identify those areas requiring sweeping. Using the cone of 30° beneath the echo sounder, depths less than 1.85 times the distance between the sounding lines may require sweeping. For example, if the sounding lines are 50 m (165 ft) apart, depths less than 93 m (305 ft) will have gaps. At the discretion of the surveyor, extra sounding lines may be used to sound the gaps, using the same position-fixing techniques employed during the survey, as illustrated in solution (b) of Figure 14.15. A cost-effective compromise involves running the vessel at higher speeds parallel with the shoreline in sounding lines out to the depths where coverage is assured. See solution (c) of Figure 14.15. Note the location of any depth anomalies, and position fix only these unusual occurrences.

14.7 SOUNDING PLAN

The most economical coverage of the seabed is achieved through a series of equally spaced sounding lines over the survey area. Specific considerations are as follows:

1. Appropriate scale of the survey.
2. Spacing between the sounding lines and their orientation with the shoreline.
3. Interval between fixes along a sounding line.
4. Speed of the vessel.
5. Direction in which sounding lines are run.

The scale of the survey is set by the degree of thoroughness and precision of soundings. Therefore, the scale determines the number of sounding lines and fixes along each line. Depending on the preceding factors, scales for surveys within 5 km (approximately 3 miles) from the shoreline range between 1 : 1000 (1 in. = ±83 ft) and 1 : 20 000 (1 in. = ±1666 ft).

The distance between the sounding lines is based on the rule (accepted in Canada, the United States, and Britain) that this should not exceed 10 mm or 1 cm (approximately 0.4 in.) *on the drawing.* Therefore, at a scale of 1 : 5000, the lines should not be greater than 5000 × 0.01 m or 50 m apart.

The intervals between fixes along a line should never be further than 25 mm apart on the scale of the survey. If the seabed topography is highly irregular, this figure could be reduced to as low as 15 mm. However, this is seldom the case as current action tends to fill in topographic depressions with sediments. Assuming normal seabed topography and a scale of 1 : 5000, the maximum spacing between fixes would be 5000 × 25 mm or 125 m.

The speed of the vessel during sounding is determined by the realistic assumption that the time interval between fixes will be a minimum of 1 minute. For a scale of 1 : 5000, the speed of the vessel would be 7 km/h (approximately 4 knots).

The sounding lines are run in a direction to be nearly at right angles with the direction of the depth contours. The effects of geological conditions, offshore bars, and the like, as discussed in Section 14.3.5, should be considered in determining the most efficient and economical direction for the angle of the sounding lines with the shoreline.

The sounding plan shown in Figure 14.16 illustrates these principles. In addition, the location of shore stations using the double-angle sextant method (see Section 14.6.1) is shown, as well as the required ties to the horizontal control network, discussed in Section 14.8.

14.8 HORIZONTAL CONTROL

The specific considerations of target design and location are discussed in Section 14.3.4. The overall control survey system should meet the specifications set out in Chapter 11. If a provincial or state control survey system has been established, the hydrographic shore stations should be tied to the overall system using trilateration, triangulation, or supplementary traverses. For example, in Figure 14.16 shore station numbers 2, 5, and 10 should be accurately tied to control stations *A, B,* and *C,* respectively. Shore stations 2, 5, and 10 should be monumented to provide for hydrographic resurvey possibilities in the future. This low cost of monumenting will pay large dividends versus having to repeat the connections with the overall control system.

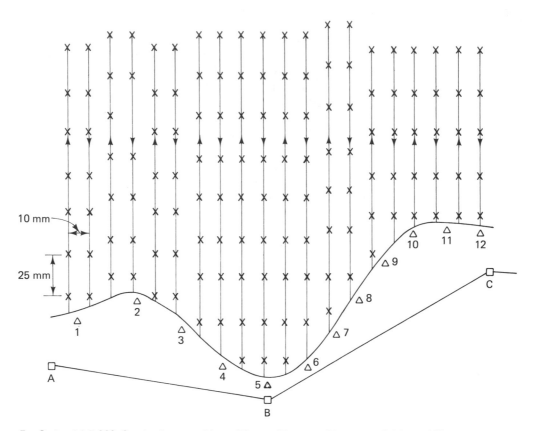

For Scale of 1:5,000, Spacing between Line = 50 m and between Fixes on each Line = 125 m

△ Shoreline Stations Placed to Satisfy Sextant Angle Requirements from Each Fix: Also in Linear Pattern along Shoreline to Minimize Occurrence of Danger Circles in Offshore Sounding Area.

□ Overall Horizontal Control System: May be Established Provincial/State Coordinate System Monuments or Traverse set up for Particular Hydrological Survey being Conducted. Shoreline Stations Tied to Control System and Coordinated to Required Accuracy.

FIGURE 14.16 Sounding plan and shore stations for double sextant method.

The other shoreline stations should be tied to stations 2, 5, and 10 using less accurate survey methods. Normally 1 : 5000 or 1 : 3000 will suffice.

14.9 PROCESSING AND PRESENTATION OF DATA

Prior to commencing the fieldwork, the sounding plan is drawn on Mylar to the desired scale, similar to that illustrated in Figure 14.16. This plan should accurately show the location of all shore stations and each sounding line. Ideally, as previously discussed, the plotting of individual fixes and depth should take place in the vessel immediately after the information has been acquired.

During the fieldwork, the time at the instant of the fix, the position number of the fix, and the lane number, including the shore stations observed, are recorded. The fix is plotted and joined by a straight line to the previous fix to determine the efficiency of the vessel track and compare it to the original sounding plan. If a constant-recording echo sounder is used, the instant of the fix is marked on the trace.

Tidal variations above or below the chosen datum are record-ed against time to provide the information necessary in obtaining the reduced soundings. The soundings are then corrected on the Mylar and the reduced sounding shown in brackets beside the field sounding. If a constant-recording echo sounder is used, the tidal variations are marked directly on the trace and the reduced soundings subsequently read directly from the trace. If depth con-tours are required, these are obtained through interpolation be-tween the individual reduced soundings and plotted on the Mylar.

Due to the number of steps involved, it is necessary to pro-duce a fair tracing and properly drafted plan for presentation to the client. A number of automatic plotting instruments such as that illustrated in Figure 14.17 have been developed. These are computer-based data acquisition and processing systems for off-

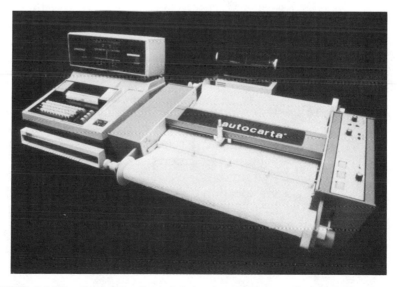

FIGURE 14.17 Computer-based data acquisition and processing system including real-time plotting. (Courtesy Racal Decca Survey, Inc., Houston, Texas)

shore survey operations. The degree of sophistication varies considerably. Some systems store the data on tape to be plotted at a later time. Experience has shown that if this system malfunctions some or all the data may be lost. Consequently, it is important to use a system incorporating a plotter as illustrated in Figure 14.17. A visual plot of all information before leaving the survey area is essential. It is also recommended that the proper functioning of the plotter be checked against firm data before, during, and after the survey.

PROBLEMS

14-1 Calculate the reduced sounding for each of the conditions tabulated below,

Depth using echo sounder	Tidal height above (+) or below (−) chart datum	Error due to weather conditions above (+) or below (−) tidal water level at time of sounding
a) 10.2 m	+1.5 m	−0.3 m
b) 15.7 m	−0.8 m	+0.5 m
c) 32.7 ft	+3.8 ft	−1.4 ft
d) 50.2 ft	−2.3 ft	−1.8 ft

14-2 Using a transparent overlay, plot the locations of shoreline stations 3, 4, 5, 6, and 7 as shown on Figure 14.16. Draw the danger circle for stations 4, 5, and 6. Identify the sounding locations, from those shown on Figure 14.13, that are on or close to the danger circle. Number the sounding locations affected and tabulate the three shore stations that should be used to position fix each of these points.

14-3 The reduced soundings for each of the offshore sounding points illustrated in Figure 14.10 are tabulated below in meters (the equivalent depth in feet is given in brackets.

Location	Depth	Location	Depth	Location	Depth
A1	1.3 (4.3)	C1	0.8 (2.6)	E2	0.6 (2.6)
A2	3.0 (9.9)	C2	2.0 (6.6)	E3	1.1 (3.6)
A3	5.0 (16.5)	C3	3.3 (10.8)	E4	0.9 (3.0)
A4	7.0 (23.1)	C4	3.5 (12.5)	E5	1.2 (3.9)
A5	8.7 (28.5)	C5	3.5 (12.5)		
B1	1.0 (3.3)	D1(Q)	0.6 (2.6)		
B2	2.5 (8.2)	D2	1.5 (4.9)		
B3	4.0 (13.2)	D3	2.3 (7.6)		
B4	5.4 (17.8)	D4	2.0 (6.6)		
B5	5.6 (18.4)	D5	1.8 (5.9)		

a) If the horizontal distance between adjacent points (eg. B1-B2) on the sounding grid is 50 m, plot the soundings at a scale of 1:2000. (If the Imperial system is being used, use a horizontal distance between points of 150 ft and plot at a scale of 1 in = 200 ft.)

b) Determine the location of the depth contours through interpolation and plot these, assuming the water elevation to be zero. Use 1 m contour intervals (3 ft. for the Imperial system).

14-4 For sweeping an area with additional soundings, calculate the maximum depth at which there would be no gaps in the original soundings if the sounding lines are 30 m (100 ft) apart.

14-5 Calculate the minimum distances between sounding lines and the intervals between fixes along a line for each of the following drawing scales.

a) 1:2000

b) 1 in = 100 ft.

14-6 State which type of electromagnetic position-fixing system that you would use for the following circumstances and briefly give three main reasons for your selection.

Accuracy requirements: ± 0.25 m (0.8 ft)
Maximum distance offshore: 10 km (6 mi)

15 Photogrammetry and Airphoto Interpretation

15.1 INTRODUCTION

This chapter deals with the application of aerial photographs to surveying. Under the proper conditions, cost savings for survey projects are enormous. Consequently, it is critical that the surveyor be capable of identifying the situations where the use of aerial photographs may be of benefit. This chapter discusses the basic principles required to use intelligently aerial photographs, the terminology involved, the limitations of their use, and specific applications to various projects, including examples from the authors' experience.

It is imperative that the reader clearly understand the differences between photogrammetry and airphoto interpretation. *Photogrammetry* is the science of making measurements from aerial photographs. Measurements of horizontal distances and elevations are the backbone of this science. These capabilities result in the compilation of *planimetric maps* showing the horizontal locations of both natural and cultural features and *topographic maps* showing spot elevations and contour lines. The preparation of *mosaics,* which involves an assembly of vertical airphotos to produce a photographic map, is also considered to be an aspect of photo-

715

grammetry as certain scientific adjustments are sometimes required. Photogrammetry is truly a science, resulting in quantitative measurements from aerial photographs.

Airphoto interpretation involves the assessment of physical and biological terrain conditions from airphotos. This information is used for engineering, geological, and forestry investigations. It is a science from the standpoint that well-established techniques are available to examine the patterns on the photographs and to outline areas having common terrain characteristics, such as landforms, topography, soil types, drainage systems, erosional patterns, and vegetation. This information is used to determine preliminary highway route locations; areas suitable for residential development; site selection for marine facilities, based on erosion and sedimentation patterns; areas to be conserved for reasons of vegetation types and/or wildlife habitat; base data for environmental assessments; and to identify geological features such as lineaments, fractures, faults, and rock types for mineral explorations. Evidence of original township boundary lines also shows on airphotos although no indications of these lines appear on the ground. In addition, accumulations of alluvium or removal of material along shorelines through erosional processes can be assessed and measured, including the causes. Airphoto interpretation is scientifically based, but it also relies on keen observation of subtle features on the airphotos to solve the survey problem at hand. The information is therefore more qualitative in nature, as precise measurements are not required.

15.2 CAMERA SYSTEMS

Two main types of camera systems are commonly used for acquiring aerial photographs. The first is the single camera illustrated in Figure 15.1, which uses a large-format negative, usually 9 in. (22.5 cm) by 9 in. (22.5 cm). This camera is used strictly for aerial photography and is equipped with a highly corrected lens to minimize distortion. This system is required to obtain aerial photographs for photogrammetric purposes due to the stringent accuracy requirements.

The second system consists of one or more cameras using a smaller photographic format (negative size), such as the common 35 mm used for ground photographs. These smaller-format systems do not have the high quality for lens that is required to meet the normal measurement accuracies for the production of stand-

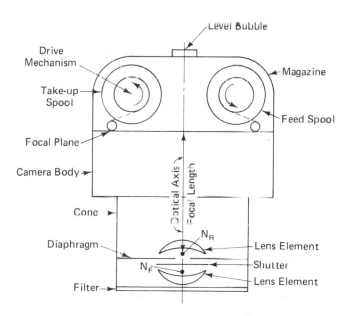

FIGURE 15.1 Components of aerial survey camera (large format).

ard maps using photogrammetry. Smaller-format camera systems are very useful and inexpensive for updating land-use changes and for the acquisition of special types of photography to enhance particular terrain aspects, such as vegetative health and algae blooms on lakes. Both types are discussed in the following sections.

15.2.1 Single Camera with Large Format

The aerial camera illustrated in Figure 15.1 is distinguished by the complexity and accuracy of its lens assembly. Although the lens is shown in simplified form in Figure 15.1, it is actually composed of several elements involving different types of glass with different optical characteristics. This is necessary to satisfy the requirements of high resolution and minimal distortion of the image created on the film by the passage of the light rays through the lens. The quality of the lens is the most important consideration and hence represents the greatest cost factor for this type of camera. The complexity and size of the lens are illustrated by the comparison of this lens with the 35-mm camera lens in Figure 15.2.

The drive mechanism is housed in the camera body, as shown in Figure 15.1. This is motor driven, and the time between exposures to achieve the required overlap is set based on the pho-

FIGURE 15.2 Comparison between large-format aerial camera lens and 35-mm camera lens. (Courtesy Wild Leitz Canada Ltd.)

tographic scale and the ground speed of the aircraft. The film is thus advanced from the feed spool to the take-up spool at automatic intervals. The focal plane is equipped with a device to hold the film flat at the instant of exposure. This is necessary to achieve sharp focusing. The lower air pressures outside the aircraft tend to pull the film away from the focal plane toward the lens, resulting in incorrect focusing.

The camera mount, illustrated in the lower left of Figure 15.3, permits flexible movement of the camera for leveling purposes. Using the level bubble mounted on the top of the camera body as the indicator, every attempt is made to have the camera as level as possible at the instant of exposure. This requires constant attention by the operator, as the aircraft will be subject to pitching and rolling, resulting in a tilt when the photographs are taken. The viewfinder, mounted vertically to show the area being photographed at any time, is illustrated in the upper right of Figure 15.3.

Two points, N_F and N_R, are shown on the optical axis in Figure 15.1. These are the front and rear *nodal* points of the lens system, respectively. When light rays strike the front nodal point (N_F), they are refracted by the lens so that they emerge from the rear nodal point (N_R) parallel with their original direction. The focal length (f) of the lens is the distance between the rear nodal point and the focal plane along the optical axis, as shown in Figure 15.1. The value of the focal length is accurately determined through calibration for each camera. The most common focal length for aerial cameras is 6 in. (15 cm).

FIGURE 15.3   Aerial camera mount and viewfinder. (Courtesy Wild Leitz Canada Ltd.)

As atmospheric haze contains an excessive amount of blue light, a filter is used in front of the lens to absorb some of the blue light, thus reducing the haze on the actual photograph. A yellow, orange, or red filter is used depending on atmospheric conditions and the flying height of the aircraft above mean ground level.

The shutter of a modern aerial camera is capable of speeds ranging from $1/50$ to $1/2000$ s. The range is commonly between $1/100$ to $1/1000$ s. A fast shutter speed minimizes blurring of the photograph, known as *image motion,* which is caused by the ground speed of the aircraft.

Figure 15.4 illustrates the complexity of the single-camera, large-format system. Note that six different filters and five different lenses are shown. Two different mounts for the same camera are illustrated. The one used depends on the size and shape in the opening or *port* in the bottom of the aircraft. Four ports of two sizes are shown in the underside of the aircraft in Figure 15.5.

FIGURE 15.4 Variety of filters and lens available for single-lens, large-format aerial camera. Note filters in foreground, lens in middle foreground, mounts in right background, and viewfinders in left background. (Courtesy Wild Leitz Canada Ltd.)

FIGURE 15.5 Underside of aircraft used for aerial photography. (Courtesy Wild Leitz Canada Ltd.)

15.2.2 Smaller-Format Camera

Cameras having smaller formats, usually either 70 mm (2.5 in.) or 35 mm (1.25 in.), are commonly used for nonphotogrammetric purposes. These systems consist of one of the following:

1. A single camera.
2. Individual cameras placed closely together in a supporting frame, usually up to 4 in total, which are capable of operating individually or simultaneously using different film-filter combinations in each. This is commonly known as a *camera pack*.
3. Multilens cameras, resulting in either four simultaneous exposures using two film types or nine simultaneous exposures using three different films.

A single camera system, usually 35 mm, is commonly used to acquire either vertical or oblique aerial photographs. The aircraft used may be as small as a four-seater, which can be rented at a reasonable rate. This photography is used for purposes of recording land-use changes (usually vertical airphotos) and making photographs suitable for illustrating a particular terrain or cultural situation in an engineering/surveying report. The former use is of more interest to the surveyor as it is useful to update existing land use plans economically. This is true only where the additional information required does not have to be plotted at standard aerial mapping accuracies, and points or areas on the ground are recognizable on both the original and newly acquired photography, thus facilitating plotting of the changes.

A fundamental rule for acquiring aerial photographs must be emphasized at this point. If vertical photos are to be taken through either a hatch (opening) in the bottom of the aircraft or by opening the passenger's door and leaning out (well strapped in and/or with a friend holding your feet), peculiar environmental conditions exist. First, outside temperatures lower considerably, at a rate of approximately 5°F (3°C) per 1000 ft (300 m) of altitude change. Some cameras, designed and lubricated for operation at normal room temperatures, will seize under these conditions. Second, lower air pressure outside the aircraft will tend to pull the film away from the focal plane, thus causing focusing problems. Consequently, the rule is *always to take more than one camera.* Also, take a complete set of photographs with each camera, a minimum of two complete sets. The major cost of acquiring aerial photography is mobilizing the aircraft to have it available near the site. In addition, actual flying time during the photography acquisition and paying for "downtime" when the weather is unsuitable are also high-cost items. For small-format cameras, extra film is cheap.

Individual cameras placed in a supporting frame are termed a *camera pack,* as illustrated in Figure 15.6. Each camera is operated by a central control mechanism, termed in *intervalometer.* The cameras are usually installed inside the aircraft, and the photographs are taken through a port in the underside of the aircraft, as shown in Figure 15.5. Under certain conditions when the only available aircraft does not have a port or hatch, the camera pack can be mounted on a wing strut if this is permitted by aircraft administration regulations.

The cameras used in these packs invariably have a 70-mm format and always have a mechanized film drive. As many of the camera models used for this purpose were not designed for taking aerial photographs, certain precautions must be observed. The cameras should be lubricated for cold temperature conditions for the reasons previously described. In addition, aircraft vibration will cause the adjustable focusing setting on the cameras to revolve out of position during flight. As the proper focus setting for all aerial photographs is infinity, it is wise to tape firmly the focusing adjustment in this position prior to takeoff.

Acquiring airphotos using a camera pack is usually done for two reasons: (1) For selected terrain data, certain film–filter combi-

FIGURE 15.6 Camera pack and mount. (Courtesy Ontario Ministry of Transportation and Communications, Toronto)

nations, particularly using color and infrared color films, are more sensitive to certain aquatic or terrestrial features than the panchromatic (black and white) film used for most photogrammatic mapping missions. For example, color photography has greater water penetration and permits easier differentration of tree species. Infrared color photography is sensitive to small variations in soil moisture and allows vegetative health to be evaluated. (2) New aerial photography is required for purposes such as land-use updating and the detection of vegetation boundary changes. The advantage of a camera pack under these circumstances is reliability. If one or more cameras malfunction during the flight, at least one camera will still be operational. One example from the authors' experience stands out. Special photography was required to examine a severe landslide in a remote area in northern Canada. The nearest camera repair facilities were in excess of 500 miles (800 km) away. Before the mission had been completed, three of the four cameras in the pack had malfunctioned. The remaining single camera permitted the completion of the work. Temporary abandonment of the project for camera repairs would have resulted in an additional cost of at least $6000 Canadian ($7500 U.S.). Also, the timing of the work would have been seriously delayed, another critical factor. Consequently, the availability of several cameras was invaluable.

Multilens cameras are illustrated in Figure 15.7. Either four (Figure 15.7a) or nine (Figure 15.7b) simultaneous exposures of

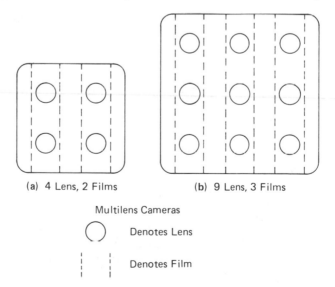

(a) 4 Lens, 2 Films   (b) 9 Lens, 3 Films

Multilens Cameras

◯   Denotes Lens

┆ ┆   Denotes Film

FIGURE 15.7 Multilens cameras: (a) four lens, two films; (b) nine lens, three films: O denotes lens, ¦¦ denotes film.

the same ground area are made, each on a 70-mm format. The film–filter combinations are limited by the number of lenses and films shown in Figure 15.7. The film for a four-lens camera advances two frames after each exposure, and three frames after each exposure for a nine-lens camera.

The primary application of multilens cameras involves the detection of specific terrain conditions, as described previously for the camera pack. This system does not have the flexibility of the camera pack because one mechanical drive operates all the films at once. Hence, a malfunction of the drive mechanism would abort the mission.

15.3 PHOTOGRAPHIC SCALE

The scale of a photograph is the ratio between a distance measured on the photograph and the ground distance between the same two points.

The features, both natural and cultural, shown on a photograph are similar to those on a planimetric map. There is one important difference. The planimetric map has been rectified through ground control so that the horizontal scale is consistent among any points on the map. The airphoto will contain scale variations unless the camera was perfectly level at the instant of exposure and the terrain being photographed was also level. As the aircraft is subject to tip, tilt, and changes in altitude due to updrafts and downdrafts, the chances of the focal plane being level at the instant of exposure are minimal. In addition, the terrain is seldom flat. As illustrated in Figure 15.8, any change in elevation will cause scale variations. The basic problem is transferring an uneven surface like the ground to the flat focal plane of the camera.

In Figure 15.8a, points A, O, and B are at the same elevation and the focal plane is level. Therefore, all scales are true on the photograph as the distance $AO = A'O'$ and $OB = O'B'$. A', O', and B' are points on a level reference datum that would be comparable to the surface of a planimetric map. Therefore, under these unusual circumstances, the scale of the photograph is uniform as the ratio $ao : AO$ is the same as the ratio $ob : OB$.

Figure 15.8b illustrates a more realistic situation, as the focal plane is tilted and the topographic relief is variable. Points A, B, O, C, D, and E are at different elevations. It can be seen visually that, although $A'B'$ equals $B'O'$, the ratio $ab : bo$ is far from equal.

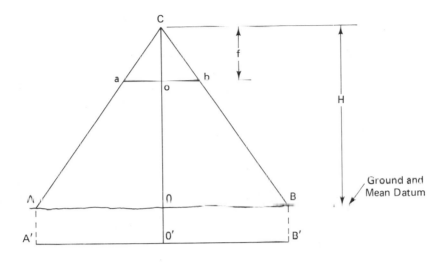

(a) Level Focal Plane and Level Ground

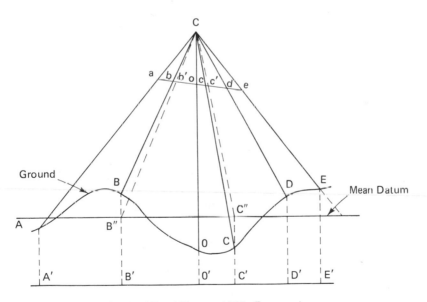

(b) Tilted Focal Plane and Hilly Topography

FIGURE 15.8 Scale differences caused by tilt and topography: (a) level focal plane and level ground; (b) tilted focal plane and hilly topography.

Therefore, the photographic scale for points between a and b will be significantly different than that for points between b and o. The same variations in scale can also be seen for the points between c and d and between d and e on the photograph.

The overall average scale of the photograph is based partially on the elevations of the *mean datum* shown in Figure 15.8b. The mean datum elevation is intended to be the average ground elevation. This is determined by examining the most accurate available contour maps of the area and selecting the apparent average elevation. Distances between points on the photograph that are situated at the elevations of the mean datum will be at the intended scale. Distances between points having elevations above or below the mean datum will be at a different photograph scale, depending on the magnitude of the local relief.

The scale of a vertical photograph can be calculated from the focal length of the camera and the flying height above the mean datum. Note that the flying height and the altitude are different elevations having the following relationships:

$$\text{Altitude} = \text{flying height} + \text{mean datum}$$

By similar triangles, as shown in Figure 15.8a,

$$\frac{ao}{AO} = \frac{Co}{CO} = \frac{f}{H}$$

where

AO/ao = scale ratio between the ground and the photograph
f = focal length
H = flying height above mean datum

Therefore, the scale ratio is

$$\text{SR} = \frac{H}{F} \qquad (15.1)$$

For example, if $H = 1500$ m and $f = 150$ mm,

$$\text{SR} = \frac{1500}{0.150} = 10,000$$

Therefore, the average scale of the photograph is 1 : 10000.

In units of the British system, the scale would be stated as 1 in. = 10000 ÷ 12, or 1 in. = 833 ft. The conversion factor of 12 is required to convert both sides of the equation to the same unit. For example, 1 in. = 833 ft expressed as a scale ratio would be 1 in. = 833 × 12 in., or 1 in. = 10000 in., or 1 : 10000.

15.4 FLYING HEIGHTS AND ALTITUDE

When planning an airphoto acquisition mission, the flying height and altitude must be determined. This is particularly true if the surveyor is acquiring supplementary aerial photography using a small-format camera.

The flying height is determined using the same relationships discussed in Section 15.3 and illustrated in Figure 15.8. Using the relationship in Eq. (15.1), $H = SR \cdot f$.

Example 15.1

If the desired scale ratio (SR) is 1 : 10000 and the focal length of the lens (f) is 150 mm, then $H = 10000 \times 0.150 = 1500$ m (4920 ft $\pm$).

Example 15.2

If the desired scale ratio (SR) is 1 : 5000 and the focal length of the lens is 50 mm, then $H = 5000 \times 0.050 = 250$ m (820 ft $\pm$).

The flying heights calculated in the examples are the vertical distance that the aircraft must fly above the mean datum, illustrated in Figure 15.8. Therefore, the altitude at which the plane must fly is calculated by adding the elevation of the mean datum to the flying height. If the elevation of the mean datum were 330 ft (100 m), the altitudes for Examples 15.1 and 15.2 would be 1600 m (5250 ft $\pm$) and 350 m (1150 ft $\pm$), respectively. These are the readings for the aircraft altimeter throughout the flight to achieve the desired average photographic scale.

If the scale of existing photographs is unknown, it can be determined by comparing a distance measured on the photograph with the corresponding distance measured on the ground or on a map of known scale. The points used for this comparison must be easily identifiable on both the photograph and the map, such as road intersections, building corners, and river or stream intersections. The photograph scale is found using the following relationship:

$$\frac{\text{Photo scale}}{\text{Map scale}} = \frac{\text{photo distance}}{\text{map distance}}$$

As this relationship is based on ratios, the scales on the left side must be expressed in the same units. The same applies to the measured distances on the right side of the equation. For example, if the distance between two identifiable points on the photo-graph is 5.75 in. (14.38 cm) and on the map it is 1.42 in. (3.55 cm), and the map scale is 1 : 50000, the photo scale is

$$\frac{\text{Photo scale}}{1:50000} = \frac{5.75 \text{ in.}}{1.42 \text{ in.}}$$

$$\text{Photo scale} = \frac{5.75 \times {}^{1}/_{50000}}{1.42}$$

$$= 1:12{,}349$$

If the scale is required in inches and feet, it is calculated by dividing the 12,349 by 12 (number of inches per foot), which yields 1029. Therefore, the photo scale is 1 in. = 1029 ft.

15.5 RELIEF (RADIAL) DISPLACEMENT

Relief displacement occurs when the point being photographed is not at the elevation of the mean datum. As previously explained, and as illustrated in Figure 15.8a, when all ground points are at the same elevation, no relief displacement occurs. However, the displacement of point b on the focal plane (photograph) in Figure 15.8b is illustrated. Because point B is above the mean da-tum, it appears at point b on the photograph rather than at point b', which would be the location on the photograph for point B'', which is on the mean datum.

Relief displacement depends on the position of the point on the photograph and the elevation of the ground point above or be-low the mean datum. Referring to Figure 15.8b, the following should be noted:

1. The displacement at the center or *principal point*, represent-ed by O on the photograph, is zero.
2. The farther that the ground point is located from the princi-pal point, the greater the relief displacement. The displace-ment cc' is less than bb', even though the ground point C is farther below the mean datum than point B is above it.
3. The greater the elevation of the ground point above or below the mean datum, the greater is the displacement. If the ground elevation of point B was increased so that it was far-ther above the mean datum, the displacement bb' would in-crease correspondingly.

Relief displacement is radial to the principal point of the pho-tograph, as illustrated in Figure 15.9. The direction of the relief

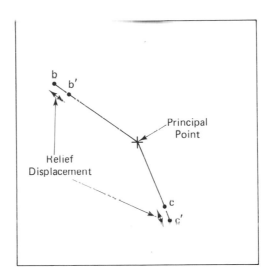

FIGURE 15.9 Direction of relief displacement (compare with Figure 15.8b)

distortion on the photograph is shown for photo points *b* and *c* in Figure 15.8b.

The practical aspects of relief displacement relate primarily to the assemblage of mosaics. A *mosaic* is a series of overlapping aerial photographs that form one continuous picture (see Section 15.8). This involves matching the terrain features on adjacent photographs as closely as possible. As the relief displacement of the same ground point will vary substantially, both in direction and magnitude for the reasons discussed previously, this presents a real difficulty in matching identical features on adjacent photographs when assembling a mosaic. A partial solution to this problem involves using only the central portion of each photograph in assembling the mosaic, because relief displacement is greatest near the photograph edges. This problem was particularly evident in a recent project carried out by the authors in the Rocky Mountains. The relief displacement was so large that mosaic assembly, even when using only the central portion of the photographs, was extremely difficult.

15.6 FLIGHT LINES AND PHOTOGRAPH OVERLAP

It is important for the reader to understand the techniques by which aerial photographs are taken. Once the photograph scale, flying height, and altitude have been calculated, the details of implementing the mission are carefully planned. Although the planning process is beyond the scope of this text, the most significant factors are as follows:

1. A suitable aircraft and technical personnel must be arranged for, including their availability if the time period for acquiring the photography is critical to the project. The costs of having the aircraft and personnel available, known as *mobilization,* are extremely high.

2. The study area must be carefully outlined and means of navigating the aircraft along each flight line, using either ground features or magnetic bearings, must be determined.

3. The photographs must be taken under cloudless skies. The presence of high clouds above the aircraft altitude is unacceptable because of the shadows cast on the ground by the clouds. Therefore, the aircraft personnel are often required to wait for suitable weather conditions. This downtime can be very expensive. Most aerial photographs are taken between 10 A.M. and 2 P.M. to minimize the effect of long shadows obscuring terrain features. Consequently, the weather has to be suitable at the right time.

To achieve photogrammetric mapping and to examine the terrain for airphoto interpretation purposes, it is essential that each point on the ground appear in two adjacent photographs along a flight line. Figure 15.10 illustrates the relative locations of flight lines and photograph overlaps both along the flight line and between adjacent flight lines. An area over which it has been decided to acquire airphoto coverage is called a *block.* The block is outlined on the most accurate available topographic map. The locations of the flight lines required to cover the area properly are then plotted, such as flight lines A and B in Figure 15.10a. The aircraft proceeds along flight line A, and airphotos are taken at time intervals calculated to provide 60% forward overlap between adjacent photographs. As illustrated in Figure 15.10a, the format of each photograph is square. Therefore, the hatched area represents the forward overlap between airphotos 2 and 3, flight line A. The minimum overlap to ensure that all ground points would show on two adjacent photographs would be 50%. However, at least 60% forward overlap is standard as the aircraft is subject to altitude variations, tip, and tilt as the flight proceeds. The extra 10% forward overlap allows for these. The airphoto coverage of flight line B overlaps that of A by 25%, as illustrated in Figure 15.10a. This not only ensures that no gaps of unphoto-

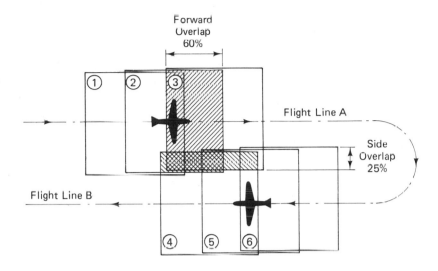

(a) Photographic Overlap

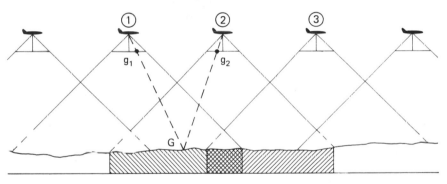

(b) Overlap Along Flight Line

FIGURE 15.10 Flight lines and photograph overlap: (a) photographic overlap; (b) overlap along flight line.

graphed ground exist, but it is also necessary to extend control between flight lines for photogrammetric methods.

The flight line in profile view is illustrated in Figure 15.10b. The single-hatched areas represent the forward overlap between airphotos 1 and 2 and photos 2 and 3. Due to the forward overlap of 60%, the ground points in the double-hatched area will appear on each of the three photographs.

The problems of mosaic assemblage caused by relief displacement, discussed in Section 15.5, are partially solved by increasing the forward overlap to as high as 80%. Although this results in roughly twice the total number of airphotos at a similar scale, a smaller portion of the central portion of each photo, which is least affected by relief displacement, can be used to assemble the mosaic. Consequently, the surveyor should consider increasing the forward overlap in areas of high topographic relief.

The number of airphotos required to cover a block or study area is a very important consideration. It should be kept in mind that each airphoto has to be cataloged and stored. Most importantly, these photographs have to be used individually and collectively and/or examined for photogrammetric mapping and/or airphoto interpretation purposes.

All other factors being equal, such as focal length and format size, the photographic scale is the controlling factor regarding the number of airphotos required. The approximate number of airphotos required to cover a given area stereoscopically (every ground point shows on a least two adjacent photos along the flight line) can be easily calculated. The basic relationships required for this computation are set out next for a forward overlap of 60% and a side overlap of 25%.

For a photographic scale of 1 : 10000, the area covered by one photograph, accounting for loss of effective area through overlaps, is 0.4 square miles or 1 km^2. Therefore, the number of airphotos required to cover a 200-square-mile (500 km^2) area is $200 \div 0.4 = 500$, or $500 \div 1 = 500$.

It is important to realize that the approximate number of photographs varies as the *square* of the photographic scale. For example, if the scale is 1 : 5000 versus 1 : 10000 in Figure 15.10, the aircraft would be flying at one-half the altitude. Consequently, the ground area covered by each airphoto would be reduced in half in *both* directions. This means that twice the number of airphotos would be required along each flight line and that the number of flight lines required to cover the same area would be doubled. The following examples illustrate the effect on the total number of airphotos required, based on the coverage required for a 200-square-mile (500 km^2) area.

1. For a scale of 1 : 5000, the number of photographs is $500 \times (10000/5000)^2 = 2000$.
2. For a scale of 1 : 2000, the number of photographs is $500 \times (10000/2000)^2 = 12500$.
3. For a scale of 1 : 20000, the number of photographs is $500 \times (10000/20000)^2 = 125$.

Thus, it can be seen that the proper selection of scale for the mapping or airphoto interpretation purposes intended is critical. The scale requirements for photogrammetric mapping depend on the accuracies of the analytical equipment to be used in producing the planimetric maps. For general airphoto interpretation purposes, including survey boundary line evidence (cut lines, land-use changes), photographic scales of between 1 : 10000 and 1 : 20000 are optimum based on the authors' experience in over 500 projects. Scales lower than 1 : 10000, for example, 1 : 5000, do not provide the combination of overview or regional coverage and detail required.

The surveyor or engineer should realize (1) that the highest cost factor in acquiring airphotos is mobilization and availability of the aircraft and (2) that film is relatively cheap. Consequently, the temptation to acquire more airphotos for a given area than are required is always present. If 400 airphotos are acquired from a flight rather than 100 at almost the same total cost, the cost per airphoto appears to be about 25% by comparison. When dealing with airphotos, any consideration of cost per photo is complete folly. The *least* number of airphotos required to complete the project satisfactorily should be the sole objective. The authors vividly recall the difficulty of convincing a detail-minded client that 100 airphotos would adequately do the job versus the 10000 airphotos required for his proposed scale. Do not allow yourself to be drawn into this trap.

15.7 GROUND CONTROL FOR MAPPING

As stated previously, the aerial photograph is not perfectly level at the instant of exposure. As a result, ground control points are required to manipulate the airphotos physically or mathematically before mapping can be carried out.

Two situations occur where ground control is required. One involves the establishment of control points where *existing* photography is to be used for the mapping. The other requires the establishment of ground control points *prior* to the acquisition of the airphoto. Although the principles for both are similar, each technique is described in the following subsections.

Ground control is required for each of the following applications. The accuracy with which the measurements must be made varies in each case depending upon the following final product requirements:

1. Measurements of distances and elevations, such as building dimensions; highway or road locations; and cross sectional information for quantity payments for cut and fill in construction projects.

2. Preparation of planimetric and topographic maps, usually including contour lines at a fixed interval.

3. Construction of controlled mosaics.

4. Construction of orthophotos.

Acquisition of ground control data is the most costly aspect of map preparation using airphotos. The surveyor should therefore give considerable thought and planning to every detail of ground control requirements.

15.7.1 Existing Photographs

When mapping of an area is required and airphotos having a suitable scale are available, by far the most economical procedure involves using the existing photographs. The minimum ground control for one pair of overlapping airphotos is three points. As leveling of the stereoscope model (area within the overlap) is the objective, one can equate the requirements with the minimum number of legs required on an ordinary table in order for it to stand by itself. Vertical control (elevations) is required for all three points and at least one horizontal distance between two points. Normally, as all three control points must be accessed on the ground, the normal procedure involves the acquisition of north and east coordinates, as well as the elevation of each point.

The selection of ground control points must be based upon the following criteria:

1. *Well spread out in the overlap area.* If the points are well separated, the model will be more stable and therefore the results will be more accurate. Using the preceding analogy, the table will be better supported if the legs are near the edges rather than grouped in the center.

2. *Easily identifiable on both adjacent airphotos.* If this is not the case, the control point will be useless. As the photographs are taken at an angle that varies directly with the distance from the principal point or airphoto center, features such as trees or shadows from adjacent buildings can obscure the control point on one of the airphotos. This difficulty can be avoided by closing one eye and then the other while viewing the overlaps stereoscopically. If the control point disappears on either photograph, it is not acceptable.

3. *Permanent significant changes may have taken place in the overlap area since the existing photographs were taken.*

Therefore, the ground points should be selected on the assumption that these will still exist when the ground survey is carried out. Points such as building corners, main road intersections, angles or corners of year-round docking facilities, and fence intersections clear of overhanging trees are suitable. Natural features generally do not provide good control points, as they are subject to removal by erosion, landslides, and cultural activities such as timbering. A clearly defined intersection of rock fracture lines would be satisfactory as permanency is virtually guaranteed.

4. *Accessible.* The surveyor should consider ease of access to all control points to minimize open-ended traverse lines, particularly over heavily wooded terrain having high topographic relief. The traverse or trilateration lines to be run should be anticipated, and the necessary changes should be made if cost savings in the ground surveying will result.

The selected identifiable points on the photograph are termed either *photo points* or *picture points*. The horizontal control between these points is obtained by direct triangulation or trilateration between the control points, if clear lines of sight exist, or by a closed traverse connecting the points. Much of this survey work is done using a theodolite and an electronic distance-measuring device (see Chapter 8). Vertical control is obtained using vertical angles and slope distances (see Chapter 3), differential leveling (see Chapter 3), stadia (see Chapter 7), or atmospheric pressure differentials between aneroid barometers for large contour intervals.

The field method chosen depends on the required accuracy, which depends on the scale of the map or mosaic and the contour interval to be mapped.

While a minimum of three control points is required for controlling a single overlap or "model" between two adjacent photographs, each model along a flight line or within a block does not require three ground control points. This saving in ground surveying is achieved through a process termed *bridging*, which is discussed in Section 15.10.

15.7.2 New Photography

While a number of situations arise where new photography is required, the three listed are common, assuming that a high degree of accuracy is required for the photogrammetric mapping.

1. *Areas containing few existing, identifiable ground control points.* As natural features do not make good ground control points for reasons previously discussed, premarked points or *targets* must be used.

2. *Legal surveys of densely developed areas.* If the property lines in a municipality require resurveying as a group, considerable economy can be affected by placing targets over or close to all known boundary corners and subsequently obtaining the horizontal coordinates of each using photogrammetric methods.

3. *Municipal surveys of roads and services.* As discussed in Chapter 13, municipalities often require accurate maps showing the location of both aboveground and underground services. Targets are easily set by painting the appropriate symbol on existing roads and/or sidewalks. This ground control network can be surveyed at convenient times of low traffic volumes. Aerial photography is obtained under low traffic conditions, and features such as manhole covers, hydro lines, roads, and sidewalks can be recognized on the photographs and accurately mapped.

The targets for each of these situations must be placed prior to the acquisition of the airphotos. This should be done as close to the anticipated flight time as possible, particularly in populous areas. The targets are an attraction of sorts, and people remove them readily for whatever purpose. For example, a property boundary survey for a small town was carried out by the authors using photogrammetric methods. A total of 650 targets were placed during the early morning of the same day as the flight, which was carried out the same afternoon. Within this short space of time, about 50 targets did not show on the airphotos.

The targets are used in several shape configurations. For the preceding example, small (1 ft or 0.3 m) square plywood targets were centered over the monuments. In other situations, such as higher-altitude photography, a target in the form of a cross is common. Painted targets on roads and sidewalks for municipal surveys are more flexible with regard to shape. In addition, they cannot be easily removed and are often used for more than one flight.

The photographic tone of the terrain on which the target is placed is a critical consideration. The camera records only *reflected* light. For example, a black or dark gray asphalt highway will commonly appear light gray to white on the photograph because of its smooth reflective surface. Therefore, a white target on what

appears to be a dark background will disappear on the airphotos.
Too much contrast between the target and the background will result in *lateral image spread*. This is caused by a gradient in the
film density from the light object to the dark. This condition is
primarily caused by scattering of light in the film emulsion at the
edge of the target. The effect of this phenomenon can render the
target unidentifiable at worst or cause difficulties in locating
the center point at best. The surveyor is well advised to consider
the following:

1. Examine previous airphotos of the area, usually available
   from the appropriate government agency or a private aerial
   survey company, to determine the relative gray tones of various backgrounds, such as asphalt roads, gravel roads, grass,
   and cultivated fields.
2. Determine the best gray tone of the targets for each type of
   background to achieve the delicate balance of enough contrast but not too much. For example, a medium to dark gray
   target on a white background (gravel road) would be logical.
   Also, a white target on a medium gray background (grass)
   would be suitable.
3. Bare or recently disturbed soils, a situation created by digging to uncover an existing monument, usually photograph
   as a light tone, even though they may appear medium to
   dark gray from ground observation. Therefore, a medium to
   dark gray target is required for identification on the airphoto.
4. Ground control points are normally targeted using a different
   configuration from the property corners for ease of recognition during the photogrammetric mapping process.

15.8 MOSAICS

A mosaic is an assembly of two or more airphotos to form one
continuous picture of the terrain. Mosaics are extremely useful for
one or more of the following applications because of the wealth of
detail portrayed.

1. Plotting of ground control points at the optimum locations to
   ensure the required distribution and *strength of figure* (see
   Section 11.10).
2. A map substitute for field checkpoint locations and approximate locations of natural and cultural features. A mosaic is
   not an accurate map because of relief displacement (see Sec-

tion 15.5) and minor variations in scale due to flying height differences during the flight.

3. A medium for presenting ground data. Using standard photographic procedures, a copy negative is produced. It is then common practice to produce a *cronaflex*, which is a Mylar transparency with an airphoto background. Economical whiteprints can be easily produced from this transparency using standard blue or white printing equipment. If the contrast in the airphoto background is reduced through a photographic process called *screening*, sufficient terrain detail will still show on the print. The advantage of screening is that valuable information can be drafted onto the Mylar transparency and still be clearly read, as it is not obscured by the darker-toned areas of the airphoto background.

After photograph film has been processed, each negative of a flight line is numbered consecutively. The flight line number is also shown. Other information that may also be shown is the roll number and the year that the photographs were taken. As these numbers are always shown in one corner of the photograph prints, it is useful to construct a mosaic by laying down the photographs in order and matching the terrain features shown on each so that all numbered information is visible. These mosaics, termed *index mosaics*, are useful to determine the photograph numbers required to cover a particular area. These mosaics are often photographically reproduced in a smaller size for ease of storage.

For projects where having photo numbers on the finished product is not a concern, every second photograph is laid down. For example, in Figure 15.10, photos 1 and 3, flight line *A*, and photos 4 and 6, flight line *B*, would form the mosaic. Photo 2, flight line *A*, and photo 5, flight *B*, are then available to permit stereoscopic viewing of the mosaic by simply placing the single photos properly on top of the mosaic. Information can then be transferred directly onto the mosaic during the stereoscopic viewing process.

This type of mosaic and the index mosaic are uncontrolled. The only practical way to adjust the overall mosaic scale involves photographing it and producing a positive print to the desired scale. If the mosaic is constructed using alternative prints and is to be used for stereoscopic viewing (see Section 15.9), the scale *cannot* be adjusted, as this will render stereoscopic viewing impossible using the single photos not forming part of the mosaic.

As it is often illogical to take the original mosaic to the field for on-site investigations because of possible damage or loss, a positive print at the same scale can be made on photographic paper at a nominal cost. Thus the original mosaic consisting of the overlapping airphotos can be left in the office. The single photos not used in the mosaic, but necessary for stereoscopic viewing, can be taken to the field with the positive print of the mosaic, thus permitting stereo viewing in the field.

If an uncontrolled mosaic is to be used for graphical presentation purposes and/or as a base for mapping terrain information, it is necessary to *feather* the photograph to avoid shadows along the edges of the overlapping airphotos, as well as to improve the appearance of the mosaic. This is accomplished by cutting through the emulsion with a razor-edge knife and pulling the outside of the photograph toward the photo center, thus leaving only the thin emulsion where the photographs join. The overlapping photograph edges are matched to the terrain features on both adjoining photos as accurately as possible. The best means of permanently attaching the adjacent photos together is the use of a special adhesive wax applied to the underside of the overlapping photograph by a hot roller. Forms of rubber cement are satisfactory for this work, but the photo edges tend to curl. The joins between the photos are then taped securely on the back of the mosaic using masking tape. The mosaics may be constructed by pasting the photographs to a mounting board such as masonite. Two facts should be kept in mind if mounting boards are to be used: (1) portability for field use is limited and (2) requirements for storage space increase substantially.

Where a mosaic is required both as a pictorial map and for accuracy of horizontal measurements, it must be constructed to conform with a combination of ground and photo control points. The density of these points and accuracy of their location depend upon the required accuracy of the mosaic. The photo points are located by *radial line plotting*. The description of this method is beyond the scope of this text, and the reader is referred to books written solely on photogrammetry. The control can also be provided by mounting a planimetric map on the mounting board to the same scale as the mosaic being constructed. The airphotos are laid down on the planimetric map and the scales are matched as closely as possible. This is time consuming and therefore costly. It is also very difficult if the average scale of the airphotos varies from that of the planimetric map or from airphoto to airphoto along the flight line. In these cases, it is necessary to reprint the airphotos from the original negatives, causing the scale problems and to increase or decrease their scale to suit. Tip and tilt at the instant of the airphoto exposure result in scale variations between opposite sides or corners of the airphoto. To compensate for these effects as well as flying height variations, an enlarger whose negative holder can be tilted, known as a *rectifier,* is used.

The advantages and disadvantages of mosaics versus maps prepared by ground survey methods are listed next.

Advantages

1. The mosaic can be produced more rapidly, because the time requirements to carry out the ground surveys and to plot the related information on a map are extensive.

2. The mosaic is less expensive, even if the cost of acquiring the airphotos is included.

3. The mosaic shows more terrain detail, as all natural and cultural features on the ground surface show clearly on the airphoto. Ground surveys are carried out to locate only the features specified by the contract and/or those that can be shown by standard symbols in the legend.

4. For airphoto interpretation purposes, subtle terrain characteristics such as tone, texture, and vegetation must be visible. Therefore, the use of mosaics for these purposes is essential.

Disadvantages

1. Horizontal scale measurements between any two points on a mosaic, regardless of the degree of ground and photo control employed, are limited in accuracy primarily due to relief displacement.

2. Mosaics are not topographic maps and therefore do not show elevations.

As previously mentioned, the degree of local ground relief is the main factor in deciding whether the central portion of each airphoto is used to assemble the mosaic or whether every second airphoto will suffice. If every photo is used, subsequent stereoscopic examination using the photos in the mosaic is both time consuming and confusing because two complete sets of airphotos are required.

15.9 STEREOSCOPIC VIEWING AND PARALLAX

Stereoscopic viewing is defined as observing an object in three dimensions. To achieve stereoscopic vision, it is essential to have two images of the same object taken from different points in space. The eyes thus meet this requirement. A person with only one eye cannot see stereoscopically.

On two adjacent airphotos, because of forward overlap, over half the ground points on both photos are imaged from two different points in space. For example, the image of ground point G in Figure 15.10b is located at g_1 and g_2 on photos 1 and 2, respectively. If the observer looks at point g_1 on photo 1 with the left eye and at point g_2 on photo 2 with the right eye, point G will be seen stereoscopically.

The eyes are used to converging on an object and therefore resist diverging. Divergence is necessary if each eye is to focus on the images of the same point on two adjacent photographs, as the airphotos have to be separated by between 2 in. (5 cm) and 3 in. (7.5 cm). This is illustrated in Figure 15.11a, where a *pocket* or *lens* stereoscope with two-power magnification is placed over two adjacent airphotos. The stereoscope assists in allowing the eyes to diverge. Pocket stereoscopes are easily portable for fieldwork and also are inexpensive. A *mirror* stereoscope is often used for office work, as the internal mirror system allows the adjacent airphotos to be placed a greater distance apart, as illustrated in Figure 15.11b. As a result, the observer is able to view the total area covered by the overlap of the two photos. This permits the stereoscopic examination to take place more rapidly. Also, the degree of magnification can be easily varied by substituting the removable binoculars shown in Figure 15.11b. The eye base for both types of stereoscopes is adjustable to suit the observer. Also, the adjacent airphotos must be adjusted slightly until the images coincide to provide stereoscopic viewing.

Figure 15.12 provides an excellent example of an area of high topographic relief. The mountain near coordinates K5 is approximately 800 ft (270 m) high. The two images of the same area on these adjacent airphotos are set at an average spacing between identical points of 2.4 in. (6 cm). A pocket or lens stereoscope can be placed over this stereopair and the high relief will be clearly seen. The distance between the lenses, known as the *interpupillary range*, is adjustable from 2.2 in. (55 mm) to 3.0 in. (75 mm). This adjustment will assist the reader in viewing any stereopair stereoscopically without eye strain.

As the ratio between the distance between principal points, *PP*1 and *PP*2 on Figure 15.13, to the flying height is much greater than the ratio of the interpupillary distance to the distance from the stereoscope lens to the photo, combined with long-range focusing when using a stereoscope, the phenomenon of *vertical exaggeration* is introduced. This means that the height of a feature appears higher through stereoscopic viewing of airphotos than the

(a)

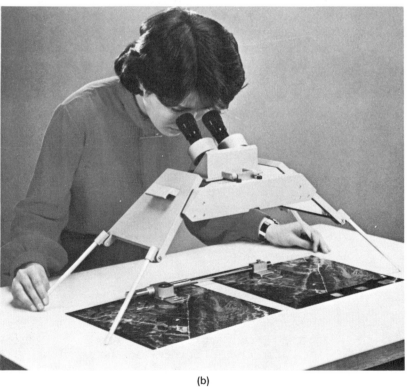

(b)

FIGURE 15.11 Lens and mirror stereoscopes: (a) lens (pocket) stereoscope; (b) mirror stereoscope. (Courtesy Cansel Survey Equipment Limited [Canada], Toronto)

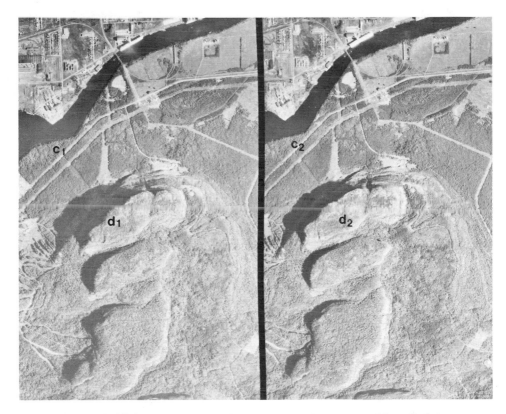

FIGURE 15.12 Stereopair illustrating high topographic relief. (Courtesy Bird & Hale Limited, 1982)

actual height. This exaggeration is usually by a factor of between 2 :1 and 2.5 :1, depending on the interpupillary distance of the individual. This also applies to the examination and estimation of terrain slopes, an important factor in airphoto interpretation. Vertical exaggeration is useful for both photogrammetry and airphoto interpretation as it emphasizes ground elevation differences, thus rendering these easier to observe and measure. However, this exaggeration must be kept in mind when estimating slopes for terrain analysis purposes.

In Figure 15.13, the mountain shown on the stereopair in Figure 15.12 has been photographed from two consecutive camera stations, $PP1$ and $PP2$. The images for both the bottom and top of the mountain are designated c_1 and d_1 for the left photo (photo 1) and c_2 and d_2 for the right photo (photo 2). The same designations are shown on both Figures 15.12 and 15.13.

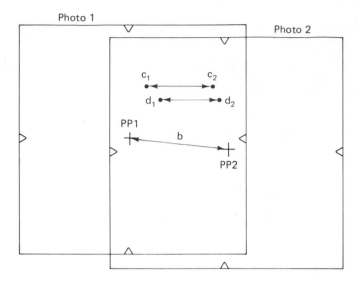

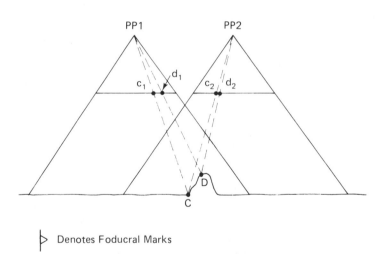

▷ Denotes Foducral Marks

FIGURE 15.13   X parallax along the flight line.

Parallax is the displacement along the flight line of the same point on adjacent aerial photographs. For example, the difference between the distances c_1c_2 and d_1d_2 in Figure 15.13 is the displacement in the X direction (the direction along the flight line) or the difference in parallax between the image points c and d. This difference in parallax is a direct indication of the elevation difference of the height of an object. If points c and d were at the same elevation (not in this example), the difference in parallax would be zero.

The relationship between difference in elevation is

$$dh = \frac{dp\,H}{dp + b}$$

where dh = difference in elevation
dp = difference in parallax
H = flying height
b = photo base (distance between the two adjacent principal points)

The following examples illustrate the calculation of dh using both measurement systems. All pertinent data given are illustrated in Figures 15.12 and 15.13, and the calculations determine the dh between points c and d (the height of the mountain). The data supplied or measured are

$$H = 18000 \text{ ft (5472 m)}$$
$$b = 3.00 \text{ in. (76.2 mm)}$$
$$dp = 0.17 \text{ in. (4.32 mm)}$$

Note that dp is determined in this example by measuring $c_1 c_2$, and $d_1 d_2$, and taking the difference: $2.41 - 2.24 = 0.17$ in. (4.32 mm) (see Figure 15.13).

Example 15.3

$$dh = \frac{0.17(18\ 000)}{0.17 + 3.00} = 965 \text{ ft}$$

Example 15.4

$$dh = \frac{4.32(5472)}{4.32 + 76.2} = 294 \text{ m}$$

The following procedure is carried out to determine b and dp.

1. The airphotos are aligned so that (a) the principal points shown in Figure 15.14 form a straight line as illustrated and (b) the airphotos are the proper distance apart for stereoscopic viewing by a mirror stereoscope, as shown in Figure 15.11b.

2. The airphotos are taped in position and an instrument known as a *parallax bar* is used to measure the distances

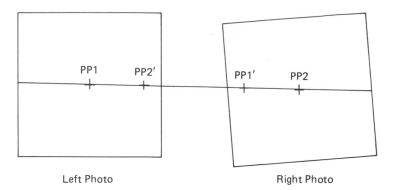

FIGURE 15.14 Positioning of adjacent photos for parallax measure-ments.

along the flight line between identical image points for which differences in elevation are required. As illustrated in the previous examples, dp is obtained by calculating the difference between these two measurements.

3. The principal points of each airphoto are determined by the intersection of lines between opposite *fiducial marks* (see Figure 15.13). These marks show the midpoint of each side of the airphoto and are permanently set in the focal plane of each aerial camera. As shown in Figure 15.14, the principal points of each adjacent airphoto are plotted on the other photograph by transference of the ground point through stereoscopic viewing. The value of b is determined by averaging the distances $PP1$ to $PP2'$ and $PP1'$ to $PP2$, thus canceling the effects of relief displacement.

The parallax bar is designed so that a small dot is placed on the bottom side of the plastic plates at each end, as illustrated in Figure 15.11. The adjustment dial at the left end of the bar on this figure allows the distances between the dots to be adjusted. When viewed stereoscopically together with two adjacent airphotos, the dots at each end of the bar appear as one *floating mark*. The bar is adjusted until the floating mark appears to be at the same elevation as the point being examined, and the distance is read from the bar scale. This process is repeated for the other point involved in the difference of elevation calculation. The two measurements are then subtracted to calculate dp.

The parallax bar is a useful piece of equipment, and the theory of parallax is one of the keys of photogrammetric measurement. The use of the parallax bar to measure difference of parallax is, in the authors' experience, reliable only when carried out by an experienced person. Initially, it is difficult to determine when the floating mark is directly on top of a terrain feature. Personal studies by the authors have shown variations in difference in elevation calculations between the same two points of plus or minus 30% among groups of 30 students having apparently equal stereoscopic viewing capabilities. The reliability of difference in elevation using a parallax bar is a true example of "practice makes perfect." The inherent danger of any calculation carried out by inexperienced people is not recognizing the accuracy of the data input. In addition, not enough attention is paid to significant figures, the tendency being to record all the figures (up to eight decimal places!) in the final answer. This is particularly true since the advent of electronic calculators and computers. Forewarned is forearmed.

15.10 PHOTOGRAMMETRIC PLOTTING INSTRUMENTS

The techniques and systems presented in this chapter up to this point have dealt only with photographic paper prints. The airphotos have been assumed to be vertical for most purposes, the exception being adjustments for tip and tilt relating to controlled mosaic production.

To compile accurate maps, the photograph image must be produced on glass plates known as *diapositives* to provide dimensional stability, thus minimizing shrinkage and expansion due to temperature and humidity changes. In addition, the adjustments to correct for tip and tilt must be carried out using precision equipment. This requires a stereoscopic plotting instrument to plot planimetric and topographic maps to the required scale using photographs taken with a precise, calibrated aerial camera. The precision to which ground elevations can be determined depends on the precision of the stereoscopic plotter or stereoanalyzer (computerized plotter). This error for both elevations and contours ranges between 1/500 of the flying height to 1/5000 times the flying height. The type of ground cover existing at the time of airphoto acquisition considerably affects accuracy.

All plotting instruments have the following features:

1. *Projection system*. Light sources project the images of two adjacent photographs in the proper plane where these images can be viewed stereoscopically. The diapositive images

are projected in the same way as ordinary positive slides projected on a screen. The focal length of the projector lens is designed to produce the image at the correct location for viewing and analysis.

2. *Viewing system.* Most modern stereoplotters and stereoanalyzers use binocular viewing systems, as illustrated in Figures 15.15 and 15.16b, which are similar to the system used for mirror sterescopes.

3. *Measuring system.* The space rods shown in Figure 15.16a are the key parts of a very finely machined system. Each rod positions the same photo point on each of the two adjacent airphotos. When the tracing table is moved, the movement of the space rods is accurately recorded using a rectangular coordinate system. Movement along the flight line is termed the X axis and movement perpendicular to the flight line is called the Y axis. The elevation of the floating mark observed through the viewing system is termed the Z axis.

4. *Plotting system.* The plotting can be carried out directly below the tracing table (see Figure 15.16) or with a pantograph as shown on the right side of Figure 15.15. Each movement of the tracing table either is directly plotted or is recorded and stored on a computer in terms of X, Y, and Z coordinates. For example, if a contour line on the stereoscopic model is to be traced, the elevation of the tracing table is adjusted to the

FIGURE 15.15 Components of a stereo plotter. (Courtesy Wild Heerbrugg)

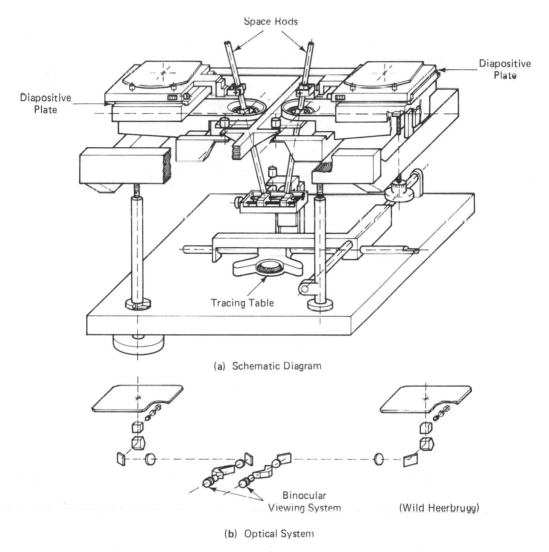

(a) Schematic Diagram

(b) Optical System

FIGURE 15.16 (a) Schematic diagram and (b) optical system for a stereoplotter.

required elevation. The floating mark is placed on top of the ground, and the tracing table is moved along the contour line by rotating the X and Y control knobs shown in the operator's hands in Figure 15.17.

This section is divided into four parts: (1) relative and absolute orientation, (2) analytical stereoplotter, (3) stereoplotting using photo prints, and (4) digital mapping.

FIGURE 15.17 Computer-assisted plotting system. (Courtesy Wild Leitz Canada, Toronto)

15.10.1 Relative and Absolute Orientation

Relative orientation refers to the coincidence of points in a sterescopic model, thus involving the complete matching of all ground points with each other in adjacent airphotos. The model can be visualized as a three-dimensional view that is not oriented toward any ground control. It is therefore independent of any control and can be manipulated in any direction as long as both adjacent airphotos are adjusted in the same direction by exactly the same amount.

Absolute orientation refers to matching of the sterescopic model through relative orientation to the terrain surface by means of ground control points. A minimum of three control points is required for the orientation of a stereoscopic model. This orientation is carried out by matching the coordinates of the ground control points to the same coordinates in the stereoscopic model. The stereo plotter operator sets the coordinates of the ground control points in the plotter and adjusts the model to fit these, thus correcting for tip, tilt, and variations in flying height. This can be accomplished mechanically or through storage in a computer of the coordinates after relative orientation of the ground control points in the model and the actual ground control point coordinates. The computer thus makes the appropriate corrections for each elevation or contour location subsequently measured.

15.10.2 Analytical Stereoplotter

The operation of the stereoplotter has been discussed previously. In addition, it is important to realize that this equipment is capable of *bridging* or *aero-triangulation*, a process that reduces the number of ground control points required. As the acquisition of ground control coordinates is the most costly part of any photogrammetric mapping operation, a reduction in the total number of points has great significance.

The process of aero-triangulation can be visualized using the following example situation. Consider that mapping is required for a flight line or strip consisting of 10 photographs and therefore 9 stereoscopic models. The photographs are numbered consecutively from 1 to 10. Ground control points are available for stereoscopic models for photos 1 and 2 and photos 9 and 10. Assume, for simplicity, that all adjustments required for absolute orientation of the stereoscopic models are made by mechanically adjusting the positions of the diapositives. Also, a computer forms part of the system, thus permitting reading and storage of X, Y, and Z coordinates. The following procedure is carried out to bridge from model 1, 2 (stereoscopic model for photos 1 and 2) to model 9, 10 (stereoscopic model for photos 9 and 10).

1. Model 1, 2 is absolutely oriented in the plotter. This means that both photos are in the identical position regarding tip, tilt, and flying height at the instant they were taken in the aircraft.

2. Photo 1 is removed and photo 3 is substituted for it in the plotter. Photo 3 is oriented to photo 2. As photo 2 is absolutely oriented, the same is now true of photo 3. Therefore, model 2, 3 is absolutely oriented.

3. The operator selects three or more *photo* control points for model 2, 3, using the same criteria discussed in Section 15.7. The coordinates of each of these points are entered into the computer. The physical location of each point is carefully marked and numbered on a paper print of either photo 2 or photo 3.

4. Photo 2 is removed and photo 4 is substituted for it. The process described in step 2 is carried out to achieve absolute orientation for model 3, 4. Photo control points are selected, coordinated, and plotted on the photo print.

5. The preceding steps are continued for the remainder of the flight line. As model 9, 10 has established ground control points, the photo coordinates for these same points are obtained using the plotter. If the photo coordinates agree with the ground coordinates within the specified tolerances, the bridging process is complete. If the number of airphotos in the flight lines is large, factors such as earth curvature

must be considered. Therefore, an adjustment is required, by which each of the photo coordinates for models 2, 3 to 8, 9 inclusive are corrected by a computer having the necessary program or software.

This procedure means that any portion of the flight line can be mapped since either photo or ground control points exist for each stereoscopic model. Using a similar process, it is possible to bridge between adjacent flight lines due to the 25% overlap discussed in Section 15.6. Some reduction in accuracy of measurements occurs because of the bridging process. However, as the ground coordinates are determined to a high degree of accuracy and the stereoplotter used for this process is a high-precision instrument, the reductions in accuracy are not critical, provided that the bridging is not carried across an excessive number of airphotos and/or flight lines.

15.10.3 Stereoplotting Using Photo Prints

Plotters are available that use photographic prints rather than glass or stable-base dispositives as the basis for photogrammetric measurements. These measurements are not as accurate, because the photo prints do not have the dimensional stability of the diapositives.

Several instruments are available that have the capabilities of measuring from photo prints. Some, such as the KEK plotter or the Wernstedt-Mahan plotter, provide for mechanical adjustment of the prints for tip and tilt corrections for orientation, although the solutions are approximate. Other instruments, such as the Zeiss Stereocord, shown in Figure 15.18, use electronic computers to carry out the orientation. These computers correct for model deformation, relief displacements, and tip and tilt. Elevations are measured using the parallax bar concept, and the resulting data are provided to very close approximations. The stage on which the photographs are mounted (see right side of Figure 15.18) is moved manually under the stereoscope and parallax bar. The digitizer coordinates the points along the travel path and these locations are digitally plotted.

15.10.4 Digital Mapping

The traditional map conveys a pictorial representation of the terrain and portrays terrain details with high geometric accuracy so that measurements taken from the map are accurate within the

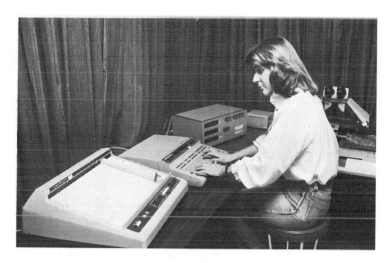

FIGURE 15.18 Computer-assisted plotter for use with photo prints. From left to right: plotter; computer; digitizer; and binocular system for viewing stereoscopic model.

limitations imposed by the map scale. These maps also present a vast array of different types of information on the same sheet. The latter presents problems of data differentiation, particularly for municipal surveys, as discussed in Section 13.6. Digital mapping systems assist substantially in solving these difficulties.

In digital mapping, the stereo aerial photographs are viewed on high-precision television circuitry. Each photograph is scanned by an electric mechanical scanner. The overlapping area of the two photographs is divided into the same square segments, known as *patches*. Video representations of the left and right patches are electronically sent to a correlator, where the analog signals are converted to digital signals. The correlator analyzes the parallaxes for each pair, calculates their height, and alters the scanning pattern. This process is repeated until both images merge into a single orthographically, correctly oriented image.

A matrix of heights is computed for each patch. A grid of about 1 million elevation points per stereomodel is produced and registered on magnetic tape. This digital terrain model gives elevations in a grid pattern of 0.18-mm intervals at the photo scale. If the photo scale is 1: 55000, this represents a grid pattern at 10-m intervals on the ground. This scanning system involving the individual grid units can be repeated for selected features such as buildings, roads, hydro lines, manholes, and property boundaries, each of which is individually coded for identification. All the information is stored on magnetic tape.

Digital mapping systems are extremely flexible for the following reasons:

1. The required data for a particular mapping requirement can be viewed on a video display, and both the scale of the product required and the information to be presented can be specified before the printout is ordered.
2. As each information type is coded separately, not only can individual information sets be recalled when required, but also additions or deletions due to land-use changes, topographic changes due to grading or excavations, and additional installation of services such as roads and pipelines can be easily made.
3. Rather than having to scale horizontal distances to the best of one's ability from a dimensionally unstable paper map, these distances can be accurately and quickly calculated by the computer.
4. Storage space can be substantially reduced, as all the data are stored on tapes and can be recalled when required either on the video display unit or as a map printout.

For the reasons given, this newly developed technology appears to have a bright future. The capital costs of installation are presently high, but these will decrease as demand increases.

15.11 ORTHOPHOTOS

Relief, tip, and tilt affect the positions of points on airphotos, as previously discussed in this chapter. The most direct effect of these displacements relates to the construction of mosaics. Even for a controlled mosaic (see Section 15.8), the overall accuracy for scaling horizontal measurements will be good. This is not true for scaling the distance between two points relatively close together, particularly if their ground elevations are significantly different.

Horizontal measurements made from an absolutely oriented stereoscopic model are very accurate because they are made from orthogonal projections to form a map beneath the model, as illustrated in Figure 15.19. An orthophoto combines the accuracy of scaling, such as from a map, with the detailed photographic representation of the mosaic. The orthophoto is produced by manipulating the images on the photograph using mechanical or electronic techniques to eliminate the adverse effect on scale of tip, tilt, and relief displacement. Although there are several methods for preparing orthophotos, the simplest to understand is the *fixed-line-element* rectification method. The principles of this method are described next.

The stereoscopic model is absolutely oriented, as illustrated by the terrain model in Figure 15.19. The film stage shown in the figure is covered by a layer of orthochromatic film, which is not

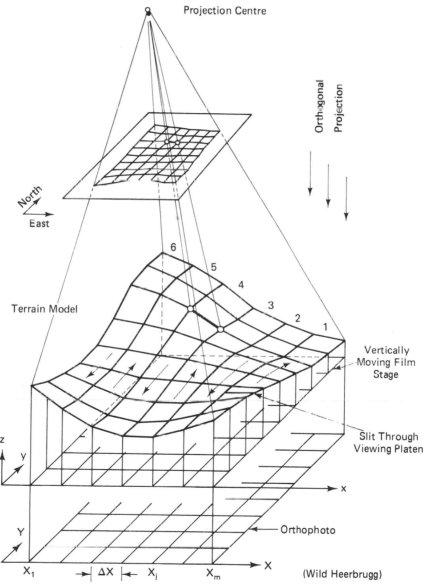

FIGURE 15.19 Relations among photograph, terrain model, and orthophoto.

sensitive to red light. The terrain model is divided into strips for purposes of the process, six strips in Figure 15.19. A slit through the viewing platen is moved at constant speed in the direction shown in strip 1. The film stage is moved up and down vertically as the slit scans the strip so that the film on top of the stage is constantly in contact with the surface of the terrain model. The only light projected onto the film is that passing through the slit. This is achieved through keeping the rest of the film covered by movable curtains. The slit scans along strip 1 and then returns along strip 2 in the opposite direction, as illustrated. The remaining strips are then scanned until the model is completed.

The net result is an orthographic projection, which is developed as an orthophoto, shown graphically below the film stage in Figure 15.19. It can be seen that the distortions due to relief displacement, tip, and tilt have been largely removed. Minimal distortions still occur within each strip because the film stage is moved to match the *average* terrain elevation across the strip. Therefore, the number of strips is increased and the width of each is subsequently decreased to reduce these minor distortions. However, this would increase the cost of the production process.

Figure 15.20 shows the strip boundaries on the terrain model superimposed on the original airphoto in the upper-right corner. The effect of relief displacement is particularly obvious in the hilly area in the upper-left portion of this airphoto. The rest of the figure is the orthophoto with distortions largely removed.

Strips of adjacent orthophotos are easily matched to each other, thus resulting in a true scale mosaic. Also, contour lines obtained from the stereoplotting process can be superimposed on the orthophotos with reliability, as their horizontal location on the photos will be correct, at least from strip to strip.

15.12 PHOTOGRAMMETRIC MAPPING: ADVANTAGES AND DISADVANTAGES

The advantages of photogrammetric mapping over traditional ground methods are discussed next.

Advantages

1. Cost savings. This is related to the size of the area to be mapped, assuming that the ground is not obscured from view by certain vegetation types such as coniferous trees. The authors' experience indicates that the break-even point be-

FIGURE 15.20 Comparison between orthophoto and original airphoto (upper-right corner). (Courtesy Wild Heerbrugg, Switzerland)

tween photogrammetric and ground methods lies between 100 acres (40 ha) and 500 acres (200 ha), depending primarily on the availability of existing photography, the season of the year in which it was taken, and the distance to and accessibility throughout the area to be mapped. Within these limits, the surveyor is advised to cost the mapping using both methods and to select the most economical. As the size of the mapping area increases, photogrammetric methods become rapidly cheaper on a per acre or per hectare basis.

2. Reduction in fieldwork required. It is generally recognized that fieldwork is a very high cost component of any surveying project. It is also access and weather dependent. The relatively few control points required for photogrammetric mapping thus reduce the field problems and reduce the time of data acquisition substantially.

3. Speed of compilation. The time required to prepare maps using a stereoplotter is minimal compared with the time required to carry out the ground survey, plot the spot elevations, interpolate the contours, and prepare a final draft of the manuscript.

4. Freedom from adverse weather conditions and inaccessible terrain conditions.

5. Provision of a constant record. This is useful for checking the photogrammetric maps. The airphotos also provide an accurate record of the terrain features at that instant in time. This is useful for direct comparisons with photography taken at other times to record either subtle or major changes in the landscape.

6. Flexibility. Photogrammetric mapping can be designed for any map scale, provided that the proper selection of flying heights, focal lengths, ground control point placement, and plotting instruments is made. Scales vary from 1: 200 upward to 1: 250000. Contour intervals as small as 0.5 ft (0.15 m) can be measured.

Disadvantages

1. Viewing terrain through dense vegetative cover. If the airphotos have been acquired under full-leaf cover and/or if the vegetative cover is coniferous, the ground is not visible in the stereo model. This problem can only be overcome by obtaining leaf-free airphotos in spring or fall if the tree cover is deciduous or by supplementing the mapping with field measurements if the vegetation is coniferous. Responsible

photogrammetric mapping companies identify contours over areas where the ground cannot actually be seen on the airphotos as dashed lines and/or appropriate notations on the map. Unfortunately, some companies do not. Consequently, the user of the map assumes that the contours represent the ground surface whereas, in fact, these contours represent variations in tree heights in wooded areas. This misrepresentation has caused serious problems on certain projects within the authors' experience, where preliminary planning or engineering design was required based on these erroneous data. Therefore, if the engineer or surveyor is confronted with a contour map of an area partially covered by dense vegetation, and no indication of approximate contours over these areas is given on the map, he or she should acquire the following information.

   a. Whether the contours were prepared photogrammetrically or using ground methods?

   b. If photogrammetrically, in what month and year was the photography obtained (leaf-free or not)?

   c. If not leaf-free photography and/or considerable coniferous cover exists, was supplementary fieldwork carried out to complete the ground mapping in vegetated areas?

      The answers to these questions from the mapping agency will determine the reliability of the mapping before considerable time has been wasted working with incorrect data.

2. Contour line locations in flat terrain. It is difficult to place accurately the floating mark in the plotter on areas of flat terrain. Consequently, it is sometimes necessary to carry out additional field measurements in such areas.

3. The site must be visited to determine type of roads and surfacing, locations of certain utility lines not easily visible on the airphotos, and road and place names. This can usually be achieved effectively and is usually combined with spot checks of the mapping to ensure that the relief is properly represented.

15.13 PRINCIPLES OF AIRPHOTO INTERPRETATION

In the use of aerial photographs, the interpreter is interested in all the environmental characteristics that are recorded or can be inferred, including specific boundary line evidence for the surveyor.

Because the environmental characteristics include most of the physical sciences, the interpreter must have an understanding of the basic concepts of climatology, geomorphology, geology, pedology, geography, ecology, hydrology, and civil engineering. For this reason, multidisciplinary teams of scientists and engineers trained in photo interpretation are frequently used in airphoto studies.

The photo interpretation technique is based on three fundamental assumptions:

1. The aerial photograph is a record of the results of longtime natural and man-made processes that are reflected on the photograph as surface features.
2. The surface features on the airphoto can be grouped together to form patterns that are characteristic of particular environmental conditions.
3. The environmental conditions and their reflected airphoto patterns are repetitive within major climatic zones; that is, similar environments will produce similar airphoto patterns, whereas different environments will usually produce different airphoto patterns.

The terrain elements that collectively produce patterns on aerial photographs include the topography (geometry of the surface), regional drainage, local erosion, vegetation, and cultural features. All these elements are essentially interrelated, but they have been arbitrarily separated in accordance with a classification designed to facilitate description, analysis, and evaluation.

15.14 PHOTO PROPERTIES

Before attempting to interpret terrain, the interpreter must know something of the properties inherent in the photographs themselves. The conventional black and white aerial photo, used as a record of surface conditions, is created by the reflection of certain spectra of electromagnetic energy from the surface of objects. A panchromatic photo records the surface only as tones of gray, and not, as does the human eye, in tones of gray and variations of color. Thus, two widely different colors that would be readily differentiated by the human eye may have exactly the same gray tone on a photograph. In addition, the photo is a much-reduced image of the actual situation; that is, it is a *scaled* representation. Accordingly, familiar pattern elements may become so reduced in size as to present an unfamiliar appearance on an aerial photograph, for example, the familiar linear pattern of a plowed field becomes an even gray without perceptible pattern when the scale of the photo is very small.

In general, photographs have two basic properties: (1) *photo tone*, which is the variation in tones of gray on a photograph, and (2) *photo texture*, which is the combination of tones of gray on an exceedingly small scale, such that individual features are close to the limit of resolution, but in which an overall pattern of variation can still be detected.

15.14.1 Tone

Photo tone responds to variations in a wide range of factors. Many of these are not related to surface conditions, which makes it imperative that the interpreter understand and compensate for these variations.

One factor that produces endless variation is *reflectance geometry* (i.e., the relation between light source, reflecting surface, and camera). Because of this factor, tone in a given area may change from flight line to flight line, even from photograph to photograph, because in each successive exposure the light from any given point is reflected to the camera at a slightly different angle. As an extreme example of the effect of this factor, a freshly plowed field in cohesive soil might photograph either very light or very dark, depending on the reflectance geometry. When the light is received by the camera primarily from the "polished" side of the soil ridges, the photo tone might be light gray; when it is received from the tumbled and crumbly face, the photo tone might be dark gray. A variation of this condition occurs with topographic position as a result of complete or partial shadowing.

The altitude at which a photograph is taken also affects tone. In general, the higher the altitude, the lighter will be the tone. This effect is produced by the increasing atmospheric scattering of light with altitude.

Despite the variations in tone produced by reflectance geometry, processing, and altitude, the properties of the ground surface itself produce the major differences in tone. In general, the conditions resulting in tone variation are differences in (1) the texture of materials, (2) the water content of surface materials, and (3) the intrinsic colors of materials. In general, these variables affect the photo tone as follows:

1. The finer the grain size, the lighter the tone is. Tone is the result of the sum of the reflectance from all surfaces; a large

proportion of the total area of very fine grained materials is composed of reflecting surfaces.

2. The higher the moisture content, the darker the tone is. Water readily absorbs light energy and reflects very little.

3. The darker the color of the soil or rock, the darker the photo tone is.

4. It is most important to note that photo tones are frequently a function of vegetative cover. Care must be taken to ensure that soil tones are not confused with vegetative tones, which tend to be slightly to significantly darker.

15.14.2 Texture

There is no clear line of demarcation between photo tone and photo texture, nor is there one between texture and pattern. Photo texture is a pattern produced by innumerable repetitions of very tiny features, each of which has a distinct appearance. When these features become so small that the camera and film combination will no longer resolve them, the texture vanishes and the area becomes a single tone of gray.

Variations in photo texture are difficult to classify, and even more difficult to describe adequately. However, for this text five fundamental distinctions have been made. These are linear textures, in which the sense impression is one of closely spaced parallax lines; botryoidal, in which the sense impression is one of randomly arranged hemispheres of mixed sizes; flecked, in which the sense impression is one of a spotty surface of variable tones; mottled, in which gradual tonal changes randomly occur; and smooth, in which the impression is one of an even and featureless surface.

1. *Linear.* Cultivation very commonly results in distinctly linear photo textures. Row crops, such as corn and beans, produce very markedly linear textures. For example, see Figure 15.21. Many different factors can create an overall linearity. In drumlin fields, for example, the topography frequently controls field shapes and drainage and results in lineal parallel features (Figure 15.22). In raised beaches, lineal or parallel arcuate texture is caused by vegetative species variation between the wetter soils and drier ridges (Figure 15.23).

2. *Botryoidal.* Forests, especially of broadleaf trees, photograph with a botryoidal texture. The wooded areas in Figure 15.21 illustrate this texture.

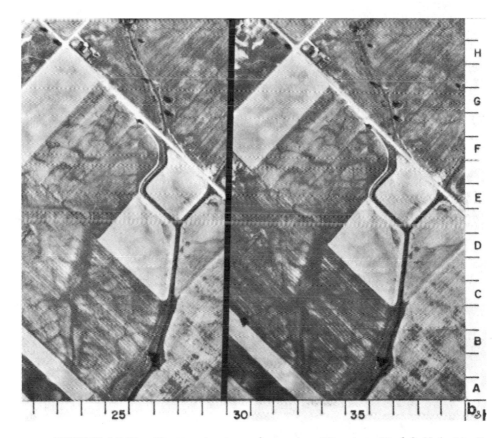

FIGURE 15.21 Stereopair: ground moraine. (Courtesy Bird & Hale Limited, 1982)

3. *Flecked.* A variable thickness of grass cover or pasture lands produces this texture. It may also be produced by crop stubble. This texture is also produced by certain types of surface configurations; any combination of circumstances that produces randomly arranged flecks of either light or dark against an otherwise uniform background will result in this texture, such as in Figure 15.24 (I 5).

4. *Mottled.* Mottled textures are very common. It is, in fact, difficult to find a photograph without some trace of mottling. Normally, in vertical photographs it is produced directly by variations in soil moisture, color, or grain size, and indirectly by vegetative differences induced by such variations. For example, Figure 15.21 (G,30) displays mottled textures.

5. *Smooth.* Smooth textures are produced by featureless surfaces. Surfaces that usually photograph as featureless are water surfaces without wave action, snow, and bare sand

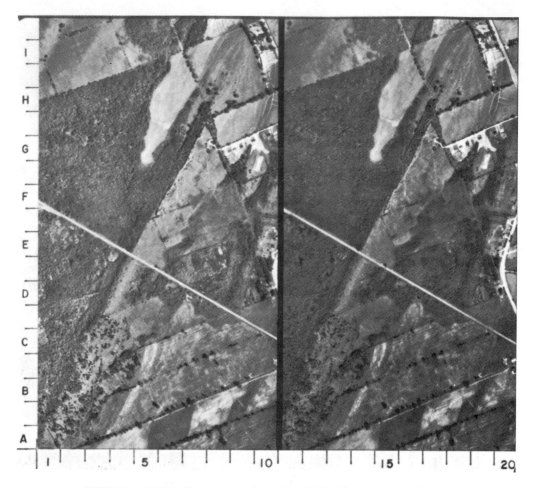

FIGURE 15.22   Stereopair: drumlin field. (Courtesy Bird & Hale Limited, 1982)

areas that have not been eroded by the wind. In Figure 15.25 (J, 22), the soils are well drained and have uniform moisture content; the vegetation is almost uniform; and the resulting texture is smooth.

15.15 TERRAIN (PATTERN) ELEMENTS

Terrain elements are the characteristics of the features and materials composing a landscape. For purposes of systematic and reasonable objective description, each fundamental component of the landscape must be classified in consistent and rigorously defin-

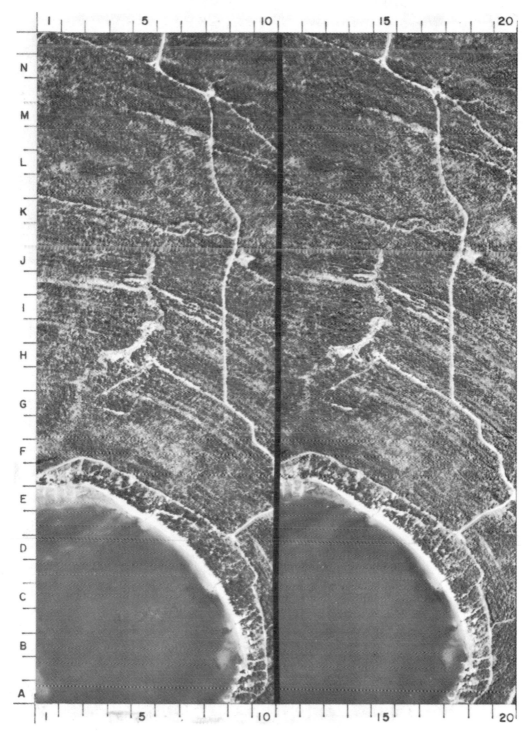

FIGURE 15.23 Stereopair: beach ridges.

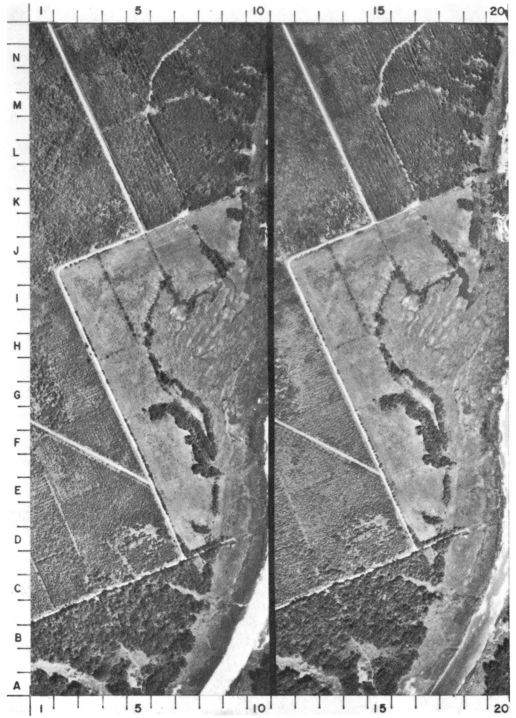

FIGURE 15.24 Stereopair: lacustrine plain and landslides. (Courtesy Bird & Hale Limited, 1982)

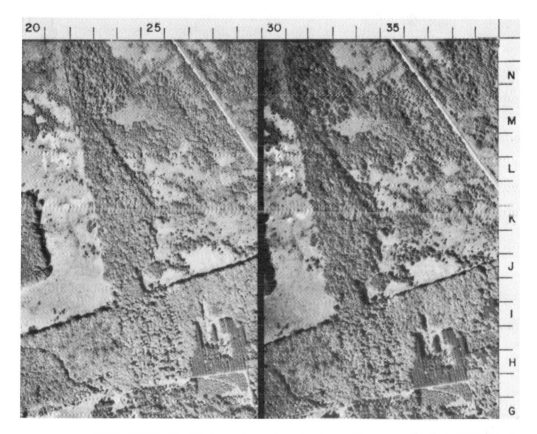

FIGURE 15.25 Stereopair: outwash plain. (Courtesy Bird & Hale Limited, 1982)

able terms. In this text, the fundamental components of a landscape are considered to be regional drainage, topography, local erosion, natural vegetation, the works of people, parent materials, and soils.

The topography (i.e., the geometry of the earth's surface) is primarily the result of the interaction of erosional and depositional agents, the nature of the rocks and soils, the structure of the earth's crust, the climatic regime, the vegetation, and the works of people. The topographic surface is, in effect, a synthesis of all environmental elements into a single expression. As such, it plays an important role in the evaluation of an area, because it provides a key for deducing the properties and processes at work in the region.

The topographic surface is composed of two fundamental elements: the *plan arrangement* and the *cross-sectional shape*. The plan arrangement of topographic surfaces is produced by the arrangement of terrain elements of different elevations (i.e., the topographic highs and lows) with respect to each other. The cross-sectional shape is the figure that would be in evidence if, for example, a topographic high were sliced vertically from crest to drainageway on either side.

In general, the plan arrangement of the topographic surface is most readily defined in terms of the regional drainage pattern, because the drainageways automatically define the low topographic positions, and the higher topographic positions must necessarily occupy the spaces between.

A *region* is any area that is characterized by repetitive features *at the scale of examination*. For example, on photographs at small scale, an area might display only two detectably different patterns as the result of differences in landscape features. At this scale, only two regions could be delineated because such regions can only be established from the distinctive patterns that are recognized. However, when large-scale photographs of the same area are examined, additional patterns may become evident. Thus, each of the two original regions might be subdivided into secondary regions. This sliding scale of definition can theoretically be carried to the maximum level of detail that is significant for the purpose at hand. There is accordingly *no quantitative implication* as to the size of the area defined by the term *region.*

15.15.1 Regional Drainage

Major drainage patterns (i.e., regional drainage) are the result of either present or past erosion or deposition. These patterns subdivide areas into regions throughout which the plan arrangements of features are essentially similar. As used in this text, regional drainage is not intended to imply that water will be found in the drainageways at all times or even at *any* time.

Many attempts have been made to classify regional drainage patterns. Probably the most thorough system is that shown in Figure 15.26. Like the shapes of uplands, stream patterns are infinitely variable. Thus, although the types are easy to identify in their extreme (i.e., *type*) form, they grade into each other in so complex a fashion that countless borderline cases occur. Nevertheless, when they can be identified, they constitute an invaluable tool in the hands of the photo interpreter.

A single drainage system may display several drainage types simultaneously. For example, the gross drainage pattern of a region may be parallel, while the low-order stream patterns (see Figure 15.22 at H+, 9 for illustration) may be pinnate. This situation

FIGURE 15.26 Drainage pattern types.

is very common in drumlin fields for example. In this connection, it is principally the low-order drainageways that are affected by those considerations that are of greatest interest to the analyst, that is, the type and structure of the parent materials and the soil types. High-order streams are generally much less delicately controlled by subtle variations in parent materials and soils and are therefore much less useful in the analysis.

15.15.2 Topography

Topography or the cross-sectional shape of uplands can be fully appreciated only by stereoscopic examination of airphotos. Furthermore, the size and the angles of component slopes are commonly necessary for adequate interpretation; therefore, the photo interpreter must know the basic mechanics of photogrammetry. For the purposes of this text, four variations of cross-sectional shape are recognized: undulating, rolling, crested, and blocky. Analysis of cross-sectional shape is important because each type has certain general implications as to underlying structure, parent materials, erosional or depositional history, and climatic regime. Although the infinite variation of cross-sectional shapes makes precise definition difficult, the following generalizations have been employed in this chapter.

1. *Undulating* shapes are composed of smooth sigmoidal curves of very small amplitude (Figures 15.21 and 15.25).
2. *Rolling* shapes are composed of smooth sigmoidal curves of relatively small amplitude (Figure 15.27).
3. *Crested* shapes are composed of relatively straight lines meeting at an apex. The topographic lows (i.e., the areas between the upland) need not be V-shaped. Crested shapes may occur with any amplitude (Figure 15.28).
4. *Blocky* shapes are composed of relatively straight lines with the tops of the topographic highs flat or nearly so. Such shapes may occur with any amplitude (Figure 15.24, near I, 14).

In addition to terms describing cross-sectional shape, certain general descriptive terms frequently employed in the literature are used in this text in a somewhat restrictive sense. These are valley, basin, lowland, plain, plateau, hill, mountain, ridge, and terrace. The terms are defined as follows:

1. *Valley*: a depression in the surface drained by a surface channel. No implication of scale or shape is intended.

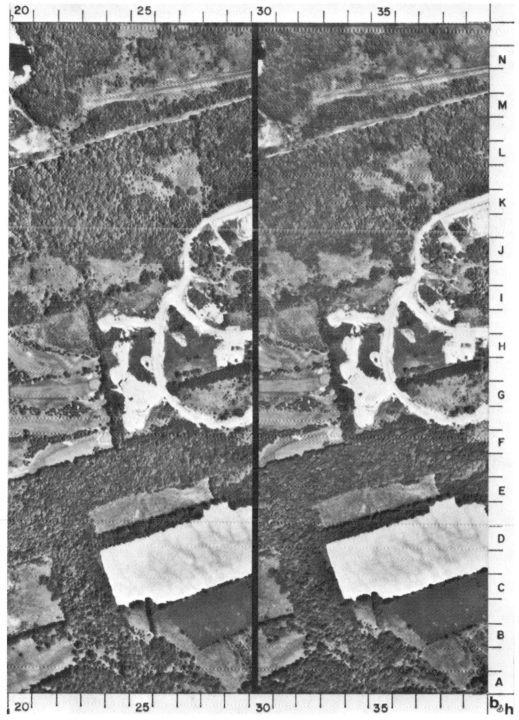

FIGURE 15.27 Stereopair: ridge moraine in upper portion. (Courtesy Bird & Hale Limited, 1982)

771

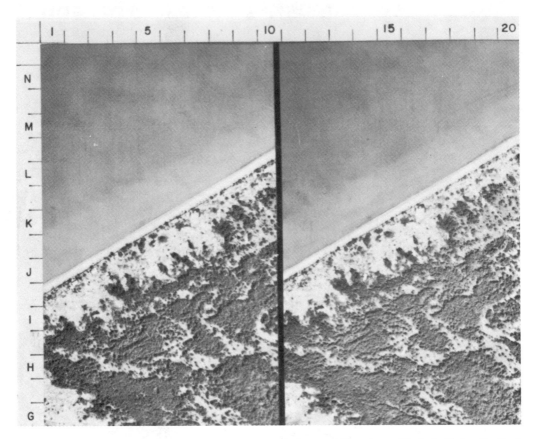

FIGURE 15.28 Stereopair: sand dunes. (Courtesy Bird & Hale Limited, 1982)

2. *Basin*: a depression in the surface not drained externally by a surface drainage channel. No implication of scale or specific shape is intended. Thus Death Valley and the small closed depressions common on outwash surfaces are both *basins*. It should be noted that valleys may (and commonly do) occur within basins, and vice versa.

3. *Lowland*: the portion of a valley that has been formed primarily by fluvial depositional processes.

4. *Plain*: a surface of less than 500 ft (150 m) of local relief and formed on essentially horizontal or very gently inclined materials.

5. *Plateau*: a surface of more than 500 ft (150 m) of local relief and formed on essentially horizontal or very gently inclined materials.

6. *Hill*: a surface of less than 1000 ft (300 m) of local relief and formed on (or from) materials having inclined bedding or having no bedding. This term might be used to describe such diverse features as dunes, pyroclastic cones, linear moraines, or surfaces eroded from intrusive igneous rocks.

7. *Mountain*: a surface of more than 1000 ft (300 m) of local relief and formed on (or from) inclined or nonstratified materials.

8. *Ridge*: an elongate topographic high. No implication of size is intended.

9. *Terrace*: an essentially horizontal surface of relatively low relief bounded on one or more sides by a topographic low and sometimes on the remaining side or sides by a topographic high.

15.15.3 Erosion

The minor features of the landscape produced by extremely localized concentrations of erosive energy are termed *local erosion* in this text. Three principal types are important to the photo interpreter: water erosion, wind erosion, and gravity erosion. The features produced by these three processes, as recorded on airphotos, are important aids in interpretation because they are commonly diagnostic of surface soil textures and, to a lesser extent, of soil profile characteristics and soil moisture conditions. The specific implications of each type are discussed in the following paragraphs.

Wind Erosion. Evidences of wind erosion are chiefly the following: blowouts, which are normally smoothly rounded and irregularly shaped depressions, and sand streaks, which are commonly light-toned and poorly defined patches. An example of a blowout is shown in Figure 15.25 at L, 20.

Water Erosion. Moving water is the major agent active in developing the surface configuration of the earth. Despite its awesome power in extreme forms, such as floods and tidal waves, it is also so delicately responsive to variations in environment that even modest changes in the material being eroded or the climatic regime can profoundly modify the surface expressions produced. As a result, the landscape patterns produced through the agency of moving water are of great importance to the photo interpreter. However, the interpreter must have at least a basic knowledge of the interrelations among climate, surface materials, surface configuration, and vegetation. This is because the relative importance of the factors influencing runoff varies depending on the specific environmental conditions that occur in any given area. Accordingly, then, interrelations can be stated only as general principles, which can be employed only with the clear understanding that the validity of each statement is sometimes modified by other conditions. These fundamental principles are as follows:

1. The amount and intensity of rainfall determine the amount of runoff. A large amount of rainfall of short duration may produce more runoff than the same amount over a longer period of time.
2. The amount of runoff is dependent on the amount of moisture in the soil prior to rainfall. A given rainfall on wet soil will produce more runoff than the same rainfall on dry soil, because a certain proportion of the incident water will be stored by the dry soil, whereas the wet soil has much less available storage capacity.
3. A noncohesive soil is removed more readily by runoff (i.e., eroded more readily) than a cohesive soil.
4. The greater the permeability of the soil, the less the surface runoff is.
5. In general, the greater the density of vegetation, the less the runoff is for a given quantity of incident water because of wetting of the vegetation and litter.
6. The steeper the slope is, the greater the runoff.

In general, there are three relatively distinct processes by which moving water erodes material and in so doing modifies the shapes composing a landscape. These processes are called *sheet, rill,* and *gully* erosion. While a discussion of all three is beyond the scope of this text, gully erosion is the best indicator of soil types and the most visible on the airphotos, as discussed next.

Gully erosion consists of the incipient or poorly developed channels tributary to the main channels. Gully erosion is the advanced stage of rill erosion, which usually follows sheet erosion. These incipient channels, the headmost expression of the entire drainage system, are known by a variety of local names, gully being perhaps the most common. The shape of a gully in plan, profile and cross section is closely related to the materials in which the gully is being formed. It is influenced by the texture, cohesiveness, and permeability of both the surface soil and subsoils.

The basic gully shapes, classified according to the gully cross section and gradient, are illustrated in Figure 15.29. Although

Symbol Used in Text	Generalized Sketch	Gradient	Climate	Parent Materials	Symbol Used in Text	Generalized Sketch	Gradient	Climate	Parent Material
		Simple types			C		Gentle	Humid	Unconsolidated: fine-grained; impermeable in both soil zone and sub-soil poor; commonly associated with sheet erosion; in general, the broader the gully the more impermeable the materials. Most common materials: clay, silty clay.
\	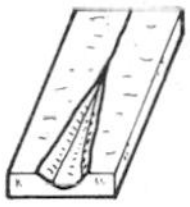	Steep	Humid	Unconsolidated; relatively homogeneous; coarse-grained; permeable more common in non-cohesive materials, but may also occur in some cohesive materials; commonly associated with rill erosion. Most common material: sand. Consolidated: virtually all materials					
							Compound Types		Unconsolidated: thin, impermeable, cohesive, fine-grained stratum overlying coarse-grained, permeable, noncohesive stratum; vertical sides may develop in coarse-grained, permeable, noncohesive material if closely bound by root mat. Consolidated: thin, cohesive strata, grain size not significant overlying rock.
⊢		Gentle	Humid	Unconsolidated: thick; relatively homogeneous; cohesive; fine-grained; permeability of soil zone fair to good, but poor in subsoil, very common in thick loess. See Fig. 20. Most common material: silt. Consolidated: rare; sometimes developed along joints in limestone.	FV		Steep	Humid	
					CF	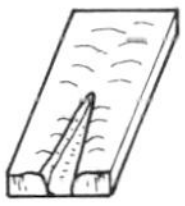	Gentle	Humid	Unconsolidated: very fine-grained (usually clay), cohesive, impermeable, soil zone overlying fine-grained, (usually silt), impermeable, cohesive stratum. Consolidated: fine-grained, impermeable, cohesive but unconsolidated surface stratum overlying rock; usually strongly angular in plan.

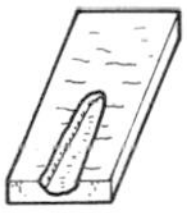

FIGURE 15.29 Gully characteristics. (Courtesy U.S. Army Corps of Engineers, Vicksburg, Mississippi)

gullies are important indicators of soil characteristics, they should not be used alone to interpret the properties of soils. This is because gully forms are to some extent controlled by vegetative cover, topography, and land use. For example, the dense root mat of some sod-forming grasses may bind sandy soils so tightly that gullies may be the FV type (see Figure 15.29), even though the soils are sandy and well drained throughout. In Figure 15.30 at F, 15, there is an example of gully type V in coarse-grained soils.

Gravity Erosion. Large-scale features caused by mass-wasting or gravity erosion may be discernible locally and yet have regional importance since they may reflect general environmental condi-

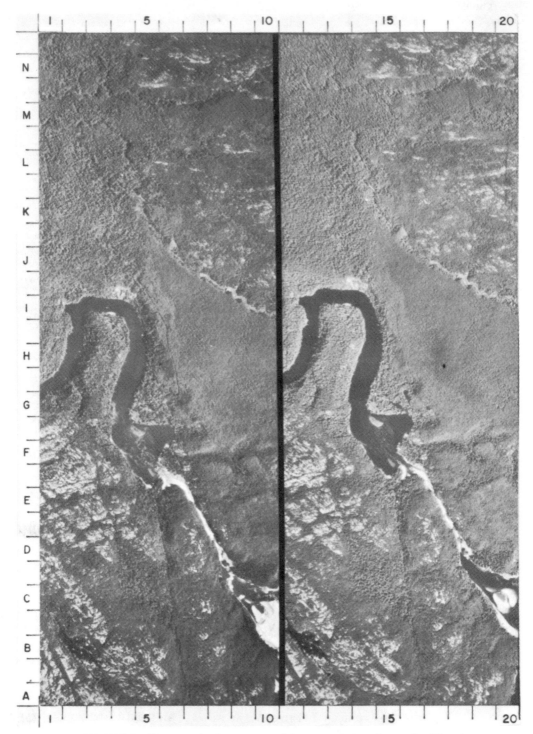

FIGURE 15.30 Stereopair: terraces and igneous intrusive rock. (Courtesy Bird & Hale Limited, 1982)

tions, including geology, vegetation, climate, and physiography. In this category, landslides, rock glaciers, talus material (due to avalanches and rock falls), and land subsidence phenomena can be included. These features may be obvious from airphotos simply because of their aerial extent and should be considered important, especially if hazard lands and land-use activities are targets of airphoto or field-based studies. For example, parts of the landscape around the St. Lawrence River are dominated by a thick surface deposit of marine clay (Leda clay). Due to the highly unstable and sensitive internal structure of these materials, under current environmental conditions the materials are highly susceptible to failure by liquefaction and sliding. In several locations, the potential hazard is indicated by large bowl-shaped slide scars along valley walls. These features may result from periods of heavy rainfall or simply the vibrations produced by heavy vehicles. Similar slide features occur along the coast of Alaska and owe their presence to earthquake activity. Figure 15.24 illustrates the pattern of a previous landslide throughout the area around H, 16.

15.15.4 Vegetation

Vegetation is commonly a delicate indicator of environmental conditions and is therefore often of great value to the photo interpreter. However, it is affected by such a multiplicity of environmental factors that interpretation on the basis of vegetation alone must be made with great caution. The interpreter must have an extensive background knowledge of the region, especially of its climatic regime.

In many cases, individual plant species are indicators of specific and very narrow ranges of environmental conditions, such as soil type and moisture relations. However, species cannot normally be identified from airphotos except at very large scales, although notable exceptions occur. For example, a distinctive shape of crown may positively identify a particular species in some places. Inability to identify particular species forces the interpreter to rely on general patterns. These are so numerous and so widely variable from region to region that all are not described in this chapter. However, some notable examples of general patterns are the following:

1. The concentric zones of vegetation around lakes and ponds, especially in cold humid regions. For example, a common zonation around a lake is, from the water outward, the sedge zone, heath zone, and spruce zone. This zonation indicates successive reductions in waterlogging of the soil and is also

to a degree a measure of the amount of organic material in the soil. The sedges may actually be floating, the heaths normally rest on peat, and the spruce grows in highly organic silty soils or peat.

2. A combination of tall grass or sedges and a reticulate drainage pattern commonly indicates fine-grained, highly organic soils subject to repeated flooding.

3. In the hardwood forest belt, the oaks, such as scarlet oak and scrub oak, that hold their leaves in winter are indicative of dry soils such as sands. Stands of poplar, birch, maple, walnut, and red and white oaks indicate moist, fine-textured soils. Sycamore, cottonwood, and willow indicate wet soils. Large-scale photographs are necessary to identify tree species.

4. Organic terrain can readily be differentiated as a contrast with the surrounding forested or cultivated land. Tree heights become uniform and a single species dominates on deep organic soil.

In general, the photo interpreter should be alert for any change or modification of vegetation pattern, because this is almost invariably an indication of changing environment that may influence engineering considerations.

15.15.5 Cultural Influences

The practices of people in adjusting themselves to their environment can be important in evaluating conditions from aerial photographs. Terrace farming, dead furrowing, and gully erosion control usually signify a friable soil on less pervious subsoil. Windbreaks and strip cropping are man-made methods of controlling wind erosion. The various pits and excavations made by people may be significant to interpretation of parent material types and soil textures if other features are visible that signify methods or needs for processing. Although practices of people are a part of or are related to the natural elements of the environment, the individual study of the cause and effect of such practices may have an important bearing on interpreting soil conditions.

15.15.6 Seasonal Influences

The comparison of any one photograph or photo mosaic with another of the same area taken in different years or at different times of the year may, at first glance, suggest completely different areas. This effect is most pronounced in areas having climatic regimes involving significant seasonal changes. Thus, the effect may not be detectable in the humid tropics, and it may be very minor in temperate-zone deserts and even in the boreal forests, where dense stands of relatively unchanging evergreen trees mask the ground. However, in humid-temperate regions, and especially in those under intense cultivation, not only the vegetation and soil moisture patterns may change, but even the outlines of the fields themselves. The following paragraphs discuss seasonal photo variations and the effects of scale that are applicable, in general, only to regions having humid-temperate climates.

1. *Summer.* Photographs are dominated by dark tones produced by intensely green, growing crops and heavily foliaged trees. Soil moisture content is normally relatively low, so bare soil tends to photograph light gray. Field patterns are somewhat subdued because fields are green.

2. *Autumn.* Field patterns are relatively distinct because of various stages of crop development and harvesting. Variations in tone resulting from variations in soil moisture content are subdued.

3. *Winter.* Photos are drab and dull, with some field patterns indistinct. Mottling due to variations in soil moisture is practically nonexistent, and bare ground tends to photograph dark because soil moisture content is uniformly high. Low angle of incident light makes sharp shadow patterns in wooded areas, producing a distinctive form of flecked texture.

4. *Spring.* Field patterns are sharp and distinct due to differences in the state of tillage and crop development. Mottled textures due to differences in soil moisture content are very distinct; high topographic positions, even those only a few meters or centimeters above the adjacent low positions, tend to photograph light, while low topographic positions photograph dark because of large local variations in soil moisture. This is the best season for obtaining soil pictures because all differences in the relations between soil types and soil moisture are emphasized.

Seasonal variations can be observed at large scales as well as small scales, but the features by which the seasonal changes are identified are quite different. For example, the subdued field patterns observed in winter photographs at small scales are produced in part by the lack of foliage on the trees, but the individual trees cannot be seen. On the other hand, the leaflessness is clearly apparent on individual trees in photos taken at large scales, and the resultant flecked pattern is distinctive.

15.15.7 Elements Derived by Deduction

In general, the elements just discussed are those that permit a direct interpretation of the patterns or textures on an airphoto. However, certain attributes of the landscape do not permit this direct interpretation, but are nevertheless important. In fact, the photo and terrain elements just discussed are valuable chiefly because they provide the data from which deductions can be made of the nature of the factors directly affecting terrain characteristics. The aspects of terrain important to engineers are those that permit identification of soil characteristics.

To deduce such factors, an intervening step between the directly interpretable patterns and textures (the photo and terrain elements) and those factors identified purely by deduction (soil characteristics) must be established. This step consists of the identification of parent materials. The materials from which the soils of a given landscape were derived are the parent materials of that landscape (Tables 15.1 and 15.2). These materials are of fun-

TABLE 15.1 *Classification of Unconsolidated Parent Materials*

Eolian materials (stratified or nonstratified)	Lacustrine materials (well stratified)
Coarse-grained: Sand	Coarse-grained
Fine-grained: Silt (loess)	Nonorganic: Sand
Glacial materials (mostly nonstratified)	Organic: Sand
Coarse-grained: Gravel (boulders, cobbles, sand)	Fine-grained
	Nonorganic: Silt, clay
Fine-grained: Silt, clay	Organic: Silt, clay
Fluvial materials[a] (crudely stratified)	
Coarse-grained	Littoral deposits (crudely to well stratified)
Nonorganic: Gravels, sand	Coarse-grained
Organic: sand	Nonorganic: Gravel, sand
Fine-grained	Fine-grained
Nonorganic: Silt	Nonorganic: Silt
Organic: Silt, clay	Organic: Silt

[a]Fluvial materials, as used in this text, are the many materials deposited from streams or streamlike water bodies. They thus include the so-called "glacial-fluvial" deposits, which are simply stream or flood deposits formed downgrade from an ice mass.

TABLE 15.2 *Classification of Consolidated Parent Materials*

Sedimentary Rocks	Igneous Rocks	Metamorphic Rocks
Stratified[a]	Stratified	Stratified
Limestone	Pyroclastics	Quartzite
Dolomite	Rhyolites	Slate
Shale	Basalt	Marble
Sandstone		
Evaporite		
Nonstratified	Nonstratified	Nonstratified
Limestone	Granitelike	Quartzite
Dolomite	Gabbrolike	Marble
Shale		Gneiss
Sandstone		Schist
Evaporite		

[a]Stratification is used in this text in a sense applicable specifically to photo interpretation. In this context, a stratified material is one that exhibits traces of bedding within the region under examination. Thus, if the total relief of the area under examination is 150 ft (45 m.) and the bedding planes of the rock are 200 ft (60 m.) the rock may exhibit no trace of stratification in the area under examination. In this text, such rock is designated "nonstratified."

damental importance because they control to a considerable degree, or at least strongly influence, the characteristics of the soil. Knowledge of the nature of the parent materials, coupled with subsidiary information on climatic regime, structure of the crust, erosive history, and cultural history, makes it possible in many cases to estimate the resulting soil properties with considerable precision. Thus every effort should be made to establish the nature of the parent materials of a region before an attempt is made to interpret the photos. Failing this, parent materials must be established by deduction by applying the implications of topographic shape, drainage properties, and the like. The terrain or pattern elements discussed in this subsection are tabulated in Table 15.3.

15.16 LANDSCAPES IN UNCONSOLIDATED MATERIALS*

These landscapes are in general composed of materials that are unconsolidated. Most, but by no means all, of them owe their present configuration chiefly to depositional processes. Classifica-

---

*In this text, unconsolidated materials are those that can be broken or deformed by hand.

Table 15.3 Generalized Correlations of Landscape and Interpretive Elements (Vegetation and cultural practices not included)

Representative Landscape	Drainage	Topography	Local Erosion	Parent Materials	Soil Profile
Unconsolidated Materials					
Dune fields	Drainageless	Crested, rolling; many basins uplands asymmetric; lee slope 34°	Gullies absent; blowouts, sand smears	Eolian sand	None
Loess surfaces	Regionally dendritic, locally pinnate	Rolling, blocky; relief less than 150 m	F-Type gullies, flat grade	Eolian silt	30 to 60 cm thick; uniform, fine-grained
Old ground moraines (Illinoian stage)	Dendritic	Undulating; relief less than 12 m	Gullies C-type, flat grade; white-fringe	Glacial till; mostly fine-grained	0.3 to 5 m thick; mostly fine-grained organic in lows in some places
Young ground moraines (Wisconsin stage)	Dendritic, some kettle hole and deranged	Undulating; relief less than 15 m	Gullies C-type, flat grade	Glacial till; mostly fine-grained, heterogeneous	0.6 to 1.5 m thick; mostly fine-grained; organic soils in lows
Ridge moraines	Dendritic or deranged	Rolling or crested: random; some basins: relief less than 60 m	Gullies V-type, steep grade; saucer-shaped, flat grade	Glacial drift; heterogeneous	1.2 to 2.5 m thick; heterogeneous
Kettle-kame moraines	Kettle hole or deranged; many small lakes	Rolling; basins numerous; random: relief less than 30m	Gullies rare; V-type, steep grade	Fluvially deposited glacial materials; coarse-grained, crudely stratified	0.3 to 1 m thick; mostly coarse-grained; organic soils in kettles
Drumlin fields	Deranged; commonly ditched	Uplands rolling, parallel, drop-shaped; lowlands undulating random	Gullies rare, C-type	Glacial drift; uplands mostly clay; lowlands fine-grained	0.6 to 1.2 m thick; fine-grained

Landform	Drainage	Topography	Gullies	Material	Thickness
Kame	Usually drainageless	Crested: conical uplands	Gullies rare; V-type steep grade	Fluvially deposited glacial material; coarse-grained, crudely stratified	0.3 to 1 m thick; mostly coarse-grained
Eskers	Usually drainageless	Crested: sinuous ridge	Gullies rare; V-type, steep grade	Fluvially deposited glacial material; coarse-grained, crudely stratified	0.3 to 1 m thick; mostly coarse-grained
Alluvial fans and aprons	Fan distributaries, radial parallel; parallel; channels braided	Blocky; regional slope 5° to 7°	U-shaped, V-type; saucer-shaped gullies; eolian erosion	Alluvial materials; coarse-grained near upper edge; fine-grained near lower; crudely stratified	None or very thin; desert pavement in some places; sand and gravel in washes
Outwash surfaces (glacial-fluvial)	Regionally crudely parallel; locally dendritic or kettle hole	Undulating; relief less than 15 m	Gullies rare; V-type steep gradients; rare deflation basins	Fluvial materials; coarse-grained; crudely stratified	0.6 to 1.8 m thick; coarse-grained in uplands; fine-grained and organic in lowlands
Floodplain (ridge and swale)	Collinear, arranged in concentric areas	Undulating; relief less than 15 m	Gullies absent	Fluvial materials; coarse-grained in uplands, fine-grained in lowlands	Thin; coarser on uplands; finer and organic on lowlands
Floodplain (natural levee)	Parallel or absent	Undulating or blocky; relief less than 6 m	Gullies V-type steep grade	Fluvial sand, silty sand	Thin, sandy silt and silty clay
Floodplain (backswamp)	Dendritic, reticulate, deranged; channels meandering, contorted, lakes common	Undulating, nearly plane	Gullies rare; C-type, flat-grade	Fluvial and lacustrine silt, clay	Thin; organic, fine-grained, impermeable
Floodplain	Channels meandering, oxbow lakes common		Gullies absent	Fluvial and lacustrine silt, clay	Thin; organic, fine-grained, impermeable *(continued)*

Table 15.3 *(Continued)*

Representative Landscape	Drainage	Topography	Local Erosion	Parent Materials	Soil Profile
Floodplain (high-level floodplain)	Dendritic, crudely collinear or parallel	Undulating, blocky; relief less than 9 m	Gullies common; V-type, steep grade	Fluvial materials; sand, silt, clay	0.6 to 1.8 m thick; variable; mostly fine-grained; organic
Deltaic surfaces	Distributary; channels widely variable resembles floodplain types	Undulating; resembles floodplain types	Gullies absent or rare	Fluvial materials; resembles floodplain	Thickness variable; resembles floodplain types
Terraces	Drainageless; some swallow hole; anomalous	Undulating; small basins common; relief commonly less than 9 m	Gullies rare; V-type, steep grade; F-type, moderate grades	Commonly coarse-grained; crudely-stratified; fine-grained in some places	0.6 to 1.5 m thick; mostly fine-grained; organic in basins
Coastal alluvial plains (coastal swamps)	Dendritic, some deranged and reticulate	Undulating; relief less than 15 m	None	Fluvial silts and clay; stratified	Thick: organic and fine-grained; impermeable
Coastal alluvial plains (marine terraces)	Regionally crudely parallel, locally dendritic	Undulating or blocky; relief less than 15.2 m	Gullies rare; F-type, moderate grade; V-type, steep grade	Gravel, sand, silty clays; crudely stratified	Thick; coarse- to fine-grained; commonly includes hardpan
Coastal alluvial plains (undifferentiated surfaces)	Dendritic	Updulating, rolling, or blocky; relief less than 30 m	Gullies of all types common	Variable; partly to completely hardened in some places	Thick; commonly fine-grained; may include hardpan
Beds of perennial lakes	Ditches or tiled; major stream channels commonly meandering	Undulating; relief less than 3 m	Gullies C-type flat grade; some sheetwash	Lacustrine materials; fine-grained, stratified	0.6 to 1 m thick; fine-grained, impermeable; organic in some places
Playas	Channelless	Undulating; relief less than 1.5 m	Gullies absent; deflation hollows rare	Lacustrine materials; silt, clay, evaporites; stratified	Absent to thick; mostly fine-evaporites in places

Tidal flats	Reticulate, channels meandering, contorted	Undulating; relief less than 1.5 m	Ephemeral runoff channels	Littoral materials; sand, silt, silty sand	Thin or absent; organic sands or silts
Beach ridges	Channelless basins; collinear or parallel	Undulating; sinuous ridges; relief less than 9 m	Gullies absent; deflation hollows common	Littoral materials; coarse-grained crudely stratified	Absent to 1.5 m thick; coarse-grained on uplands; fine-grained
Pyroclastic cones grades	Parallel radial	Crested; relief up to several hundred feet	Gullies V-type, steep	Volcanic ejecta; ash, lapilli, cinders, blocks	Absent to very thick; coarse- to fine-grained, stratified
Consolidated Materials					
Limestone plains	Swallow hole; locally dendritic; rare through-flowing streams	Undulating or blocky; basins rare to common; relief less than 15 m	Gullies common; C-type or V-type	Limestone; strata horizontal	None to 5 m thick; fine-grained, permeable if undisturbed
Shale plains in humid regions	Regionally dendritic; locally pinnate; channels closely spaced	Rolling; slopes smoothly sigmoid; relief less than 120 m; valley bottoms flat	Gullies common; C-type if relief low, V-type if relief high	Shale; strata horizontal	0.6 to 1 m thick; fine-grained
Shale plains in arid and semiarid regions	Regionally dendritic or parallel; locally pinnate	Crested; slopes straight, up to 30°; relief less than 150 m; valley bottoms flat	Gullies incredibly numerous; V-type in arid regions. C-type in semiarid	Shale; strata horizontal	Absent or very thin; fine-grained
Sandstone plains	Regionally dendritic; locally rectangular or trellis	Rolling; valley bottoms slightly concave upward; relief less than 150 m	Gullies rare; V-type, steep grade; rare deflation basins	Sandstone; strata horizontal	0.3 to 3 m thick; mostly coarse-grained, permeable

(continued)

Table 15.3 *(Continued)*

Representative Landscape	Drainage	Topography	Local Erosion	Parent Materials	Soil Profile
Limestone-shale plains	Regionally dendritic; slightly pinnate locally	Blocky; "scarplets" on valley sides; relief less than 150 m; valley bottoms flat	Gullies have "stepped" profiles; V-type or C-type with C-type generally on upper topographic high surfaces	Shale and limestone; strata horizontal	0.6 to 2 m thick; fine-grained
Sandstone-shale plains	Regionally dendritic; locally pinnate rectangular	Blocky; "scarplets" on slopes; relief less than 150 m	Gullies V-type, two-layered	Shale and sandstone; strata horizontal	0.6 to 2 m thick; coarse- to fine-grained, relatively impermeable
Limestone plateaus	Swallow hole on uplands; lowlands have through-flowing streams	Blocky, crested; basins common; relief exceeds 150 m; steep valley walls; valley floors flat	Gullies rare; C-type, rockfalls common	Limestone; strata horizontal	0.6 to 5 m thick on uplands; fine-grained, permeable if undistrubed; thin or absent on valley side
Sandstone plateau	Regionally dendritic; locally dendritic, rectangular, parallel	Blocky, crested; valley bottoms slightly concave upward; valley cross section usually opened U- or V-shaped	Gullies rare; V-type, steep grade	Sandstone; strata horizontal	1 to 1.5 m thick; coarse- to fine-grained
Sandstone-shale plateaus	Regionally dendritic, locally dendritic, rectangular parallel	Blocky, crested; "scarplets" on valley sides; slopes steep	Gullies V-type; steep grades; "stepped" profiles; sheet and rill erosion common	Interbedded shales & sandstone; strata horizontal	0.6 to 2 m thick on uplands; coarse-grained; thin to absent on valley sides
Limestone hills (humid climate)	Regionally trellis; locally trellis, dendritic, swallow hole, parallel	Rolling; small elongate basins; ridges mostly asymmetric	Gullies rare; C-type; rockfalls common at base of valley sides	Limestone; may include thin shale or sandstone beds; strata inclined	Thin on uplands; thick on lowlands fine-grained, permeable if undisturbed

Limestone hills (arid climate)	Regionally trellis; locally trellis, dendritic, parallel; channels braided	Crested: ridges steep, rough, "broken", mostly asymmetric	Gullies V-type, steep; "stepped" profiles; rockfalls common	Limestone; thin shale or sandstone beds; strata inclined	Thin or absent on uplands; stony
Dolomite hills (humid climate)	Trellis: locally parallel	Rolling; ridges mostly asymmetric	Gullies rare; C-type; rockfalls common on steep slopes	Dolomite; interbedded shale or sandstone; strata inclined	Thin on uplands; stony; thick on lowlands, silty, clayey
Shale hills (humid climate)	Trellis	Rolling; ridges subdued	Gullies generally C-type; some broadly V-type	Shale; thin interbeds of sandstone or limestone; strata inclined	Thin on uplands; fine-grained; includes shale chips; 0.6 to 1.5 m thick on lowlands; fine-grained
Shale hills (arid climate)	Trellis	Crested; slopes straight, less steep than limestone	Gullies generally C-type; some V-type, steep grade	Shale; thin interbeds of sandstone or limestone; strata inclined	Thin or absent; mostly shale chips
Sandstone-shale hills	Regionally trellis; dendritic on valley floors	Crested: ridges minutely serrated at crest	Gullies rare; C-type shaped or broadly F-type; rockfalls common on steep slopes	Mostly sandstone in ridges; mostly shale in lowlands	Thin to thick; coarse-grained on uplands; fine-grained on lowlands
Limestone-shale hills	Trellis: dendritic on floors of large valleys	Rolling, crested; valley floors undulating	Gullies rare on uplands V-type; profiles commonly "stepped," rockfalls common on steep slopes; gullies on lowlands C-type	Ridges mostly limestone; valleys mostly shale	Thin on uplands; stony; 1 to 2 m thick on lowlands; fine-grained
Basalt plains and plateaus	Swallow hole, deranged; through-flowing streams deeply incised	Regionally blocky; locally rolling, many basins; surface rough, botryoidal; large valleys steep-sided; columnar jointing common	Gullies rare; V-type, grades moderate	Basalts, andesites, rarely rhyolites; strata horizontal	0.5 to 3 m thick; sand, sandy silt, silty

(continued)

Table 15.3   (Continued)

Representative Landscape	Drainage	Topography	Local Erosion	Parent Materials	Soil Profile
Granite hills	Dendritic, rectangular; low-order channels "surround" uplands	Rolling, valley sides sigmoid; uplands randomly arranged	Gullies mostly compound; C-type near heads; V-type near mouths; sheet erosion common	Granite and granite-like rocks; nonstratified	2 to 3 m thick; silty sand; clayey sand; finer grained in lowlands
Slate hills	Dendritic; trellis	Rolling, crested; valley floors flat	Gullies C-type on gentle slopes; F-type on steep slopes	Slates, phyllites; strata horizontal	0.6 to 1.2 m thick; fine-grained; relatively impermeable
Gneiss hills	Dendritic; strongly resembles pattern typical of granitic areas; "enclosed" areas oval, rectangular	Rolling; valley sides sigmoid; tendency toward preferred orientation	Gullies compound; C-type near heads; F-type near mouths	Gneiss and gneiss-like rocks; nonstratified	2 to 5 m thick; silty, sand; finer grained in lowlands
Schist hills	Regionally dendritic; locally rectangular, dendritic, trellis	Rolling where relief is low; crested where relief is high; uplands tend to be angular	Gullies V-type	Schistose rocks; nonstratified; chlorite, talc, hornblende schists if uplands rounded and subdued; quartzose schists if uplands angular	2 to 4 m thick; coarse- to fine-grained; finer grained in lowlands

tion criteria are, first, the mode of deposition of the materials composing the component landforms and, second, the topographic configuration (i.e., the shapes of the landforms and the ways in which they are arranged with respect to each other within the landscape).

Although the configurations on which the categories in this group of landscapes have been based are chiefly the result of depositional processes, almost all have been modified to a greater or lesser degree by erosional processes. For example, in the case of loess surfaces, the existing configuration is almost entirely the result of erosional processes, whereas in the case of natural levees the configuration is almost entirely depositional.

In the paragraphs that follow, each landscape is described in terms of landscape elements and interpretive elements. Landscape elements are discussed for regional drainage, topography, local erosion, natural vegetation, land use, and construction practices. Interpretive elements are discussed for parent material.

15.16.1 Landscapes in Eolian Materials

Eolian landscapes are those formed from unconsolidated materials deposited by wind action.

Dune Fields. Major areas of sand dunes occur on every continent and in virtually every climatic region. Although the largest areas are associated with deserts, extensive dune belts are also found along many coastlines, on the shores of large lakes, and in some instances even along large streams. Dunes are exceedingly variable in size, ranging from a few meters to as much as 600 ft (180 m) in height, and in area from a few hundred square meters (usually isolated barchans) to huge dune fields covering hundreds of square kilometers. Surface drainage within the dunes is generally absent. The complete absence of scour channels is a primary recognition feature of all types of dunes, whether active or fixed. Dunes occur in a wide variety of types, with a corresponding variety of topographic configurations. The most common types are the following:

1. Isolated barchans, in which each dune occurs independently of all others.
2. Parabolic dunes, which occur as hairpin or crescent shaped.
3. Transverse dunes, which occur as series of ridges at right angles to the direction of the prevailing wind.
4. Longitudinal dunes, which occur as a series of ridges parallel to the direction of the prevailing winds.

5. Star dunes, which are normally very large peaks from which four or more ridges radiate. The cross-sectional shapes may include combinations of rolling and crested.

6. Complex dunes, in which dunes of indeterminate or gradational forms occur in such proximity that the forms touch or interpenetrate.

Regardless of type, a primary recognition feature is the lee slope, which remains virtually constant at approximately 61%, which corresponds to the angle of repose of dry sand (34°). Water erosion normally does not occur, because infiltration and internal drainage are so good that surface flows cannot exist. Dunes are, of course, exceedingly susceptible to wind erosion. All active dunes owe their shape to a combination of eolian erosion and deposition. Any break in the vegetation cover of dunes stabilized by vegetation may result in rapid eolian erosion. These areas, called *blowouts*, almost always photograph in brilliant white tones, as illustrated in Figure 15.28 (G, 2), contrasting sharply with the gray tones of the fixed dunes.

Even areas of active dunes are rarely entirely devoid of vegetation. Note that the obviously active dune illustrated in Figure 15.28 (G, 2) actually supports a few scattered shrubs and trees. However, the overwhelming impression is one of extreme barrenness. The excellent internal drainage results in a general lack of soil moisture, so dunes tend to be grass-covered even when adjacent areas of other soil types are tree-covered.

In humid and subhumid climates, dune areas are frequently used as pastures. Dune fields are also frequently used for recreational purposes. Evidence of hiking and off-road vehicle trails are often seen as white lines under the tree canopy.

Dune sands are usually uniform in grain size. In general, about 90% of the material passes the No. 40 sieve (0.42 mm), and from 20% to 10% passes the No. 200 sieve (0.074 mm). Active dunes have no developed soil profiles; the materials are essentially uniform from the surface to the base of the dune.

Low topographic positions in fixed dunes in some instances contain minor amounts of organic material. Low topographic positions in dune areas may also have high water tables during wet periods.

15.16.2 Loess Surfaces

Large loess areas occur in the United States in Mississippi, Illinois, Iowa, Kansas, Nebraska, and Washington. The material is apparently related to large streams such as the Mississippi, Missouri, and Snake rivers. Elsewhere in the world, extensive areas of loess occur in Argentina, across central Europe and extending into southern Russia, and across south-central Asia and into northern China.

Loess is exceedingly variable in thickness, varying from a few inches to over 60 m in a few places. In general, it must be several meters thick before it produces a distinctive landscape pattern. There is no correlation between loess and the materials on which it lies. Therefore, no further discussion of this landscape will be undertaken in this text.

15.16.3 Wisconsin Ground Moraines

This type of ground moraine is widely distributed in the northern United States and Canada. The deposits vary in thickness from a few centimeters to several hundred meters, and in many places the larger streams have cut entirely through the glacial deposits and flow in bedrock channels.

The gross drainage pattern is chiefly dendritic. Low-order drainage patterns are commonly dendritic or crudely developed centripetal types around lakes. Often the drainageways are widely spaced and poorly developed. Both surface and internal drainage are so poor in many places that it has been necessary to improve them by either ditches or tiles or both (see Figure 15.21). A *phantom* drainage pattern frequently develops in low-order drainage swales. This feature is particularly noticeable in Figure 15.21.

In the upper reaches of the watersheds, surface water penetrates the *A* and *B* horizons and moves along the top of the impervious *C* horizon. There is a concentration of moisture in the soils of the lows, and no crosion channel develops. Rather, an interconnected swale system is developed by sheet erosion in the dendritic pattern. The result is a hazy, gray-toned dendritic pattern, which is phantom drainage.

The cross-sectional shape of ground moraines is somewhat variable. The most common expression is undulating. It generally contains undrained basins and gives an overall visual impression of randomness. In general, the local relief is very small, and rarely exceeds a few meters or few tens of meters. The most common type of gully is *C*-type with very flat gradients. Other young ground moraine areas, however, may have few or no gullies. The combination of sheet erosion and poor surface and subsurface drainage results in the typically mottled pattern (or texture) illustrated in Figure 15.21.

Ground moraines are almost entirely under cultivation. In northern areas, the poor site drainage resulted in the accumulation of extensive organic soil deposits called *fens*, which range in depth from less than 1 m to more than 6 m. Where the demand for agricultural land is high and the organic soil is shallow, it can be removed and the underlying ground moraine tilled. Field boundaries are normally rectangular and generally pass indiscriminately across both drainage and topographs. In the more humid regions (e.g., the Great Lakes basin), the crops are diversified, with resulting wide differences in photo tone and texture, although there is generally at least some trace of a linear texture developed as a result of plowing and cultivation, as illustrated in Figure 15.21.

Roads normally are independent of drainage or topography. Fields may be ditched and/or tiled in some regions to improve drainage. Figure 15.21 has both natural and manmade drainage features, including tile drainage at C, 24.

The parent materials are highly variable both vertically and horizontally. In most areas, the most common type is a *till* (i.e., a heterogeneous mixture composed of clay and silt with varying admixtures of sand and gravel). In places, the parent material is considerably coarser, being chiefly sand and gravel with minor amounts of silt and clay. There is a general correlation between parent material type and both topographic expression and drainage pattern. In general, basined surfaces or deranged drainage patterns imply coarse-grained materials, and valleyed surfaces and dendritic drainage patterns imply fine-grained materials.

The contrast between high and low topographic positions is usually strikingly displayed by variations in tone on the photographs. The organic and silty or clayey soils of the low positions are both darker in color and hold moisture longer than the slightly more coarse-grained soils of the high positions. They therefore photograph dark, resulting in the typical mottled pattern or texture illustrated in Figure 15.21.

15.16.4 Ridge Moraines

As used in this study, a ridge moraine is an accumulation of glacially deposited materials that forms a band or belt of hills or ridges that rise appreciably above the adjacent landscape. It includes terminal, lateral, interlobate, and any other type of moraine made of soil that was released from the ice through ablation.

These landscapes are exceedingly variable in both surface expression and internal composition. The same moraine system may exhibit several variations in shape and parent materials within the space of a few miles.

The gross drainage pattern is normally kettle hole, as illustrated in Figure 15.31 (N, 15). Low-order drainage patterns are commonly dendritic, but well-developed radial patterns may occur around individual topographic highs.

The soils in ridge moraines are frequently generally permeable, resulting in a near horizontal regional water table. The result is that lake and organic terrain surfaces usually appear at about the same elevations throughout the landscape.

Where the parent materials are coarse grained, the cross-sectional shape is usually rolling and basins are relatively frequent, although frequency varies widely from place to place. Slopes are generally fairly steep, and in some cases reach approximately 60%. The moraine illustrated in Figure 15.31 is relatively coarse grained. Local relief of ridge moraines may range from 8 to about

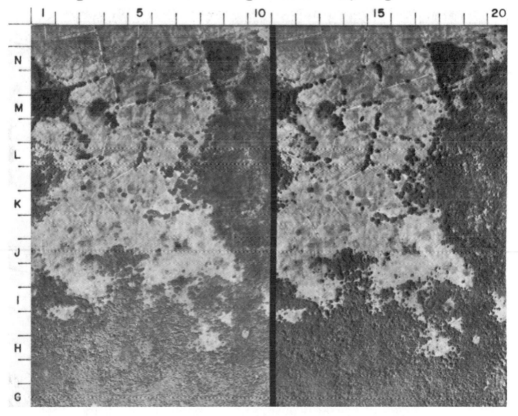

FIGURE 15.31 Stereopair: ridge moraine. (Courtesy Bird & Hale Limited, 1982)

100 m, with the coarser-grained varieties tending to exhibit more visible relief.

Where the parent materials are generally coarse grained, gullies are V-type with steep gradients. In some places, where coarse-grained parent materials exhibit such excellent internal drainage that surface flows do not develop, even relatively steep slopes may show no trace of gullying.

Moraines in humid climates are, in general, tree covered. Where the parent materials are coarse grained and internal drainage is good, the rapid loss of soil moisture results in relatively open woodlands and, in extreme cases, open grasslands with only an occasional tree. In subhumid climates, the vegetation of moraines is normally grass, with trees only in the wetter ravines and valleys. Where the parent materials are coarse-grained, normally only scattered fields are cultivated; the remainder of the land is generally used as pasture or timberland. Field boundaries are, to a considerable degree, independent of topography or drainage, but frequently the field shapes are partially controlled by topography. Where the parent materials are finer grained, much more of the surface is normally cultivated, with field patterns being essentially independent to topography. Figure 15.31 shows a particularly bouldery till. Attempts at improving the land for pasture are evidenced by the stone fences.

Roads tend to cross moraine systems on slightly sinuous paths. Road cuts generally have steeper banks where parent materials are coarse grained than in areas where the materials are fine grained.

The parent materials of this landscape are generally heterogeneous mixtures of sand, gravel, and silt. Virtually all possible combinations occur, with the nature of the materials varying rapidly both vertically and horizontally. There is a general relation between drainage types and grain size of the parent materials. Very wide spacings between drainageways, the local presence of deranged patterns, and an abundance of kettle holes generally indicate coarse-grained materials. The complete absence of drainageways normally indicates very sandy or gravelly parent materials.

15.16.5 Kettle-kame Moraines

As described in this text, kettle-kame moraines are any accumulations of glacial deposition that result in landscapes that rise appreciably above the adjacent terrain, that are generally not arranged in linear bands or belts, and that are composed of crudely stratified parent materials. The division between kettle-kame moraines and ridge moraines is entirely arbitrary; in nature they grade into each other. Kettle-kame moraines occur in scattered regions throughout areas which that been subjected to glaciation.

The gross drainage pattern is chiefly of the kettle-hole type, as in Figure 15.27, but perhaps the most striking characteristic is the almost complete absence of developed scour channels over fairly large areas, as illustrated in Figure 15.32. Low-order drainage patterns are generally weakly developed dendritic or centripetal.

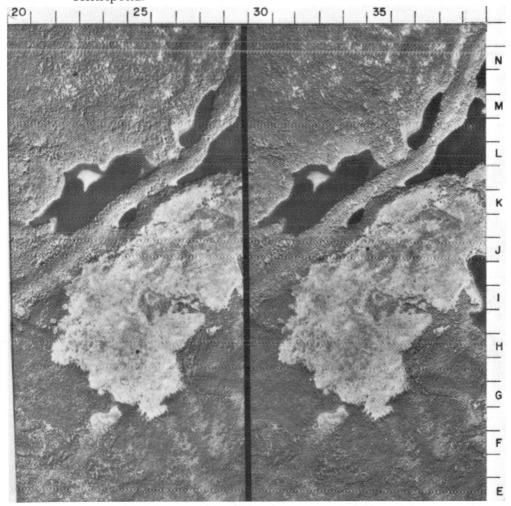

FIGURE 15.32 Stereopair: esker and kames. (Courtesy Bird & Hale Limited, 1982)

The cross-sectional shape is almost invariably rolling, as illustrated in Figure 15.32. A typical kame is at H, 34 in Figure 15.32. This type of surface is frequently referred to as *hummocky* topography. Slopes are widely variable and may range up to as steep as 60%. The overall visual impression is one of extreme randomness. Gullies are rare or absent. Where they do occur, they are almost invariably V-type with steep gradients. Their presence generally indicates a somewhat cohesive topsoil underlain by noncohesive materials.

The uplands of most kettle-kame moraines are now so completely modified by cultural practices that little or no trace of the natural vegetation remains. In humid areas the uplands were chiefly covered by forest consisting of relatively widely spaced trees. Pure stands of jack pine are common in northern regions, along with mixed forests of poplar, birch, and white spruce. The field boundaries are generally rectangular, but they are normally somewhat modified by topographic or drainage considerations. Irregular wooded areas are common and are readily identifiable by their generally dark tone and botryoidal texture. Reforestation may be necessary for the control of runoff-related flooding and wind erosion.

Road alignments are basically rectangular, but in most areas at least some of them have been forced into sinuous paths to avoid steep slopes or kettles. Road cuts commonly display side slopes of approximately 60%. Gravel and sand pits are present; their presence is an indication of relatively clean sand or gravel.

The parent materials of kettle-kame moraines are chiefly relatively clean, crudely stratified sands and gravels. However, there is an appreciable variation within relatively short distances in many places. In general, the following correlations are usually apparent:

1. The greater the number of kettles is, the coarser the parent material.
2. The absence of developed drainage channels, including gullies, implies coarse-grained and relatively clean parent materials.
3. A deranged drainage pattern implies silty or clayey parent materials.
4. Deflation hollows and sand smears imply relatively fine sand.
5. Consistently steep slopes imply coarse-grained parent materials; consistently gentle slopes imply silty or clayey materials.
6. Crested ridges imply coarse-grained parent materials.

The nature of the soils in the kettles and other depressions is to a large extent dependent on internal drainage. In kettles where such drainage is good, the soils are only slightly finer grained and more organic than those on the adjacent uplands.

15.16.6 Eskers

Eskers are frequently associated integrally with kames, and the differentiation between the two landforms is more of academic than practical interest. Kames are frequently no more than widenings in an esker, and the boundary between the two is difficult to define.

Eskers and kames are normally too restricted in area to develop gross drainage patterns. The presence of eskers and kames may, however, interrupt other patterns, producing a special variety of deranged patterns. In esker fields (i.e., in places where several eskers occur close together) the trunk streams generally flow roughly parallel to the esker ridges, thus producing a crudely parallel drainage pattern (see Figure 15.32).

There is no clear line of demarcation between eskers and kames. In general, eskers are sinuous ridges displaying a crested cross section while kames are conical hills displaying the same basic surface expression. However, eskers may occur in groups or fields, in which case there is normally lineation of the landscape parallel to the trend of the ridges. Furthermore, eskers frequently degenerate into lines of isolated hills or kame fields. However, in most instances kames occur apparently at random on other forms of surfaces.

The slopes reflect the soil texture. Narrow, sharply crested, and steep-sided features, as at L, 28 in Figure 15.32, invariably contain coarse material.

Gullies are rare; when present, they are V-type with steep gradients. Deflation hollows may develop where the soil is sand and the protective vegetation is removed.

Eskers and kames are generally uncultivated. The granular soils and excellent internal drainage, plus the typically steep slopes, combine to reduce their agricultural utility. As a result, kames and eskers are commonly used only as pasture. Rectangular field patterns often terminate at the bases of eskers and kames. In some places secondary roads may follow the crest of esker ridges, especially if the soils of the surrounding regions have poor bearing capacity or are highly frost susceptible. Gravel and sand pits are so prevalent in many places that they constitute a primary recognition feature.

The parent materials of eskers and kames consist chiefly of crudely stratified clean sand, gravel, and mixtures of sand and gravel. A few silt lenses occur in some places.

15.16.7 Drumlin Fields

Drumlin fields occur in widely scattered areas throughout glaciated regions. In North America, the chief areas are in New England, New York, Ontario, Michigan, and Wisconsin.

The gross drainage pattern is occasionally of the deranged type (Figure 15.22), but in most places there is a marked tendency for scour channels to occur roughly parallel to the long axes of the drumlins. Interdrumlin areas frequently accumulate significant depths of organic material because of the high runoff from the hills and the low regional gradients of the interdrumlin plain.

Drumlin fields exhibit a rolling cross-sectional shape. A drumlin field is composed essentially of groups of smoothly rounded elongate hills (the drumlins) superimposed on a flat or undulating surface. Drumlins are drop-shaped, streamlined forms, and in any given area the steep end of the form is consistently oriented in one direction and the long axes of the forms are approximately parallel. Drumlins range in height from 10 ft (3 m) to about 100 ft (30 m), and in length from a few tens of meters to 3000 ft (900 m) or more. Slopes rarely exceed 25%, and most are considerably less. The flanks of most of the larger drumlins are scarred by subparallel C-type gullies.

The natural vegetation of drumlins consists chiefly of dense forest. Drumlin fields are, however, generally so intensively cultivated that little or no trace of the original vegetation remains. Field patterns are generally rectangular, but shape and orientation may be modified by topographic and drainage influences, as shown in Figure 15.22. The steep side slopes of the drumlins are commonly not worked but are instead used as pasture or woodlots. Crops are generally diversified, resulting in widely variable tones and textures, as illustrated in Figure 15.22.

Road networks are normally rectangular, following the original survey fabric. There is a marked tendency for minor roads and driveways, or northern bush roads, to be crudely aligned with the long axes of the drumlins (Figure 15.22).

The parent material of drumlins is invariably fine-grained plastic soils, although it may be coated in some places with thin deposits of coarser-grained material. Some drumlins contain rock cores at shallow depths; such cores are occasionally exposed where the thin surficial clays have been breached by erosion.

Drumlin fields tend to be associated with areas of shale and lime stone bedrock.

The parent material of the surface on which the drumlins are emplaced (i.e., the interdrumlin surface) is commonly a heterogeneous mixture of clay and silt with but minor admixtures of coarse-grained material. Boulder piles and fences are an indication not only of local farming practices, but also of the boulder and silt content of the soils.

15.16.8 Outwash Surfaces

Outwash surfaces are formed by the deposition of materials from perennial (or nearly perennial) streams. The materials forming the surface may be derived from mountains or some other type of highland or from glaciers. In this chapter, only outwash surfaces composed of materials derived from continental glaciers are discussed and illustrated. However, such features as the High Plains region of the North American Great Plains are also outwash plains, differing only in that the materials of which they are composed were derived from the Rocky Mountains by stream erosion.

The gross drainage pattern is roughly parallel. The channels are usually very widely spaced. Drainage is almost entirely internal in such areas. Internal drainage over large areas is common, as illustrated in Figure 15.25. Areas of kettle-hole drainage are common; in many places the kettles are distinguished by an almost complete lack of drainage channels leading to them.

Outwash surfaces almost invariably exhibit undulating cross-sectional shapes (Figure 15.25). Relief is very low, rarely exceeding a few meters. It is usually too low to be detected stereoscopically. The outwash in Figure 15.30 around H, 17 exhibits virtually no topographic relief on the ground. The uniform canopy height of a pure jack pine stand reflects this characteristic.

Relief is generally too subdued to permit gully formation except in a few places along major streams. Where present, gullies are V-type with very steep gradients. Deflation hollows may develop where vegetation has been removed from sandy soils. Blowouts are present in Figure 15.25.

Vegetation is largely dependent upon climatic regime. Most glacial outwash plains in the Great Lakes region of North America were forested. The trees were either hardwoods or conifers, depending primarily upon climate. Many outwash plains have been reforested or have been logged over, so little trace remains of the original vegetation.

In cultivated areas, field patterns are rectangular and almost completely independent of topography. Common and easily recognized agricultural uses include fruit orchards and tobacco farms. Many former cleared lands have been reforested in pine to protect the topsoil from wind erosion and to provide an economic resource for the future (see Figure 15.25). Relatively large areas of forest in an otherwise cultivated outwash plain commonly indicate relatively sandy soils.

Roads are in a rectangular pattern and independent of topography. Shallow quarries where sand and gravel are being extracted are not unusual, with the depth of face being an indication of the depth of outwash over another landform or the depth to the water table.

Parent materials are relatively clean, crudely stratified sands and gravels, with occasional silt lenses of limited aerial extent. Surface soils are widely variable, depending primarily on the nature of the parent material and the topographic position. The most striking soil differences occur between the topographic highs and lows, even though the relief may be only a few inches. The lows may be subsidence hollows, or they may be the filled channels or abandoned glacial streams.

15.16.9 Floodplains

Floodplains are surfaces that are adjacent to streams and flooded by the stream more or less regularly. No restrictions on size are implied; in this text the floodplain may consist only of bars exposed in the channel at low water, as well as the vast expanses of plains flooded by such streams as the Mississippi. The stream with which the plain is associated need not be perennial; in this text the alluvial fill in the valleys of such streams is regarded as floodplain if the stream normally floods it from time to time. This general category provides no clear visualization of the complexly associated environments of many floodplains. For this reason, the descriptions of the various factors have been subdivided into several of the more important (i.e., widespread) environments. Those specifically described are ridge and swale topography, natural levees, backswamps, abandoned courses, and high-level floodplains. It must, however, be emphasized that not all these environments occur in every floodplain. Many floodplains are composed of only one or two.

Regional Drainage

1. *Ridge and swale.* The gross drainage pattern is basically dendritic, but is commonly very distorted; it may form a spe-

cial type of deranged pattern. The channels may be broadly meandering, and in some cases there will be small-scale meanders superimposed on much larger loops. The examples of floodplain ridge and swale are all of limited extent, consist of coarse material, and a gross drainage pattern has not developed. The beach ridge and swale, as shown in Figure 15.23, illustrates a similar pattern.

2. *Natural levee.* These landscapes are generally too restricted in area to develop gross drainage patterns, but may be characterized by low-order drainage in usually the parallel type, with the channels draining the gentle reverse slope (i.e., away from the river of the ridge), and thus oriented roughly at right angles to the trend of the ridge and the river, as illustrated along the upper river bank in Figure 15.33.

3. *Backswamp.* The gross drainage pattern is normally dendritic, but also may exhibit some of the features of the reticulate

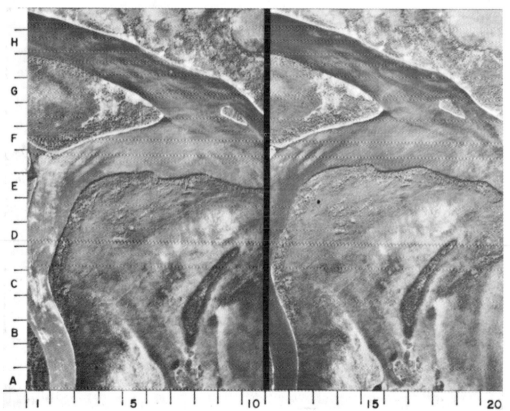

FIGURE 15.33 Stereopair: floodplain. (Courtesy Bird & Hale Limited, 1982)

type (i.e., channels bifurcate and rejoin in complex patterns very similar to those found on deltaic plains, as in Figure 15.33). Low-order channels are basically dendritic, but they may be entirely concealed in places where the swamps are forest covered.

4. *Abandoned course.* These landscapes are generally too restricted in area to develop gross drainage patterns. Low-order drainage is normally parallel to, and remains within, the abandoned course. Channels are commonly meandering. Lakes of the *oxbow* type usually fill or partially fill the abandoned course (see Figure 15.34).

5. *High-level floodplain.* Both gross and low-order drainage patterns are predominantly dendritic, but the landscape may exhibit traces of patterns inherited from earlier times; thus, it may be crudely collinear or parallel, as shown in Figure 15.34.

Topography

1. *Ridge and swale.* The cross-sectional shape is undulating. The ridges and swales are concentric and usually arcuate. Local relief is rarely more than a few meters and may be so subdued as to be indetectable by stereoscopic examination.

2. *Natural levee.* The cross-sectional shape is generally undulating or blocky. The levees consist of asymmetric ridges of very low relief parallel to a stream channel. The steep slope (frequently an abrupt cut bank) faces the water, and the very gentle reverse slope faces away from the stream. Relief may be from a few centimeters to as much as 10 to 30 ft (3 to 9 m), though these extremes are rare. The ridges range in width from a few meters to several kilometers. Changes in river courses may leave natural levees abandoned and far from the present course.

3. *Backswamp.* The cross-sectional shape is undulating and is commonly very nearly flat; relief is normally only a few centimeters or at most a meter. Areas of backswamp are extremely variable in size even in the same floodplain; they may range from a few hundred square meters to several square kilometers. They are normally bounded on at least one side, and frequently on all, by natural levees. To this extent they are very shallow closed basins, although usually there is at least one drainage exit.

4. *Abandoned course.* This surface normally consists of a relatively narrow band or belt bounded by either (or both) natu-

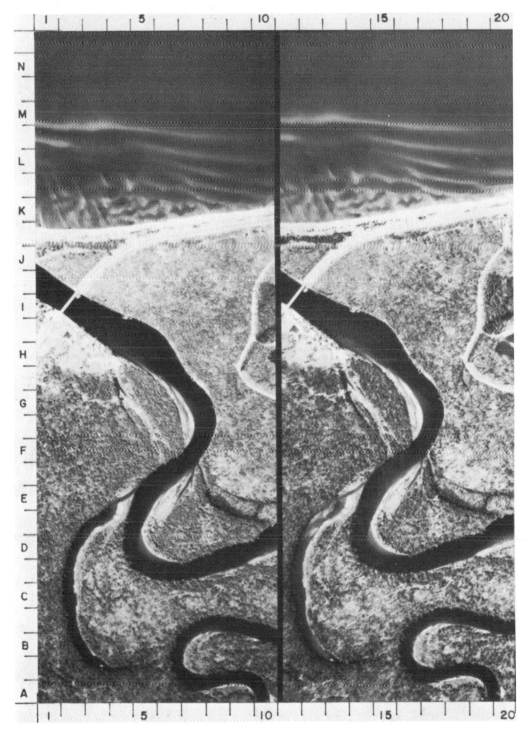

FIGURE 15.34 Stereopair: floodplain. (Courtesy Bird & Hale Limited, 1982)

803

ral levees or ridge and swale topography. It may range in width from a few meters to a kilometer or more, depending on the size of the stream that once flowed therein. The cross-sectional shape is generally undulating and of very low relief. It may in some places contain closed basins.

5. *High-level floodplain.* This surface is normally flooded only during extremely high water stages. The cross-sectional shape is generally undulating or blocky (see Figure 15.35, H33). Relief is normally only a fraction of a meter or, at most, 33 ft (10 m.). This surface commonly occurs as discontinuous segments along the edges of the floodplain, but it may also occur as outliers within the low-level floodplain. Sizes are variable, ranging from a few hundred square meters to several square kilometers.

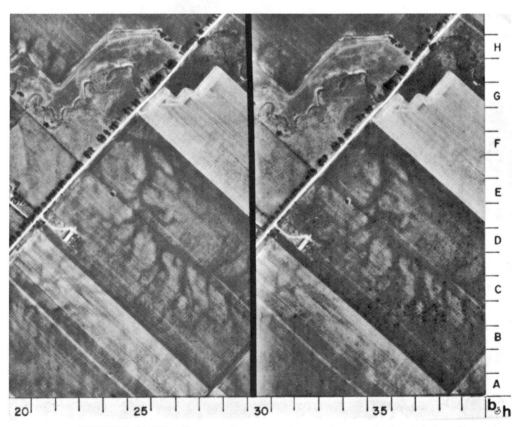

FIGURE 15.35 Stereopair: ground moraine and floodplain. (Courtesy Bird & Hale Limited, 1982)

Local Erosion

1. *Ridge and swale.* This surface rarely if ever exhibits gullies. Where the surface is newly formed (i.e., along active channels), deflation hollows may develop.

2. *Natural levee.* Most natural levees exhibit little or no gullying, chiefly because local relief is inadequate for gully development. However, where such levees have been abandoned, the steep face of the ridge may develop gully systems. In general, these are V-type with steep gradients.

3. *Backswamp.* These surfaces rarely exhibit evidence of local erosion.

4. *Abandoned course.* These surfaces rarely exhibit evidences of local erosion of any type.

5. *High-level floodplain.* Gullies are relatively common on these surfaces, especially near the margins. Because the parent materials and soils of such features are quite variable, the gully forms are also variable. Perhaps the most common is the V-type with steep gradients, but virtually any type may occur.

Natural Vegetation

Floodplain vegetation in general is dependent on climatic regime, but to a lesser extent than that of most other landscapes. Water is normally adequate for abundant plant growth, even though rainfall may be very low. Nevertheless, there are significant variations between tropic, desert, temperate, and arctic climatic types. The following descriptions apply only to humid-temperate climates.

1. *Ridge and swale.* When ridges and swales are newly formed, vegetation may be absent. These surfaces are then colonized by small herbs and grasses, then shrubs, and finally by dense forest. The contrast between the dark-toned botryoidal texture of the trees on the ridges and the smooth, generally light toned grasses of the swales is often striking.

2. *Natural levee.* These surfaces were normally forested, and many have been subsequently cultivated. When such an area is forest covered, the trees may mask the typical streaked appearances.

3. *Backswamp.* These surfaces were normally forested, although in some places extensive areas of marsh, grass, and reeds occur, especially in and near the lakes, very similar to those occurring in deltaic plains.

4. *Abandoned course.* These surfaces were normally forested. However, near oxbow lakes, especially at their ends, extensive

areas of grass or reeds may occur (see Figure 15.34). En-croachment of vegetation into stagnant oxbow lakes often re-sults in complete closure of the surface (bog) with the same hazard as described for encroached kettle lakes.

5. *High-level floodplain.* These surfaces are normally densely forested.

Land Use and Construction Practices

1. *Ridge and swale.* Newly formed ridge and swale areas and those in the process of forming (i.e., active point bars) are rarely cultivated. Old surfaces of this type are commonly cultivated. In such places, the normally sharp distinction between drainage conditions and soil types from ridge to swale commonly results in a field pattern consisting of elongate rectangles oriented parallel to the trend of the topography.

2. *Natural levee.* This landscape is generally the most thoroughly cultivated of the floodplain environments. The soils are usually well drained, encouraging diversified agriculture. Field patterns are often rectangular and tend to be aligned parallel to the typical streaked pattern.

3. *Backswamp.* The soils of backswamps are generally poorly drained and difficult to work. As a result, this environment is commonly left in timber; where cleared, it is frequently used for grazing.

4. *Abandoned course.* The soils of abandoned courses are generally poorly drained and difficult to work. As a result, this environment is left in timber or used for grazing.

5. *High-level floodplain.* These surfaces are commonly cultivated. The field patterns tend to be rectangular and independent of topography.

Parent Materials

1. *Ridge and swale.* The parent materials of ridge and swale surfaces are generally sand or gravelly on the ridges, with finer-grained materials forming the swales. In some places this distinction becomes extreme, with sand composing the ridges and clays and silty clays composing the swales.

2. *Natural levee.* This landscape is generally composed of silt or silty sand.

3. *Backswamp.* The parent materials of backswamps are chiefly silt, clay, and mixtures of both.

4. *Abandoned course.* The parent materials of this landscape are chiefly clay and silty clay. Oxbows frequently contain deep organic soil.

5. *High level flood plain.* The parent materials are highly variable, depending on surrounding land forms.

15.16.10 Deltaic Surfaces

There is no clear line of distinction between a delta and a floodplain; the delta is in a sense only the floodplain extended into the terminal water body. However, because the deltaic surface is formed by the complex interaction of littoral, lacustrine, marine (if formed at sea coasts), and fluvial processes, the features typical of floodplains are somewhat modified, and several features not normally found in floodplains may occur.

The deltaic surface is typically a very gently undulating surface slightly above the level of the terminal water body. The surface normally consists of complex arrangements of natural levees, distributary channels, abandoned beach ridges, and swamps and marshes, but any one or several of these components may be absent in some places.

Deltaic surfaces of some type tend to form at the mouth of nearly every stream that discharges into a lake or sea. Some are destroyed by wave and current action as rapidly as they are formed, but the remainder develop to a greater or lesser extent, depending on complex factors of the environment. Three fundamental delta types are recognized: digitate, arcuate, and estuarine. There is no clear distinction between the forms; a very large number of intermediate forms occur. Deltas range in size from a few square meters to hundreds of square kilometers.

The gross drainage pattern of most deltas consists of a basic distributary (i.e., delta) system through which the delta-forming stream discharges. The channels of such distributaries may be of any type, depending on such factors as stream gradient, tidal range of the water body into which the stream discharges, and the transported materials. The major channels of the digitate type are usually relatively straight and simple. Those of arcuate deltas may be straight or meandering and are commonly complexly interconnected. It should be emphasized that all delta types grade into each other, so occasionally one portion of a delta may be typically digitate, another arcuate, and still another estuarine. True estuarine deltas are restricted by definition to tidal waters. Estuarine distributaries may be of almost any type; meandering, straight, and braided channels are all common.

The cross-sectional shape is chiefly undulating, with local relief rarely exceeding a few meters. Many aligned forms, such as ridge and swale surfaces, are common on deltas.

The relief of deltas is generally too low to permit gully erosion. Deltas, being normally well watered, are even less subject to eolian erosion than are floodplains, although such erosion does sometimes occur in arid climates. Even in semiarid areas, deltas are commonly forested.

Deltas are used in a multitude of different ways. They may be intensively cultivated, as are those of the Nile or Ganges, or they may be relatively untouched, as are those of the Yukon, Niger, and Amazon, or they may be combination of both, such as the Mississippi.

Where deltas are cultivated, drainage works such as ditches and tiles are prevalent. Extensive drainage canals have been constructed in many places as public health measures.

As with floodplains, delta materials vary widely in type, depending on the nature of the depositing stream and to some extent on the nature of the terminal water body. Streams with steep gradients tend to form deltas composed of relatively coarse grained materials, while streams with gentle gradients tend to deposit finer-grained materials.

15.16.11 Terraces

Terraces are found along the valleys of most large and many small streams. In this text, the landscape is a terrace if it is not inundated by floodwaters more or less regularly. The surface of a terrace may range from a few meters above the present floodplain to heights of 500 ft (150 m) or more above it.

Terraces are generally too restricted in area to develop gross drainage patterns. In many places they are devoid of all drainage channels, even low-order ones.

The cross-sectional shape is generally undulating. In many places a shallow depression may extend along the base of the bounding valley wall (see Figure 15.30 around J, 18).

Terraces are normally devoid of gullies except at their valleyward edge. V-type gullies with steep gradients indicate coarse-grained materials (see Figure 15.30 around F, 15), and F-type gullies with relatively slight gradients indicate silty materials.

Terraces are generally cultivated or used as grazing land. Field patterns are generally rectangular and independent of topography on the terrace surface, but they commonly terminate at odd angles at the terrace edge or at the valley wall.

Terrace materials are exploited for construction purposes, road materials, or industrial uses. The presence of a gravel or sand pit generally indicates relatively clean, coarse-grained parent materials. Terraces are frequently used as sites for airfields and towns; the generally level surface, good drainage, and good foundation conditions are ideal for these purposes.

In the Great Lakes region, where most terraces formed during glacial or postglacial times, the abundant coarse-grained material furnished by the melting ice has resulted in terraces in which materials tend to be sand and gravels.

15.16.12 Lacustrine and Littoral Deposits

Landscapes in lacustrine deposits owe their essential shape to deposition from standing water, either salt or fresh, and either perennial or ephemeral. Such surfaces are widely distributed on every continent. They are, essentially, the dry beds of lakes.

These landscapes are in general the beds of large freshwater lakes that have been drained within geologically recent time. A few surfaces of this general type are the result of deposition from brackish or saline water (see Figure 15.24). Lacustrine surfaces are most common in and adjacent to glaciated regions; excellent examples occur in the Great Lakes region, as well as in northern Europe and Asia.

The gross drainage pattern is generally dendritic, but in cultivated regions it is usually ditched or canalized so completely as to be difficult to recognize. Low-order drainage patterns are generally of the ditched or tiled type if the region is cultivated and dendritic if it is not.

The cross-sectional shape is generally undulating. Beds of perennial lakes are commonly so flat that no relief can be seen with a stereoscope.

Where gullies occur, they generally are very broad C-type, with very flat gradients. A lack of gullies normally indicates a lack of available relief, rather than a lack of susceptibility of the soils to erosion. Where there is sufficient relief, such as along major river valleys, gullies are very common. Slope instability is dramatically demonstrated by the large slump features in Figure 15.24.

Most lake basins are now so intensively cultivated that little or no trace of the original vegetation remains. In northern regions, extensive fens with black spruce forests are common.

Field patterns are almost entirely rectangular and independent of topography or drainage. Most lacustrine plains in humid climatic regimes are intensively cultivated.

Most lake beds are composed of considerable thicknesses of stratified materials. Materials range from heavy clay through silt–clay mixtures to silts and even fine sands.

Fine soils are frequently varved. Marine soils are usually sensitive, such as the Leda-type clays in Figure 15.24.

The generally low relief is often an inhibition to both internal and external drainage, so soils of lake basins frequently retain high moisture content throughout the year, promoting the accumulation of organic soils.

15.16.13 Beach Ridges

Beach ridges occur along most shorelines of low relief that are subject to wave attack. They occur around large lakes as well as around oceans and seas. They commonly form the margins of the larger lacustrine plains. Beach ridge complexes are widely variable in area, ranging from narrow bands a few meters wide to areas covering several square kilometers. This landscape type is usually too restricted in area to develop gross drainage patterns.

Relatively young beach ridges are commonly of the channel-less-basin type, the basins being elongate, roughly parallel swales (see Figure 15.23). Old beach ridges may be of this type, or they may develop scour channels through the swales, resulting in collinear patterns.

The cross-sectional shape is frequently undulating. The arcuate or sinuous ridges may range in height from a fraction of a meter to 33 ft (10 m), but the normal range is less than 10 ft (3 m.). The parallel ridges are separated from each other by swales (see Figure 15.23). The following generalizations can be made, but it must be recognized that many exceptions occur.

1. The coarser the material is, the steeper the slopes. Gravel ridges may have slopes as steep as 60% in some cases, although most are much less.
2. The older the ridge is, the more gentle the slopes. Those ridges closest to the shore (i.e., the youngest) are commonly appreciably steeper than those farther from the shore (i.e., the oldest).
3. Ridges formed on lakeshores or coastlines subjected to ice push are often steeper than those formed by wave action alone.

The farthest inland penetration of ancient water bodies is frequently marked by a shore cliff such as that shown in Figure 15.23 at N, 17.

Beach ridges rarely if ever exhibit gullies. Where vegetation is absent, they are subject to wind erosion.

Along exposed coasts, beach ridges are commonly covered with sparse grass and shrubs. In humid regions, areas of beach ridges far from present shorelines may be forest covered, and, in such cases, the subdued topography may be partially masked by the trees. This situation is also common in areas of beach ridges bordering large lakes or lacustrine plains.

In some places, the markedly different soil properties between swales and ridges result in field patterns aligned with the trend of the ridges. The resulting alternations of light (the ridges) and dark (the swales) photo tones are sometimes striking.

However, if the alignment of the ridges happens to trend in roughly the direction of a road or railroad, the ridges are not uncommonly used as a ready-made embankment (Figure 15.23). Beach ridges are sometimes used as sources of sand and gravel for industrial purposes or road materials.

The parent materials are almost entirely coarse grained, ranging from gravel to find sand. The materials are generally almost entirely free of fine-grained materials.

The swales (i.e., the topographic lows) contain somewhat organic soils. In regions where climatic and drainage conditions are conducive to the development of organic soils, mucks and peats of considerable depth may be found.

In general, the coarse-grained beach ridge soils tend to support tree growth, resulting in a dark tone. The swales, with grasses, moss, and sparse tree growth, photograph in a lighter tone. The resulting streaked pattern of alternating dark and light in concentric arcuate or sinuous patterns can commonly be used to assist in the identification of beach ridges even when the relief is so subdued as to be undetectable stereoscopically.

15.16.14 Organics

Organic deposits may be confined to a small area in the order of several square meters, or they may be extensive, covering hundreds of square kilometers. The soil consists of partially decayed organic matter such as the leaves, branches, trunks, and roots of trees.

Gross drainage patterns are dendritic (Figure 15.33), but in cultivated regions, canals are prominent. Low-order drainage will be ditches or dendritic streams.

The cross-sectional shape is very flat because the surface is controlled in effect by the water table. Because of uniform canopy heights, fens and bogs occasionally appear to be topographically lower near their center. In fact, the ground surface is virtually flat and horizontal. Fens and bogs invariably occupy topographic depressions.

No evidence of wind, water, or gravity erosion will be seen within organic terrain. Major streams will flow on the underlying landform, reflecting its erosion features.

Shallow organic terrain, and that which has developed a firm base, will support tree growth. In general, the deeper the peat is, the more sparse and stunted the tree vegetation.

Agricultural uses are common close to market areas. Peat is extracted by mining and processed into potting soil or fuel. In populated areas, marshes, bogs, and fens are frequently protected for their value as habitats for wildlife.

Organic terrain is normally avoided where possible because of the adverse engineering properties of the soil. Where unavoidable, special construction techniques are used. These include floating foundations for buildings and roads, extensive filling using toe-blasting, and sand drains to resolve short- and long-term consolidation problems.

Peat ranges in depth from a few centimeters to 10 or more meters.

15.17 LANDSCAPES IN CONSOLIDATED MATERIALS

As used in this text, consolidated materials are those that are so strong they cannot be broken or deformed by hand. Hereafter these materials are called rocks.

Landscapes in consolidated materials are strongly controlled by both the nature of the materials from which the form is being sculptured and by the gross structure (i.e., the attitude of the component rock masses with respect to the horizontal) of the rocks or other materials. Accordingly, the classification used in this text is primarily based on three factors: (1) stratification or its absence; (2) structure, or attitude of the beds, if any; and (3) lithology, or the nature of the component rocks or other materials.

For the purpose of this text, it will be sufficient to differentiate between major classes of bedrock and to assess certain engineering parameters within them. The three classes are (1) igneous intrusive, igneous extrusive, and metamorphic rock with no apparent bedding, banding, or stratification; (2) metamorphic rock with parallel topographic features; and (3) horizontally stratified interbedded sedimentary rock landscapes.

15.17.1 Igneous Intrusive, Igneous Extrusive, and Metamorphic Rock with No Apparent Bedding, Banding, or Stratification

In the sense used in this text, nonstratified rocks are those that do not exhibit topographic or other characteristics that normally develop in a landscape because of bedding. Thus, the igneous rocks included in this group are the plutonic rocks that form in large masses (i.e., batholiths, laccoliths, etc.).

Granite and related rocks occur in every continent, chiefly in the "shield" areas and in the central portion of mountain ranges. The pertinence of the discussion that follows is confined to granitic areas of roughly less than 500 ft (150 m) of local relief and to regions having humid climatic regimes.

Both gross and low-order drainage patterns are generally either dendritic or rectangular. Where the rectangular pattern is evident, there are normally at least two well-developed joint systems at angles of approximately 60° to 90° as illustrated in Figure 15.30, upper right.

The cross-sectional shapes in granite hills are generally rolling. The slopes from the topographic highs to the topographic lows are usually sigmoidal, although some streams flow in broadly V-shaped valleys. There is normally a distinct sense of randomness; the ridges between drainageways are short, often tortuous, and randomly oriented.

The soils of granite regions are peculiarly susceptible to gully erosion due to steep topographic slopes. Where no gully control measures are taken, gullies are numerous. Talus slopes are frequent.

This landscape is usually covered with dense forest or else devoid of vegetation, except for organic accumulations in the low areas. Occasionally, isolated rock outcrops are visible through dense forest cover.

Logging, mining, and recreational uses are most common in this rock type. Large tracts of land with no apparent human activity are common.

Granites are extensively quarried in some places; the quarries are generally typified by smooth, stepped working faces, with the risers of the steps vertical and the treads nearly or perfectly flat. Steep rock cuts for highway construction are commonly seen in these areas. Road alignments are frequently controlled by bedrock topography and the availability of construction material in nearby landforms.

The residual soils of these landscapes are usually fairly thin or nonexistent. Where the bedrock is heavily vegetated, there is usually a thin veneer of lodgement-type till, less that 6 ft (2 m) thick, which is best described as a bouldery silt till. The depth of the soil can be estimated using drainage, vegetation, and topography as indicators.

15.17.2 Landscapes in Metamorphic Rocks with Parallel Topographic Features

Gneisses, schists, and related rocks occur on every continent, chiefly in shield areas and in regions that have been subject to intense tectonic activity. The sites considered in this text have less than 500 ft (150 m) of local relief. The emphasis of the discussion that follows is confined to metamorphic areas of roughly this order of relief and to regions having humid climatic regimes.

The gross drainage pattern is usually rectangular and strongly resembles the features exhibited by granite hills (see Figure 15.36, upper). The local drainage is normally parallel. Thin, parallel lakes are common.

The cross-sectional shape is rolling. It strongly resembles granitic surfaces of comparable relief, but is more ordered; that is, there is usually a definite tendency for the generally short ridges to exhibit a preferred orientation, as in Figure 15.36 (upper), depending on the degree of metamorphism. The general appearance could be described as ridge and swale.

Gneisses are very similar to granites in their susceptibility to erosion. Gullies are compound, with saucer-shaped segments near their heads and F-type plus V-type near their mounts, with a sharp break in gradient at the transition.

The ridges and swales can frequently be observed, not because of topographic differences, but by vegetation changes. The ridges are either bare rock or treed, while the wetter swales have different tree species or organic soils.

As in granitic glaciated landscapes, the residual soil was removed by the glaciers, and the rock was either left bare or covered with glacial features. Shallow soils preclude farming activities.

Gneisses are less extensively quarried than granite; quarry faces may be either straight and stepped (dimension-stone quarried) or curved and irregular (road material and the like). Roads are sinuous in most places and tend to follow ridge lines.

Parent materials may be any one of the many varieties of gneiss or schist. The rocks are commonly strongly jointed, usually with at least two sets at more or less right angles to each other.

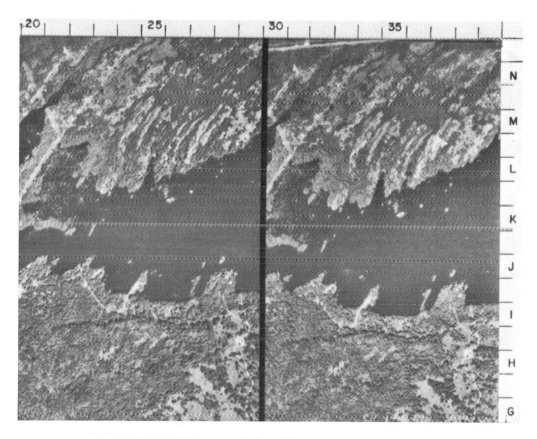

FIGURE 15.36 Stereopair: banded metamorphic rocks. (Courtesy Bird & Hale Limited, 1982)

15.17.3 Landscapes of Low Relief in Interbedded, Horizontally Stratified, Sedimentary Rocks

These landscapes are normally called *plains*. They have a local relief of less than 500 ft (150 m). The component rocks may be of any lithologic type, excluding igneous or metamorphic rocks. For the purposes of this text, *horizontal* implies that the bedding planes are not inclined at angles of more than a few degrees. The pattern elements may change slightly depending on the nature of the uppermost stratum. Where differences are notable, they are described.

Many rock sequences consist of complexly interbedded sandstones, shales, and limestones. In this text, interbedded refers to the variety and complexity of different sedimentary rocks that may be exposed and seen on airphotos, rather than to the specific geological implications of the term. In general, the landscapes that develop on such sequences are complex combinations of the forms typical of all three types.

The drainage patterns are normally typical of either shale or sandstone, since the drainage features typical of limestone are rarely developed. As a result, the gross drainage pattern is normally either dendritic or rectangular, and commonly a combination of both. For example, note that the gross drainage pattern in Figure 15.37 is basically dendritic, but with a marked angularity that suggests the rectangular type. The low-order drainage pattern is mostly dendritic.

The landscapes are often stepped. The general impression is one of blocky cross-sectional shapes (Figure 15.37), but very commonly isolated crests of blocky masses will rise above the general upper surface of the topographic highs. These features are well illustrated in Figure 15.12.

Like all other topographic expressions of this landscape type, the gullies are widely variable. Those that form in predominantly shale parent materials are normally C-type, while those in predominantly sandstone materials tend to be V-type.

In humid areas, natural vegetation is normally relatively dense forest. Growth tends to be more dense on shales and sandstones than on limestones, Cedar commonly grows on limestone and can frequently be identified by its triangular shadow. Bedrock and boulder fences and piles are common, as are scattered cedar trees. Figure 15.37 illustrates typical farming practices.

In regions of low relief, the roads tend to be independent of topography, but where relief is relatively high, the roads tend to be sinuous and to follow either ridgelines or valley bottoms. Since the ridges are usually irregular and discontinuous, road alignments are not uncommonly marked by many cuts and fills. Clusters of farm buildings tend to be on the ridge tops, and quarries are numerous in many places.

The parent materials may be any combination of interstratified limestone shale and sandstone. In such sequences the individual rock types are often impure; thus, the limestones tend to be shaley or even sandy, and the sandstones are often shaley or limey. These gradational types make it even more difficult to determine either the precise sequence or the exact type of rock underlying any given soil, because the resulting topographic expressions also tend to be gradational. The most common sequences are, from top to bottom, limestone, shale, and sandstone, or the reverse. In glaciated areas, it is common to have a thin covering of a glacial or periglacial material on the bedrock.

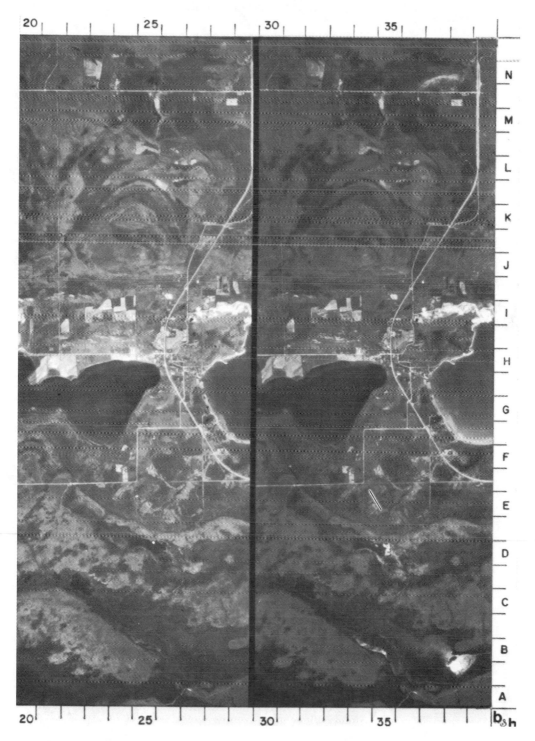

FIGURE 15.37 Stereopair: horizontally bedded sedimentary rocks. (Courtesy Bird & Hale Limited, 1982)

General statements concerning the characteristics of soils from topographically high or low sites are impossible to formulate because of the widely variable rock types from which they may have developed. Each must be interpreted in terms of the probable sequence in the area being studied, rather than in terms of the landscape type as a whole.

There is a great deal of difficulty encountered when trying to predict soil depths over bedrock in this landscape. The regularity of the rock surface results in a general lack of topographic evidence of the proximity of bedrock. Other indicators, such as stream rectangularity, profusion of boulders, and land uses, can occasionally lead to general ideas on soil depth, but field evidence is almost a necessity, and extrapolation from information like well logs is unreliable.

15.18 APPLICATIONS OF AIRPHOTO INTERPRETATION FOR THE ENGINEER AND THE SURVEYOR

Airphoto interpretation is used for a wide variety of projects. The main advantages of this technique are as follows:

1. The identification of landforms and consequently site conditions, such as soil type, soil depth, average topographic slopes, and soil and site drainage characteristics, is made before going to the field to carry out either engineering or surveying field work. When the field work begins, there is a strong feeling of having been at the site before, owing to the familiarity achieved through careful examination of the airphotos.

2. The surveyor can examine the topographic slopes, areas of wet or unstable ground, and the density and type of vegetative cover. Therefore, he or she can become familiar with the ease of or difficulties to be expected in carrying out the field survey. The surveyor can also determine the location of property or section boundaries from the airphotos, some of which are extremely difficult to see on the ground.

3. Airphotos provide an excellent overview of the site and the surrounding area, which cannot be achieved through ground work alone. For example, if evidence of soil movement such as landslides is indicated on the airphotos (see Figure 15.24), the surveyor would avoid placing monuments in such an

area as they would be subject to movement with the soils. The engineer would avoid using this area for any heavy structures owing to the potential of gradual or sudden soil failure.

4. As soil test holes should always be used to verify the results of the airphoto interpretation, these can be carefully preselected on the airphoto prior to doing the field work. Existing road or stream bank cuts can be preidentified for use as field checkpoints, thus saving on drilling costs. Based on the authors' experience in using airphoto interpretation for over 500 projects, the average savings in the amount of field work required is between 70% and 80%.

A few words of caution are appropriate at this point. The information presented in this chapter represents, owing to space limitations, only an introduction to airphoto interpretation. A much more detailed publication entitled *Airphoto Interpretation of the Physical Environment,* by G. Bird and I. Hale,* leads to a thorough understanding of the techniques. In addition to pursuing the subject more thoroughly, one should develop the habit of personally field checking one's own airphoto interpretation whenever possible. To become proficient in this technique requires considerable practice.

Two examples of property boundary line determinations from the airphotos are as follows:

1. In Figure 15.27, a number of existing property lines through heavily wooded areas are evident. One example of several is the boundary line from K, 37 to M, 36.
2. Figure 15.38 illustrates the location of an original boundary by intermittent narrow cut lines from E, 9 to N, 2. This is a much more subtle example than that given for Figure 15.27. Therefore, it is much more useful, particularly because of the remoteness of the area.

---

*Available from Bird and Hale Limited, 1263 Bay Street, Toronto, Ontario, Canada M5R 2C1

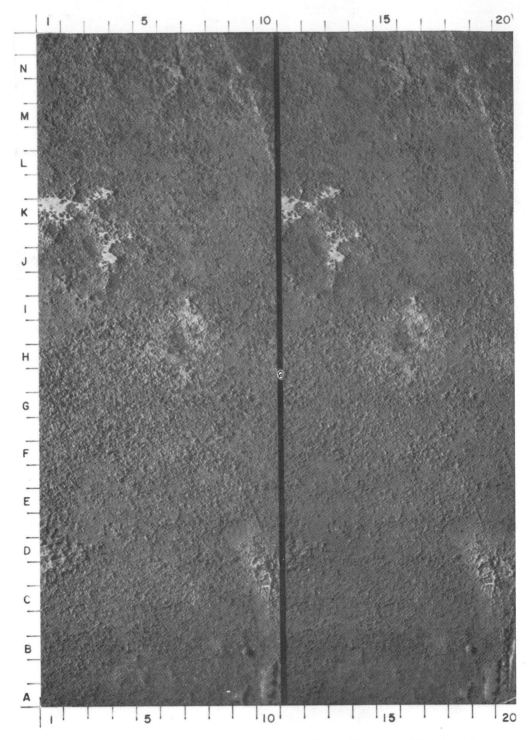

FIGURE 15.38 Stereopair: original survey lines. (Courtesy Bird & Hale Limited, 1982)

PROBLEMS

15-1 Calculate the flying heights and altitudes, given the following information.

a) Photographic scale — 1 : 20000
Lens focal length = 153 mm
Elevation of mean datum = 180 m
b) Photographic scale 1 in = 20,000 ft
Lens focal length = 6.022 in.
Elevation of mean datum = 520 ft.

15-2 Calculate the photo scales, given the following data.

a) Distance between points A and B on topographic map (scale 1: 50000) = 4.75 cm. Distance between same points on airphoto — 23.07 cm.
b) Distance between points C and D on topographic map (scale 1: 100000) = 1.85 in. Distance between same points on airphoto = 6.20 in.

15-3 Calculate the approximate numbers of photographs required for stereoscopic coverage (60% forward overlap and 25% side overlap) for each of the conditions set out below.

a) Photographic scale = 1: 30000
Ground area to be covered is 30 km by 45 km.
b) Photographic scale 1: 15000
Ground area to be covered is 15 miles by 33 miles
c) Photographic scale is 1 in = 500 ft.
Ground area to be covered is 10 miles by 47 miles.

15-4 Calculate the dimensions of the area covered on the ground in a single stereo model having a 60% forward overlap, if the scale of the photograph is

a) 1: 10000
b) 1 in. = 400 ft.

15-5 a) Calculate how far the camera would move during the exposure time for each of the conditions below.

(i) Ground speed of aircraft: 350 km/hr
Exposure time: $1/100$ S
(ii) Ground speed of aircraft: 350 km/hr
Exposure time: $1/1000$ S
(iii) Ground speed of aircraft: 200 miles per hour
Exposure time: $1/500$ S

b) Compare the effects on the photograph of all three situations in (a).

15-6 Does a longer focal length camera increase or decrease the relief displacement? Briefly explain the reasons for your answer.

15-7 Would a wide-angle lens increase or decrease relief displacement? Briefly explain the reason for your answer.

15-8 State the best alternative of those shown below for the assemblage of an uncontrolled mosaic, using alternative photographs, over rolling terrain having vertical relief differences of up to 50 ft. (15 m). Give the main reasons for your choice.

a) focal length = 150 mm, scale = 1: 20000
b) focal length = 150 mm, scale = 1: 5000
c) focal length = 12 in (300 mm), scale 1 in = 1000 ft.
d) focal length = 12 in (300 mm), scale 1 in = 2000 ft.

15-9 Calculate the time interval between airphoto exposures if the ground speed of the aircraft is 100 miles per hour (160 km per hour), if the focal length is 50 mm (2 in.); the format is 70 mm (2.8 in.); and the photographic scale is 1: 500 (approximately 1 in. = 40 ft). Forward overlap = 60%.

15-10 a) What problem does the channel straightening from F,36 to C,37 in Figure 15.21 indicate about the drainage of this area?

b) Briefly state the negotiations that should be carried out with the land owners adjacent to this straightening regarding construction costs, loss of land due to excavation, and benefits per acre (hectare) for each owner.

15-11 Are the soils on the drumlin in Figure 15.22 well drained internally? State the main reason for your answer.

15-12 a) State the main concern regarding the permanency of survey monumentation in the area around H,4 on Figure 15.24.

b) Judging by the vegetation which has regenerated in the area of the landslide (around H, 5 to H, 9), estimate roughly the number of years since the landslide occurred.

15-13 See Figure 15.25.

a) What terrain characteristics do the light toned areas around M, 21 indicate?

b) What is the angle, using a protractor, between the well-defined original survey lines intersecting just above I, 33?

c) List the pattern elements (see Table 15.3) that identify this area as an outwash plan.

15-14 See Figure 15.27.
   a) The area around I, 35 is under construction. Why do the soils appear so light in tone?
   b) What is the resurvey significance around G 22 and H 22?

15-15 See Figure 15.28.
   Sand dunes are located behind the shoreline bluff. List the pattern elements (see Table 15.3) that confirm the area as sand dunes.

15-16 See Figure 15.30.
   a) It is intended to subdivide the right hand portion of the river, on the right side between G, 5 and J, 5. Give the coordinates of the ends of the cut line along the rear lot line of the subdivision.
   b) State the type of regional drainage of the main river.

15-17 See Figure 15.31.
   a) Give the corrdinates of the evidence of extension of the property lines indicated by the tree line around N, 7.
   b) List the pattern elements (see Table 15.3) that identify the arca around M, 5 as a ridge moraine.

15-18 See Figure 15.32.
   a) Give the coordinates of each end of the esker.
   b) What do you think is the cause of the cleared area around J, 26?

15-19 See Figure 15.33.
   a) What is the probable soil type around D, 8?
   b) List the terrain conditions that must be considered in giving an estimate of costs for running a survey line from A, 9 to H, 7.

15-20 See Figure 15.35.
   Briefly explain the cause of the lineal texture around G, 28.

15-21 See Figure 15.36.
   a) Identify the landform around H, 28.
   b) What type of bedrock underlies this landform?

15-22 See Figure 15.37.
   What photo pattern element is the best indicator of property lines in the area around L, 26?

16 Computerized Surveying Data Systems

16.1 GENERAL

The impact of computer technology on the survey sciences has been notable. The techniques of ground data collection, processing, and graphics portrayal are presently undergoing revolutionary changes. Figure 16.1 gives a schematic summary of an integrated surveying system.

The chief characteristics of a survey data gathering system include the *automatic recording and transfer to the computer of ground data* collected by tacheometers, the *high-precision digitizing* capable with modern digitizers and stereo analyzers, and the *wealth of data* available from analyzed satellite imagery.

Once the data have been stored in the computer, the system permits interaction with the data by way of computer terminals for problem solving and adjustments, and the high-resolution graphics screen permits data manipulation, editing and design by precise graphics techniques. The output portion of the system includes various printing and plotting devices for both text and graphics.

The chief advantages of such a system include speed of operation at all stages, fewer mistakes, higher precision, and lower costs.

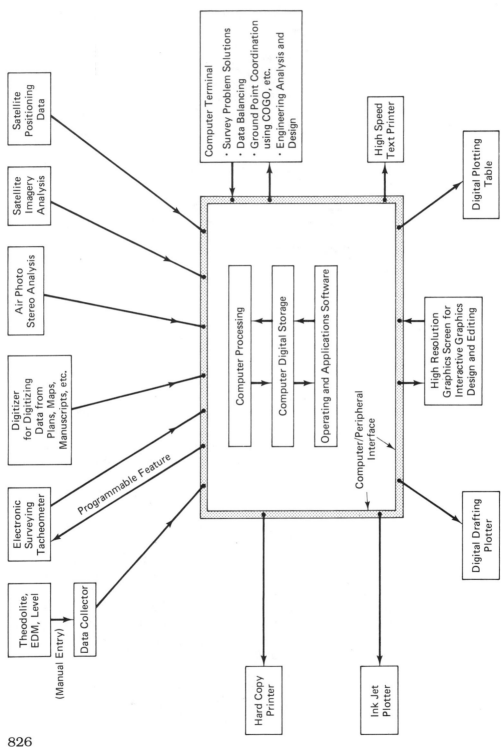

FIGURE 16.1 Computerized surveying data systems: the computer core with the input and output peripherals.

16.2 FIELD SURVEYING COMPONENTS OF THE SYSTEM

Electronic tacheometers were introduced in Chapter 8. The *data collectors* (Figure 16.2) used with the electronic tacheometers can also be used for any type of field operation utilizing traditional EDMs, theodolites, and levels. The traditional methods of note keeping described in Section 1.18 must be reviewed to determine reliable techniques for various forms of electronic note keeping. Complete sketches are required to show the location and relationship of the entered data. Extra care and precautions are advised because much of the data are lost to view once they have been entered into the data collector making it impossible to scan the data for errors or omissions.

FIGURE 16.2 Kern E-1 electronic theodolite combined with a Kern DM 502 infra-red EDM. A Hewlett Packard 41 Cv is used as a data collector and for preliminary computations. The H.P. 41 Cv itself can be interfaced with a Digital Corp. LSI-II (also shown) for further data processing. (Courtesy Seneca College)

Some knowledge of the total surveying data system is required by the field surveyor to ensure that the field coding of data is optimal for all system's requirements. Figure 16.3 illustrates a set of pipeline survey field codes used by a large Canadian surveying company. It can be seen that many of the codes can be used for most preliminary surveys, but that some of the codes apply only to pipeline surveys. Codes can be designed for any specific survey purposes and are chosen to describe suitably the terrain conditions, both for survey data purposes and for subsequent plan and manuscript presentations. A chief advantage noted in the previous section was speed; this advantage is particularly relevant to the use of electronic surveying tacheometers, where many applications now permit the field survey to be cost competitive once again with aerial surveys. Another advantage in the use of data collectors is the mistake-free, high-speed transfer of collected ground data to the data base.

INFORMATION FIELD

1 · B	THESE FEATURES ARE TO BE USED IN INFO I FIELD WITH PRIMARY AND SECONDARY CODES MARKED ●
2 · E	
3 · STELC	
4 · IPSCO	
5 · EDGE	
6 · CL	

J. D. BARNES LIMITED

PRIMARY AND OR SECONDARY CODES

10 OPEN END ●	11 TOPP (TOP OF PIPE)	12 SBRT (SIDE BEND RT)	13 SBLT (SIDE BEND LT)	14 OB (OVER BEND)
15 SAG	16 WELD	17 TSTL (TEST LEAD)	18 TEE (TIE OVER WELD)	19 PLCR (PIPELINE CROSSING)
20 DRIP ●	21 PADU ● (SAND PAD UPPER)	22 PADL ● (SAND PAD LOWER)	23 PADB ● (SAND PAD BOTH)	24 WELD C.O. (CUT OUT WELD)
25 WELD T (TIE WELD)	26 WELD FT (FINAL TIE WELD)	27 BORE	28 ANOD (ANODE)	29 FPAD ● (FOAM PAD)
30 OTCR ● (OUTCROP ROCK)	31 ROCK ●	32 SDTY ● (SEDIMENTARY ROCK)	33 CLAY ●	34 CLWT (CLAY WET)
35 SAND ●	36 PEAT ●	37 BLDR ● (BOULDERS)	38 CASG ● (CASING)	39 GRVL (GRAVEL)
40 SNDW (SAND WET)	41 POLE	42	43 BURC (BURIED CABLE)	44 FPL (FOREIGN PIPE)
45 DRTL ● (DRAINAGE TILE)	46 STSW ● (STORM SEWER)	47 SASW ● (SANITARY SEWER)	48 WTMN ● (WATERMAIN)	49 SLEW
50	51 SCOR ● (STRUCTURE CORNER)	52 OHTC (OVERHEAD TELE)	53 OHHC (OVERHEAD HYDRO)	54 ADPL (ADJ PIPE LINE)
55 BERM	56 VENT	57 VALVE	58 MLV (MAIN LINE VALVE)	59 TRAL (TRAIL)
60 GRRD ● (GRAVEL ROAD)	61 PVMT ● (PAVEMENT)	62 DRRD ● (DIRT ROAD)	63 FRML (FARM LANE)	64 FNCE (FENCE)
65 BUSH	66 PLND ● (PLOUGHED LAND)	67 PAST (PASTURE)	68 NATG (NATURAL GROUND)	69 RTKS (RAILWAY TRACKS)
70 RIVR (RIVER)	71 DTCH (DITCH)	72 T BNK (TOP OF BANK)	73 CRK (CREEK)	74 B BNK (BOTTOM OF BANK)
75 SWMP ● (SWAMP)	76	77	78 TREE	79 WATR ● (WATER)
80 TIE ● (TIE IN)	81 BAGS ● (SAND BAGS)	82 BRKR ● (BREAKER)	83 BOWT ● (BOLT ON RIVER WEIGHTS)	84 SDWT (SADDLE WEIGHT)
85 ANCH ● (ANCHOR)	86 RSHD ● (ROCK SHEILD)	87 SLAB ● (CONCRETE SLAB)	88 WRAP ●	89 CCC (CONCRETE COAT)
90 I BAR (IRON BAR)	91 LSM (LEGAL SURVEY MON)	92 CSM (CONTROL SURVEY MON)	93 HSP (HORIZ SURVEY POINT)	94 BM (BENCH MARK)
95 TBM (TEMP BENCH MARK)	96 WS (WOOD STAKE)	97 RP (REFERENCE POINT)	98 PRLN (PROPERTY LIMIT)	99 ROWL (RIGHT OF WAY LIMIT)
100 CMFG (CHANGE MANUFACTURER)	101 CHWT (CLASS I)	102 CHWT (CLASS 2)	103 CHWT (CLASS 3)	104

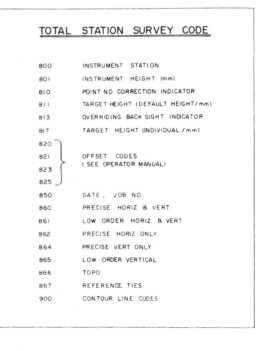

TOTAL STATION SURVEY CODE

800	INSTRUMENT STATION
801	INSTRUMENT HEIGHT (mm)
810	POINT NO CORRECTION INDICATOR
811	TARGET HEIGHT (DEFAULT HEIGHT/mm)
813	OVERRIDING BACK SIGHT INDICATOR
817	TARGET HEIGHT (INDIVIDUAL /mm)
820 821 823 825	OFFSET CODES (SEE OPERATOR MANUAL)
850	DATE, JOB NO
860	PRECISE HORIZ & VERT
861	LOW ORDER HORIZ. & VERT
862	PRECISE HORIZ ONLY
864	PRECISE VERT ONLY
865	LOW ORDER VERTICAL
866	TOPO
867	REFERENCE TIES
900	CONTOUR LINE CODES

FIGURE 16.3 Field data coding for use with a Wild Tachymat. (Courtesy J. D. Barnes Ltd., Toronto)

16.3 REMOTE SENSING COMPONENTS OF THE SYSTEM

The *stereo analysis* of air photos (Chapter 15) is not a new technique, although the technique itself has been continually refined over the years. Its value to the computerized surveying data system is its ability to identify and digitize (*x*, *y*, and *z* coordinates) large areas of terrain data. All these data are an important component of the area data base.

Satellite imagery received from Land Sat 1V can be processed by a *digital image analysis system* that classifies terrain into categories of soil and rock types and vegetative cover, and can provide moisture regime analysis; these data can be digitized and added to the data base of that area. In addition to providing terrain data, satellites are also being utilized to provide position coordinates (Chapter 11).

16.4 DIGITIZER COMPONENT OF THE SYSTEM

The remote-sensing techniques just described have the inherent capability of digitizing the data after they have been analyzed. When a data base is being prepared for a specific area, in addition to available remote-sensing data, one would want to include all the relevant data that had previously been collected on that area. These data presently exist in the form of plans, maps, and map manuscripts produced by many different agencies and probably at several different scales.

The *digitizer* shown in Figure 16.1 and Figure 16.4 has the capability of coding and digitizing various forms of terrain data from existing plans, maps, and map manuscripts, thus increasing and enhancing the data base of the area. Manual digitizing is time consuming, but if the data so obtained are preserved in a data base for unlimited future use, the technique is cost effective.

16.5 CENTRAL COMPONENT OF THE SYSTEM: THE COMPUTER

As can be seen from Figure 16.1, the computer accepts and processes the data, and then stores them in the data base. The stored information is categorized into layers of homogenous information.

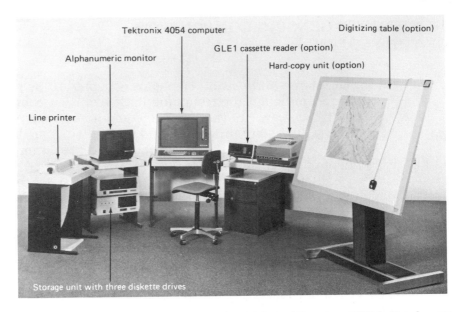

FIGURE 16.4 Wild Geomap work station. (Courtesy Wild Heerbrugg Canada Ltd.)

For example, the elevation and contour data will be kept in the same layer (storage location). The soils types, ground cover, drainage, utilities, and the like are kept in their own layers so that specific information on an area can be retrieved and portrayed or a map can be produced showing *all* the stored data. See Figures 16.5 and 16.6 for additional integrated systems.

In addition to portraying (e.g., by ditigal plotters) stored layer data, the system has the capability of analyzing the stored data and producing derived maps and plans. For example, if a complete data base existed on an undeveloped area, one could, by relating contour values to drainage patterns, delineate flood-risk areas and other areas unsuitable for development; an analysis of vegetative and soils characteristics could result in a map identifying potentially good cattle grazing areas or land that had potential for various crops. Satellite imagery has the additional advantage of being updated on a regular basis so that time effects (seasons, cycles, etc.) can also be portrayed and evaluated.

In addition to accepting digital data, the computer will also accept descriptive nongraphic data. In urban areas the data base of a neighborhood, in addition to storing locations and descriptions of all utilities, property lines, building locations, and the like, will also include attribute data such as owner's names, assessment, and type of dwelling.

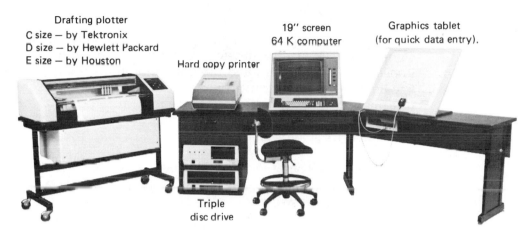

Drafting plotter
C size — by Tektronix
D size — by Hewlett Packard
E size — by Houston

Hard copy printer

19" screen
64 K computer

Graphics tablet
(for quick data entry).

Triple
disc drive

FIGURE 16.5 Computer-aided drafting system: features a 64K Computer, disk drives, 19″ display screen, graphics tablet and a computer-driven plotter (C, D or E size). (Courtesy Keuffel & Esser Co.)

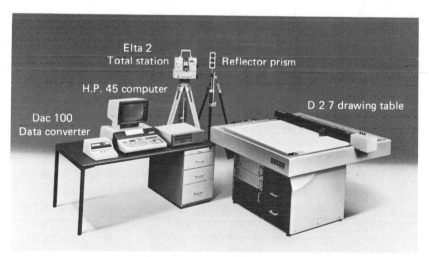

Elta 2
Total station   Reflector prism

H.P. 45 computer

Dac 100
Data converter

D 2 7 drawing table

FIGURE 16.6 Automatic surveying and plotting systems. (Courtesy Carl Zeiss Canada Ltd.)

The computer has other important functions beyond the storage, sorting, and retrieval of data. The computer can be used to solve problems, adjust data, and generate derived data. A particularly useful computer program for surveyors and highway designers is COGO (coordinate geometry). Even with a limited knowledge of computers and programming, the surveyor can in a few hours learn to access a computer and effect complex solutions utilizing the COGO program. For example, in the case of generating coordinates for inclusion in a data base, the surveyor need only follow these typical few steps:

1. Access the computer by name and password.
2. Type in the direction for the creation of a COGO data file (e.g., COGO TABLE).
3. Define the table size and identification number when prompted by the computer (e.g., 100 stations, table 1).
4. Type in the program name, COGO.
5. Store the coordinates of the first (1) point (N, E).
6. Enter bearing and distance for points 1 to 2.
7. The coordinates of point 2 will appear immediately.

All subsequent points are similarly entered. Points on curves can also be entered by following the simple format in the COGO manual. All the coordinate points can be stored indefinitely for later use.

The computer–peripheral interface shown in the schematic (Figure 16.1) refers to connecting devices such as the R.S.232C, IEEE, etc., together with handshake software programs designed to effect the data transfer; most peripheral manufacturers provide the interfaces necessary for connection to the system.

16.6 HIGH-RESOLUTION GRAPHICS SCREEN COMPONENT

The *high-resolution graphics screen* can be used in two ways. First, it can be used as a graphic output to display the stored data; color graphics can be very useful in analyzing the graphic display. Second, the graphics screen can be used to interact with the stored data. For example, if part of a subdivision plan is shown on the screen (see Figure 16.7), a lot line can be erased, moved slightly, and reentered. The revised display can be recorded via the hard-copy printer so that an ongoing record of revisions can be

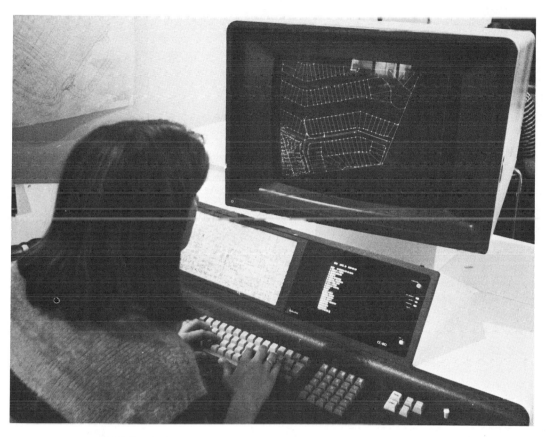

FIGURE 16.7 Land division design and editing being performed on an Autotrol Interactive Graphics work station. (Courtesy Seneca College)

kept. The screen interaction can constitute actual design (CAD; computer-assisted design), or it can be used simply for editing purposes. An example of editing would be when, in comparing the contours on two adjacent urban map sheets, the contours on one sheet border do not precisely match up with the contours on the adjacent sheet border (failure to match is a result of interpolating from different data on each sheet). The problem is solved by viewing both sheets on the screen simultaneously, and then by using an electronic stylus, simply to erase the noncompatible sections and redraw them on the screen so that they are compatible.

A much prized feature of interactive graphics is that a large drawing (say, a subdivision property plan) can be viewed at different scales, and one small section can be "windowed" and viewed in isolation at a scale suitable for interactive design, right on the screen (see Figures 16.8a and b).

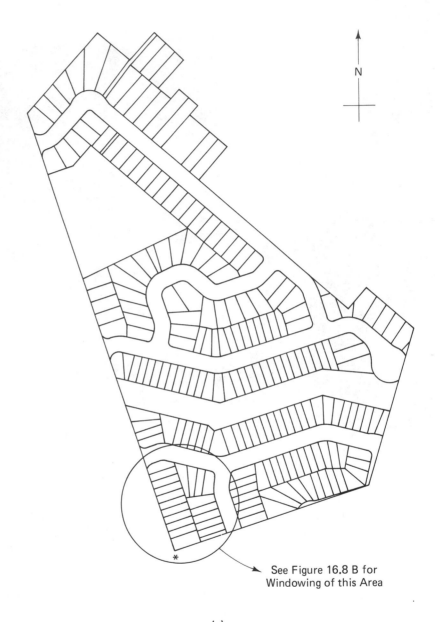

N

*

See Figure 16.8 B for
Windowing of this Area

(a)

FIGURE 16.8 (a) Reduced-scale plan of subdivision. Data generated at the Autotrol Interactive Graphics work station and plotted by the Interfaced Calcomp plotter; (b) "windowed" portion of the subdivision plan shown in part (a), showing tangent and chord distances, bearings grid coordinates, etc. At this scale (1:1000) design can be effected on the graphics screen and later printed or plotted.

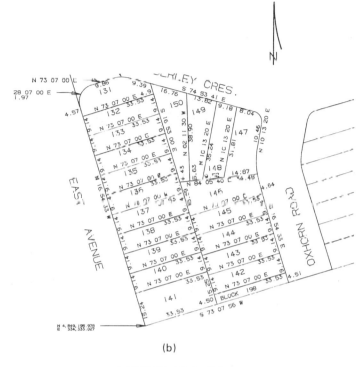

(b)

FIGURE 16.8 (b)

16.7 OUTPUT COMPONENTS OF THE SYSTEM

The *hard-copy printer,* already mentioned in regard to interactive graphics, can be supplied for black and white or color output. Once again, it is an extremely useful record in the interactive design process.

The *high-speed printer* is useful in providing the field surveyor with layout data in the form of both polar and rectangular coordinates (see Figure 16.9a). For the sake of clarity, it is very useful to have the layout point numbers added to the construction plan (Figure 16.9b). Polar coordinates (azimuth and distance) are listed for instrument setups at specific coordinated control monuments. Data for several reference control stations are included in the printout so that reference azimuths can be established and verified before the actual laying out commences.

THE MUNICIPALITY OF METROPOLITAN TORONTO - DEPARTMENT OF ROADS AND TRAFFIC

W.R. ALLEN ROAD FROM SHEPPARD AVENUE TO STANSTEAD DRIVE

ENGINEERING STAKEOUT

FROM STATION 269+80.00   TO STATION 271+30.00

BASE LINE

INSTRUMENT ON CONST CONTROL MON CX-80

AZIMUTH 213-57- 1

SIGHTING   RAP #2 - ANTENNA C.F.B.

POINT NO.	STATION	DESCRIPTION	OFFSET FROM BASELINE		AZIMUTH DEG-MIN-SEC	DISTANCE	CLOCKWISE TURN ANGLE	ELEVATION	COORDINATES NORTH	EAST
876	269+89.355	CONST CONTROL MON CX-76	22.862	LEFT	170-49-16	80.430	316-52-15	0.0	4845374.460	307710.370
885	271+29.785	CONST CONTROL MON CX-85	22.857	LEFT	350-49- 8	60.000	136-52- 7	0.0	4845513.091	307687.967
877	269+95.164	CONST CONTROL MON CX-77	22.861	RIGHT	139-19-24	87.513	285-22-24	0.0	4845387.490	307754.580
878	269+97.098	CONST CONTROL MON CX-78	38.095	RIGHT	130-50-11	94.861	276-53-11	0.0	4845391.830	307769.310
879	270+27.530	CONST CONTROL MON CX-79	38.081	RIGHT	115-33-22	74.155	261-36-21	0.0	4845421.870	307764.440
881	270+69.932	CONST CONTROL MON CX-81	22.862	RIGHT	80-38- 4	45.719	226-41- 3	0.0	4845461.300	307742.650
958	290+51.899	RAP #3 - RADIO TOWER	294.749	LEFT	344-19-40	1990.850	130-22-40	0.0	4845520.531	307159.747
884	271+29.932	CONST CONTROL MON CX-84	22.862	RIGHT	28- 3-30	75.551	174- 1-29	0.0	4847370.697	307733.077
959	0+00.000	RAP #4 - CN TOWER	0.0		152-46-14	******.***	298-49-14	0.0	4833410.793	313894.638
933	270+16.535	CONST CONTROL MON CX-133	61.272	RIGHT	113- 9- 1	99.566	259-12- 1	0.0	4845414.717	307789.088
956	271+72.134	RAP #1 - BILLBOARD FRAME	471.682	RIGHT	69- 7-32	505.019	215-10-31	0.0	4845633.810	308169.411
960	269+67.759	CONTROL MON MTR77-6119	9.329	RIGHT	153-18-17	106.983	299-21-32	196.768	4845358.277	307745.594
957	268+98.575	RAP #2 - ANTENNA C.F.B.	183.253	LEFT	213-57- 1	234.606	0- 0- 0	0.0	4845259.249	307566.519
483	269+83.555	BC CORNER ROUND	25.637	LEFT	172-39-51	86.193	318-42-51	0.0	4845368.437	307708.556
753	269+83.622	CATCH BASIN GUTTER	25.142	LEFT	172-20-12	86.687	318-23-11	0.0	4845368.437	307709.034
485	269+84.988	PI CORNER ROUND	13.500	LEFT	164-31-15	85.312	310-34-15	0.0	4845371.643	307716.388
486	269+88.075	MP CORNER ROUND	16.973	LEFT	166-41-56	81.922	312-44-55	0.0	4845374.136	307716.309
446	269+88.654	PI CORNER ROUND	13.500	RIGHT	146-40-47	88.906	292-43-46	0.0	4845379.569	307746.378
2103	269+90.000	C/L OF CONSTRUCTION	0.0		154-49-55	82.995	300-52-54	196.352	4845378.744	307732.836
444	269+90.625	BC CORNER ROUND	28.986	RIGHT	137-35-48	94.626	283-38-47	0.0	4845383.986	307761.351
754	269+91.351	CATCH BASIN GUTTER	34.690	RIGHT	134-33- 3	97.281	280-36- 2	0.0	4845385.613	307766.866
461	269+91.604	BC CORNER ROUND	36.674	RIGHT	133-51-51	98.267	279-34-50	0.0	4845386.179	307768.784
434	269+92.355	BULLNOSE TOP OF CURB	1.500	RIGHT	153-21-22	81.172	299-24-21	196.425	4845381.308	307733.941
437	269+93.105	BC BULLNOSE TOP OF CURB	2.250	RIGHT	152-11-17	80.687	298-44-17	196.395	4845382.168	307734.562
435	269+93.105	CP BULLNOSE TOP ISLAND	1.500	RIGHT	153-11-45	80.456	299-14-44	196.410	4845382.048	307733.821
436	269+93.105	EC BULLNOSE TOP OF CURB	0.750	RIGHT	153-42-23	80.233	299-45-22	196.425	4845381.929	307733.081
447	269+93.947	MP CORNER ROUND	18.162	RIGHT	142-24-37	86.221	288-27-36	0.0	4845385.538	307750.135
482	269+97.210	CENTER PT CORNER ROUND	27.250	LEFT	174-16-54	72.708	320-19-53	0.0	4845381.513	307704.785
484	269+97.210	EC CORNER ROUND-TOP CURB	13.500	LEFT	163-28-17	73.176	309-31-16	196.290	4845383.707	307718.358
755	269+98.000	CATCH BASIN TOP OF CURB	13.500	LEFT	163-23-29	72.392	309-26-28	196.270	4845384.488	307718.232
2107	270+00.000	C/L OF CONSTRUCTION	0.0		163-10-51	73.410	309-13-51	196.036	4845386.462	307717.913
2106	270+00.000	C/L OF CONSTRUCTION	0.0		152-40-57	73.433	298-43-56	196.156	4845388.616	307731.240
443	270+04.265	CENTER PT CORNER ROUND	27.250	RIGHT	133-24-39	82.485	279-27-38	0.0	4845397.175	307757.460
445	270+04.265	EC CORNER ROUND-TOP CURB	13.500	RIGHT	141-47-32	74.932	287-50-31	196.050	4845394.981	307743.887
756	270+05.265	CATCH BASIN TOP OF CURB	13.500	RIGHT	141-25- 0	74.059	287-28- 0	196.110	4845395.968	307743.727
2111	270+10.000	TOP OF CURB	13.500	RIGHT	139-30-47	69.973	285-33-46	195.854	4845400.642	307742.971
2110	270+10.000	TOP OF CURB	13.500	LEFT	161-55-21	60.513	307-58-20	195.854	4845396.334	307716.317
2109	270+10.000	C/L OF CONSTRUCTION	0.0		149-53-42	64.006	295-56-41	195.974	4845398.488	307729.644
2113	270+20.000	TOP OF CURB	13.500	LEFT	160-10-24	50.657	306-13-23	195.686	4845406.206	307714.722

CITY OF TORONTO

MANAGEMENT SERVICES DEPARTMENT DATA PROCESSING CENTRE

POINT NO.	STATION	DESCRIPTION	OFFSET FROM BASELINE	AZIMUTH DEG-MIN-SEC	DISTANCE	CLOCKWISE TURN ANGLE	ELEVATION	COORDINATES NORTH	EAST
2114	270+20.000	TOP OF CURB	13.500 RIGHT	134-40-41	61.648	280-43-40	195.685	4845410.514	307741.376
2116	270+20.000	C/L OF CONSTRUCTION	0.0	146-9-26	54.782	292-12-25	195.805	4845408.360	307723.049
2117	270+30.000	TOP OF CURB	13.500 RIGHT	128-23-45	53.896	274-26-44	195.532	4845420.386	307733.780
2116	270+30.000	TOP OF CURB	13.500 LEFT	157-34-58	40.871	303-37-57	195.532	4845416.078	307713.126
2115	270+30.000	C/L OF CONSTRUCTION	0.0	140-56-24	45.884	286-59-23	195.652	4845418.232	307725.453
2120	270+40.000	TOP OF CURB	13.500 RIGHT	120-8-38	47.000	266-11-37	195.392	4845430.258	307735.184
2119	270+40.000	TOP OF CURB	13.500 LEFT	153-22-39	31.221	299-25-38	195.392	4845425.949	307711.530
2118	270+40.000	C/L OF CONSTRUCTION	0.0	133-18-55	37.545	279-21-54	195.512	4845428.104	307724.857
757	270+42.210	CATCH BASIN TOP OF CURB	13.500 LEFT	152-4-27	29.120	298-7-26	195.363	4845428.131	307711.178
758	270+47.500	CATCH BASIN TOP OF CURB	13.500 RIGHT	112-53-36	42.907	258-55-35	195.373	4845437.168	307737.067
2123	270+50.000	TOP OF CURB	13.500 RIGHT	109-22-22	41.392	255-25-21	195.267	4845440.130	307736.589
2122	270+50.000	TOP OF CURB	13.500 LEFT	145-30-23	21.887	291-33-22	195.267	4845435.821	307709.935
2121	270+50.000	C/L OF CONSTRUCTION	0.0	121-41-51	30.231	267-44-50	195.387	4845439.975	307723.262
433	270+51.605	EC BULLNOSE TOP OF CURB	2.250 RIGHT	116-43-38	30.999	262-46-37	195.473	4845439.918	307725.227
431	270+51.605	CP BULLNOSE TOP OF ISLAND	1.500 RIGHT	117-33-23	30.395	263-36-23	195.488	4845439.799	307724.486
432	270+51.605	BC BULLNOSE TOP OF CURB	0.750 RIGHT	118-25-9	29.797	264-28-8	195.503	4845439.679	307723.746
429	270+52.355	TOP OF CURB	13.500 RIGHT	106-25-57	40.320	252-28-57	195.296	4845442.454	307736.213
430	270+52.355	BULLNOSE TOP OF CURB	1.500 RIGHT	116-24-24	29.952	262-27-23	195.479	4845440.539	307724.367
2125	270+60.000	TOP OF CURB	13.500 LEFT	127-5-55	13.559	273-8-54	195.155	4845445.693	307708.339
2126	270+60.000	TOP OF CURB	13.271 RIGHT	95-58-24	37.430	242-1-23	195.160	4845447.847	307734.757
2124	270+60.000	C/L OF CONSTRUCTION	0.0	103-59-39	24.864	250-2-38	195.275	4845447.847	307721.666
2128	270+70.000	TOP OF CURB	13.500 LEFT	79-30-15	9.360	225-33-14	195.057	4845455.565	307706.743
2129	270+70.000	TOP OF CURB	12.971 RIGHT	80-23-30	35.829	226-31-29	195.068	4845459.789	307732.875
2127	270+70.000	C/L OF CONSTRUCTION	0.0	80-15-49	22.859	226-19-48	195.177	4845457.719	307720.070
2132	270+80.000	TOP OF CURB	12.671 RIGHT	64-43-40	36.967	210-49-39	194.950	4845469.613	307730.933
2131	270+80.000	TOP OF CURB	13.500 LEFT	33-18-38	13.853	179-21-37	194.914	4845465.437	307705.148
2130	270+80.000	C/L OF CONSTRUCTION	0.0	56-44-21	25.036	202-47-20	195.064	4845467.591	307718.475
426	270+82.355	TOP OF CURB	13.500 LEFT	27-29-6	15.670	173-2-5	194.955	4845467.761	307704.772
759	270+87.500	CATCH BASIN TOP OF CURB	13.371 LEFT	18-59-12	20.095	165-2-11	194.923	4845472.862	307704.078
760	270+87.500	CATCH BASIN TOP OF CURB	12.446 RIGHT	54-10-15	39.498	200-13-14	194.947	4845476.981	307729.564
2135	270+90.000	TOP OF CURB	12.371 RIGHT	50-58-12	40.616	197-1-11	194.927	4845479.437	307729.091
2133	270+90.000	TOP OF CURB	13.309 LEFT	16-6-9	22.356	162-9-8	194.908	4845477.463	307703.741
2134	270+90.000	C/L OF CONSTRUCTION	0.0	39-19-46	30.514	185-22-45	195.024	4845477.463	307705.879
2137	271+00.000	C/L OF CONSTRUCTION	0.0	8-47-10	31.764	154-50-9	194.858	4845485.251	307702.392
2136	271+00.000	TOP OF CURB	12.071 RIGHT	39-57-24	46.183	186-0-23	194.877	4845489.261	307727.199
2138	271+10.000	C/L OF CONSTRUCTION	0.0	27-55-33	37.987	173-58-32	194.969	4845487.335	307715.283
2140	271+10.000	TOP OF CURB	12.809 LEFT	4-50-51	41.451	150-53-51	194.821	4845495.163	307701.043
2141	271+10.000	TOP OF CURB	11.771 RIGHT	31-33-58	53.059	177-35-57	194.842	4845499.085	307725.307
2139	271+10.000	C/L OF CONSTRUCTION	0.0	20-24-30	46.257	166-28-53	194.927	4845497.207	307713.688
2143	271+20.000	TOP OF CURB	12.559 LEFT	24-24-30	51.250	148-27-29	194.799	4845505.075	307699.694
2142	271+20.000	TOP OF CURB	11.471 RIGHT	25-10-33	60.827	171-13-32	194.821	4845508.909	307723.416
2142	271+20.000	C/L OF CONSTRUCTION	0.0	15-17-35	55.172	161-20-34	194.900	4845507.079	307712.092
428	271+27.355	TOP OF CURB	11.250 RIGHT	21-27-48	66.914	167-30-47	194.814	4845516.134	307722.024
2147	271+30.000	TOP OF CURB	11.250 RIGHT	20-20-48	69.203	166-23-47	194.812	4845518.746	307721.602
2146	271+30.000	TOP OF CURB	12.309 LEFT	0-45-17	61.132	146-48-16	194.720	4845514.986	307638.345
2145	271+30.000	C/L OF CONSTRUCTION	0.0	11-36-18	64.407	157-39-17	194.857	4845516.950	307710.496

FIGURE 16.9 (a) Computer printout of layout data

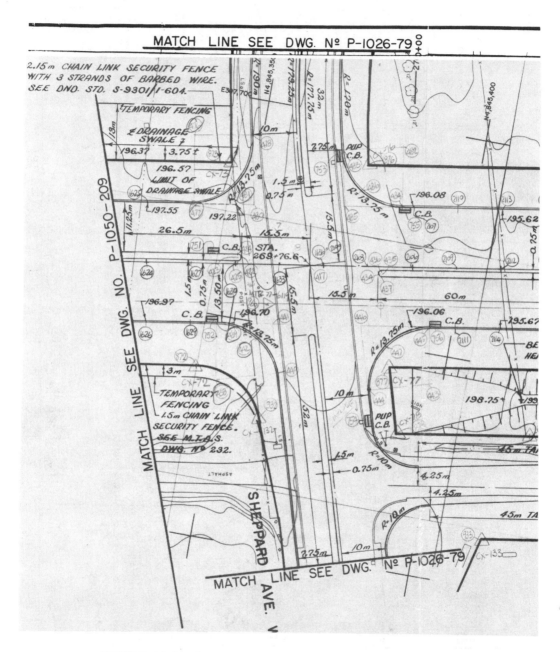

FIGURE 16.9 (b) Portion of a construction plan showing layout points. Courtesy Municiplity of Metropolitan Toronto-Department of Roads.

Layout data can also be communicated to the field surveyor by *directly loading the layout coordinates into electronic surveying tacheometers* that have this storage capability: in addition to storing layout data, some tacheometers have programmable capabilities (e.g., Zeiss ELTA-2).

The *digital drafting plotter* will plot (in a matter of seconds) entire plans of subdivisions that have been previously coordinated (see Figure 16.10). The plotter can work with four to eight pens, which gives it the capability of working with different line weights or different colors. These plotters are now widely used in the preparation of property plans and engineering plans. Having the data in digital form (instead of just graphical) ensures that items recorded close together do not disappear or become confused.

The *digital plotting table* also plots points that have been coordinated; it will accept scribing and pen or pencil tools, which makes it an ideal instrument for the preparation of plans and maps. This device is used a great deal in conjunction with stereo analyzers for the preparation of maps and plans for municipal engineering purposes (see Figures 16.11 and 16.12).

The *ink jet plotter* is a relatively new (1983) device that produces color graphics in a wide variety (256) of colors and shades. Prototype applications are being investigated for mapping purposes.

FIGURE 16.10 Land division plot on a Calcomp plotter as designed on the Autotrol work station. (Courtesy Seneca College)

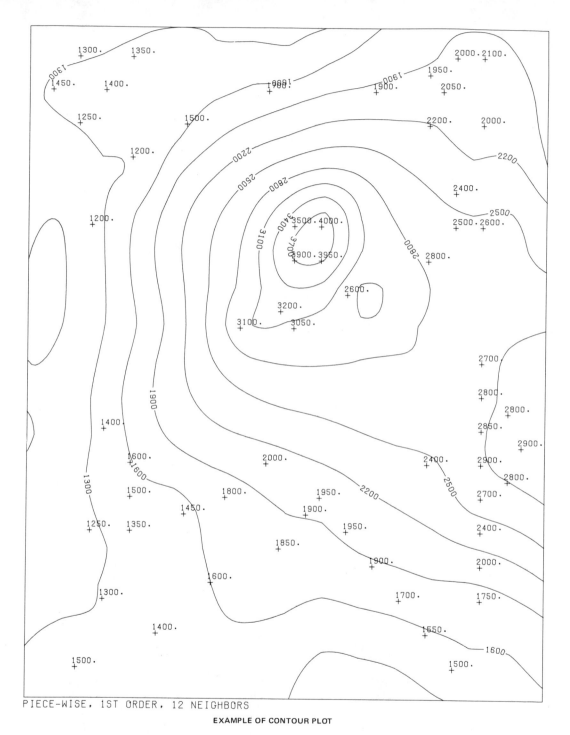

PIECE-WISE, 1ST ORDER, 12 NEIGHBORS

EXAMPLE OF CONTOUR PLOT

FIGURE 16.11 Contour plot by programmed digital plotter. (Courtesy California Computer Products, Inc.)

FIGURE 16.12 Subdivision plot by programmed digital plotter. (Courtesy California Computer Products, Inc.)

16.8 THE SYSTEM

Large systems such as described here would be effective at the state or provincial level. With all the data being fed to a central location, accessibility for all interested concerns would be optimized. All that would be required would be first to consult the data directory and identify the data layer required, and then arrange to have the desired data transferred to one's own computer. A centralization of the data would go a long way in reducing needless duplication of both surveying activities and map and plan preparations. The centrally stored data base would be equally useful to all agencies and private concerns, regardless of specific data requirements and proposed scale of drawings. Central storage would also facilitate utilization of the state or provincial grid coordinate system as a reference fabric.

The data, once transferred to an agency or company computer, would form a local data base for use in specific projects. The data in the data base would be up to date as all input data from various government agencies would be continually added, a truly dynamic system. Depending on the size of the operation and the intended use of the data, a wide variety of computers and system hardware (shown in Figure 16.1) could be utilized.

The central computer for a state or provincial system, even for a large-volume agency or company, would be a large-capacity *mainframe computer;* smaller concerns could utilize *minicomputers* of varying capacity and some, probably not all, of the peripheral hardware. Smaller concerns could well utilize a *microcomputer* to process and store field data and even to drive a small plotter (see Figure 16.13). If progress in this area continues at the present rate, problems requiring a mainframe computer now could well be handled by micro computers in a year or two. The technology described in this section is relatively new: it is improving at an incredible pace. Hardware that is too expensive to purchase today will perhaps be commonplace tomorrow. The surveyor of today cannot afford to dismiss these technological innovations as being unattainable; he or she will doubtless be using improved versions in the very near future. Although individual hardware will change, the concept of a central data base at the core of a computerized surveying data system is time valid.

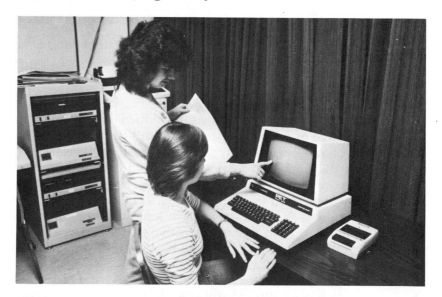

FIGURE 16.13 Personal computer (Commodore Pet) being used for traverse computations. Mini-computer (Digital LSI-11) in background is used for more complex applications.

16.9 DATA BASE

Complete data on the physical attributes of an area constitute a *digital terrain model (DTM)*. Elevation-only data on the same area would constitute a *digital elevation model (DEM)*. As noted in an earlier section, the data are sorted and stored in *layers* (i.e., a layered data base) See Figure 16.14. Layers common to different usages (e.g., mapping, municipal) would be assembled in *files*; files, however, could be cross accessed to provide tailored data or packages. The immediate problem, which is being worked on now by federal, state, and provincial agencies, is a standardization of geocodes and layer characteristics. An example is the National Digital Topographic Data Base, Ottawa (NDTDB).

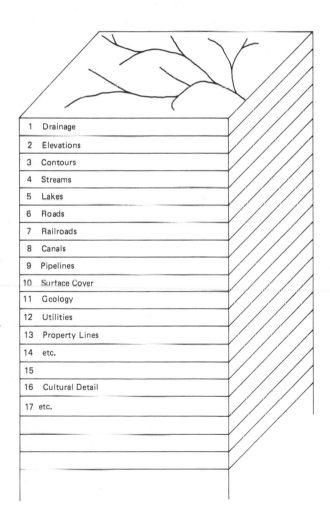

1 Drainage
2 Elevations
3 Contours
4 Streams
5 Lakes
6 Roads
7 Railroads
8 Canals
9 Pipelines
10 Surface Cover
11 Geology
12 Utilities
13 Property Lines
14 etc.
15
16 Cultural Detail
17 etc.

FIGURE 16.14 Schematic showing various data layers for a data base layer file.

16.10 SUMMARY OF CHARACTERISTICS OF A COMPUTERIZED SURVEYING DATA SYSTEM

1. Needless duplication of surveying and mapping by many agencies will be eliminated.

2. Data in the data base can be instantly accessed, with maps and plans being produced at any desired scale.

3. Existing plans and maps can be quickly revised at the same scale or at another scale. A data search would quickly indicate whether or not a sufficient amount of new data was in the data base to justify a revision.

4. The layered storage of data permits quick access to only that information desired.

5. Costs are reduced by employing digital plotting instruments that can produce whole maps in minutes instead of days, or even weeks.

6. Nongraphic data (names, descriptions, etc.) can also be stored and plotted by the digital plotters.

7. The cost effectiveness of computer-aided design (CAD) can be realized using this system. Route location and design of highways can be accomplished at large savings, even if the data base has to be assembled; if a data base already exists, savings in costs are phenomenal.

8. Large-scale planning projects from one municipality to another are automatically tied together by the state or provincial coordinate grid.

9. Accuracy of most data is improved due to automatic transcription and by the high-precision scaling techniques employed in the system.

10. Productivity is vastly improved in a wide variety of interrelated government agencies and private concerns.

11. Flexibility is possible as instruments of various levels of sophistication can be employed in a whole or part "system." Further flexibility is possible by utilizing off-line microcomputers to preprocess data and effect trial designs on desk-top plotters.

12. Drafting will be uniform and of high quality.

13. Profiles and cross sections can be generated from the data base for any specific lines or areas. These data may be sufficient for proposed uses or may require minor additional field surveying.

Appendix

A Random Errors and Survey Specifications

A.1 Random Errors

When a very large number of measurements are taken in order to establish the value of a specific dimension, the results will be grouped around the true value, much like the case of a range target as illustrated in Figure A.1. When all systematic errors and mistakes have been removed from the measurements, the residuals between the true value (dead center of the bull's-eye) and the actual measurements (shot marks) will be due to random errors.

The rifle target shown in Figure A.1 illustrates some of the characteristics of random errors:

1. Small random errors (residuals) occur more often than large random errors.
2. Random errors have an equal chance of being plus (right) or minus (left).

The number of rifle shots hitting the left and right side of each ring are shown in the ring frequency summary. These results are then plotted directly below in the form of a bar chart called a histogram.

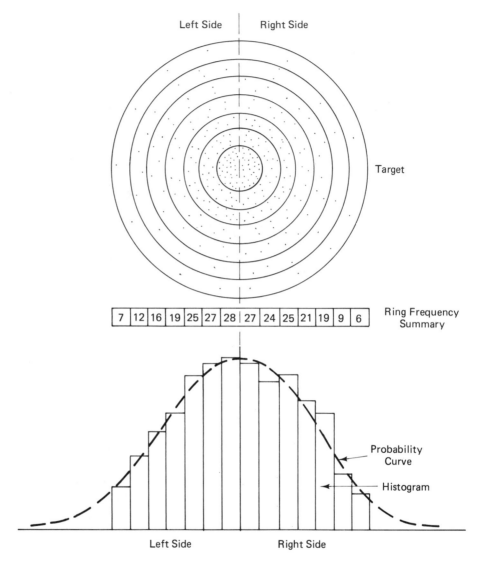

FIGURE A.1   Range target: 265 shots on target.

It will be seen that the probability of any target shot hitting a ring (or half-ring) is directly proportional to the area of the histogram rectangles. For example, the probability of one of the target shots hitting the bull's-eye (for a specific rifle and specified conditions) is $(28 + 27)/265 = 0.21$ (or 21%).

Table A.1 shows that the total of the ring probabilities is, of course, unity. Geometrically, it can be said that the area under the probability curve is equal to 1, and the probability that an error

TABLE A.1   *Target Ring Probabilities*

Ring Number	Probability
1 (Bull's-eye)	$\dfrac{28 + 27}{265} = 0.208$
2	$\dfrac{27 + 24}{265} = 0.192$
3	$\dfrac{25 + 25}{265} = 0.189$
4	$\dfrac{19 + 21}{265} = 0.151$
5	$\dfrac{16 + 19}{265} = 0.132$
6	$\dfrac{12 + 9}{265} = 0.079$
7	$\dfrac{7 + 6}{265} = 0.049$
	Total 1.000

(residual) falls within certain limits is equal to the area under the curve between those limits.

The probabilities for the hits in each ring are shown calculated in Table A.1. For example, given the *same* conditions as were present when the shots in Figure A.1 were taken (i.e., same precision rifle and same range conditions), one would expect that 40% of all future rifle shots would hit the middle two rings (0.208 + 0.192 = 0.400).

Any discussion of probability and probable behavior implies that a very large number (infinite) of observations have been taken; the larger the number of observations, the closer the results will conform to the laws of probability. In the example used, if the number of rifle shots were greatly increased and the widths of the rings greatly narrowed, the resultant plot would take the form of a smooth, symmetrical curve, known as the probability curve. This curve is shown superimposed on the histogram in Figure A.1.

In surveying, one cannot take a large number of repetitive measurements, but if *survey techniques* are used that normally give results which when plotted take the form of a probability curve, it is safe to assume that the errors associated with the survey measurements can be treated using random error distribution techniques.

Before leaving the example of the rifle target shot distribution, it is worthwhile considering the effect of different precisions on the probability curve. If a less precise rifle (technique) is used, the number of shots hitting the target center will be relatively small, and the resultant probability curve (see Figure A.2a) will be relatively flat. On the other hand, if a more precise rifle (technique) is used, a larger number of shots will hit the target center and the resultant probability curve will be much higher (see Figure A.2b), indicating that all the rifle shots are grouped more closely around the target center.

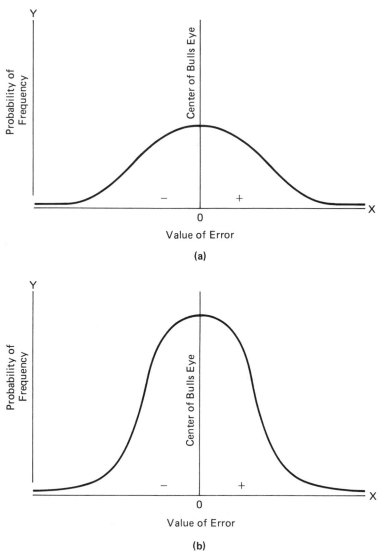

FIGURE A.2 Precision comparisons: (a) probability curve for target results from a low precision rifle: (b) probability curve for target results from a high precision rifle.

Note: If the sights of the high precision rifle being used were out of adjustment, the target hits would consistantly be left or right of the target center. The shape of the resulting probability curve would be similar to that shown in Figure A.2(b), except that the entire curve would be shifted left or right of the target center plot point. This situation illustrates that precise methods can give inaccurate results if the equipment is not adjusted properly.

A.2 Probability Curve

The probability curve shown in Figures A.1 and A.2 has the equation

$$y = \frac{1}{\sigma\sqrt{2\pi}}\, e^{\frac{-v^2}{2\sigma^2}}$$ (A.1)

where y is the ordinate value of a point on the curve (frequency of a residual of size v occurring); v is the size of the residual; e is the base of natural logarithms (2.718); and σ is a constant known as the standard deviation or standard error, a measure of precision. Since σ is associated with an infinitely large sample size, the term standard error (SE) will be used when analyzing finite survey repetitions.

A.3 Most Probable Value

In the preceding section, the concept of a residual (v) was introduced. A residual is the difference between the true value or location (e.g., bull's-eye center) and the value or location of one occurrence or measurement. When the theory of probability is applied to survey measurements, the residual is in fact the error (i.e., the difference between any field measurement and the "true" value of that dimension).

In Section 1.14 the topic of errors was first introduced, and it was noted that the "true" value of a measurement was never known, but that for the purpose of calculating errors, the *arithmetic mean* was taken to be the "true," or *most probable*, value.

Since we do not have large (infinite) numbers of repetitive measurements in surveying, the arithmetic mean will itself contain an error (discussed later as the error of the mean).

$$\text{Mean:}\quad \overline{X} = \frac{\Sigma x}{n}$$ (A.2)

where Σx is the sum of the individual (x) measurements, and n is the number of individual measurements.

A.4 Standard Error

We saw in Figure A.2 that precision could be depicted graphically by the shape of the probability curve. In statistical theory, precision is measured by the *standard deviation* (also called *standard error*). Theoretically,

$$\sigma = \pm \sqrt{\frac{\Sigma v^2}{n}} \qquad (A.3)$$

where σ is the *standard deviation* of a very large sample, v is the true residual, and n is the very large sample size. Practically,

$$SE = \pm \sqrt{\frac{\Sigma v^2}{n-1}} \qquad (A.4)$$

where SE is the *standard error* of a set of repetitive measurements; v is the error $(x - \bar{x})$, and n is the number of repetitions. Since the use of $\bar{x}$ instead of the "true" value always results in an underestimation of the standard deviation, $(n - 1)$ is used in place of n. The term $(n - 1)$ is known in statistics as *degrees of freedom* and represents the number of extra measurements taken. That is, if a line were measured 10 times, it would have 9 $(10 - 1)$ degrees of freedom. Obviously, as the number of repetitions increases, the difference between n and $(n - 1)$ becomes less significant.

The concepts just described are being used more and more to define and specify the precision of various field techniques.

As noted earlier, the arithmetic mean contains some uncertainty; this uncertainty can be expressed as the *standard error of the mean* (SE_m).

The laws of probability dictate that the error of a sum of identical measurements be given by the error multiplied by the square root of the number of measurements. SE sum $= SE \sqrt{n}$. The mean (standard error) is given by the sum divided by the number of occurrences; therefore,

$$SE_m = \frac{SE \sqrt{n}}{n} = \frac{SE}{\sqrt{n}} \qquad (A.5)$$

This expression shows that the standard error of the mean is inversely proportional to the square root of the number of measurements. That is, if the measurement is repeated by a factor of 4, the standard error of the mean is cut in half. This relationship **demonstrates that, beyond a realistic number, continued repetitions of a measurement do little to reduce uncertainty.**

Some instrument manufacturers use the concept of the standard error of the mean to specify instrument precision and attainable accuracies. The standard error of the mean (SE_m), when used in this context, is sometimes referred to as the "standard error". The reader is cautioned that the term standard error could be referring to the standard deviation or the standard error of the mean, and that clarification is required before the values can be used.

A.5 Measures of Precision

Figure A.3 shows the SE, 2SE and 3SE plotted under the probability curve. It can be shown that the area under the curve between the limits shown is as follows:

Mean $\pm$ 1SE = 68.27% of area under curve
Mean $\pm$ 2SE = 95.46% of area under curve
Mean $\pm$ 3SE = 99.74% of area under curve

We have seen in a previous section that the area under the probability curve is directly proportional to the probability of expected results.

The preceding relationship can be restated by noting that the

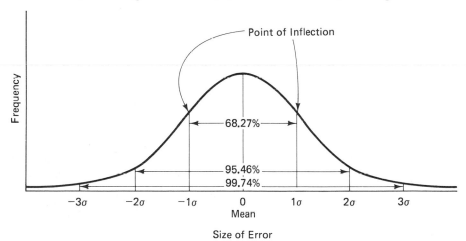

FIGURE A.3 Graph of probability curve showing standard errors.

probability of measurements deviating from the mean is as follows:

$$
\left.\begin{array}{l} 68.27\% \\ 95.46\% \\ 99.74\% \end{array}\right\} \text{of all measurements will be in range of} \quad \begin{array}{l} \bar{x} \pm SE \\ \bar{x} \pm 2SE \\ \bar{x} \pm 3SE \end{array}
$$

A term used in the past, *probable error*, was the 50% error (i.e., the limits under the curve representing 50% of the total area). Those limits are $\pm 0.6745SE$. Today's surveyors are more interested in the concept of maximum anticipated error, which varies from the 90% to the 95% probability limits.

Practical Precision Parameters

Error	% Certainty	
Probable (0.6745 SE)	= 50%	(A.6)
Standard (SE)	= 68.27%	
90% (1.6449 SE)	= 90%	(A.7)
95% (1.9599 SE)	= 95%	(A.8)

A.6 Illustrative Problem

To illustrate the concepts introduced thus far, consider the data in Table A.2. The results of 15 measurements of a survey base line are shown together with the probability computations. It is assumed that all systematic errors have already been removed from the data. The following observations are taken from the data in Table A.2.

1. The most probable distance is 266.308 m.

2. The standard error of any one measurement is ± 0.005 m.

3. The standard error of the mean is ± 0.001 m. That is, there is a 68.27% probability that the true length of the line lies between 266.308 ± 0.001 m; there is a 90% probability that the true length of the line lies between 266.308 ± 0.002 m 5 (i.e., $0.0013 \times 1.6449 = 0.002$); there is a 95% probability that the true length of the line lies between 266.308 ± 0.003 m (i.e., $0.0013 \times 1.9599 = 0.00254$).

4. With the probable error (50%) at 0.0034, it is expected that

half of the 15 measurements will lie between 266.308 ± 0.003 (i.e., from 266.305 to 266.311). In fact, only 5 (1/3) measurements fall in that range.

5. With the 90% error at ±0.008, it is expected that 90% of the measurements will lie between 266.308 ± 0.008 (i.e., from 266.300 to 266.316). In fact, 93% (14 out of 15) fall in that range.

6. With the 95% error at ±0.098, it is expected that 95% of the measurements will lie between 266.308 ± 0.0098 (i.e., from 266.298 to 266.318). In fact, all the measurements fall in that range.

TABLE A.2 *Analysis of Random Distance Errors*
$(v = x - \bar{x})$

n	Distance X (m)	Residual(v)	v^2
1	266.304	−0.0035	0.0000123
2	266.318	+0.0105	0.0001103
3	266.312	+0.0045	0.0000203
4	266.304	−0.0035	0.0000123
5	266.313	+0.0055	0.0000303
6	266.307	−0.0005	0.0000003
7	266.309	+0.0015	0.0000023
8	266.303	−0.0045	0.0000203
9	266.301	−0.0065	0.0000423
10	266.305	−0.0025	0.0000063
11	266.302	−0.0055	0.0000303
12	266.310	+0.0025	0.0000063
13	266.314	+0.0065	0.0000423
14	266.307	−0.0005	0.0000003
15	266.303	−0.0045	0.0000203
	$\Sigma X = 3994.612$	$\Sigma v = -0.0005$	$\Sigma v^2 = 0.0003565$

1. Mean (most probable value), $\bar{x} = \dfrac{\Sigma x}{n} = \quad 3994.612/15 \quad =$ 266.3075 m. $\qquad$ (A.2)

2. Standard error, $SE = \sqrt{\Sigma v^2/(n-1)} = \sqrt{0.0003565/14} = 0.0050$ m. $\qquad$ (A.4)

3. Standard error of the mean, $SE_m = SE/\sqrt{n}, = 0.0050/\sqrt{15} = 0.0013$ m. $\qquad$ (A.5)

4. Probable error (50% error) $= 0.6745\,SE = 0.0034$ m. $\qquad$ (A.6)

5. 90% error $= 1.6449SE = 0.0082$ m. $\qquad$ (A.7)

6. 95% error $= 1.9599SE = 0.0098$ m. $\qquad$ (A.8)

It should be noted that expected frequencies are entirely valid only for randomly distributed data. When the number of observations is small, as is the case in surveying measurements, one can often encounter actual data that are marginally inconsistent with predicted frequencies. If, however, the field data differ significantly from probable expectations, it is safe to assume that the data are unreliable, due either to untreated systematic errors or to undetected mistakes.

As a rule of thumb, measurements that fall outside the range of $x \pm 3.5$ SE (Figure A.3) either are rejected from the set of measurements or are repeated in the field.

A.7 Propagation of Errors

This section deals with the arithmetic manipulation of values containing errors (e.g., sums, products, etc.)

A.7.1 Sums of varied measurements

The sum of any number of measurements that have individual mean and SE values is determined as follows. If distance K is the sum of two distances A and B,

$$\text{SE}_K = \sqrt{\text{SE}_A{}^2 + \text{SE}_B{}^2} \qquad (A.9)$$

Example 1: If distance A were found to be 101.318 ± 0.010 m and distance B were found to be 87.200 ± 0.008 m, what is the distance K $(A + B)$?

$$\text{SE}_K = \sqrt{0.010^2 + 0.008^2} = 0.013 \text{ m} \qquad (A.9)$$
$$K = 188.518 \pm 0.013 \text{ m}$$

Example 2: If from the preceding data distance L is the *difference* in the two distances A and B,

$$\text{SE}_L = \sqrt{\text{SE}_A^2 + \text{SE}_B^2},$$
$$SE_L = \sqrt{0.010^2 + 0.008^2} = 0.013 \text{ m} \qquad (A.9)$$
$$L = (101.318 - 87.200) \pm 0.013$$
$$= 14.118 \pm 0.013 \text{ m}$$

Example 3: If the difference in elevation between two points is determined by taking two rod readings, each having a SE of 0.005 m, what is the SE of the difference in elevation?

$$\text{SE (diff. of elev.)} = \sqrt{0.005^2 + 0.005^2} = 0.007 \text{ m} \qquad \text{(A.9)}$$
$$\text{SE (diff. of elev.)} = 0.005 \sqrt{2} = 0.007 \text{ m (See below)} \qquad \text{(A.10)}$$

A.7.2 Sums of identical measurements

The sum of any number of measurements each one having the same SE is as follows:

$$\text{SE}_{\text{sum}} = \sqrt{n \times \text{SE}^2}$$
$$= \text{SE} \sqrt{n} \qquad \text{(A.10)}$$

Example 1. A distance of 700.00 ft is laid out using a 100.00-ft steel tape that has an SE = 0.02 ft. In this case the SE of the 700.00-ft distance is $0.02\sqrt{7} = 0.05$ ft $\qquad$ (A.10)

A.7.3 Products of measurements

The product of any number of measurements that have individual SE values can be given by the relationship

$$\text{SE}_{\text{product}} = \pm \sqrt{A^2\text{SE}_B^2 + B^2\text{SE}_A^2} \qquad \text{(A.11)}$$

where A and B are the dimensions to be multiplied.

Example 1. Consider a rectangular field having $A = 250.00$ ft $\pm$ 0.04 ft and $B = 100.00$ ft $\pm$ 0.02 ft.

$$\text{SE}_{\text{product}} = \pm \sqrt{250^2 \times 0.02^2 + 100^2 \times 0.04^2} \qquad \text{(A.11)}$$
$$= \pm 6.40 \text{ ft}$$
$$\text{Area of field} = 250 \times 100 = 25\ 000 \pm 6 \text{ ft}^2$$

A.8 Weighted Observations

If the reliability of different sets of measurements varies one to the other, then equal consideration cannot be given to those sets. Some method (weighting) must be used to arrive at a best value. For example, measurements may be made under varying conditions, utilizing varying levels of skills, and repeated a varying number of times.

A.8.1 Weight by number of repetitions

The simplest concept of weighted values can be illustrated by the following example where the weighted mean is calculated. A distance was measured six times; the values obtained were

$$6.012 \text{ m}$$
$$6.011 \text{ m}$$
$$6.012 \text{ m}$$
$$6.012 \text{ m}$$
$$6.011 \text{ m}$$
$$6.013 \text{ m}$$

The value of 6.012 was observed three times; 6.011, two times; and 6.013, one time.

Distance (x)	Weight (w)	x · w
6.012	3	18.036
6.011	2	12.022
6.013	1	6.013
	$\Sigma w = 6$	$\Sigma xw = 36.071$

$$\text{Weighted mean} = \frac{36.071}{6} = 6.012 \text{ m}$$

That is
$$\bar{x}_w = \frac{\Sigma xw}{\Sigma w} \qquad (A.12)$$

where $\bar{x}_w$ is the weighted mean, x is the individual measurement, and w is the weight factor.

If the distance had been measured six times and six different results occurred, each measurement would have received a weight of one, and the computation would simply be the same as for the arithmetic mean.

A.8.2 Weight by standard error of the mean SE$_m$

The standard error of the mean SE$_m$ was introduced in Section A.5. This measure tells us about the reliability of a measurement set and supplies a weight for the mean of a set of measurements. A set with a small SE$_m$ should receive more weight than a set with a large (less precise) SE$_m$.

$$SE_m = \frac{SE}{\sqrt{n}} \qquad (A.5)$$

It is seen that the error varies inversely with the square root of the number of measurements; it is also true that the number of measurements varies inversely with the SE_m^2.

We saw in the previous section that weights were proportional to the number (n) of measurements. That is, generally,

$$W_K \, \alpha \, \frac{1}{SE_{K^2}} \tag{A.13}$$

A.8.3 Adjustments

When the absolute size of an error is known, and when weights have been assigned to measurements having varying reliabilities, corrections to the field data will be made in such a way that the error is eliminated by applying corrections that reflect the various weightings. It is obvious that measurements having large weights will be corrected less than measurements having small weights. (The more certain the measurement, the larger the weight.) It follows that correction factors should be in inverse ratio to the corresponding weights.

Example 1. The angles in a triangle were determined with A being measured three times, B being measured two times, and C being measured once.

It is noted that the correction factor is simply the inverse of the weight, and that the actual correction for each angle is simply the ratio of the correction factor to the total correction factor, all multiplied by the total correction.

Angle	Mean Value	Weight	Correction Factor	Correction	Adjusted Angle
A	45°07'32"	3	$\frac{1}{3} = 0.33$	$0.33/1.83 \times 20 = +\ 4$	45°07'36"
B	71°51'06"	2	$\frac{1}{2} = 0.50$	$0.50/1.83 \times 20 = +\ 5$	71°51'11"
C	63°01'02"	1	$1 = \underline{1.00}$	$1/1.83 \times 20 = \underline{+11}$	$\underline{63°01'13"}$
	179°59'40"		1.83	20	179°59'60" = 180°00'00", adjusted

Error	$= -20"$
Correction	$= +20"$

Example 2. Consider the same angles, except in this case the weights will be related to the SE characteristics of three different theodolites.

Angle	Mean Value	SE	(SE²) Correction Factor	Correction	Corrected Angle
A	45°07'34"	±0.50"	0.25	0.25/29.25 × 20 = 0	45°07'34"
B	71°51'06"	±2.00"	4.00	4/29.25 × 20 = 3	71°51'09"
C	63°01'00"	±5.00"	25.00	25/29.25 × 20 = 17	63°01'17"
	179°59'40"		29.25	20	179°59'60"
				= 180°	

Error   = −20"
Correction = +20"

In this case Eq. (A.13) was used.

Example 3   Adjustment of a Level Loop. Errors are introduced into level surveying each time a rod reading is taken. It stands to reason that corrections to elevations in a level loop should be proportional to the number of instrument setups (i.e., the weights should be inversely proportional to the number of instrument setups). Furthermore, since there is normally good correlation between the number of setups and the distance surveyed, corrections can be applied in proportion to the distance surveyed. (If a part of a level loop were in unusual terrain, corrections could then be proportional to the number of setups).

Consider the following example, where temporary bench marks (TBM) are established in the area of an interchange construction. See Figure A.4.

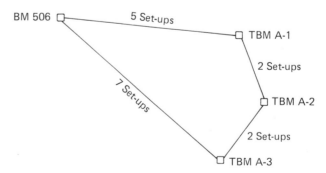

FIGURE A.4   Level loop adjustment.

Station	Elevation	No. of Setups between Stations	Correction Factor	Corrected Elevation
BM #506	172.865 (fixed)			172.865
TBM #A-1	168.303	5	$5/16 \times .011 = .003$	168.306
TBM #A-2	168.983	2	$7/16 \times .011 = .005$	168.988
TBM #A-3	170.417	2	$9/16 \times .011 = .006$	170.423
BM #506	172.854	7	$16/16 \times .011 = .011$	172.865
		16 setups		

$$\text{Error} = 172.865 - 172.854 = -0.011 \text{ m}$$
$$\text{Correction} = +0.011 \text{ m}$$

A.9 Principle of Least Squares

Reference to the data in Table A.2 will show that the smaller the sum of the squared errors, $\Sigma(x - \bar{x})^2$, the more precise will be the data (i.e., the more closely are the data grouped around the mean $\bar{x}$).

If such data were to be assigned weights, it is logical to weight each error $(x - \bar{x})$ such that the sum of the squares of the errors is a minimum. This is the principle of least squares: $\Sigma \, Wa \, (x - xa)^2$ is a minimum.

The development of the principle of least squares and the adjustments based on that principle can be found in texts on surveying adjustments. In order that a least squares adjustment be valid, a reliable extimate of the SE's of various measuring techniques must be available to properly identify the individual weights. Section A.11.4 contains some examples.

A.10 Two-Dimensional Errors

The concept of two-dimensional errors was first introduced in Section 6.7 and Figure 6.12. When the concept of *position* is considered, two parameters (i.e., x and y or r and θ) must be analyzed.

In Chapter 8 it was shown that traverse closures were rated with respect to relative accuracies (i.e., $1/5000$, or $1/10000$, etc.). Surveys are often specified (see later in this section) by stating that the standard error be less than $1/500000$, $1/250000$, and so on.

Figures A.5a and A.5b show the two-dimensional concept with an area of uncertainty being generated because of the uncertainty in distance ($\pm E_d$) in combination with the uncertainty (E_a) resulting from the uncertainty in angle $\pm\Delta\theta$.

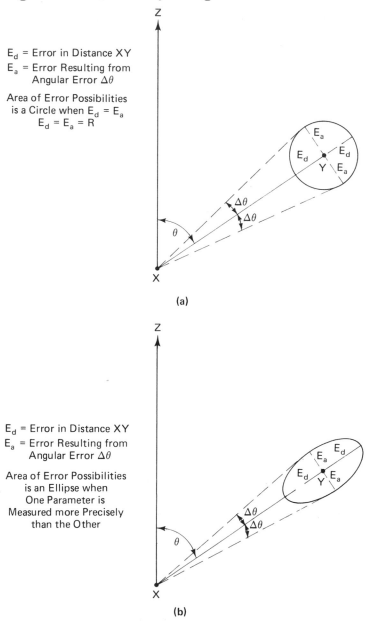

FIGURE A.5   Area of uncertainty; (a) Error circle; (b) error ellipse.

The figure of uncertainty is usually an ellipse with the major and minor axes representing the standard errors in distance and direction. When the distance and the direction have equal standard errors, the major axis equals the minor axis, resulting in a circle as the area of uncertainty, where $E_d = \sigma x$ and $E_a = \sigma y$. $r^2 = \sigma x^2 + \sigma y^2$ is the equation of such a circle of uncertainty.

In the previous discussion of one-dimensional accuracy, the probability that the true value was within $\pm 1\sigma$ was 68%. In the case of the standard ellipse ($r^2 = \sigma x^2 + \sigma y^2$), the probability that the true value is within the ellipse is 39%. If a larger probability is required, a constant K is introduced so that

$$(Kr)^2 = \sigma x^2 + \sigma y^2$$

represents the larger area. See Figure A.6.
Values for K are shown in Table A.3.

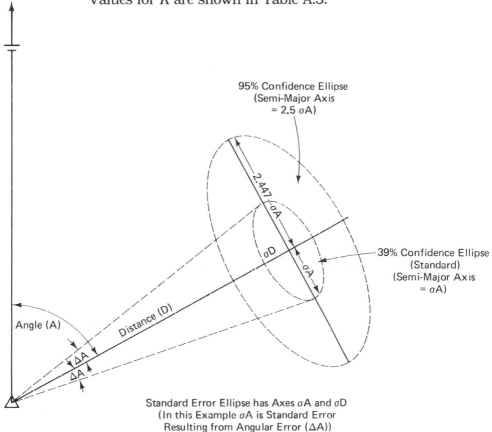

95% Confidence Ellipse
(Semi-Major Axis
= 2.5 σA)

2.447 σA

σD

σA

39% Confidence Ellipse
(Standard)
(Semi-Major Axis
= σA)

Angle (A)

Distance (D)

ΔA

ΔA

Standard Error Ellipse has Axes σA and σD
(In this Example σA is Standard Error
Resulting from Angular Error (ΔA))

FIGURE A.6   Standard error ellipse and the 95% confidence ellipse.

TABLE A.3

Probability, P (%)	K
39.4	1.000
50.0	1.177
90.0	2.146
95.0	2.447
99.0	3.035
99.8	3.500

Example: Assume a station (A) is to be set for construction control that is 450.00 ft from a control monument with a position accuracy of ± 0.04 ft. What level of accuracy is indicated for angle and distance?

$$r = \pm 0.04 = \sqrt{\sigma x^2 + \sigma y^2}$$

Since this case represents a circle, $\sigma x = \sigma y$:

$$\pm 0.04 = \sqrt{2\,\sigma x^2}$$
$$\sigma x = \frac{0.04}{\sqrt{2}} = 0.028 \text{ ft}$$

Therefore, for an error due to distance measurement of 0.028, a accuracy of

$$\frac{0.028}{450} = \frac{1}{16\,100}$$

is indicated. And since the error due to angle measurement is also 0.028, the allowable angle error can be determined as follows:

$$\frac{0.28}{450} = \text{Tan of the angle error}$$

Therefore, the allowable angle error is $\pm 0°00'13''$.

Furthermore, at the 39.4% probability level, the distance must be measured to within ± 0.028 ft; however, if we wish to speak in terms of the 95% level of probability that our point is within ± 0.04 ft, we must use the K factor (Table A.3 and Figure A.6) of 2.45. Thus, the limiting error is now

$$\frac{0.028}{2.45} = = 0.011 \text{ ft} \quad \text{for both distance and angle}$$

The angle limit is given by $0.011/450 = \tan$ angle error. Therefore, the maximum angular error is now $\pm 0°00'05''$.

One further consideration involves the normality of the distribution of errors associated with the survey techniques. We have assumed thus far that we have "good estimates" of the standard errors associated with the survey measurements, "good estimates" in this case being equated with an infinitely large number of measurement repetitions. If such "good estimates are suspect for a specific type of operation, then the sample size, given by the actual degrees of freedom, becomes a limiting factor.

Table A.4 gives factors by which the axes of the standard ellipse are to be multiplied to obtain the 95% confidence region. The table shows that a K value of 2.45 (preceding example) is valid only in the theoretical circumstance of an infinite number of measurements. For ten repetitions of a measurement, K = 2.86.

Note: For the development of error ellipse theory and a rigorous treatment of two dimensional error concepts, the reader is referred to a text on Survey Adjustments (See Appendix C).

TABLE A.4 *Probability Confidence Region Factors*

f	90%	95%	99%
1	9.95	19.97	99.99
2	4.24	6.16	14.07
3	3.31	4.37	7.85
4	2.94	3.73	6.00
5	2.75	3.40	5.15
6	2.63	3.21	4.67
7	2.55	3.08	4.37
8	2.50	2.99	4.16
9	2.45	2.92	4.00
10	2.42	2.86	3.89
11	2.37	2.82	3.80
12	2.37	2.79	3.72
13	2.35	2.76	3.66
14	2.34	2.73	3.61
15	2.32	2.71	3.57
16	2.31	2.70	3.53
17	2.30	2.68	3.50
18	2.29	2.67	3.47
19	2.28	2.65	3.44
20	2.28	2.64	3.42
25	2.25	2.60	3.34
30	2.23	2.58	3.28
40	2.21	2.54	3.22
48	2.20	2.53	3.19
60	2.19	2.51	3.16
80	2.18	2.49	3.12
120	2.17	2.48	3.09
∞	2.15	2.45	3.03

f = degrees of freedom in the measurement (practically = $n - 1$).
From "Specifications and Recommendations for Control Surveys and Survey Markers," Surveys and Mapping Branch, Department of Energy, Mines, and Resources —Canada, gives factors by which the axes of the standard ellipse are to be multiplied to obtain the 95% confidence region.

A.11 Specifications for Horizontal Control Surveys: Canada

A.11.1 General Specifications, 1973

In 1973, control survey specifications were changed by the Surveys and Mapping Branch, Department of Energy, Mines, and Resources—Ottawa, from the one-dimensional (linear) concept (i.e., 1/100000) to a two-dimensional concept stated in terms of the 95% confidence error ellipse. A survey station of a network is classified according to whether the semimajor axis of the 95% confidence region with respect to other stations of the network is equal to or less than

$$r = cd \qquad (A.14)$$

where r (centimeters) is the radius of a circle containing the 95% confidence region error ellipse, d is the distance in kilometers, and c is a factor assigned according to the order of the survey.

TABLE A.5   *Classification of Horizontal Control Surveys*

Order	c	r (ppm)
First	2	20
Second	5	50
Third	12	120
Fourth	30	300

Surveys and Mapping Branch, Department of Energy, Mines, and Resources—Ottawa, 1973.

According to Table A.5, for a first-order survey between two control stations 10 km apart, in order that these stations be classified as first order, the semimajor axis of the 95% confidence region of one station relative to the other must be equal to or less than 20 cm.

$$r = 2d = 2 \times 10 = 20 \text{ cm} \quad (0.200 \text{ m})$$

$$\frac{0.200}{10000} = \frac{1}{50000}$$

The specifications prior to 1973 (1961 specifications) used "maximum anticipated error" as a criterion for accuracy; maximum anticipated error could be interpreted, in the linear sense, as being 99% probability, or 2.5 times the standard deviation. The 95% confidence region ellipse can be related to standard deviation since its axes are approximately 2.5 times (2.447) longer than those of the standard error ellipse.

The term "maximum anticipated error" has been retained in wide popular use even though it now refers to 95% probability instead of the pre-1973 99% probability.

A.11.2 Specifications for Short Lines, 1977

A greater interest in urban control surveys and engineering works control surveys fostered a need for specifications that allowed for error propagation normally associated with short line surveys. A control survey network that includes lines of less than 3 km is classified according to whether the semimajor axis of the 95% confidence region for each station with respect to all other stations of the network is less than or equal to

$$r = c(d + 0.2) \tag{A.15}$$

where r = radius in centimeters
   d = distance in kilometers to any station
   c = factor assigned according to the order of the survey

Table A.6 shows values of r (cm) for various short distances. The table also shows comparative values for $d = 3.0$ km for the original specification $r = cd$ (1973) and the modified specification $r = c(d + 0.2)$ (1977). The modified specification reflects the fact that many errors (e.g., instrumental, centering, effects of network configuration) are simply not proportional to distance. The 1977 modification indicates that beyond a distance of 3 km the disproportionate characteristics (with respect to distance) have largely been dissipated.

In addition to the specifications issued by the Survey and Mapping Branch, local and provincial agencies have also published suggested specifications. In many jurisdictions, the maximum allowable error for legal (property) surveys is set by law to be $1/5000$. This could be stated as

$$r = 10(2d + 0.1) \text{ cm} \tag{A.16}$$

where r (cm) is the maximum allowable error (MAE) and d is the distance in kilometers. For example, for 3 km,

$$r = 10(6.1) = 61 \text{ cm}$$

$$= 1/4900$$

For urban development, it has been suggested that the following formula apply:

$$r = 12(d^{2/3} + 0.1) \text{ cm} \tag{A.17}$$

TABLE A.6 Accuracy Standards for Horizontal Control Surveys with Short Lines

| | | Semimajor Axis of 95% Confidence Region, $r = c(d + 0.2)$; d is the distance between points | | | | | | | | | | | | | | | $r = cd$ (1973 specs) | | |
| | | d = 0.1 km | | | d = 0.3 km | | | d = 1.0 km | | | d = 2.0 km | | | d = 3.0 km | | | d = 3.0 km | | |
Order	C	cm	ppm	Ratio	cm	ppm	Ratio	cm	ppm	Ratio	cm	ppm	Ratio	cm	ppm	Ratio	cm	ppm	Ratio
1	2	0.6	60	1/16 700	1.0	33	1/30 000	2.4	24	1/41 700	4.4	22	1/45 500	6.4	21	1/46 900	6.0	20	1/50 000
2	5	1.5	150	1/6700	2.5	83	1/12 000	6.0	60	1/16 700	11.0	55	1/18 200	16.0	53	1/18 800	15.0	50	1/20 000
3	12	3.6	360	1/2800	6.0	200	1/5000	14.4	144	1/6900	26.4	132	1/7600	38.4	128	1/7800	36.0	120	1/8300
4	30	9.0	900	1/1100	15.0	500	1/2000	36.0	360	1/2800	66.0	330	1/3000	96.0	320	1/3100	90.0	300	1/3300

Source: Surveys and Mapping Branch, Department of Energy Mines and Resources, Ottawa, Canada, 1977.

thus, for 3 km,

$$r = 12(3^{2/3} + 0.1) = 26 \text{ cm}$$

$$= {}^1/_{11500} \quad \text{Class 3}$$

For intensely developed urban areas, it has been suggested that the following formula apply:

$$r = 5(d^{2/3} + 0.1) \text{ cm} \tag{A.18}$$

for 3 km,

$$r = 5(3^{2/3} + 0.1) = 10.9 \text{ cm}$$

$$= {}^1/_{27500} \quad \text{Class 2}$$

Formulas A.16, A.17, and A.18 are included for illustrative purposes only—they have no official standing.

A.11.3 Measurement Guidelines

The accuracy of a horizontal control survey can best be assessed properly from the results of a rigorous least squares adjustment of the measurements. Since this assessment can only be made after the fieldwork has been completed, something more helpful is needed for those who wish to design networks and require some reasonable assurance that a particular order of accuracy will be obtained when the fieldwork is done.

The best solution is to simulate the proposed network in a suitable computer program such as GALS* using a priori estimates for the standard deviations of the proposed measurements. The results of such a simulation study, tempered with the wisdom of practical experience, will usually provide a reliable indication of the accuracy likely to be obtained in the field.

For those not able to conduct computer simulation studies, the guidelines in Table A.7 may prove useful. These guidelines are based on practical experience and the results of computer simulation studies of somewhat idealized networks. *At best they are a general guide only and must be treated with caution.* They have been arrived at by considering networks with the following characteristics.

Triangulation

First Order. Single chain; average length 20 km; all angles and

---

*A Geodetic Survey of Canada computer program.

TABLE A.7 Measurement and Internal Check Guidelines[a]

Characteristics	Order			
	1	2	3	4
Measurements				
Directions:				
Least count of theodolite in seconds	0.2	1.0	1.0	10.0
Minimum number of sets (positions)	24 (4 × 6)	6	2	2
Standard deviation (mean of all sets) in seconds	0.7	2.0	4.0	8.0
Lengths:				
Standard deviation (complete determination)				
Triangulation and traverse (ppm)	4	10	20	40
Trilateration (ppm)	3	5	10	20
Azimuths:				
Minimum number of sets	32	12	4	4
Standard deviation (mean of all sets) in seconds	0.4	1.5	3.0	6.0
Maximum number of courses between control azimuths				
[a]Triangulation	10	12	15	20
[a]Trilateration	6	8	10	12
[a]Traverse	5	6	9	12
Checks				
Directions:				
Standard deviation of a triangle misclosure in seconds	1.7	5.0	10.0	20.0
Directions vs. lengths:				
Consistency checks	See explanatory notes on pages 21 and 22 of *Specifications and Recommendations for Control Surveys and Survey Markers,* Surveys and Mapping Branch, Ottawa			
Azimuths:				
Maximum difference between control azimuths	2"/N	5"/N	10"/N	20"/N

[a]These guidelines can be applied with reasonable confidence only to networks having characteristics similar to those depicted in Figures A.8 to A.13.

sides measured; 10 courses between azimuth controls (see Figure A.7).

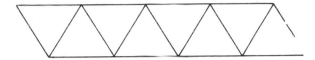

FIGURE A.7 Positional uncertainty.

Second, Third, and Fourth Orders. Single chain; average lengths 15 km, 10 km, and 5 km, respectively; all angles measured; one side measured every fourth triangle; 12, 15, and 20 courses, respectively, between azimuth controls (see Figure A.8).

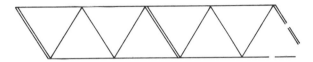

FIGURE A.8 Triangulation, single chain network, first order.

Cross-braced quadrilaterals; average lengths 15 km, 10 km, and 5 km, respectively; all angles measured; one length measured every second quadrilateral; 12, 15, and 20 courses, respectively, between azimuth controls (see Figure A.9).

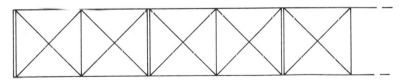

FIGURE A.9 Triangulation, single chain network, second to fourth order.

Trilateration

All Orders. Patterns shown in Figures A.10 and A.11; one redundant line per point; average lengths 20, 15, 10, and 5 km, respectively; 6, 8, 10, and 12 courses, respectively, between azimuth controls.

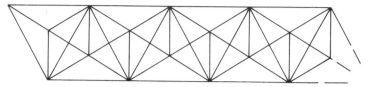

FIGURE A.10 Triangulation, cross-braced quadrilateral network, second to fourth order.

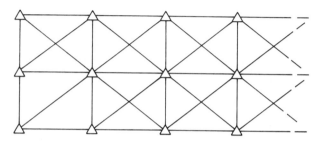

FIGURE A.11 Trilateration network, first to fourth order.

Traversing

All Orders. Simple, reasonably straight traverses as shown in Figure A.12; average lengths 20, 15, 10, and 5 km, respectively; 5, 6, 9, and 12 courses, respectively, between azimuth controls.

FIGURE A.12 Trilateration network, first to fourth order.

A.11.4 Examples* of Standard Deviations That Can Be Expected under Normal Circumstances from Various Instruments

TABLE A.8 *Distance by Tape*

Method	Standard Deviation (m)	Notes
Invar tape	$\sqrt{0.003^2 + (0.3L\ 10^{-6})^2}$	Techniques described in Geodetic Survey Pub. 73 L, meters
Steel tape	$\sqrt{N[0.002^2 + (40P\ 10^{-6})^2]}$	N = no. of tape lengths P = length of each tape Very careful slope, sag and temp. corrs. applied
Steel tape	$\sqrt{N[0.006^2 + (80P\ 10^{-6})^2]}$	N and P as above Clinometer used for vertical angles up to 5°, alignment by picket, air temp. used for corrections; tension handle used for spans over 30 m
Steel tape	$\sqrt{N[0.01^2 + (200P\ 10^{-6})^2]}$	N and P as above, no tension handles, nominal temp. correction
Subtense bar	$\sqrt{2(0.001)^2(2.5L^2\ 10^{-6})^2}$	Angle measured to accuracy approx. 1″, 1 mm for plumbing at each end of line

*Surveys and Mapping Branch, Department of Energy, Mines, and Resources, Ottawa, 1977.

TABLE A.9 *Examples of Standard Deviations for Various Types of EDM Instruments*

Type	Standard Deviation (m)	Standard[a]	Notes
Mekometer	$\sqrt{0.0002^2 + (2.0L\ 10^{-6})^2}$	M3	Forced centering
Short-range infrared	$\sqrt{0.005^2 + (3L\ 10^{-6})^2}$	M2	Mean of 8 determinations anytime
	$\sqrt{.008^2 + (4L\ 10^{-6})^2}$	M5	Mean of 2 determinations anytime
Laser	$\sqrt{0.005^2 + (2L\ 10^{-6})^2}$	M2	Mean of 8 determinations anytime
	$\sqrt{0.008^2 + (5L\ 10^{-6})^2}$	M5	Mean of 2 determinations anytime
Visible light	$\sqrt{0.010^2 + (1L\ 10^{-6})^2}$	M1	Mean of day and night sets, each of 6 observations
(Model 6 Geodimeter)	$\sqrt{0.015^2 + (5L\ 10^{-6})^2}$	M5	Mean of 2 determinations anytime
Microwave	$\sqrt{0.05^2 + (5L\ 10^{-6})^2}$	M1	Tell. MRA 2, 31 cavities, three pairs
	$\sqrt{0.03^2 + (5L\ 10^{-6})^2}$	M1	Tell. MRA 3, 13 cavities, three pairs
	$\sqrt{0.01^2 + (3L\ 10^{-6})^2}$	M1	Tell. MRA 4, 10 cavities, three pairs
	$\sqrt{0.10^2 + (10L\ 10^{-6})^2}$	M4	Tell. MRA 2, 10 cavities
	$\sqrt{0.5^2 + (10L\ 10^{-6})^2}$	M4	Tell. MRA 3, 10 cavities
	$\sqrt{0.02^2 + (7L\ 10^{-6})^2}$	M4	Tell. MRA, 4, 10 cavities

Notes for Microwave rows 1–3: Each pair alternating; master end separated by at least 4 hours

Notes for Microwave rows 4–6: Measured once in one direction

[a]Standards as listed in Tables A.10 and A.11

Note: The proportional part of the accuracy of all EDM measurements is limited by the accuracy of the modulation frequency, by the accuracy of the meteorological measurements, and by the reliability of the samples measured as representative of the conditions along the measured line. The nonproportional part is dependent on the accuracies of plumbing and of the ''zero'' corrections of the instrument and reflector and the resolution of the instrument. The modulation frequency and zero correction of the instrument must be checked whenever the instrument is repaired or after rough treatment, as well as on a regular schedule. Accuracy standards referred to in this table, for recording meteorological measurements and frequency of calibration tests are given in Table A.10.

TABLE A.10 *Standards for Meteorological Measurements and Calibration Tests*

Standard	Temp. at One or Both Ends	Calibration Schedule (additional to calibration after repair or rough handling)		Pressure (mm)	Standard Deviations		
		Zero	Frequency		Dry Bulb Temp. (°C)	Wet Bulb Temp. (°C)
M1	2	4 mo.	2 wks.	3	0.1	0.1
M2	2	4 mo.	1 yr.	3	0.5	—
M3	2	2 mo.	2 mo.	3	0.5	—
M4	2	1 yr.	1 yr.	3	0.1	0.1
M5	1	6 mo.	1 yr.	3	1	—
M6	1	4 mo.				

The accuracy of the sea level and slope corrections is limited by the accuracy of the mean elevation H and elevation difference (h). Accuracy standards for determining the mean elevation and elevation differences are given in Table A.11.

TABLE A.11 *Accuracy Standards for Determining Elevation and Elevation Differences for a Line L Meters Long*

Standard[a]	Standard Deviation of Measurements for:	
	Elevation Difference h (m)	Mean Elevation H (m)
M1	$10^{-6}\,L/h$	2
M2	$10^{-6}\,L/h$	2
M3	$10^{-6}\,L/h$	2
M4	$2 \times 10^{-6}\,L/h$	10
M5	$2 \times 10^{-6}\,L/h$	10

[a]These standards correspond to the standards listed in Table A.10 and are referred to in Table A.9.

TABLE A.12 *Directions*

Instrument Least Count	Standard Deviation		Notes
	Seconds	*Meters Perpendicular to L*	
0."2	$\sqrt{\dfrac{3 \times 10^5\,P^{\,2}}{L} + (0.6)^2}$	$\sqrt{2P^2 + (3.01L \times 10^{-6})^2}$	P is standard deviation of plumbing of instrument and target. Normal values 0.0014 m. Four groups of 6 sets during 2 nights. Starting time of each set differing by at least 2 hours.
1."0	$\sqrt{\dfrac{3 \times 10^5\,P^{\,2}}{L} + 1.5^2}$	$\sqrt{2P^2 + (7.5L \times 10^{-6})^2}$	Mean of 6 sets
1.0"	$\sqrt{\dfrac{3 \times 10^5\,P^{\,2}}{L} + 2.0^2}$	$\sqrt{2P^2 + (10L \times 10^{-6})^2}$	Mean of 4 sets
1."0	$\sqrt{\dfrac{3 \times 10^5\,P^{\,2}}{L} + 4^2}$	$\sqrt{2P^2 + (20L \times 10^{-6})^2}$	Mean of 2 sets
10."0	$\sqrt{\dfrac{3 \times 10^5\,P^{\,2}}{L} + 8^2}$	$\sqrt{2P^2 + (40L \times 10^{-6})^2}$	Mean of 2 sets
20.0"	$\sqrt{\dfrac{3 \times 10^5\,P^{\,2}}{L} + 15^2}$	$\sqrt{2P^2 + (75L \times 10^{-6})^2}$	Mean of 2 sets
30.0"	$\sqrt{\dfrac{3 \times 10^5\,P^{\,2}}{L} + 21^2}$	$\sqrt{2P^2 + (100L \times 10^{-6})^2}$	Mean of 2 sets

Angles

Note: The angle between two directions with standard deviations σ_1 and σ_2 as derived above will have a standard deviation $\sqrt{\sigma_1^2 + \sigma_2^2}$. If the lengths of the lines are about the same in the most accurate case above, ($\sigma_1 = \sigma_2$), the standard deviation of the angle between them will be about

$$\sqrt{\left(\frac{4 \times 10^5\,P}{L}\right)^2 + (0.85)^2} \text{ seconds}$$

874

TABLE A.13 *Azimuths*

Instruments	Standard Deviation	Notes
First-order astronomic	1″	Azimuth and longitude observed
Second-order survey theodolite with stride level (Polaris observations)	3″	No observed longitude; nonmountainous terrain where small deflection in prime vertical can be expected and estimated or assumed zero
Second-order survey theodolite without stride level (Polaris observations)	6″	
Second-order survey theodolite with stride level (Polaris observations)	10″	No observed longitude; mountainous terrain where large deflections can be expected and only rough estimates made
Second-order survey theodolite without stride level (Polaris observations)	15″	
Gyro theodolite	3″–10″	No observed longitude; using techniques devised by L.F. Gregerson; accuracy depends on local deflection
Gyro theodolite	20″	No observed longitude; using normal techniques
Second-order survey theodolite (solar observations)	15″–25″	Depending on method

TABLE A.14 *Position Differences*

Method	Standard Deviation (m)	Notes
Doppler satellite	1.5	35 passes, broadcast or precise ephemeris; 5 to 10 days simultaneous observations at the two positions
Doppler satellite	5.0	40 passes, broadcast ephemeris, 1 to 2 days nonsimultaneous observation at the two positions
Inertial survey system	0.2–5	Depending on method and distance

A.12 Specifications for Horizontal Control Surveys: United States of America

Specifications for geodetic control surveys were developed by the Federal Geodetic Control Committee in consultation with other surveying experts in the civil engineering and surveying and mapping fields. Surveys are classified as being first, second, or third order depending on the techniques used to establish the control stations. First-order surveys are used for the primary control network and also for highly precise scientific purposes (e.g., deformation studies, crustal movements). Second-order surveys are used to densify and support the primary network and to establish control for large-scale engineering studies (e.g., marine, highways, dam construction). Third-order surveys are used for small-scale engineering works, large-scale mapping projects, and the like.

Survey classification, standards of accuracy, and general specifications are shown for triangulation (Table A.15), trilateration (Table A.16), and traverse (Table A.17).

A.13 Vertical Control

To control elevations, bench marks were established throughout the North American Continent. The reference datum originally used was mean sea level (MSL) and was referred to as the sea level datum of 1929. This datum was established by constraining the combined interconnected United States and Canadian networks of first-order leveling as it existed in 1929 to conform to mean sea level of various epochs as determined at 26 tidal stations distributed along the Atlantic, Gulf of Mexico, and Pacific coasts.

All first-order leveling, together with the tidal station data, was adjusted for orthometric and other definable systematic errors to form the National Geodetic Vertical Datum of 1929. Most bench-mark elevations in North America have now been adjusted to that datum.

Specifications for vertical control surveys are shown in Table A.18 (Canada) and Table A.19 (United States of America). New specifications may be forthcoming from both countries as work on the new North American Vertical Datum (NAVD-88) progresses.

TABLE A.15 Triangulation

Classification	First Order	Second Order		Third Order	
		Class I	Class II	Class I	Class II
Recommended spacing of principal stations	Network stations seldom less than 15 km. Metropolitan surveys 3 to 8 km and others as required.	Principal stations seldom less than 10 km. Other surveys 1 to 3 km or as required.	Principal stations seldom less than 5 km or as required	As required	As required
Strength of figure					
R_1 between bases					
Desirable limit	20	60	80	100	125
Maximum limit	25	80	120	130	175
Single figure Desirable limit					
R_1	5	10	15	25	25
R_2	10	30	70	80	120
Maximum limit					
R_1	10	25	25	40	50
R_2	15	60	100	120	170
Base measurement					
Standard error	1 part in 1 000 000	1 part in 900 000	1 part in 800 000	1 part in 500 000	1 part in 250 000
Horizontal directions					
Instrument	0".2	0."2	0".2 or 1".0	1".0	1".0
Number of positions	16	16	8 or 12	4	2
Rejection limit from mean	4"	4"	5" or 5"	5"	5"
Triangle closure					
Average not to exceed	1".0	1".2	2".0	3".0	5".0
Maximum seldom to exceed	3".0	3".0	5".0	5".0	10".0

(continued)

TABLE A.15 (continued)

Classification	First Order	Second Order		Third Order	
		Class I	Class II	Class I	Class II
Side checks					
In side equation test, average correction to direction not to exceed	0".3	0".4	0".6	0".8	2"
Astro azimuths					
Spacing figures	6-8	6-10	8-10	10-12	12-15
Number of observations/night	16	16	16	8	4
Number of nights	2	2	1	1	1
Standard error	0".45	0".45	0".6	0".8	3".0
Vertical angle observations					
Number of and spread between observations	3D/R-10"	3D/R-10"	2D/R-10"	2D/R-10"	2D/R-20"
Number of figures between known elevations	4-6	6-8	8-10	10-15	15-20
Closure in length					
(Also position when applicable) after angle and side conditions have been satisfied, should not exceed	1 part in 100 000	1 part in 50 000	1 part in 20 000	1 part in 10 000	1 part in 5000

TABLE A.16 *Trilateration*

Recommended spacing of principal stations	Network stations seldom less than 10 km. Other surveys seldom less than 3 km.	Principal stations seldom less than 10 km. Other surveys seldom less than 1 km.	Principal stations seldom less than 5 km. For some surveys a spacing of 0.5 km between stations may be satisfactory.	Principal stations seldom less than 0.5 km	Principal stations seldom less than 0.25 km.
Geometric configuration					
Minimum angle contained within, not less than	25°	25°	20°	20°	15°
Length measurement					
Standard error	1 part in 1 000 000	1 part in 750 000	1 part in 450 000	1 part in 250 000	1 part in 150 000
Vertical angle observations					
Number of and spread between observations	3D/R-10"	3D/R-10"	2D/R-10"	2D/R-10"	2D/R-20"
Number of figures between known elevations	4-6	6-8	8-10	10-15	15-20
Astro azimuths					
Spacing figures	6-8	6-10	8-10	10-12	12-15
Number of observations/night	16	16	16	8	4
Number of nights	2	2	1	1	1
Standard error	0".45	0".45	0".6	0".8	3".0
Closure in position					
After geometric conditions have been satisfied should not exceed	1 part in 100 000	1 part in 50 000	1 part in 20 000	1 part in 10 000	1 part in 5000

TABLE A.17 *Traverse*

Classification	First Order	Second Order Class I	Second Order Class II	Third order Class I	Third order Class II
Recommended spacing of principal stations	Network stations 10–15 km. Other surveys seldom less than 3 km.	Principal stations seldom less than 4 km except in metropolitan area surveys where the limitation is 0.3 km.	Principal stations seldom less than 2 km except in metropolitan area surveys where the limitation is 0.2 km.	Seldom less than 0.1 km in tertiary surveys in metropolitan area surveys. As required for other surveys.	
Horizontal directions or angles					
Instrument	0".2	0".2 or 1".0	0".2 or 1".0	1".0	1".0
Number of observations	16	8 or 12*	6 or 8*	4	2
Rejection limit from mean	4"	4" or 5"	4" or 5"	5"	5"

*May be reduced to 8 and 4, respectively, in metropolitan areas.

Classification	First Order	Second Order Class I	Second Order Class II	Third order Class I	Third order Class II
Length measurements					
Standard error	1 part in 600 000	1 part in 300 000	1 part in 120 000	1 part in 60 000	1 part in 30 000
Reciprocal vertical angle observations					
Number of and spread between observations	3D/R–10"	3D/R–10"	2D/R–10"	2D/R–10"	2D/R–20"
Number of stations between known elevations	4–6	6–8	8–10	10–15	15–20
Astro azimuths					
Number of courses between azimuth checks	5–6	10–12	15–20	20–25	30–40

Number of observations/night	16	16	12	8	4
Number of nights	2	2	1	1	1
Standard error	0".45	0".45	1".5	3".0	8".0
Azimuth closure at azimuth check point not to exceed	1".0 per station or 2"$\sqrt{N}$	1".5 per station or 3"$\sqrt{N}$. Metropolitan area surveys seldom to exceed 2".0 per station or 3"$\sqrt{N}$.	2".0 per station or 6"$\sqrt{N}$. Metropolitan area surveys seldom to exceed 4".0 per station or 8"$\sqrt{N}$.	3".0 per station or 10"$\sqrt{N}$. Metropolitan area surveys seldom to exceed 6" per station or 15"$\sqrt{N}$.	8" per station or 30"$\sqrt{N}$.
Position closure After azimuth adjustment	0.04m $\sqrt{K}$ or 1:100 000	0.08 $\sqrt{K}$ or 1:50 000	0.2m $\sqrt{K}$ or 1:20 000	0.4m $\sqrt{K}$ or 1:10 000	0.8m $\sqrt{K}$ or 1:5000

TABLE A.18 *Classification, Standards of Accuracy, and General Specifications for Vertical Control*

Classification	Special Order	First Order	Second Order (first-order procedures recommended)	Third Order	Fourth Order
Allowable discrepancy between forward and backward levelings	$\pm 3 \text{ mm } \sqrt{K}$ $\pm 0.012 \text{ ft } \sqrt{m}$	$\pm 4 \text{ mm } \sqrt{K}$ $\pm 0.017 \text{ ft } \sqrt{m}$	$\pm 8 \text{ mm } \sqrt{K}$ $\pm 0.035 \text{ ft } \sqrt{m}$	$\pm 24 \text{ mm } \sqrt{K}$ $\pm 0.10 \text{ ft } \sqrt{m}$	$\pm 120 \text{ mm } \sqrt{K}$ $\pm 0.5 \text{ ft } \sqrt{m}$
Instruments:					
Self-leveling high-speed compensator	Equivalent to 10″/2 mm level vial	Equivalent to 10″/2 mm level vial	Equivalent to level vial 20″/mm	Equivalent to sensitivity below	Equivalent to sensitivity below
Level vial telescopic magnification	10″/2 mm 40×	10″/2 mm 40×	20″/2 mm 40×	40″ to 50″/2 mm	40″to 50″/2 mm
Parallel plate micrometer	X	X			
Sun shade and instrument cover	X	X	X		
Rods:					
Invar and double scale	X	X	X		
Invar-checkerboard footplates or steel		X	X		
Graduations on wood, metal alloy, or fiber glass are satisfactory				Graduations on wood, metal alloy, or fiber glass are satisfactory	

TABLE A.18 *(continued)*

Classification	Special Order	First Order	Second Order (first-order procedures recommended)	Third Order	Fourth Order
pins for turning points					
Circular bubble attached to rod	X	X	X		
Rod supports	X X	X X	X X		
Difference between backsight and foresight distances and their total for the section not to exceed:	5 m	10 m	10 m	Balanced	Balanced
Maximum length of sight:					
1 mm-wide rod mark	50 m	X	X		
1.6 mm-wide rod mark	60 m				
parallel plate method	80 m	80 m	80 m	N/A	N/A
three-wire method	110 m	110 m	110 m	N/A	N/A

Adapted from "Specifications and Recommendations for Control Surveys and Survey Markers".
Surveys and Mapping Branch
Department of Energy, Mines and Resources
Ottawa, Canada 1973

TABLE A.19 *National Ocean Survey, U.S. Coast and Geodetic Surveys: Classification, Standards of Accuracy, and General Specifications for Vertical Control*

Classification	*First Order* Class I, Class II	*Second Order* Class I	*Second Order* Class II	*Third Order*
Principal uses: Minimum standards; higher accuracies may be used for special purposes	Basic framework of the National Network and of metropolitan area control. Extensive engineering projects. Regional crustal movement investigations. Determining geopotential values	Secondary control of the National Network and of metropolitan area control. Large engineering projects. Local crustal movement and subsidence investigations. Support for lower-order control	Control densification, usually adjusted to the National Network. Local engineering projects. Topographic mapping. Studies of rapid subsidence. Support for local surveys	Miscellaneous local control; may not be adjusted to the National Network. Small engineering projects. Small-scale topographic mapping. Drainage studies and gradient establishment in mountainous areas
Recommended spacing of lines: National Network	Net A; 100 to 300 km Class I; Net B; 50 to 100 km Class II	Secondary network; 20 to 50 km	Area control: 10 to 25 km	As needed
Metropolitan control;	2 to 8 km	0.5 to 1 km	As needed	As needed
other purposes	As needed	As needed	As needed	As needed
Spacing of marks along lines	1 to 3 km	1 to 3 km	Not more than 3 km	Not more than 3 km
Gravity requirement[a]	0.20×10^{-3} gpu	—	—	—
Instrument standards	Automatic or tilting levels with parallel plate micrometers; invar scale rods	Automatic or tilting levels with optical micrometers or three-wire levels; invar scale rods	Geodetic levels and invar scale rods	Geodetic levels and rods
Field procedures	Double-run; forward and backward, each section	Double-run; forward and backward, each section	Double- or single-run	Double- or single-run

(continued)

TABLE A.19 *(continued)*

Classification	First Order Class I, Class II	Second Order		Third Order
		Class I	*Class II*	
Section length	1 to 2 km	1 to 2 km	1 to 3 km for double-run	1 to 3 km for double-run
Maximum length of sight	50 m *Class I*; 60 m *Class II*	60 m	70 m	90 m
Field procedures[b]				
Max. difference in lengths				
Forward and backward sights				
per setup	2 m *Class I*; 5 m *Class II*	5 m	10 m	10 m
per section (cumulative)	4 m *Class I*; 10 m *Class II*	10 m	10 m	10 m
Max. length of line between connections	Net A; 300 km Net B; 100 km	50 km	50 km double-run 25 km single-run	25 km double-run 10 km single-run
Maximum closures[c]				
Section; forward and backward	3 mm $\sqrt{K}$ *Class I*; 4 mm $\sqrt{K}$ *Class II*	6 mm $\sqrt{K}$	8 mm $\sqrt{K}$	12 mm $\sqrt{K}$
Loop or line	4 mm $\sqrt{K}$ *Class I*; 5 mm $\sqrt{K}$ *Class II*	6 mm $\sqrt{K}$	8 mm $\sqrt{K}$	12 mm $\sqrt{K}$

[a]See text for discussion of instruments.
[b]The maximum length of line between connections may be increased to 100 km double run for second order, class II, and to 50 km for double run for third order in those areas where the first-order control has not been fully established.
[c]Check between forward and backward runnings where F is the distance in kilometers.

Notes:
1. K = kilometers, m = miles = the one-way distance between bench marks measured along the leveling route.
2. To maintain the specified accuracy, long narrow loops should be avoided. The distance between any two bench marks measured along the actual route should not exceed four times the straight line distance between them.
3. Branch, spur, or open-ended lines should be avoided because of the possibility of undetected gross errors.
4. For precise work, the sections should be leveled once forward and once backward independently using different instrumentmen and, if possible, under different weather conditions and at different times of the day.
5. A starting BM must be checked against another independent BM by two-way leveling before the leveling survey can commence. If the check is greater than the allowable discrepancy, both BMs must be further checked until the matter is resolved.
6. When a parallel-plate micrometer is used for special- or first-order leveling, double-scale invar rods must be used; the spacing of the smallest graduations must be equivalent to the displacement of the parallel-plate micrometer. When the three-wire method is used for first- or second-order leveling, rods with the checkerboard design must be used.
7. Line of sight not less than 0.5 m above the ground (special and first order).
8. Alternate reading of backsight and foresight at successive setups.
9. Third- and lower-order surveys should use the two-rod system; read only one wire and try to balance backsight and foresight distances.
10. Results equivalent to fourth-order spirit leveling can sometimes be obtained by measurement of vertical angles in conjunction with traverses, trilateration, or triangulation. Best results are obtained on short (<16 km) lines with simultaneous (within the same minute) measurement of the vertical angles at both ends of the line, using a one-second theodolite.

Tables A.15, A.16, A.17, and A.19 are taken from the 1974 publication *Classification, Standards of Accuracy, and General Specifications of Geodetic Control Surveys.* Detailed specifications related to these tables are available in the publication *Specifications to Support Classification, Standards of Accuracy, and General Specifications of Geodetic Control Surveys* by the Federal Geodetic Control Committee. Both publications can be obtained from U.S. Department of Commerce, National Oceanic and Atmospheric Administration, National Ocean Survey, Rockville, Maryland.

Appendix

B TRIGONOMETRIC DEFINITIONS AND IDENTITIES

Right Triangles

Basic Function Definitions

(B.1) $\operatorname{Sin} A = \dfrac{a}{c} = \operatorname{Cos} B$

(B.2) $\operatorname{Cos} A = \dfrac{b}{c} = \operatorname{Sin} B$

(B.3) $\operatorname{Tan} A = \dfrac{a}{b} = \operatorname{Cot} B$

(B.4) $\operatorname{Sec} A = \dfrac{c}{b} = \operatorname{Cosec} B$

(B.5) $\operatorname{Cosec} A = \dfrac{c}{a} = \operatorname{Sec} B$

(B.6) $\operatorname{Cot} A = \dfrac{b}{a} = \operatorname{Tan} B$

Derived Relationships

$$a = c \operatorname{Sin} A = c \operatorname{Cos} B = b \operatorname{Tan} A = b \operatorname{Cot} B = \sqrt{c^2 - b^2}$$

$$b = c \operatorname{Cos} A = c \operatorname{Sin} B = a \operatorname{Cot} A = a \operatorname{Tan} B = \sqrt{c^2 - a^2}$$

$$c = \frac{a}{\operatorname{Sin} A} = \frac{b}{\operatorname{Cos} B} = \frac{a}{\operatorname{Sin} B} = \frac{b}{\operatorname{Cos} A} = \sqrt{a^2 + B^2}$$

Algebraic Signs for Primary Trigonometric Functions

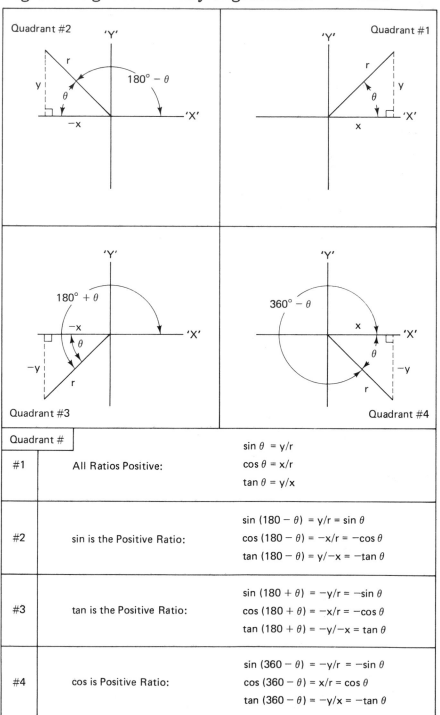

Quadrant #		
#1	All Ratios Positive:	$\sin \theta = y/r$ $\cos \theta = x/r$ $\tan \theta = y/x$
#2	sin is the Positive Ratio:	$\sin (180 - \theta) = y/r = \sin \theta$ $\cos (180 - \theta) = -x/r = -\cos \theta$ $\tan (180 - \theta) = y/-x = -\tan \theta$
#3	tan is the Positive Ratio:	$\sin (180 + \theta) = -y/r = -\sin \theta$ $\cos (180 + \theta) = -x/r = -\cos \theta$ $\tan (180 + \theta) = -y/-x = \tan \theta$
#4	cos is Positive Ratio:	$\sin (360 - \theta) = -y/r = -\sin \theta$ $\cos (360 - \theta) = x/r = \cos \theta$ $\tan (360 - \theta) = -y/x = -\tan \theta$

Note: The quadrant numbers reflect the traditional geometry approach (counter clockwise) to quadrant analysis. In surveying, the quadrants are numbered from 1 (N.E.), 2 (S.E.), 3 (S.W.) and 4 (N.W.). The analysis of algebraic signs for the trigonometric functions (as shown) remains valid. Hand held calculators will automatically provide the correct algebraic sign if the angle direction is entered in the calculator in its azimuth form.

Oblique Triangles

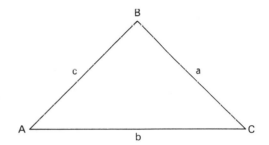

Sine Law

$$(\text{B.7}) \quad \frac{a}{\operatorname{Sin} A} = \frac{b}{\operatorname{Sin} B} = \frac{c}{\operatorname{Sin} C}$$

Cosine Law

$$(\text{B.8}) \quad a^2 = b^2 + c^2 - 2bc \operatorname{Cos} A$$

$$(\text{B.9}) \quad b^2 = a^2 + c^2 - 2ac \operatorname{Cos} B$$

$$(\text{B.10}) \quad c^2 = a^2 + b^2 - 2ab \operatorname{Cos} C$$

Given	Required	Formulas
A, B, a	C, b, c	$c = 180 - (A + B)$; $b = \dfrac{a}{\operatorname{Sin} A} \operatorname{Sin} B$; $c = \dfrac{a}{\operatorname{Sin} A} \operatorname{Sin} C$
A, b, c	a	$a^2 = b^2 + c^2 - 2bc \operatorname{Cos} A$
a, b, c	A	$\operatorname{Cos} A = \dfrac{b^2 + c^2 - a^2}{2 bc}$
a, b, c	Area	Area $= \sqrt{s(s - a)(s - b)(s - c)}$ where $s = \frac{1}{2}(a + b + c)$
C, a, b	Area	Area $= \frac{1}{2} ab \operatorname{Sin} C$

General Trigonometric Formulas

(B.11) $\text{Sin } A = 2 \text{ Sin } \frac{1}{2}A \text{ Cos } \frac{1}{2}A = \sqrt{1 - \text{Cos}^2 A} = \text{Tan } A \text{ Cos } A$

(B.12) $\text{Cos } A = 2 \text{ Cos}^2 \frac{1}{2}A - 1 = 1 - 2 \text{ Sin}^2 \frac{1}{2} A = \text{Cos}^2 \frac{1}{2}A - \text{Sin}^2 \frac{1}{2}A = \sqrt{1 - \text{Sin}^2 A}$

(B.13) $\text{Tan } A = \dfrac{\text{Sin } A}{\text{Cos } A} = \dfrac{\text{Sin } 2A}{1 + \text{Cos } 2A} = \sqrt{\text{Sec}^2 A - 1}$

Addition and Subtraction Identities

(B.14) $\text{Sin } (A \pm B) = \text{Sin } A \text{ Cos } B \pm \text{Sin } B \text{ Cos } A$

(B.15) $\text{Cos } (A \pm B) = \text{Cos } A \text{ Cos } B \mp \text{Sin } A \text{ Sin } B$

(B.16) $\text{Tan } (A \pm B) = \dfrac{\text{Tan } A \pm \text{Tan } B}{1 \mp \text{Tan } A \text{ Tan } B}$

(B.17) $\text{Sin } A + \text{Sin } B = 2 \text{ Sin } \frac{1}{2}(A + B) \text{ Cos } \frac{1}{2}(A - B)$

(B.18) $\text{Sin } A - \text{Sin } B = 2 \text{ Cos } \frac{1}{2}(A + B) \text{ Sin } \frac{1}{2}(A - B)$

(B.19) $\text{Cos } A + \text{Cos } B = 2 \text{ Cos } \frac{1}{2}(A + B) \text{ Cos } \frac{1}{2}(A - B)$

(B.20) $\text{Cos } A - \text{Cos } B = 2 \text{ Sin } \frac{1}{2}(A + B) \text{ Sin } \frac{1}{2}(A - B)$

Double-Angle Identities

(B.21) $\text{Sin } 2A = 2 \text{ Sin } A \text{ Cos } A$

(B.22) $\text{Cos } 2A = \text{Cos}^2 A - \text{Sin}^2 A = 1 - 2 \text{ Sin}^2 A = 2 \text{ Cos}^2 A - 1$

(B.23) $\text{Tan } 2A = \dfrac{2 \text{ Tan } A}{1 - \text{Tan}^2 A}$

Half-Angle Identities

(B.24) $\text{Sin } \dfrac{A}{2} = \sqrt{\dfrac{1 - \text{Cos } A}{2}}$

(B.25) $\text{Cos } \dfrac{A}{2} = \sqrt{\dfrac{1 + \text{Cos } A}{2}}$

(B.26) $\text{Tan } \dfrac{A}{2} = \dfrac{\text{Sin } A}{1 + \text{Cos } A}$

Appendix

C *References and Additional Readings*

Airphoto Interpretation of the Physical Environment. S. J. Glenn Bird and Ian M. Hale. Bird and Hale Ltd., Toronto, Canada. 1982.

Analysis and Adjustments of Survey Measurements. Edward M. Mikhail and Gordon Gracie. New York, Toronto: Van Nostrand Reinhold Co. 1981.

Department of the Army Technical Manuals: TM 5-241-4/1 *Transformation of Coordinates—Geographic to Grid.* TM 5-241-4/2 *Transformation of Coordinates—Grid to Geographic. Universal Transverse Mercator Grid Tables for Latitudes 02 to 802.* U. S. Army Map Service, Corps of Engineers.

Electronic Surveying. Simo H. Laurila. New York and Toronto: John Wiley & Sons. 1976.

4444 *Review Questions for Surveyors.* (or latest). (A comprehensive collection of numerical and written questions and answers dealing with most aspects of surveying.) Russell C. Brinker. Published by the author. 1975.

Geodetic Engineering Surveying Computations and Data Adjustments, Volume 10. State Plane Coordinates. (1927 NAD and 1983 NAD.) 2nd. Edition. Herbert W. Stoughton. Published by the author.

Journals and Technical Papers: The Canadian Institute of Surveying 1750 Courtwood Cres. Box 5378, Station "F". Ottawa, Ontario K2C 3J1.

Journals and Technical Papers: American Congress on Surveying and Mapping 210 Little Falls Street, Falls Church, VA. 22046.

The Land Surveys of Ontario 1750–1980. L. M. Sebert. Cartographica 17:3. pp. 64–105. 1980.

Manual of Instructions for the Survey of Public Lands of the United States. Bureau of Land Management. U. S. Government Printing Office. 1973.

Manual of Photogrammetry, 4th Edition. American Society of Photogrammetry. 1981.

Manual Relating to Surveys and Surveyors. Association of Ontario Land Surveyors. Toronto, Canada. 1973.

Metric Curve Tables. Circular and Spiral Curve Functions. Roads and Transportation Association of Canada. 1765 St. Laurent Blvd., Ottawa, Ontario, K1G 3V4.

Observations and Least Squares. Edward M. Mikhail and F. Ackerman. New York: IEP. A Dun-Donnelley Publisher. 1976.

Route Location and Design. 5th Edition. Thomas F. Hickerson. New York, Toronto: McGraw-Hill Book Company. 1964.

Route Surveying and Design. 5th Edition. Carl F. Meyer and David W. Gibson. New York: Harper & Row Publishers. 1980.

Solar Ephemeris. (This ephemeris is published annually and contains ephemeris data for sun and star observations.) Keuffel & Esser. 20 Whippany Road, Morristown, NJ 07960.

Surveying Instruments. (A virtual encyclopedia of surveying instruments.) 7th Edition. Fritz Deumlich. English Translation by Wolfgang Faig. New York: Walter de Gruyter Co. 1982.

Surveying—Instruments and Methods. (A look at the historical development of surveying instruments.) J. Wartnaby. London: Her Majesty's Stationery Office. 1968.

Index